Computational Studies

The book covers a diverse range of topics based on computational studies, including modeling and simulations based on quantum chemical studies and molecular dynamics (MD) simulations. It contains quantum chemical studies on several molecules, including biologically relevant molecules and liquid crystals and various aspects of superatomic clusters including superalkalis and superhalogens. It gives an overview of MD simulations and their applications on biomolecular systems such as HIV-1 protease and integrase.

Features:

- Includes first principle methods, density functional theory, as well as molecular dynamics simulations.
- Explores quantum chemical studies on several molecules.
- Gives readers an overview of the power of computation.
- Discusses superatomic clusters, superalkalis, and superhalogens.
- Covers themes from molecules, clusters, materials, as well as biophysical systems.

This book is aimed at researchers and graduate students in materials science and computational and theoretical chemistry.

Emerging Materials and Technologies

Series Editor: Boris I. Kharissov

The *Emerging Materials and Technologies* series is devoted to highlighting publications centered on emerging advanced materials and novel technologies. Attention is paid to those newly discovered or applied materials with potential to solve pressing societal problems and improve quality of life, corresponding to environmental protection, medicine, communications, energy, transportation, advanced manufacturing, and related areas.

The series takes into account that, under present strong demands for energy, material, and cost savings, as well as heavy contamination problems and worldwide pandemic conditions, the area of emerging materials and related scalable technologies is a highly interdisciplinary field, with the need for researchers, professionals, and academics across the spectrum of engineering and technological disciplines. The main objective of this book series is to attract more attention to these materials and technologies and invite conversation among the international R&D community.

Smart Micro- and Nanomaterials for Pharmaceutical Applications
Edited by Ajit Behera, Arpan Kumar Nayak, Ranjan K. Mohapatra, and Ali Ahmed Rabaan

Friction Stir-Spot Welding
Metallurgical, Mechanical and Tribological Properties
Edited by Jeyaprakash Natarajan and K. Anton Savio Lewise

Phase Change Materials for Thermal Energy Management and Storage
Fundamentals and Applications
Edited by Hafiz Muhammad Ali

Nanofluids
Fundamentals, Applications, and Challenges
Shriram S. Sonawane and Parag P. Thakur

MXenes
From Research to Emerging Applications
Edited by Subhendu Chakroborty

Biodegradable Polymers, Blends and Biocomposites
Trends and Applications
Edited by A. Arun, Kunyu Zhang, Sudhakar Muniyasamy and Rathinam Raja

Bioinspired Materials and Metamaterials
A New Look at the Materials Science
Edward Bormashenko

Computational Studies
From Molecules to Materials
Edited by Ambrish Kumar Srivastava

For more information about this series, please visit: www.routledge.com/Emerging-Materials-and-Technologies/book-series/CRCEMT

Computational Studies

From Molecules to Materials

Edited by Ambrish Kumar Srivastava

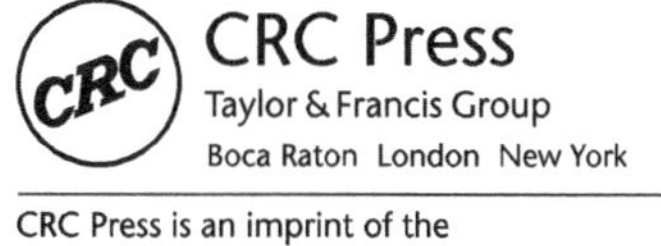

CRC Press is an imprint of the
Taylor & Francis Group, an **informa** business

First edition published 2025
by CRC Press
2385 NW Executive Center Drive, Suite 320, Boca Raton FL 33431

and by CRC Press
4 Park Square, Milton Park, Abingdon, Oxon, OX14 4RN

CRC Press is an imprint of Taylor & Francis Group, LLC

ISBN: 978-1-032-52854-0 (hbk)
ISBN: 978-1-032-57858-3 (pbk)
ISBN: 978-1-003-44132-8 (ebk)

DOI: 10.1201/9781003441328

Typeset in Times
by Apex CoVantage, LLC

Dedicated to
My Parents
(whose sacrifices turned into my achievements)

Contents

Preface

Atoms form molecules, which form bulk materials. There is also an intermediate phase between molecules and materials, known as clusters. This book deals with the computational studies of molecules to materials. Computational studies cover an important part of modern science, which plays a pivotal role in complementing experimental studies and explaining the experimental results in a very constructive way. In addition, the studies become a powerful tool in the absence of experiments. With the advent of powerful computers and advancement in theory and its implementation in the form of computer code or software, the scope of computational research expanded to a variety of fields and reached the next levels. This book, *Computational Studies: From Molecules to Materials*, aims to present readers with multidimensional forms of computational research.

The book consists of 14 chapters contributed by leading and active researchers and experts in their respective fields. The chapters are based on computational studies with a variety of themes. Chapter 1 discusses various aspects of the molecules of liquid crystals. Chapter 2 compares the spectroscopic results of some organic compounds obtained by theory with experiments. Chapter 3 offers various properties of a biologically relevant molecule using density functional theory. Chapter 4 provides insight into drug interactions and drug delivery systems using quantum chemical methods. Chapter 5 discusses graphene oxide-based nanomaterials and their applications in biomedicine. Chapter 6 details the concept of molecular dynamics and its role in biomolecular simulations. Chapter 7 offers a technique for the optimization of atomic clusters and related algorithms in detail. Chapter 8 discusses the role of a magnesium nanocluster in the purification of oxygen, nitrogen, and carbon monoxide gases. Chapter 9 reveals the effect of confinement on bonding and catalysis. Chapter 10 provides a comprehensive account of the nonlinear optical properties of molecules and clusters with excess electrons. Chapter 11 discusses the properties of molecules for various devices in organic electronics such as diodes, transistors, and solar cells. Chapter 12 presents the hydrogen storage capacity of Cu(I)-triazine-based organometallic complexes. Chapter 13 discusses the properties of pure and doped activated carbon sheets using density functional theory. In the end, Chapter 14 sheds some light on the role of quantum computing in the study of materials. All in all, the book reveals various aspects of computational research along with the latest developments. I believe that this book will benefit young researchers and scientists working at the interface of physics, chemistry, and biology. It is particularly useful for researchers in theoretical chemistry, computational chemistry, computational materials science, biophysics, biochemistry, modeling, and simulations.

Best wishes,

Ambrish Kumar Srivastava
Gorakhpur, India
July 2023

Preface

About the Editor

Ambrish Kumar Srivastava is Assistant Professor at the Department of Physics in Deen Dayal Upadhyaya Gorakhpur University, Gorakhpur, India. He was Junior Research Fellow (with All India Rank 18) and Senior Research Fellow of the Council of Scientific and Industrial Research (CSIR), India. He earned his PhD on the topic entitled "Computational Studies on Biologically Active Molecules and Small Clusters: DFT and TDDFT Approaches" from the University of Lucknow, India, and subsequently worked as National Postdoctoral Fellow of the Science & Engineering Research Board (SERB) at D.D.U. Gorakhpur University. He has published over 120 research papers in various journals of international repute with an h-index of 23 and a citation index of 1700. In addition, he has authored/edited 3 books, and 3 more books are currently in press. He is an active reviewer for various leading journals and has reviewed more than 40 research papers so far. He is an Associate Editor of Frontiers in Physics for the Chemical Physics and Physical Chemistry section and also serves on the Editorial Board of several journals. He is a member of various scientific societies and organizations including the American Chemical Society, Royal Society of Chemistry, Indian Chemical Society, Materials Research Society of India, among others. He has recently received the prestigious NASI-Young Scientist Platinum Jubilee Award–2022 from the National Academy of Sciences, India. His broad research interests include Superatomic Clusters, Computational Materials Science, and Biophysics.

Laboratory website: https://cutt.ly/CMSLab

Contributors

Ahmadi, Sara
Department of Chemistry
Islamic Azad University
Firoozabad, Iran

Bag, Abhishek
Department of Chemistry
Cooch Behar Panchanan Barma University
Cooch Behar, India

Chakraborty, Bhrigu
Department of Chemistry
Indian Institute of Technology Kharagpur
Kharagpur, India

Chattaraj, Pratim Kumar
Department of Chemistry
Birla Institute of Technology
Ranchi, India

Dash, Mrinal Kanti
School of Applied Sciences and Humanities
Haldia Institute of Technology
Haldia, India

De, Gobinda Chandra
Department of Chemistry
Cooch Behar Panchanan Barma University
Cooch Behar, India

Dehghan, Mahmood Reza
Department of Chemistry
Islamic Azad University
Firoozabad, Iran

Dubey, Kshatresh Dutta
Department of Chemistry
Shiv Nadar Institution of Eminence
Gautam Buddha Nagar, India

Giri, Santanab
School of Applied Sciences and Humanities
Haldia Institute of Technology
Haldia, India

Jha, Ruchi
Advanced Technology Development Centre
Indian Institute of Technology Kharagpur
Kharagpur, India

Joshi, Sravani
Bioinformatics Centre
CSIR–Centre for Cellular and Molecular Biology
Hyderabad, India

Kardam, Vandana
Department of Chemistry
Shiv Nadar Institution of Eminence
Gautam Buddha Nagar, India

Kargeti, Ankit
Department of Applied Sciences
BML Munjal University
Gurugram, India

Kumar, Abhishek
Department of Physics
University of Lucknow
Lucknow, India

Kumar, Ratnesh
Department of Physics
University of Lucknow
Lucknow, India

Mishra, Abhishek Kumar
Department of Physics
University of Petroleum and Energy Studies
Dehradun, India

Mishra, Ashok Kumar
Department of Physics
Dr. Shakuntala Misra National Rehabilitation University
Lucknow, India

Mishra, Soni
Department of Physics
Graphic Era Hill University
Dehradun, India

Misra, Neeraj
Department of Physics
University of Lucknow
Lucknow, India

Pal, Ranita
Advanced Technology Development Centre
Indian Institute of Technology Kharagpur
Kharagpur, India

Rasheed, Tabish
Department of Applied Sciences
BML Munjal University
Gurugram, India

Roymahapatra, Gourisankar
School of Applied Sciences and Humanities
Haldia Institute of Technology
Haldia, India

Sharma, Dipendra
Department of Physics
Deen Dayal Upadhyaya Gorakhpur University
Gorakhpur, India

Siddiqui, Shamoon Ahmad
Department of Physics
Integral University
Lucknow, India

Srivastava, Ambrish Kumar
Department of Physics
Deen Dayal Upadhyaya Gorakhpur University
Gorakhpur, India

Srivastava, Ruby
Bioinformatics Centre
CSIR–Centre for Cellular and Molecular Biology
Hyderabad, India

Sun, Wei-Ming
Department of Basic Chemistry School of Pharmacy
Fujian Medical University
Fujian, China

Tewari, Satya Prakash
Department of Physics
Dr. Shakuntala Misra National Rehabilitation University
Lucknow, India

Tiwari, Gargi
Department of Physics
Patna University
Patna, India

Tiwari, Sugriva Nath
Department of Physics
Deen Dayal Upadhyaya Gorakhpur University
Gorakhpur, India

1 DFT-Based Studies on Thermodynamic, Electronic, Optical, and Spectroscopic Aspects of Liquid Crystals

An Overview

Dipendra Sharma, Gargi Tiwari, Abhishek Kumar, Neeraj Misra, and Sugriva Nath Tiwari

1.1 INTRODUCTION

Conventionally, it is presumed that matter exists only in three states: solid, liquid, and gas. However, this is not quite true: in particular, some of the organic substances do not undergo a single transition from solid to liquid but rather a cascade of transitions involving new phases; the mechanical properties and the symmetry properties of these phases are intermediate between those of a liquid and those of a crystal. For this reason, they have often been called liquid crystals, or mesogens. A more proper name is "mesomorphic phases" (mesomorphic: of intermediate form). However, many crystals show a transition from a strongly ordered state to a phase where each molecule commutes through several equivalent orientations. The high-temperature phase is positionally ordered, but orientationally disordered. It is sometimes (loosely) called a plastic crystal. Examples of such rotational transitions are solid hydrogen, ammonium halides, and also certain types of organic molecules. In the liquid crystalline phase, molecules of certain organic liquids are aligned preferentially along one direction or in two dimensions at lower temperatures; they are positionally disordered but orientationally ordered anisotropic liquids. At higher temperatures, they undergo a transition to a traditional (isotropic) liquid phase.[1–3] A typical phase transformation for a thermotropic liquid crystal is shown in Figure 1.1.

The development of liquid crystal science has a long and colorful history. The first research on liquid crystals was discovered and published in 1888 by Austrian Botanist Friedrich Reinitzer from

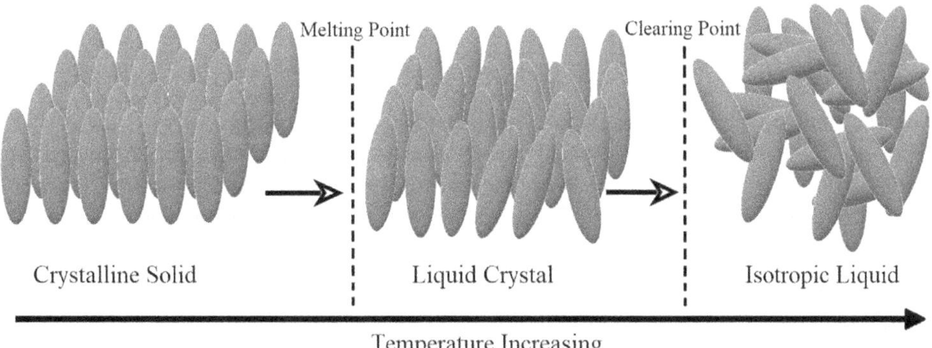

FIGURE 1.1 Arrangements of molecules in thermotropic liquid crystalline phase transitions represented by rod-like molecules.

DOI: 10.1201/9781003441328-1

Karl-Ferdinands-Universität. He discussed the melting behavior of cholesteryl benzoate, which has two melting points: at 145.5 °C, the compound changes into a murky fluid, and at 178.5 °C, it becomes clear liquid.[4] German Physicist Otto Lehmann, a specialist in polarization microscopy, was consulted by Reinitzer when he was unable to explain these data. Lehmann examined the optical characteristics of the compound, observed crystallites in the hazy fluid, and concluded that this was an intermediate phase (or "mesophase"—from Ancient Greek, Îo (mésos) meaning "middle") between the liquid and solid phases. This transitional phase exhibited birefringence while simultaneously having a liquid-like flow.[5] Such intermediate (liquid crystal) phases can also be obtained by mixing substances like sodium and potassium salts of higher fatty acids to suitable solvents.[1]

At present, liquid crystals have emerged as beautiful, mysterious, and soft condensed materials. Since the topic of liquid crystal is very interesting and it encapsulates physics, chemistry, and, in some aspect, also biology, it constitutes a very suitable content for an interdisciplinary integration of the natural science studies.[6–12]

Depending upon the nature of the building blocks of molecules and upon external parameters (temperature, solvents etc.), a wide variety of phenomena and transitions among liquid crystals can be observed. To design a liquid crystal, one must use elongated objects. Presently, at least three different ways are known to achieve this: with small organic molecules; with long helical rods that either occur in nature or can be made artificially; and with more complex units that are really associated structures of molecules and ions. Accordingly liquid crystals are classified into various types.[13]

1.2 LIQUID CRYSTAL CLASSIFICATION

Liquid crystalline phases can be obtained either by heating and/or cooling the substances or by treating the substances with water or polar solvents (aqueous medium). Depending on the way mesomorphic phases are observed, liquid crystal materials are classified as thermotropic or lyotropic. Further subgroups of liquid crystals are made depending on the molecular architecture of constituent entities.

For instance, the molecules in the mesophases can be rod-like (or calamitic), disc-like (or discotic), amphiphilic, nonamphiphilic, polymeric, etc. There is no need of a solvent in case of thermotropic liquid crystalline substances. The concentration of the solvents, on the other hand, has an impact on the behavior of the liquid crystals and aggregation in lyotropic liquid crystals.

1.3 THERMOTROPIC LCs

The order and orientation of molecular constituents in thermotropic mesophases can be used to categorize them. Broadly, thermotropic LCs are classified as nematic and smectic liquid crystals. Cholesterics are treated as a special class of nematic liquid crystals.[14] Salient features of liquid crystals are very briefly described here.

1.3.1 Nematic LCs

The nematic phase is the most straightforward mesophase that has ever been observed; there is no short-range positional order but rather long-range orientational order between mesogens. Mesogens are randomly dispersed in space, yet they can freely translate and rotate along the director field (Figure 1.2a). "Nematic" is derived from the Ancient Greek word νεμα ("nema") meaning thread. Since thread-like molecular structures are seen under a microscope, they are referred to as nematic structures.[14]

1.3.2 Cholesteric LCs

The cholesteric phase is identical to the nematic phase, with the exception that the mesogens change their orientation in a helical direction with respect to the director field. In the cholesteric phase, the director field reverses course and faces the helix in the other direction. Each layer of the helix that the molecules form, which is nonsuperimposable, gives rise to the chirality. This is why cholesteric LCs are mostly referred to as chiral nematics. The gap between two mesogenic layers that have rotated 180° to the director field is known as the half of the pitch within the cholesteric mesophase (Figure 1.2b). Further, cholesteric LCs exhibit emission of blue-violet radiations with decreasing temperature from the isotropic phase and hence give rise to a new class of LCs, commonly known as blue phase LCs.[15–17]

1.3.3 Discotic LCs

Numerous compounds made up of molecules with a disc shape have been shown in recent investigations to have stable thermotropic liquid crystalline phases. The name "discotic liquid crystals" is now used to describe them. Most of them can be divided structurally into two groups: columnar and discotic

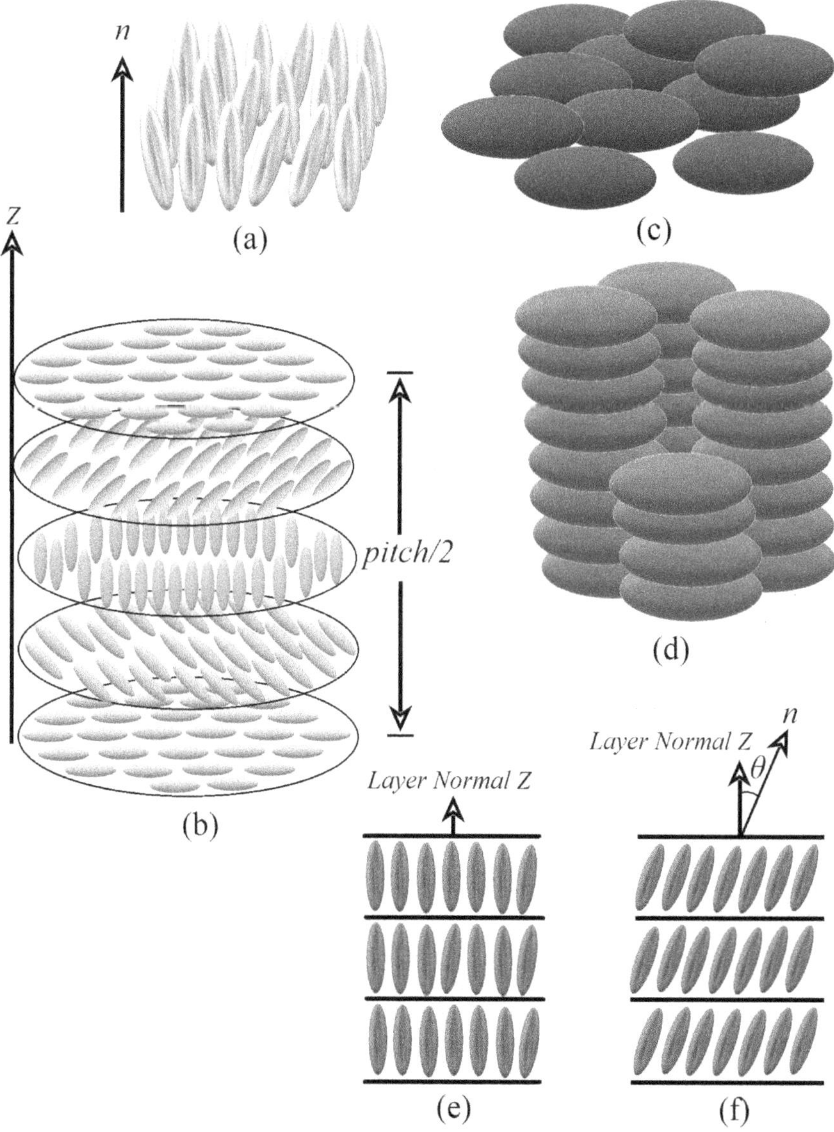

FIGURE 1.2 Different subphase alignments of molecular configurations: (a) nematic, (b) cholesteric, (c) discotic nematic, (d) discotic columnar phase, (e) smectic A, and (f) smectic C.

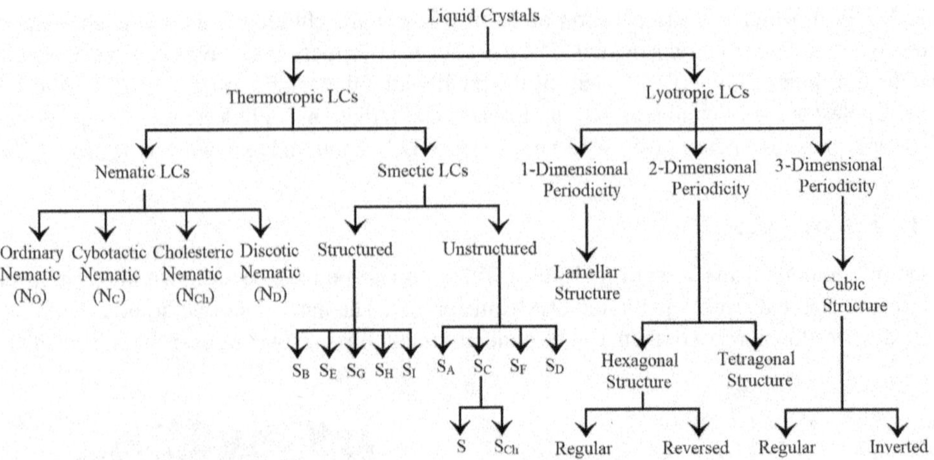

FIGURE 1.3 General classification of thermotropic liquid crystals.

nematic (Figures 1.2c and 1.2d). In its most basic form, the columnar phase is an orientationally ordered arrangement of discs without any long-range translational order, while the nematic phase exhibits liquid-like disorder in the third dimension and long-range translational periodicity in the second.

1.3.4 Smectic LCs

The word "smectic" is derived from the Ancient Greek word σμεκτοσ ("smektos") meaning soap-like). This phase was first identified in liquid crystals made up of amphiphilic molecules. The term "smectic" is now used to describe liquid crystals where the molecules are organized in layers in addition to having an orientational order. The organized layers' ability to slide past one another adds to the liquid nature of the mesomorphic phase. Numerous smectic stages have been discovered, and each of them varies in the position and orientation of the mesogens. Smectics are further designated as SmA, SmB, SmC, etc. To differentiate between the smectic phases, we may look at the molecular orientation inside the layers (Figures 1.2e and 1.2f). The smectic A phase has molecules aligned along a director field (n) and parallel to the layer normal, whereas the smectic C phase has molecules tilted at an angle away from the layer normal. Figure 1.3 shows a general classification of the various polymorphs of thermotropic liquid crystals.[2, 18] Substances that display smectic LCs are occasionally referred to as two-dimensional liquids.[10, 19]

In addition to these categories, ferroelectric and antiferroelectric liquid crystals, as well as many other mesogenic substances, constitute separate classes because of their potential applications and future prospects.[11, 12]

1.4 LYOTROPIC LCs

A lyotropic liquid crystal, on the other hand, is created when amphiphilic molecules melt in a certain type of solvents under proper pressure, temperature, and concentration conditions. By adjusting the solvent concentration, one can control the lyotropic liquid crystal combination. According to the concentration of the amphiphilic species, lyotropic liquid crystals (LLCs), which are self-assembled surfactant-solvent systems, can form a variety of mesophases, including micelles, micellar cubic, bicontinuous cubic, hexagonal, lamellar structures, and cell membrane bilayer (Figure 1.4).[6, 20, 21] Further, liquid crystalline properties in polymers, biopolymers (like protein, nucleic acids, blood cells, viruses, etc.) and other bioactive substances are well established, and they constitute another class of liquid crystals and help to mediate many living processes and biomolecular recognitions.[7, 8, 21]

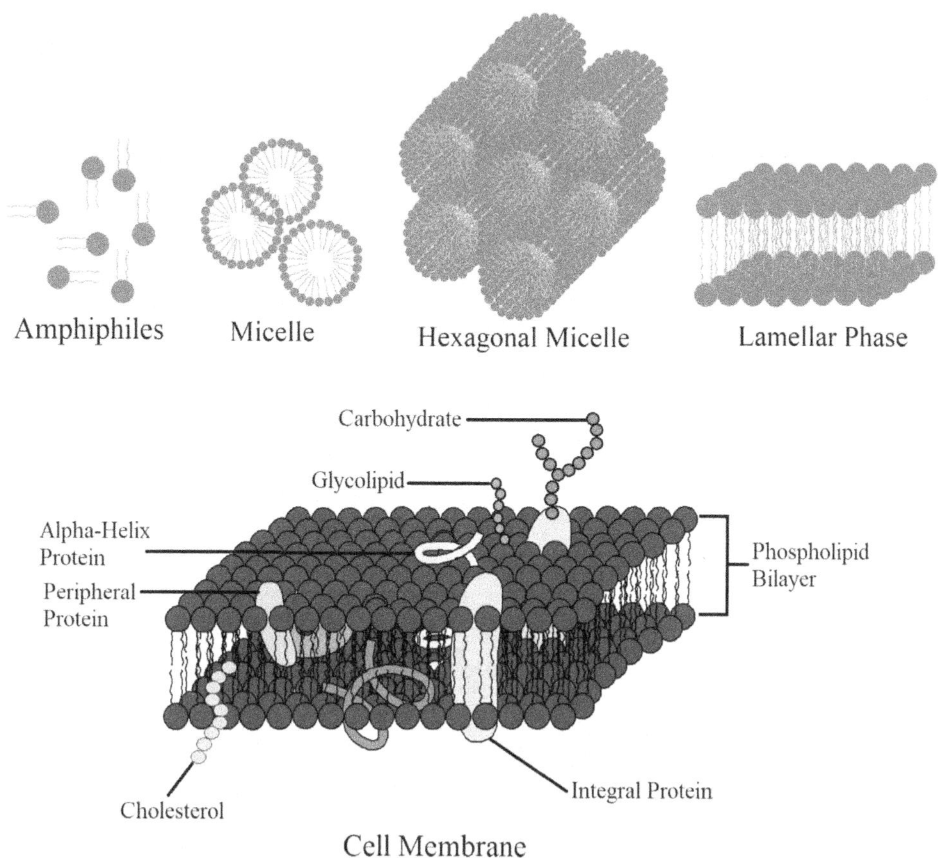

FIGURE 1.4 Representation of molecular form and the architecture of different types of lyotropic liquid crystal phases.

The order parameter is crucial for distinguishing between different liquid crystal phases. Figure 1.5 shows the order parameter (S) and an *n*OCB LC molecule exhibiting different phases. Molecular orientation research is one of the most important and inescapable concerns since the degree of order determines the anisotropy of the physical characteristics of liquid crystal substances.[1, 22]

Over the past few decades, mesogens or liquid crystals (LCs) have attracted a lot of scientific interest in the field of soft condensed matter due to their distinctive anisotropic properties, fluidity, and wide range of applications, including spatial light modulators, sensors, optical antennas, flat panel displays, and beam steering devices.[6, 23–38] The length of the aliphatic chain is one of the key molecular structure factors that affects mesomorphic behavior. Since the beginning of the LC investigation, its variation has been the simplest molecular method of regulating the capacity of mesogenic materials to self-organize. Comparing different properties (electrical, optical, and mesomorphic) of the members in a homologous series is a well established tool for gathering information on the relationship between structure and properties. Beginners in the liquid crystalline field can easily see the strong odd-even effect dependent on the length of the terminal chain or chains, etc. The spontaneous molecular ordering that is a characteristic of liquid crystalline phases has long been used in many technical applications, such as electro-optic displays. Since the beginning of display device technology, cyanobiphenyl liquid crystals have been used extensively due to their mesomorphic behavior near the room temperature.[2, 29]

In view of these facts, this chapter presents the thermodynamic, electronic, optical, and spectroscopic aspects of the members of *n*CB and *n*OCB LC homologous series that have been investigated using the DFT method.

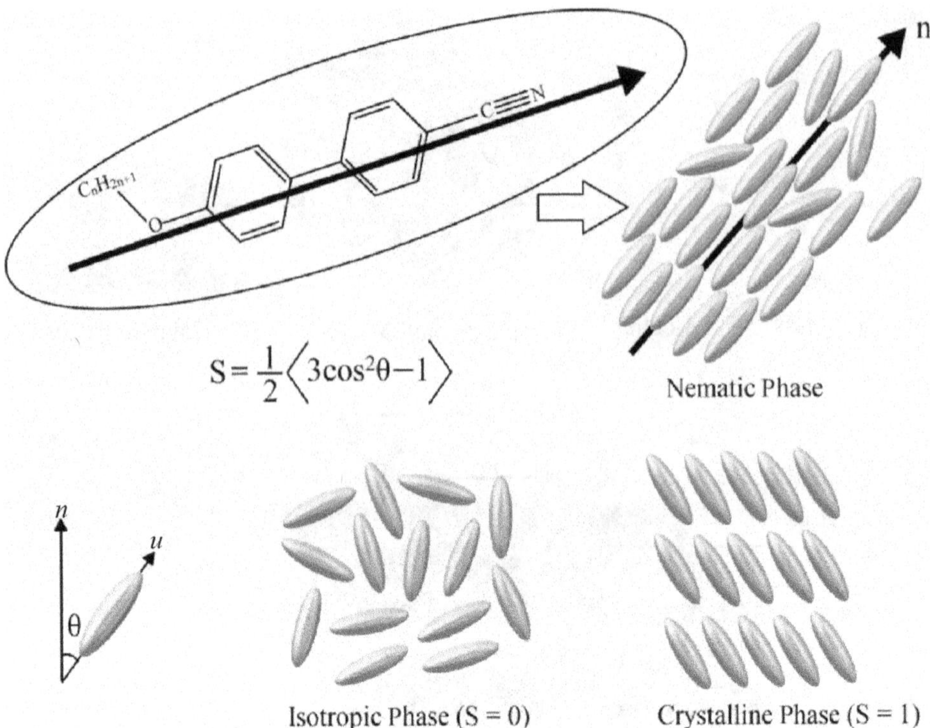

FIGURE 1.5 Order parameter (S) and dependence of order parameter on director angle (θ) defining different phases of LC material.

1.5 COMPUTATIONAL METHOD

The density functional theory (DFT) based B3LYP/6–31G(d,p) and B3LYP/6–311+G(d,p) schemes developed in the GAUSSIAN 16W program[39] were employed as the computational tool in this study. The total electronic energy of a molecular system[40] is expressed as:

$$E_{DFT} = E_n[\rho] + E_e[\rho] + E_{CR}[\rho] + E_X[\rho] + E_C[\rho] \tag{1.1}$$

where E stands for the energy functional and the subscripts n, e, CR, X, and C, respectively, refer to the nuclear repulsion energy, one electron's (kinetic + potential) energy, coulomb repulsion energy, exchange energy, and correlation energy. For $E_C = 0$, this equation is reduced to Hartree–Fock (HF) form. The electron density function (ρ) and its gradient are represented as conventional integrals of DFT functionals in equation (1.1):

$$E[\rho] = \int f(\rho(r), \nabla\rho(r))dr \tag{1.2}$$

Equation (1.2) leads to integrals that must be assessed using numerical integration because they are not directly solvable. The one aspect of DFT methods that differs between E_X [ρ] and E_C [ρ] is the function f that is used.

The hybrid functional B3LYP has shown to be a good trade-off between computing cost, coverage, and result correctness. It was initially developed to study circular dichroism and vibrational absorption.[41] Analyzing organic molecules in the gas phase with this method is increasingly widespread. The exchange correlation functional produced by this B3LYP method includes the parameterized exchange term of A. D. Becke[42] and the correlation term developed by Y. Lee, W. Yang, and R. G. Parr.[43] B3LYP is a hybrid functional with three parameters that includes a combination of Becke's exchange term, HF exchange, and Lee–Yang and Parr's correlation term. The final two elements of equation (1.1) are typically stated in B3LYP[44] as:

$$\alpha.E_X^{Slater} + (1-\alpha).E_X^{HF} + \beta.\Delta E_X^{Becke} + E_C^{local} + \gamma.\Delta E_C^{non-local} \tag{1.3}$$

where the three parameters α, β, and γ are determined by fitting to a diverse set of molecules. A number of organic compounds have already adopted this widely used method.[45–50] There are many functionals. e.g. B3PW91,[51] ωB97XD,[52] LC-ωHPBE,[53] CAM-B3LYP,[54] M062X,[55] etc. have been used in recent years due to their quantum computational accuracy.

Vibrational infrared frequencies are computed using harmonic approximation:

$$E \approx \frac{1}{2}\sum_{i=1}^{3N}\sum_{j=1}^{3N}\left(\frac{\partial^2 E}{\partial\bar{q}_i\partial\bar{q}_j}\right)_o \bar{q}_i\bar{q}_j \tag{1.4}$$

where

$$\bar{q}_i = \sqrt{m_i}(q_i - q_i^o) \tag{1.5}$$

Here m_i is the mass of atom i, and q_i represents the x-, y-, and z-coordinates for each of the N atoms in a molecule. Once vibrational frequencies are obtained, vibrational contribution (which is the most significant) of thermodynamic parameters are computed using following relations:

$$H = U = R\sum_{i=1}^{3N-6}\left[\frac{hc\tilde{v}_i}{2k_B} + \frac{hc\tilde{v}_i}{k_B(e^{hc\tilde{v}_i/k_BT}-1)}\right] \tag{1.6}$$

$$S = R\sum_{i=1}^{3N-6}\left[\frac{hc\tilde{v}_i}{k_BT(e^{hc\tilde{v}_i/k_BT}-1)} - \ln(1-e^{-hc\tilde{v}_i/k_BT})\right] \tag{1.7}$$

where H, U, and S denote enthalpy, internal energy, and entropy, respectively. Other parameters are calculated using standard thermodynamic relations.

Numerical differentiation can be used to determine the dipole moment (μ), mean polarizability (α), and anisotropy in polarizability (Δα) with an electric field magnitude of 0.001 au accordingly as:

$$\mu = (\mu_x^2 + \mu_y^2 + \mu_z^2)^{\frac{1}{2}} \tag{1.8}$$

$$\alpha = \frac{\alpha_{xx} + \alpha_{yy} + \alpha_{zz}}{3} \tag{1.9}$$

$$\Delta\alpha = \left[\frac{(\alpha_{xx}-\alpha_{yy})^2 + (\alpha_{yy}-\alpha_{zz})^2 + (\alpha_{zz}-\alpha_{xx})^2}{2}\right]^{\frac{1}{2}} \tag{1.10}$$

$$\beta = \left[(\beta_{xxx}+\beta_{xyy}+\beta_{xzz})^2 + (\beta_{yyy}+\beta_{xxy}+\beta_{yzz})^2 + (\beta_{zzz}+\beta_{xxz}+\beta_{yyz})^2\right]^{1/2} \tag{1.11}$$

Molar refractivity (MR) can be determined by the Lorenz-Lorentz formula:

$$MR = \left[\frac{n^2 - 1}{n^2 + 2}\right]\left(\frac{MW}{\rho}\right) = 1.333\pi N\alpha \tag{1.12}$$

The molar volume (MW/ρ), the Avogadro number (N), the refractive index (n), and the polarizability of the molecular system (α) are all present in this equation. For α, 1 au = 0.1482×10^{-24} esu.

Further, Koopmans theorem has been used to determine the electronic and global properties like ionization potential (I), electron affinity (A), electronegativity (χ), global hardness (η), and softness (S).[56]

According to Parr et al.,[57] the relationship between chemical potential (φ) and electronegativity (χ) is provided by the relation:

$$\varphi = \left(\frac{\partial E}{\partial N}\right)_{v(r)} = -\chi \tag{1.13}$$

where the terms denoted by $v(r)$ and μ are the exterior and electronic chemical potentials, respectively. The values of the electron affinity (A) and ionization potential (I) are given as:[58]

$$A = -E_{LUMO} \quad \text{and} \quad I = -E_{HOMO} \tag{1.14}$$

The I and A values can be used to calculate the molecule's global hardness (η)[59] and electronegativity (χ)[65]:

$$\chi = \frac{I + A}{2} \quad \text{and} \quad \eta = \frac{I - A}{2} \tag{1.15}$$

The electrophilicity index (ω), introduced by Parr et al.,[60] and chemical softness (S)[61] are given as:

$$S = \frac{1}{\eta} \quad \text{and} \quad \omega = \frac{\varphi^2}{2\eta} \tag{1.16}$$

The following equations can be used to calculate the molecules' electron donating capability (ω^-) and electron accepting capability (ω^+):

$$\omega^- = \frac{(3I + A)^2}{16(I - A)} \quad \text{and} \quad \omega^+ = \frac{(I + 3A)^2}{16(I - A)} \tag{1.17}$$

1.6 RESULTS AND DISCUSSION

1.6.1 4-Alkyl-4′-Cyanobiphenyl Series

It has been observed that there is an odd-even effect in the transition temperature for the nCB LC homologous series in the nematic region. Additionally, the odd-even effect was exclusively explored for the nematic zone.[29] Previously, the odd-even effect was not observed in smectic liquid crystals, and the transition temperature is almost linearly related to the number of carbon atoms in the alkyl chain. In this section, we discuss the electro-optical and electronic properties of the nCB series, which were studied theoretically for nematic and smectic phases of the homologues.[62–65]

1.6.2 Optimized Parameter Analysis

Figure 1.6 depicts the equilibrium geometries of the *n*CB liquid crystal series. The first methylene group's carbon (-CH_2) along the extended molecular axis lengthens the molecule but has no effect on its width, as in the case of 1CB, while the second methylene group's carbon increases the length and simultaneously widens the molecule, as in the case of 2CB. This applies to *n*CB molecules having an alkyl chain at the terminal end. As a result, the carbon in the second methylene group, that is the carbon in the odd position, increases the length but not by the same amount as the carbon of the first methylene group.[66] Depending on whether the terminal chain has an even or odd number of methyl(ene) groups, the dimensions of the *n*CB members change as the chain grows. This phenomenon, called the odd-even effect, was clearly seen in the case of the *n*CB series homologues.[29]

1.6.3 Electronic and Global Parameter Analysis

The trend of some electronic characteristics, including electron affinity (A), ionization potential (I), chemical hardness (η), and absolute-electronegativity (χ), has also been examined. These variables, often known as reactivity descriptors, describe how chemically reactive molecular systems are or how they interact with other species. Figure 1.7 displays how these parameters can vary with a homologous number of *n*CB series. In both nematic and smectic zones, it is evident that the ionization potential (Figure 1.7a) and global hardness (Figure 1.7d) vary with the number of carbon atoms and reflect the phenomena of odd-even effect; i.e., these variables are affected by the n parity value. The molecule's propensity to donate an electron or to become an electron donor and yield a cation is measured by the ionization potential. Chemical hardness gives a clear indication of how stable a molecule's electronic state is; in general, odd-number carbon chain molecules are more stable than even-number ones. Chemical hardness and electronegativity are indicators of a molecular system's stability and chemical reactivity.[67] The electron affinity of a molecule evaluates its propensity to absorb an electron, transform into an electron acceptor, and produce an anion, in contrast to ionization potential.

Strong electron affinity exists for $n = 2$ and decreases with the number of carbon atoms in an alkyl chain in the nematic zone, but it increases with the number of carbon atoms in the smectic zone, as

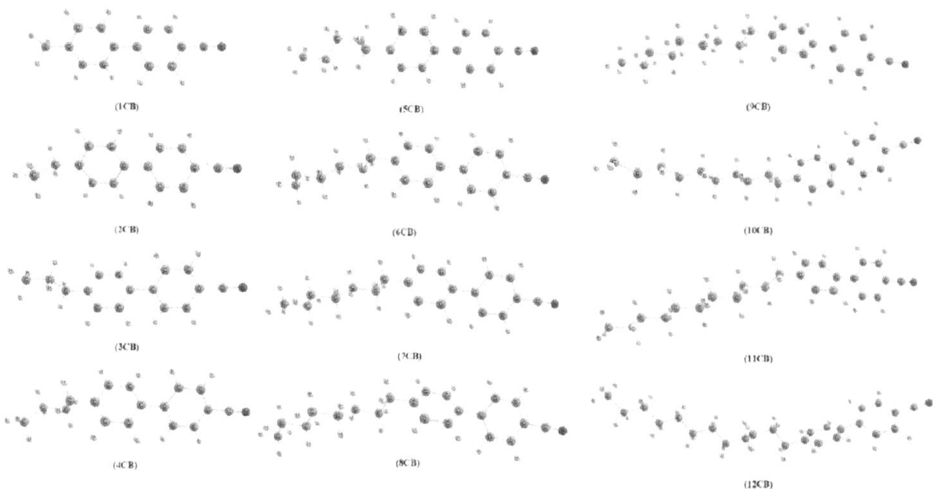

FIGURE 1.6 Optimized geometries of *n*CB liquid crystalline series calculated at the B3LYP/6–311++ G(d,p) level.

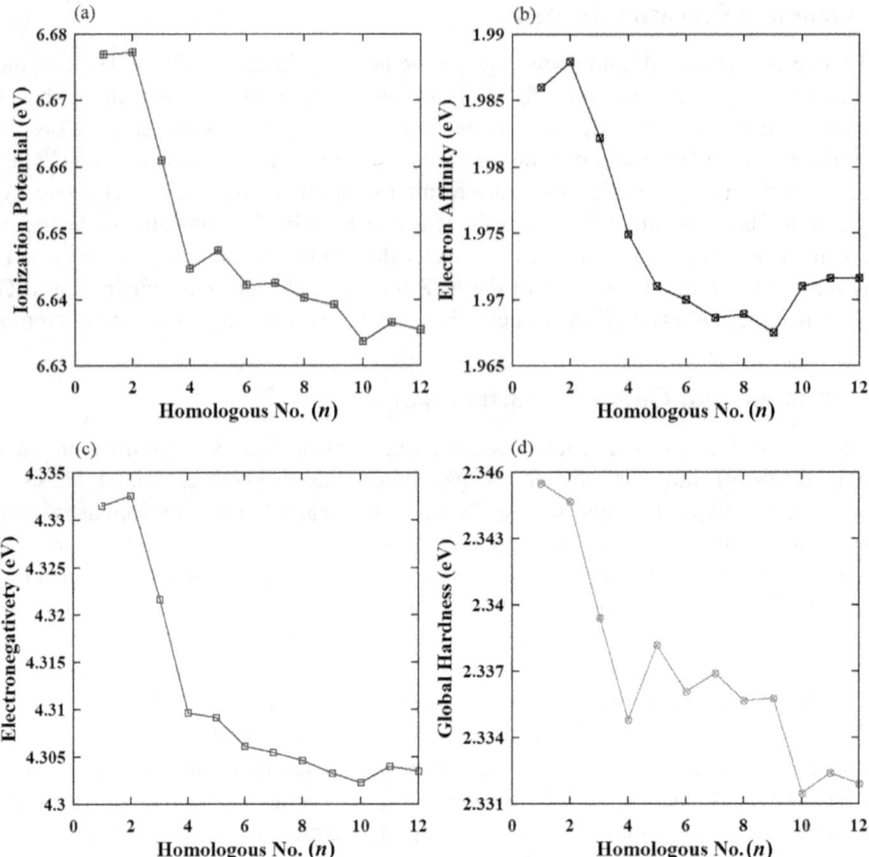

FIGURE 1.7 Variation of (a) ionization potential, (b) electron affinity, (c) electronegativity, and (d) chemical hardness with regard to the number of carbon atoms in the alkyl chain of the *n*CB liquid crystalline series.

shown in Figure 1.7b. This is the key characteristic that explains the liquid crystalline behavior of the *n*CB series. In the same way as that of the chemical hardness and ionization potential, the *n*CB's electronegativity demonstrates that this parameter is quite high for $n = 2$ and that it goes down with the increase of the number of carbon atoms in the alkyl chain (Figure 1.7c). All of the estimated characteristics, with the exception of electron affinity, indicate that the odd-even effect in the *n*CB family members persists beyond the nematic range and spreads into the smectic region. Table 1.1 provides values for various global parameters.

1.6.4 Electro-Optical Parameter Analysis

We have examined a number of electro-optical parameters, including dipole moment, mean polarizability, anisotropy in polarizability, and hyperpolarizability, to observe the odd-even effect in *n*CB liquid crystalline series. Table 1.2 contains the values of these parameters. An indication of the shape and charge distribution of a molecule can be found in its dipole moment. There is an odd-even effect in the dipole moment depicted in Figure 1.8a, which may be seen as the dipole moment of an even-number carbon atom molecule and which is greater than the dipole moment of an odd-number carbon atom of the methylene group. Moreover, as previously described[63] this variance extends into the smectic zone in addition to the nematic region. Clearly, it has to do with the molecular design and charge distribution, putting aside the intermolecular interactions that exist in bulk liquid

TABLE 1.1
Electronic and Global Reactivity Descriptor Parameters of the *n*CB (*n* = 1 to 12) Series

Homologous No. (*n*)	*I* (eV)	*A* (eV)	χ(eV)	η(eV)
1	6.677	1.986	4.331	2.345
2	6.677	1.987	4.332	2.344
3	6.661	1.982	4.321	2.339
4	6.644	1.974	4.309	2.334
5	6.647	1.971	4.309	2.338
6	6.642	1.970	4.306	2.336
7	6.642	1.968	4.305	2.336
8	6.640	1.968	4.304	2.335
9	6.639	1.967	4.303	2.335
10	6.633	1.971	4.302	2.331
11	6.636	1.971	4.304	2.332
12	6.635	1.971	4.303	2.331

TABLE 1.2
Electro-Optical Parameters of the *n*CB (*n* = 1 to 12) Series

Homologous No. (*n*)	Dipole Moment (Debye)	Mean Polarizability (α) (Bohr³)	Anisotropy in Polarizability (Δα) (Bohr³)	Hyperpolarizability (β) (Bohr⁵)
1	6.110	183.170	181.402	1511.922
2	6.131	191.823	195.274	1364.240
3	6.180	204.395	209.518	1699.821
4	6.305	211.953	222.328	1918.433
5	6.283	222.531	235.840	1926.032
6	6.376	227.689	248.500	2042.674
7	6.338	237.322	261.775	2057.229
8	6.402	238.314	274.594	2104.940
9	6.364	246.241	287.589	2112.702
10	6.426	248.307	300.581	2169.996
11	6.378	255.610	313.439	2139.265
12	6.436	256.127	326.261	2183.852

crystals. The homologous number of the alkyl chain is directly related to the anisotropy in polarizability, as seen in Figure 1.8c. It is widely acknowledged that torsional angle has the greater impact on the anisotropy in polarizability values[68] The torsional angle alters as the carbon atoms are added to the alkyl chain; which, in turn, also increases the anisotropy in polarizability. For even and odd numbers of carbon atoms, the mean polarizability and hyperpolarizability values are somewhat different (Figures 1.8b and 1.8d). The odd-even effect has been observed for $n \geq 7$ in the case of mean polarizability, and below this, polarizability grows essentially linearly. In addition to the negligible contribution from acoustic phonons, it is typically considered that the first-order hyperpolarizability of organic molecules is entirely of pure electrical origin.[69] Although the hyperpolarizability values do indeed exhibit an odd-even effect, this impact becomes less pronounced as the number of carbon atoms increases. It is anticipated that changes to the dihedral angle or deviations from planarity will cause a decrease in hyperpolarizability. It seems pertinent to note that the planarity requirement must be met for a molecule to be nonlinearly optically active. Thus the polarizabilities and

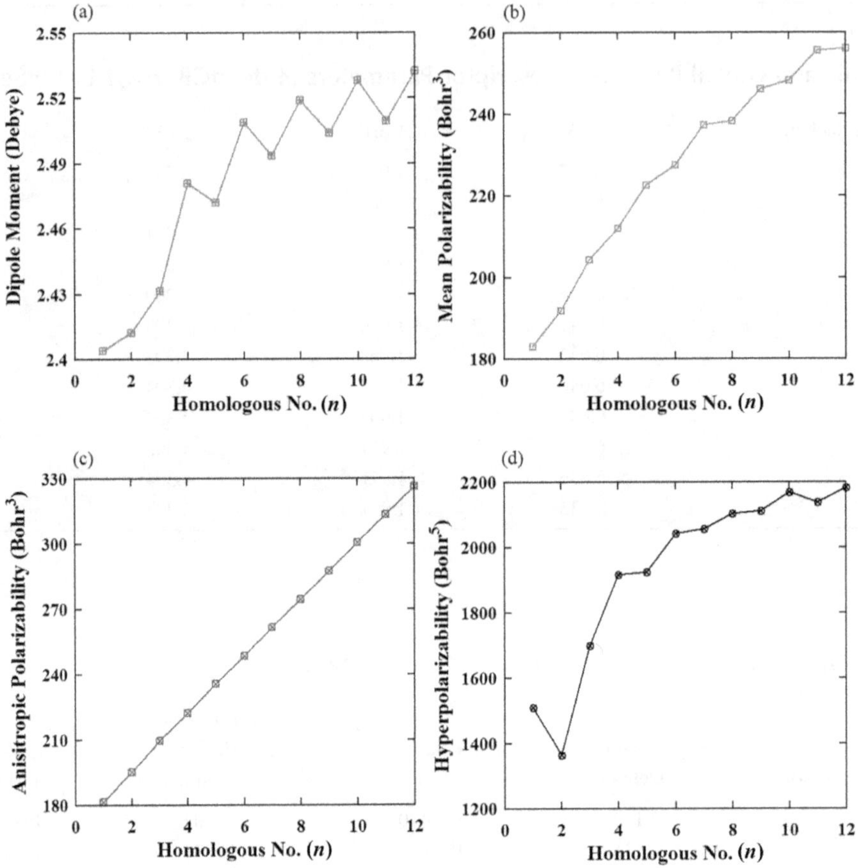

FIGURE 1.8 Variations in the (a) dipole moment, (b) mean polarizability, (c) anisotropic polarizability, and (d) hyperpolarizability of the homologues of the *n*CB liquid crystalline series with regard to the number of carbon atoms in the alkyl chain.

hyperpolarizabilities control the system's nonlinear optical (NLO) properties in addition to the cross sections of various scattering and collision processes and the strength of molecular interactions.[70]

1.7 4-*N*-ALKOXY-4′-CYANOBIPHENYL LIQUID CRYSTAL SERIES

This section covers the theoretically explored Raman spectra, UV-vis spectra, electro-optical, global (electronic), and thermal features of the *n*OCB LC series.[71] The homologues of 4-*n*-alkoxy-4′-cynobiphenyl series (*n*OCB; $n = 1$ to 12) are optimized by the DFT/B3LYP method and are shown in Figure 1.9. The behavior of *n*OCB LC molecules is nematic for $n = 1$ to 7 and smectic up to 12. The main difference between the homologous *n*OCB and *n*CB series is that the *n*OCB molecules have an oxygen atom between the biphenyl ring and the alkyl unit. Hence it becomes interesting to see how the extra oxygen atom affects the quantities of the *n*OCB LC series.

1.7.1 Thermal Parameter Variation with Homologous Number

Vibrational frequency calculations at the B3LYP/6–31G(d,p) level have been done for procuring thermal parameters after the optimization operations. Figure 1.10a shows the correlation between the number of carbon atoms present in an alkoxy chain and thermal properties such as Gibbs free

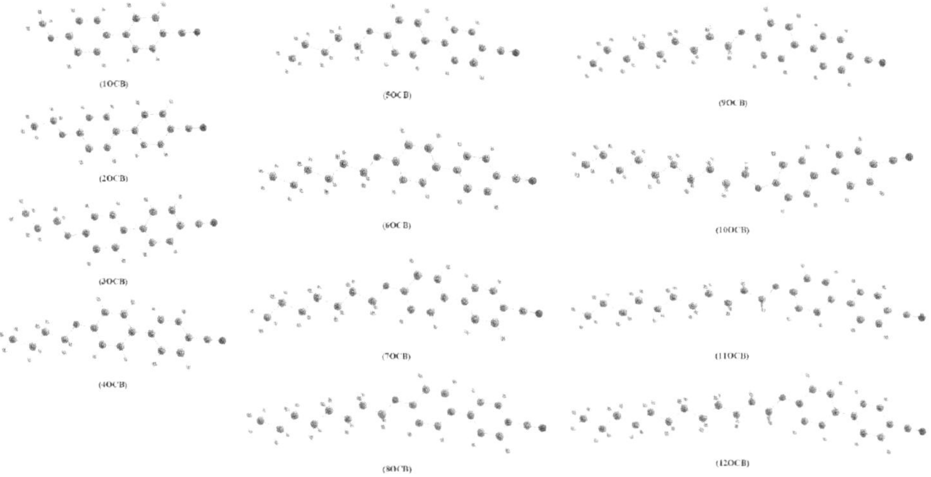

FIGURE 1.9 Equilibrium geometries of *n*OCB liquid crystal homologous series computed at the B3LYP/6–31G(d,p) level.

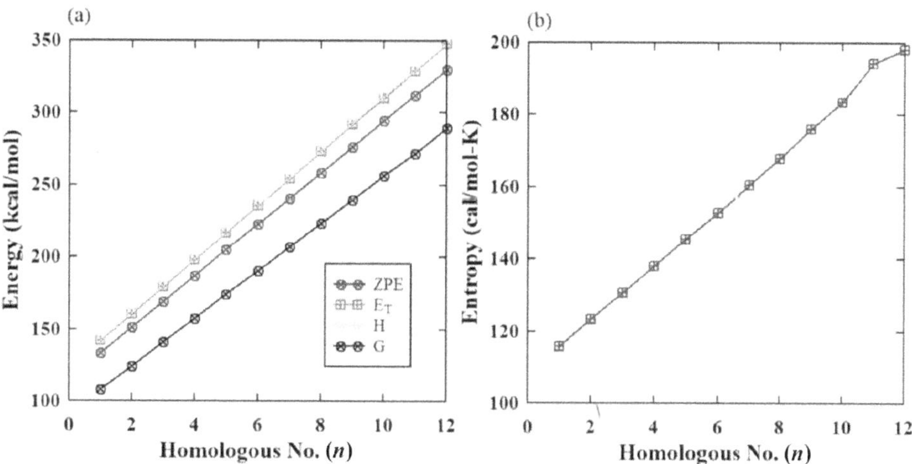

FIGURE 1.10 Variation of (a) thermal energy (*ET*), zero point energy (*ZPE*), Gibbs free energy (*G*), and enthalpy (*H*); (b) entropy with respect to homologous number for *n*OCB LC series.

energy, zero-point energy, thermal energy, and enthalpy. Figure 1.10b depicts the variance in entropy. The addition of carbon atoms in the alkoxy chain causes an increase in all the thermal parameters. Figure 1.10a vividly reflects that thermal energy (E_T) and enthalpy (H) are in close proximity for each individual homologue of the *n*OCB (Table 1.3). Further, as evident from Figure 1.10b, with an increase in the number of carbon atoms in the alkoxy chain, entropy of the individual homologue increases, leading to the inference that disorder assumes a higher value for higher homologues. This means the higher the homologue is, greater will be the disorder in the molecular system.

1.7.2 Electronic and Global Parameter Variation with Homologous Number

Figures 1.11a–c depict the energy variation of the HOMO (E_{HOMO}) and LUMO (E_{LUMO}), as well as the associated energy gap between the HOMO-LUMO (ΔE) with the homologous number for *n*OCB (n =1 to 12) LC series. Table 1.4 contains the values of these quantities. With increasing n,

TABLE 1.3
Thermal Parameters of the *n*OCB (*n* = 1 to 12) Series

Homologous No. (*n*)	ZPE (kcal/mol)	Thermal Energy ($E_{thermal}$) (kcal/mol)	Enthalpy (*H*) (kcal/mol)	Gibbs Free Energy (*G*) (kcal/mol)	Entropy (cal/mol-K)
1	133.588	141.951	142.544	108.030	115.761
2	151.337	160.589	161.181	124.414	123.318
3	169.228	179.352	179.945	140.961	130.754
4	187.133	198.113	198.704	157.532	138.093
5	205.041	216.874	217.466	174.081	145.515
6	222.926	235.618	236.211	190.641	152.842
7	240.818	254.371	254.963	207.111	160.499
8	258.692	273.106	273.699	223.644	167.886
9	276.542	291.828	292.421	239.946	176.002
10	294.408	310.558	311.151	256.452	183.458
11	312.241	329.271	329.864	271.927	194.320
12	330.254	348.017	348.609	289.496	198.266

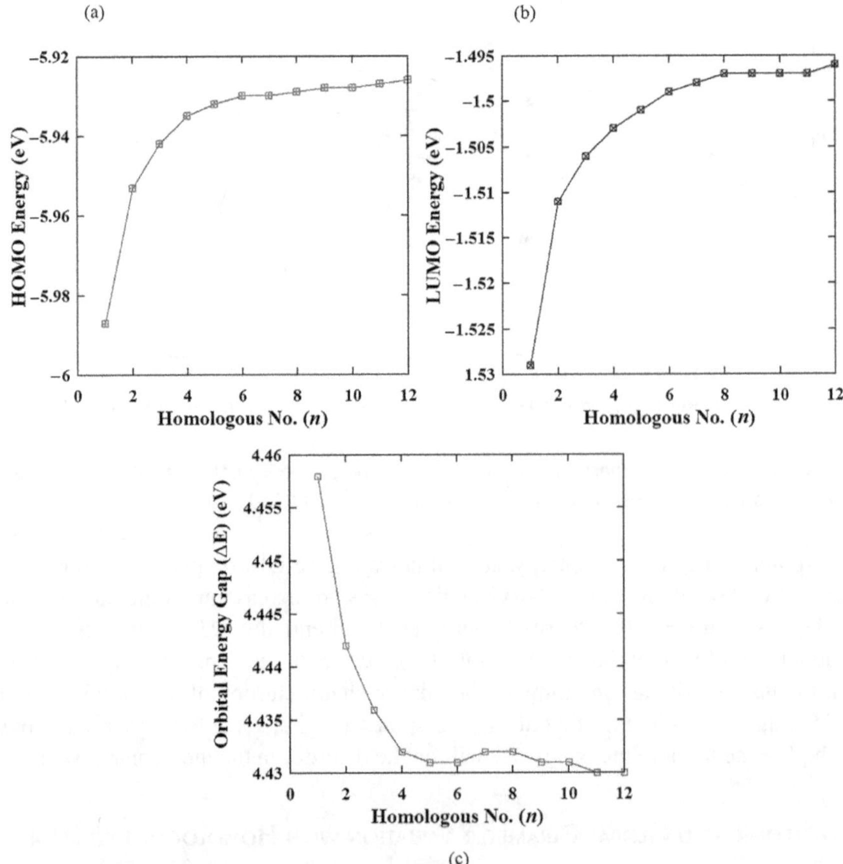

FIGURE 1.11 (a) HOMO, (b) LUMO, and (c) orbital energy gap (ΔE) variation with the homologous number for *n*OCB LC series.

TABLE 1.4
Orbital Energy Gap (ΔE), HOMO Energy, and LUMO Energy of the *n*OCB (n = 1 to 12) Series

Homologous No. (n)	E_{HOMO} (eV)	E_{LUMO} (eV)	ΔE (eV)
1	−5.987	−1.529	4.458
2	−5.953	−1.511	4.442
3	−5.942	−1.506	4.436
4	−5.935	−1.503	4432
5	−5.932	−1.501	4.431
6	−5.930	−1.499	4.431
7	−5.930	−1.498	4432
8	−5.929	−1.497	4.432
9	−5.928	−1.497	4.431
10	−5.928	−1.497	4.431
11	−5.927	−1.497	4.430
12	−5.926	−1.496	4.430

the values of HOMO, LUMO, and ΔE are poorly changed. As the homologous number increases, all these quantities descend gradually (Table 1.4). These molecules occupy the region of ultraviolet transparency, according to the energy gaps (ΔE) of the *n*OCB series.

The variation of global parameters for the *n*OCB (n = 1 to 12) series with homologous numbers is also shown in Figure 1.12. Their respective values are given in Table 1.5. These variables are known as reactivity descriptors because they describe the chemical reactivity of molecular systems. These global descriptors are followed by orbital energy gaps (ΔE), and they decrease as the homologous number increases. According to global hardness variation, lower homologues are likely to be more stable than higher ones. Similarly, the variation in electron affinities suggests that lower homologues can readily take electrons to transform themselves into electron acceptors in order to generate anions.

1.7.3 Electro-Optical Parameter Variation with Homologous Number

Figure 1.13 shows the total energy, dipole moment, mean polarizability, anisotropy in polarizability, and molar refractivity. Table 1.6 provides the values of electro-optical parameters. The addition of carbon atoms in the alkoxy chain linearly increases the total molecular energy of the *n*OCB homologues (Figure 1.13a). Thus, it can be inferred that the configurational stability of the homologues of *n*OCB (n = 1 to 12) reflects ascending values with the increase of alkoxy chain length. That is, higher homologues are more stable as compared to their lower homologues; thus 12OCB poses the most stable configuration among all the chosen molecules. Similarly, as the homologous number rises, the dipole moment also increases (Figure 1.13b). The occurrence of the odd-even effect in the dipole moment was explored by similar investigations on *n*CB series[63] Figure 1.13c illustrates the dependence of mean polarizability and anisotropic polarizability on the homologous number. Hence, anisotropic polarizability is largely influenced by the torsional angles.[68]

Due to the addition of carbon atoms in the alkoxy chain, the torsion angle of the homologues changes, leading to an increment in the value of the anisotropy in polarizability. A linear relationship for the homologous number is also indicated by the molar refractivity plot, which is depicted in Figure 1.13d. Figures 1.14a, and 1.14b, respectively, demonstrate the comparison between estimated mean polarizability and anisotropy in polarizability values of *n*OCB (n = 9 to 12) molecules with experimental data.[72] In Figure 1.13c, the components of α and $\Delta\alpha$ are reported in atomic units ($Bohr^3$), while the same (calculated values) are converted into cm^3 ($1\ Bohr^3 = 0.1482 \times 10^{-24}\ cm^3$) in Figures 1.14a and 1.14b.

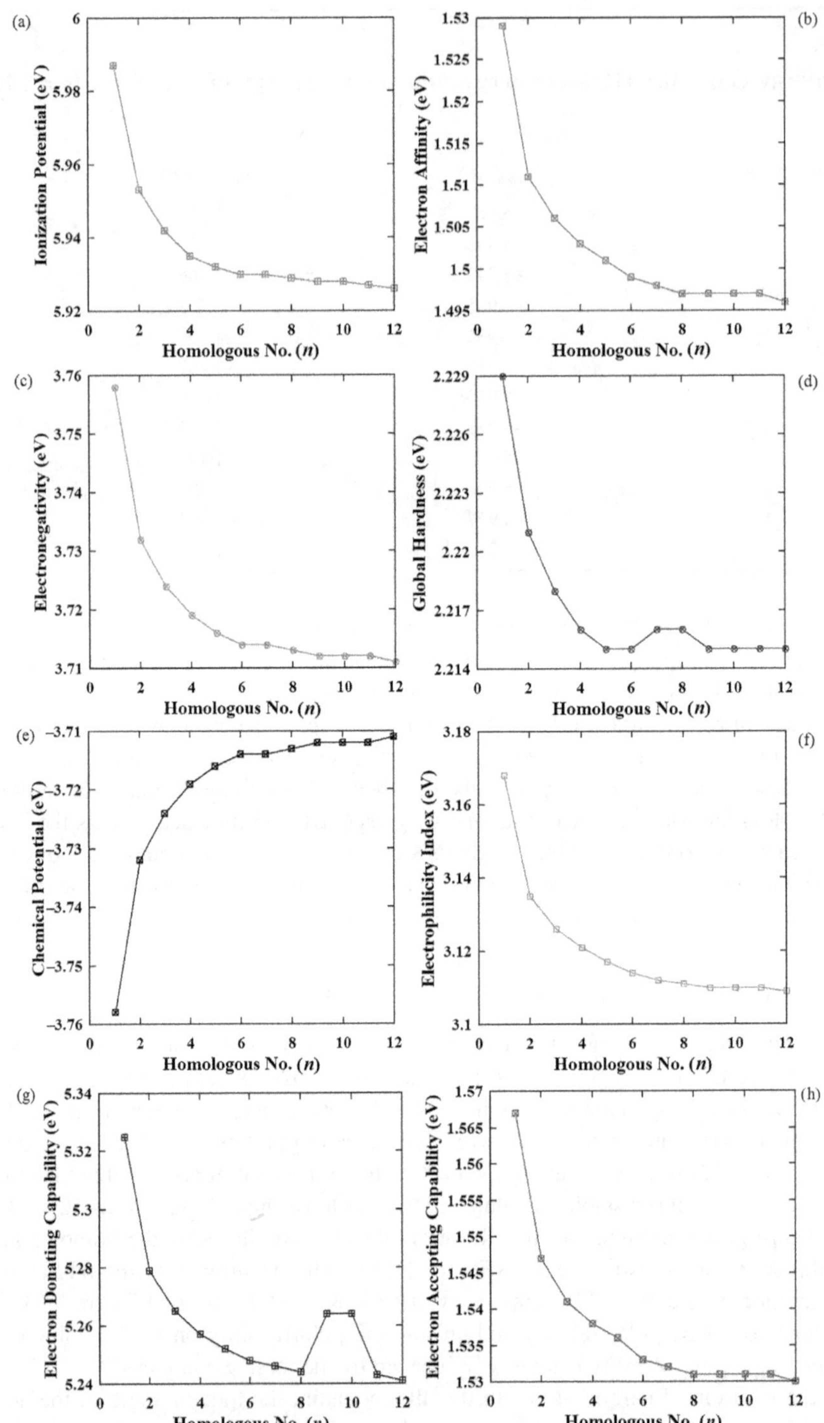

FIGURE 1.12 Dependence of parameters like (a) ionization potential, (b) electron affinity, (c) electronegativity, (d) global hardness, (e) chemical potential, (f) electrophilicity index, (g) electron donating capability, and (h) electron accepting capability upon the homologous number for *n*OCB LC series.

TABLE 1.5
Electronic and Global Parameters for the Homologues of *n*OCB (*n* = 1 to 12) LC Series

Homologous No. (*n*)	*I* (eV)	A (eV)	χ(eV)	η(eV)	*S*(eV)	φ(eV)	ω(eV)	ω^-(eV)	ω^+(eV)
1	5.987	1.529	3.758	2.229	0.448	−3.758	3.168	5.325	1.567
2	5.953	1.511	3.732	2.221	0.450	−3.732	3.135	5.279	1.547
3	5.942	1.506	3.724	2.218	0.451	−3.724	3.126	5.265	1.541
4	5.935	1.503	3.719	2.216	0.451	−3.719	3.121	5.257	1.538
5	5.932	1.501	3.716	2.215	0.451	−3.716	3.117	5.252	1.536
6	5.930	1.499	3.714	2.215	0.451	−3.714	3.114	5.248	1.533
7	5.930	1.498	3.714	2.216	0.451	−3.714	3.112	5.246	1.532
8	5.929	1.497	3.713	2.216	0.451	−3.713	3.111	5.244	1.531
9	5.928	1.497	3.712	2.215	0.451	−3.712	3.110	5.264	1.531
10	5.928	1.497	3.712	2.215	0.451	−3.712	3.110	5.264	1.531
11	5.927	1.497	3.712	2.215	0.451	−3.712	3.110	5.243	1.531
12	5.926	1.496	3.711	2.215	0.451	−3.711	3.109	5.241	1.530

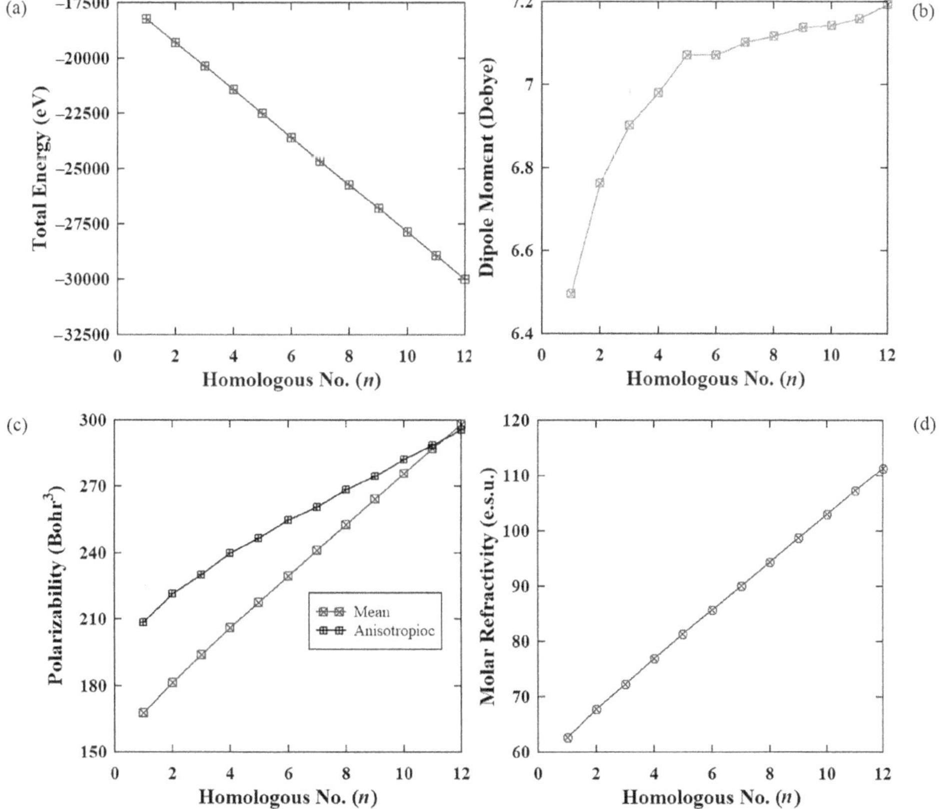

FIGURE 1.13 Variation of electro-optical parameters: (a) total energy, (b) dipole moment, (c) polarizability (mean and anisotropic), and (d) molar refractivity with the homologous number of *n*OCB LC series.

TABLE 1.6
Electro-Optical Parameters of the *n*OCB (*n* = 1 to 12) Series

Homologous No. (*n*)	Total Energy (eV)	Dipole Moment (Debye)	Mean Polarizability (α) (Bohr³)	Anisotropy in Polarizability (Δσ) (Bohr³)	Molar Refractivity (MR) in e.s.u.
1	−18234.068	6.496	167.786	208.671	62.677
2	−19304.099	6.763	181.447	221.755	67.806
3	−20373.956	6.903	193.753	230.265	72.377
4	−21443.810	6.981	205.977	239.894	76.944
5	−22513.669	7.072	217.811	246.538	81.364
6	−23583.527	7.072	229.583	254.825	85.762
7	−24653.384	7.103	241.177	260.764	90.093
8	−25723.242	7.117	252.786	268.681	94.429
9	−26793.098	7.138	264.316	274.711	98.737
10	−27862.956	7.143	275.836	282.426	103.040
11	−28932.812	7.159	287.332	288.626	107.334
12	−30002.611	7.194	298.202	295.691	111.395

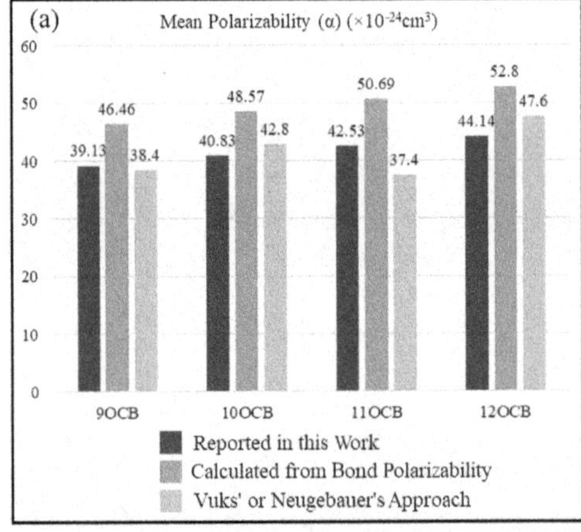

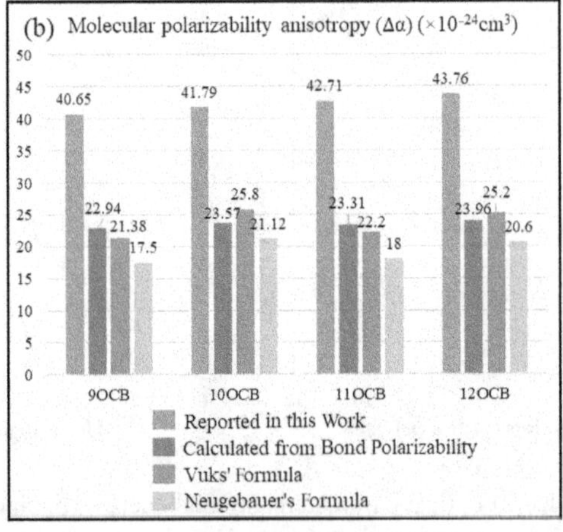

FIGURE 1.14 (a) Mean polarizability (α) and (b) molecular polarizability anisotropy (Δα) of the *n*OCB (*n* = 9 to 12) as computed by B3LYP/6–31G(d,p) method.

1.7.4 Raman and Absorption Spectra Analysis

The DFT/B3LYP/6–31G(d,p) approach has been used to examine the Raman spectra of the members of the *n*OCB LC series. Figure 1.15 displays the Raman spectra of the 8OCB LC molecule in the 400 cm^{-1} to 3500 cm^{-1} regions. The 8OCB, 9OCB, 10OCB, 11OCB, and 12OCB molecules all have six distinct strong vibrational peaks. Table 1.7 compares the calculated Raman frequencies of relevant vibrational modes for the 8OCB molecule and shows that they are in good agreement with the experimental data.[73, 74] At the wave number of 3111.19 cm^{-1} and 3037.40 cm^{-1}, peak one and peak two are connected to asymmetric and symmetric CH_3 mode stretching, respectively. Peak three is due to CN stretching, which is seen at 2344.11 cm^{-1}, and peak four is caused by C-C stretching of the biphenyl ring, which is seen at 1671.22 cm^{-1}. The fifth peak, at a wavelength of 1316.89 cm^{-1}, is a band of perpendicular deformations that combines symmetrical and asymmetrical deformations of

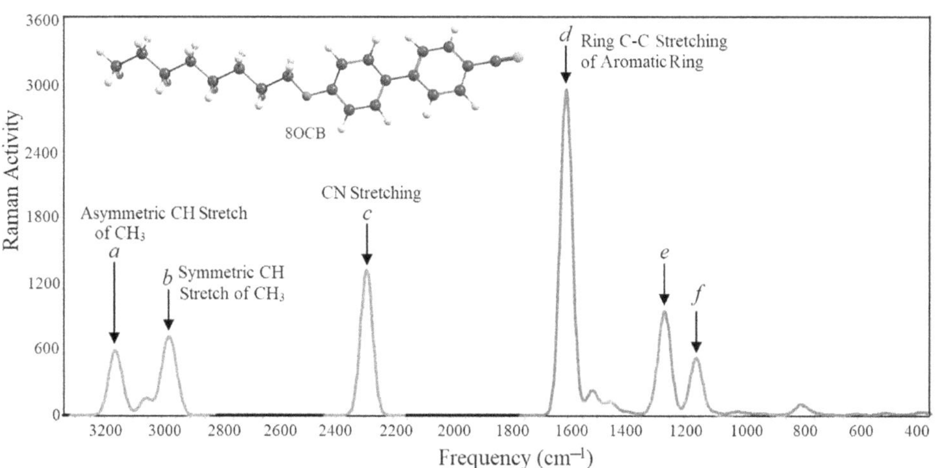

FIGURE 1.15 Raman spectrum of the 8OCB LC compound. The letters in Table 1.7 refer to distinct vibrations.

TABLE 1.7
Raman Spectrum of 8OCB LC Molecule

Raman Frequency (cm^{-1})			
Calculated by B3YP/6–31G(d,p)	**Experimental[69, 70]**	**Peaks**	**Assignment of the Raman Peaks Shown in Figure 1.2**
3111.19	2962~2988	*a*	Asymmetric CH stretch of CH_3
3037.40	2878	*b*	Symmetric CH stretch of CH_3
2344.11	2234	*c*	CN stretching
1671.22	1522	*d*	Ring C-C stretching
1316.89	1284	*e*	Combinational band of symmetric and asymmetric perpendicular deformations of CH_2 groups in the aliphatic chain
1206.49	1185	*f*	In-plane deformation of CH bonds of the biphenyl moiety

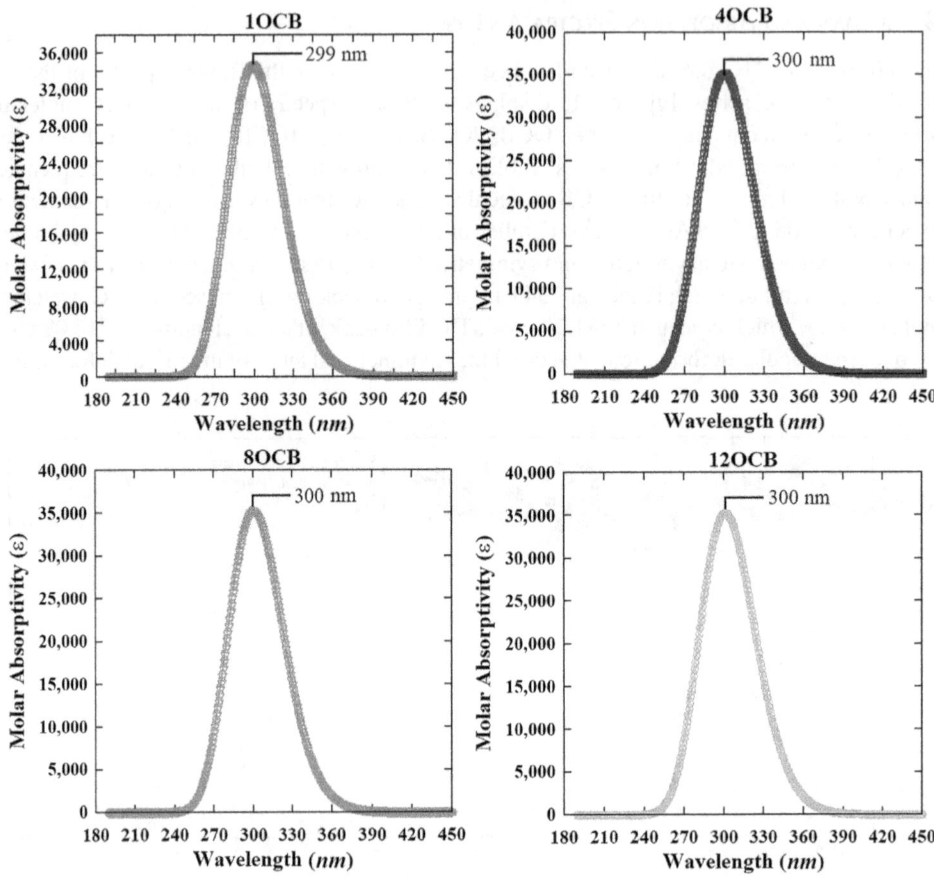

FIGURE 1.16 UV-vis spectra of *n*OCB (*n* = 1, 4, 8, and 12) liquid crystal molecules computed by the ZINDO technique.

TABLE 1.8
Oscillator Strength, Excitation Energy and Wavelength of *n*OCB LC Molecules as Computed by ZINDO Method

Molecule	Excited State	Oscillator Strength (*f*)	Excitation Energy (eV)	λ_{max} (nm)
1OCB	1	0.8324	4.138	299.61
	2	0.0233	4.302	288.23
	3	0.0015	4.355	284.68
2OCB	1	0.8396	4.131	300.13
	2	0.0238	4.298	288.41
	3	0.7965	4.031	307.61
3OCB	1	0.8444	4.126	300.45
	2	0.0239	4.298	288.48
	3	0.0014	4.352	284.89
4OCB	1	0.8470	4.126	300.52
	2	0.0239	4.297	288.49
	3	0.0013	4.352	284.90

(Continued)

TABLE 1.8 (Continued)

Molecule	Excited State	Oscillator Strength (*f*)	Excitation Energy (eV)	λ_{max} (nm)
5OCB	1	0.8485	4.126	300.52
	2	0.0239	4.297	288.50
	3	0.0014	4.352	284.90
6OCB	1	0.8496	4.126	300.50
	2	0.0239	4.297	288.50
	3	0.0014	4.352	284.89
7OCB	1	0.8501	4.127	300.44
	2	0.0240	4.298	288.49
	3	0.0014	4.352	284.87
8OCB	1	0.8508	4.127	300.45
	2	0.0240	4.298	288.49
	3	0.0014	4.352	284.87
9OCB	1	0.8514	4.126	300.47
	2	0.0240	4.298	288.50
	3	0.0014	4.352	284.88
10OCB	1	0.8519	4.126	300.50
	2	0.0240	4.297	288.50
	3	0.0014	4.352	284.88
11OCB	1	0.8526	4.125	300.55
	2	0.0240	4.297	288.51
	3	0.0013	4.352	284.90
12OCB	1	0.8489	4.115	301.28
	2	0.0250	4.288	289.23
	3	0.0009	4.349	285.04

CH_2 groups in the aliphatic chain, while the last one, at a wavelength of 1206.49 cm^{-1}, is connected to an in-plane deformation of CH bonds of the biphenyl moiety. Further, it is important to mention that all the five homologues (n = 8 to 12) of nOCB LC series furnish almost identical spectra and hence similar values of Raman frequencies.

The ZINDO technique was used to study electronic absorption spectra. The similar nature of absorption spectra is rendered by the homologues of nOCB (n = 1 to 12). The UV-vis spectra of only four homologues of the nOCB (n = 1, 4, 8, and 12) series have been shown in Figure 1.16. Table 1.8 provides a list of the excited states, oscillator strength (*f*), excitation energy (*E*), and excitation wavelength (λ_{max}). At 299.61 nm (4.138 eV) with oscillator strength of 0.8324, 1OCB experiences one strong electronic transition from HOMO to LUMO (MO contribution 86.5%). Corresponding to 288.23 nm (4.302 eV) and 284.68 nm (4.355 eV), two more electronic transitions take place, with oscillator strengths of 0.0233 and 0.0015, respectively. At a wavelength of 300.52 nm, the 4OCB molecule exhibits a strong absorption band. With 86.5% of the MO contribution, the transition from HOMO to LUMO occurs at this wavelength. The oscillator strength and the related excitation energy correspond to 4.126 eV and 0.8470, respectively. Moreover, the strong electronic transitions in the 8OCB and 12OCB molecules occur at 300.45 nm (HOMO to LUMO with MO contribution of 86.4%) and at 300.55 nm (HOMO to LUMO with MO contribution of 86.2%), respectively. The experimental values of molar absorptivity (ε) and wavelength of maximum absorption (λ_{max}) for the 8OCB LC molecule are reported to be 20.1×10^3 M^{-1} $cm^{-1}$and 297 nm, respectively,[75] whereas in this study, the molar absorptivity and maximum wavelength for the 8OCB molecule are found to be 35.3×10^3 M^{-1} cm^{-1} and 300.45 nm, respectively.

1.8 CONCLUSIONS AND VIEWPOINTS

This chapter discussed various aspects of thermodynamic, electronic, optical, and spectroscopic aspects of alkyl and alkoxy cyanobiphenyls. Both the homologous series possess odd-even effect. This investigation demonstrated the presence of phenomena of the odd-even effect in the case of the *n*CB series only, which is in conformity with experimental observations. The plots of the thermal, electronic, and electro-optical properties for the *n*OCB series elucidate a linear dependence on homologous number. While ascending the homologous series, there exists a linear increase in the values of the dipole moment, mean polarizability, anisotropy in polarizability, and molar refractivity. In both LC series, the global parameters indicate that lower homologues are more prone to anionic behavior. The higher homologues appear to be more disordered. For further research and applications of these homologues of the *n*CB and *n*OCB liquid crystalline series in fabrications of NLO materials, electro-optical parameters can offer improved insights and auxiliary information.

1.8.1 Acknowledgment

Dipendra Sharma is thankful to UGC, New Delhi, India, for providing the start-up grant.

1.8.2 Conflict of Interests

The authors declare no conflict of interest.

REFERENCES

[1] de Gennes P-G, Prost J. The Physics of Liquid Crystals [Internet] (2nd ed.). Oxford Sci. Publ. New York: Oxford University Press, 1995 [cited 2023 Apr 1]. Available from: https://global.oup.com/academic/product/the-physics-of-liquid-crystals-9780198517856.

[2] Chandrasekhar S. Liquid Crystals [Internet] (2nd ed.). Cambridge University Press, 1992 [cited 2023 Apr 1]. Available from: www.cambridge.org/core/books/liquid-crystals/7972CBEB90A0F90F0D23C9B83B6A46D2.

[3] Priestley EB, Wojtowicz PJ, Sheng P, eds. Introduction to Liquid Crystals. New York, NY: Springer, 1975.

[4] Reinitzer F. Beiträge zur Kenntniss des cholesterins. Monatshefte für Chemie [Internet]. 1888 [cited 2023 Jun 6];9:421–441. Available from: https://link.springer.com/article/10.1007/BF01516710.

[5] Lehmann O. Über fliessende krystalle. Zeitschrift für Phys Chemie [Internet]. 1889 [cited 2023 Jun 6];4U:462–472. Available from: www.degruyter.com/document/doi/10.1515/zpch-1889-0434/html?lang=en.

[6] Khoo I-C. Liquid Crystals. New Jersey, NJ: John Wiley & Sons, Inc., 2022.

[7] Petrov AG. The Lyotropic State of Matter: Molecular Physics and Living Matter Physics. Boca Raton, FL: CRC Press, 1999.

[8] Neto AMF, Salinas SRA, Press OU. Physics of Lyotropic Liquid Crystals: Phase Transitions and Structural Properties. Oxford: Oxford University Press, 2005.

[9] Collings PJ, Patel SJ. Handbook of Liquid Crystal Research [Internet]. Patel, SJ. and Collings PJ, (eds.). New York: Oxford University Press, 2006 [cited 2023 Jun 6]. Available from: www.tandfonline.com/doi/abs/10.1080/10587250008024819.

[10] Jákli A, Saupe A. One- and Two-Dimensional Fluids : Properties of Smectic, Lamellar and Columnar Liquid Crystals [Internet]. London: CRC Press; 2006 [cited 2023 Jun 6]. Available from: www.routledge.com/One-and-Two-Dimensional-Fluids-Properties-of-Smectic-Lamellar-and-Columnar/Jakli-Saupe/p/book/9780367390761.

[11] Bahadur B. Liquid Crystals—Applications and Uses. Bahadur B, (ed.). Singapore: World Scientific, 1990.

[12] Lagerwall S. Ferroelectric and Antiferroelectric Liquid Crystals, Weinheim: Wiely-VCH, 1999.

[13] Singh S. Liquid Crystals. Singapore: World Scientific, 2002.

[14] Vertogen G, de Jeu, WH. Thermotropic Liquid Crystals, Fundamentals [Internet]. Heidelberg: Springer, 1988 [cited 2023 Apr 1]. Available from: http://link.springer.com/10.1007/978-3-642-83133-1.

[15] Bahr C. Chirality Liquid Crystals. New York: Springer, 2001.

[16] Goodby J. Handbook of Liquid Crystals [Internet]. Wiley-VCH, 2014 [cited 2023 Apr 1]. Available from: www.wiley-vch.de/de/fachgebiete/ingenieurwesen/materialwissenschaften-10ms/soft-matter-10msg/handbook-of-liquid-crystals-978-3-527-32773-7.

[17] Castillo-Vallés M, Martínez-Bueno A, Giménez R, et al. Beyond liquid crystals: New research trends for mesogenic molecules in liquids. J Mater Chem C [Internet]. 2019 [cited 2023 Apr 1];7:14454–14470. Available from: https://pubs.rsc.org/en/content/articlehtml/2019/tc/c9tc04179f.

[18] Chandrasekhar S, Ranganath GS. Discotic liquid crystals. Reports Prog Phys [Internet]. 1990 [cited 2023 Jun 6];53:57. Available from: https://iopscience.iop.org/article/10.1088/0034-4885/53/1/002.

[19] Lagerwall JPF, Scalia G. A new era for liquid crystal research: Applications of liquid crystals in soft matter nano-, bio- and microtechnology. Curr Appl Phys. 2012;12:1387–1412.

[20] Dwivedi MK, Yadav RA, Tiwari SN. Liquid crystalline behaviour of N-cetyl N,N,N,-trimethyl ammonium bromide: An experimental study. Prog Cryst Growth Charact Mater. 2006;52:91–98.

[21] Brown GH, Wolker JJ. Liquid Crystals and Biological Structures. New York: Academic Press, 1979.

[22] Tsvetkov V. On molecular order in the anisotropic liquid phase. Acta Physicochim. URSS. 1942; 16:132–47.

[23] Clark NA, Lagerwall ST. Submicrosecond bistable electro-optic switching in liquid crystals. Appl Phys Lett [Internet]. 2008 [cited 2023 Apr 1];36:899. Available from: https://aip.scitation.org/doi/abs/10.1063/1.91359.

[24] Chandani ADL, Ouchi Y, Takezoe H, et al. Relation between spontaneous polarization and rotational viscosity in enantiomeric mixtures of ferroelectric liquid crystals. Jpn J Appl Phys [Internet]. 1988 [cited 2023 Apr 1];27:L276–L279. Available from: https://iopscience.iop.org/article/10.1143/JJAP.27.L276.

[25] Al-Zangana S, Turner M, Dierking I. A comparison between size dependent paraelectric and ferroelectric BaTiO3 nanoparticle doped nematic and ferroelectric liquid crystals. J Appl Phys [Internet]. 2017 [cited 2023 Apr 1];121:085105. Available from: https://aip.scitation.org/doi/abs/10.1063/1.4976859.

[26] Bubnov A, Iwan A, Cigl M, et al. Photosensitive self-assembling materials as functional dopants for organic photovoltaic cells. RSC Adv [Internet]. 2016 [cited 2023 Apr 1];6:11577–11590. Available from: https://pubs.rsc.org/en/content/articlehtml/2016/ra/c5ra23137j.

[27] Iwan A, Sikora A, Hamplová V, et al. AFM study of advanced composite materials for organic photovoltaic cells with active layer based on P3HT:PCBM and chiral photosensitive liquid crystalline dopants. http://dx.doi.org/101080/0267829220151011243 [Internet]. 2015 [cited 2023 Apr 1];42:964–972. Available from: www.tandfonline.com/doi/abs/10.1080/02678292.2015.1011243.

[28] Israelachvili JN, Mitchell DJ, Ninham BW. Theory of self-assembly of hydrocarbon amphiphiles into micelles and bilayers. J Chem Soc Faraday Trans 2 Mol Chem Phys [Internet]. 1976 [cited 2023 Apr 1]; 72:1525–1568. Available from: https://pubs.rsc.org/en/content/articlehtml/1976/f2/f29767201525.

[29] Dwivedi MK, Tiwari SN. Odd-even effect in p-alkyl-p′-cyanobiphenyl liquid crystalline series: An ab initio study. J Mol Liq. 2011;158:208–211.

[30] Yoshimi T, Shima A, Hagiwara-Norifusa S, et al. Phase Transitions of N-(4-methoxybenzylidene)-4-butylaniline (MBBA) Confined within Mesoporous Silica. Cryst [Internet]. 2020 [cited 2023 Apr 1]; 10:792. Available from: www.mdpi.com/2073-4352/10/9/792/htm.

[31] Vieweg N, Christian Jansen C, Shakfa MK, et al. Molecular properties of liquid crystals in the terahertz frequency range. Opt Express [Internet]. 2010 [cited 2023 Apr 1];18:6097–6107. Available from: https://opg.optica.org/viewmedia.cfm?uri=oe-18-6-6097&seq=0&html=true.

[32] Demus D, Goodby J, Gray GW, et al. Handbook of Liquid Crystals Set [Internet]. New York: Wiley-VCH, 1998 [cited 2023 Apr 1]. Available from: https://onlinelibrary.wiley.com/doi/book/10.1002/9783527619276.

[33] Sahu TK, Chauhan MS, Sharma D, et al. Electro-optical properties of MBBA liquid crystal surrounded by acetonitrile medium enhanced under the presence of electric field: An ab initio study. Macromol Symp [Internet]. 2023 [cited 2023 Apr 1];407:2100400. Available from: https://onlinelibrary.wiley.com/doi/full/10.1002/masy.202100400.

[34] Toyne KJ. Liquid crystal behaviour in relation to molecular structure, in Thermotropic Liq Cryst [Internet]. Gray, GW, and Society of Chemical Industry (Great Britain), (eds.). Published on behalf of the Society of Chemical Industry by John Wiley & Sons, 1987 [cited 2023 Apr 1]. p. 178. Available from: www.worldcat.org/title/15133521.

[35] Bradshaw DS, Andrews DL. Quantum Channels in Nonlinear Optical Processes. https://doi.org/101142/S0218863509004609. 2012;18:285–299.

[36] Raynes EP. Electro-optic effects in liquid crystals. Electro-optic Photorefractive Mater Springer Proc Phys [Internet]. 1987 [cited 2023 Apr 1];18:90–98. Available from: https://link.springer.com/chapter/10.1007/978-3-642-71907-3_8.

[37] Zharkova GM, Khachaturyan VM, Vostokov LA, et al. Study of liquid thermoindicators, in Advances in Liquid Crystal Research Applications. Bata, L, (ed.). Oxford: Pergamon Press; Budapest: Akademiai Kaido, 1980. pp. 1221–1239.

[38] Hollingsworth DK, Boehman AL, Smith EG, et al. Measurement of temperature and heat transfer coefficient distributions in a complex flow using liquid crystal thermography and true-color image processing. ASME Collect Pap Heat Transf Heat Transf Div. 1989;123:35–42.

[39] Frisch MJ, Trucks GW, Schlegel HB, et al. Gaussian 16, Revision B.01. Wallingford, CT: Gaussian, Inc., 2016.

[40] Hohenberg P, Kohn W. Inhomogeneous electron gas. Phys Rev [Internet]. 1964 [cited 2021 May 29]; 136:B864. Available from: https://journals.aps.org/pr/abstract/10.1103/PhysRev.136.B864.

[41] Aamouche A, Devlin FJ, Stephens PJ. Structure, vibrational absorption and circular dichroism spectra, and absolute configuration of Tröger's base. J Am Chem Soc [Internet]. 2000 [cited 2023 Jun 15]; 122:2346–2354. Available from: https://pubs.acs.org/doi/abs/10.1021/ja993678r.

[42] Becke AD. Density-functional exchange-energy approximation with correct asymptotic behavior. Phys Rev A [Internet]. 1988 [cited 2021 May 28];38:3098–3100. Available from: https://journals.aps.org/pra/abstract/10.1103/PhysRevA.38.3098.

[43] Lee C, Yang W, Parr RG. Development of the Colle-Salvetti correlation-energy formula into a functional of the electron density. Phys Rev B [Internet]. 1988 [cited 2021 May 28];37:785–789. Available from: https://journals.aps.org/prb/abstract/10.1103/PhysRevB.37.785.

[44] Becke AD. Density-functional thermochemistry. III. The role of exact exchange. Cit J Chem Phys. 1993;98:5648.

[45] Sharma D, Tiwari G, Tiwari SN. Electronic and electro-optical properties of 5CB and 5CT liquid crystal molecules: A comparative DFT study. Pramana—J Phys [Internet]. 2021;95. Available from: https://doi.org/10.1007/s12043-021-02114-z.

[46] Sharma D, Tiwari G, Tiwari SN. Thermodynamical properties and infrared spectra of 4-n-propoxy-4'-cyanobiphenyl: Hartree-Fock and density functional theory methods. Int J Electroact Mater. 2017;5: 19–30.

[47] Sharma D, Tiwari SN. Electronic structure and vibrational spectra of 1OCB liquid crystal: A DFT study. Emerg Mater Res. 2017;6:322–330.

[48] Tiwari SN, Sharma D, Dwivedi MK. Molecular structure, frontier orbital analysis and vibrational assignments of a nematic liquid crystal: 4-n-ethyl-4'-cyanobiphenyl. Int J Electroact Mater. 2017;5:31–41.

[49] Sharma D, Tiwari SN. Comparative computational analysis of electronic structure, MEP surface and vibrational assignments of a nematic liquid crystal: 4-n-methyl-4'-cyanobiphenyl. J Mol Liq. 2016;214:128–135.

[50] Tiwari SN, Sharma D. Electronic structure, FT-IR analysis and nematic behaviour studies of para-methoxybenzylidine p-ethylaniline: Ab-initio and DFT approach. J Mol Liq. 2017;244.

[51] Becke AD. Density-functional thermochemistry. III. The role of exact exchange. Cit J Chem Phys. 1993;98:5648.

[52] Chai J-D, Head-Gordon M. Long-range corrected hybrid density functionals with damped atom–atom dispersion corrections. Phys Chem Chem Phys [Internet]. 2008 [cited 2021 Aug 26];10:6615–6620. Available from: https://pubs.rsc.org/en/content/articlehtml/2008/cp/b810189b.

[53] Henderson TM, Izmaylov AF, Scalmani G, et al. Can short-range hybrids describe long-range-dependent properties? J Chem Phys [Internet]. 2009 [cited 2023 Jun 19];131. Available from: https://pubs.aip.org/aip/jcp/article/131/4/044108/838663/Can-short-range-hybrids-describe-long-range.

[54] Yanai T, Tew DP, Handy NC. A new hybrid exchange–correlation functional using the Coulomb-attenuating method (CAM-B3LYP). Chem Phys Lett. 2004;393:51–57.

[55] Zhao Y, Truhlar DG. The M06 suite of density functionals for main group thermochemistry, thermochemical kinetics, noncovalent interactions, excited states, and transition elements: Two new functionals and systematic testing of four M06-class functionals and 12 other functionals. Theor Chem Accounts 2007 1201 [Internet]. 2007 [cited 2021 Aug 26];120:215–241. Available from: https://link.springer.com/article/10.1007/s00214-007-0310-x.

[56] Geerlings P, De Proft F, Langenaeker W. Conceptual density functional theory. Chem Rev [Internet]. 2003 [cited 2023 Apr 1];103:1793–1873. Available from: https://pubs.acs.org/doi/abs/10.1021/cr990029p.

[57] Parr RG, Donnelly RA, Levy M, et al. Electronegativity: The density functional viewpoint. J Chem Phys [Internet]. 1977 [cited 2021 May 28];68:3801–3807. Available from: https://aip.scitation.org/doi/abs/10.1063/1.436185.

[58] Foresman JB, Frisch Æ. Exploring Chemistry with Electronic Structure Methods [Internet] (2nd ed.). Pittsburgh, PA: Gaussian, Inc., 1996 [cited 2022 Nov 15]. Available from: www.csus.edu/indiv/g/ghermanb/research_foo/g03_man/g_ur/refs/ref_308.htm.
[59] Senet P. Chemical hardnesses of atoms and molecules from frontier orbitals. Chem Phys Lett. 1997;275:527–532.
[60] Pauling L. The Nature of Chemical Bond (3rd ed.). Ithaca, NY: Cornell University Press, 1960. [Internet]. [cited 2021 May 28]. Available from: www.sciepub.com/reference/7634.
[61] Parr RG, Szentpály LV, Liu S. Electrophilicity index. J Am Chem Soc [Internet]. 1999 [cited 2021 May 28];121:1922–1924. Available from: https://pubs.acs.org/doi/abs/10.1021/ja983494x.
[62] Lakshmi Praveen P, Ojha DP. Molecular Structure and Conformational Behavior of a Nematogen—A Statistical Approach. http://dx.doi.org/101080/154214062011556455 [Internet]. 2011 [cited 2023 Apr 2]; 537:64–75. Available from: www.tandfonline.com/doi/abs/10.1080/15421406.2011.556455.
[63] Kumar A, Srivastava AK, Tiwari SN, et al. Evolution of anisotropy, first order hyperpolarizability and electronic parameters in p-alkyl-p'-cynobiphenyl series of liquid crystals: Odd-even effect revisited. Mol Cryst Liq Cryst [Internet]. 2019;681:23–31. Available from: https://doi.org/10.1080/15421406.2019.1641987.
[64] Mishra M, Dwivedi MK, Shukla R, et al. Study of molecular ordering in liquid crystals: EBBA. Prog Cryst Growth Charact Mater. 2006;52:114–124.
[65] Chaturvedi S, Chaturvedi N, Dwivedi MK, et al. Theoretical study of nematogenic behaviour of para-hexyl-p'-cyanobiphenyl. Indian J Phys [Internet]. 2013 [cited 2023 Apr 2];87:263–269. Available from: https://link.springer.com/article/10.1007/s12648-012-0212-1.
[66] Marčelja S. Chain ordering in liquid crystals. I. Even-odd effect. J Chem Phys [Internet]. 2003 [cited 2023 Apr 2];60:3599. Available from: https://aip.scitation.org/doi/abs/10.1063/1.1681578.
[67] Makov G. Chemical hardness in density functional theory. J Phys Chem [Internet]. 1995 [cited 2023 Apr 2]; 99:9337–9339. Available from: https://pubs.acs.org/doi/abs/10.1021/j100023a006.
[68] Alyar H, Bahat M. Torsional energy and nonlinear optical properties of 2-, 3-R-4-phenylpyridine (R = CH3, NH2 and NO2). Asian J Chem [Internet]. 2012 [cited 2023 Apr 2];24:5319–5323. Available from: https://asianjournalofchemistry.co.in/user/journal/viewarticle.aspx?ArticleID=24_11_119.
[69] Bosshard C, Spreiter R, Degiorgi L, et al. Infrared and Raman spectroscopy of the organic crystal DAST: Polarization dependence and contribution of molecular vibrations to the linear electro-optic effect. Phys Rev B [Internet]. 2002 [cited 2023 Apr 2];66:205107. Available from: https://journals.aps.org/prb/abstract/10.1103/PhysRevB.66.205107.
[70] Sun Y, Chen X, Sun L, et al. Nanoring structure and optical properties of Ga8As8. Chem Phys Lett. 2003;381:397–403.
[71] Sharma D, Sahu TK, Tiwari SN. DFT study of electro-optical, electronic and thermal properties of 4-n-alkoxy-4′-cynobiphenyl liquid crystal series. Phase Transitions [Internet]. 2021;1–11. Available from: https://doi.org/10.1080/01411594.2021.1944628.
[72] Mitra M, Gupta S, Paul R, et al. Determination of Orientational Order Parameter from Optical Studies for a Homologous Series of Mesomorphic Compounds. http://dx.doi.org/101080/00268949108030937 [Internet]. 2006 [cited 2023 Apr 2];199:257–266. Available from: www.tandfonline.com/doi/abs/10.1080/00268949108030937.
[73] Yu Y, Lin K, Zhou X, et al. New C–H Stretching Vibrational spectral features in the Raman Spectra of gaseous and liquid ethanol†. J Phys Chem C [Internet]. 2007 [cited 2023 Apr 2];111:8971–8978. Available from: https://pubs.acs.org/doi/abs/10.1021/jp0675781.
[74] Ghosh S, Roy A. Crystal polymorphism of 8OCB liquid crystal consisting of strongly polar rod-like molecules. RSC Adv [Internet]. 2021 [cited 2023 Apr 2];11:4958–4965. Available from: https://pubs.rsc.org/en/content/articlehtml/2021/ra/d0ra08543j.
[75] Özgan Ş, Okumuş M. Thermal and Spectrophotometric Analysis of Liquid Crystal 8CB/8OCB Mixtures. Brazilian J Phys 2011 412 [Internet]. 2011 [cited 2023 Apr 2];41:118–122. Available from: https://link.springer.com/article/10.1007/s13538-011-0034-1.
[76] Tiwari SN, Dwivedi MK, Sharma D. Physico-chemical properties, frontier orbitals and spectral study of a nematogen: 4-n-ethoxy-4'-cyanobiphenyl and its two constituents. Mater Today Proc [Internet]. 2020;29:987–992. Available from: https://doi.org/10.1016/j.matpr.2020.04.512.
[77] Sharma D, Tiwari G, Tiwari SN. Electronic structure and thermodynamic properties of 4-n-heptyl-4′-cyanobiphenyl: A computational study. Mater Today Proc [Internet]. 2019;15:409–415. Available from: https://doi.org/10.1016/j.matpr.2019.04.101 [53]
[78] Tiwari A, Palepu J, Choudhury A, et al. Theoretical analysis of the NH^3, NO, and NO^2 adsorption on boron-nitrogen and boron-phosphorous co-doped monolayer graphene—A comparative study. FlatChem. 2022;34:100392.

[79] Tiwari A, Chauhan MS, Sharma D. Fluorination of 2,5-diphenyl-1,3,4-oxadiazole enhances the electron transport properties for OLED devices: A DFT analysis. https://doi.org/101080/0141159420222129051 [Internet]. 2022 [cited 2023 Jun 26];95:888–900. Available from: www.tandfonline.com/doi/abs/10.1080/01411594.2022.2129051.
[80] Pandey AK, Kumar A, Srivastava AK, et al. Influence of electric field on the electro-optical and electronic properties of 4-n-alkoxy-4′-cyanobiphenyl liquid crystal series: An application of DFT. Pramana—J Phys [Internet]. 2023 [cited 2023 Jun 26];97:1–10. Available from: https://link.springer.com/article/10.1007/s12043-023-02570-9.

2 Spectroscopic Signatures of Some Organic Compounds
Theory Meets Experiment

Abhishek Kumar, Ambrish Kumar Srivastava, Ratnesh Kumar, and Neeraj Misra

2.1 INTRODUCTION

The biological importance of the derivatives of hexahydroacridine-1,8-diones, similar to the octahydroacridine-1,8-diones cannot be disputed not only for their rich and wide-ranging chemistry but also for a number of other characteristics. Anti-inflammatory and hypertensive,[1] potassium channel opener,[2] and anti-microbial-like[3] properties have been reported in the literature. Depending upon the type and placement of a substituent on the acridine core, there are a number of activities against tumor,[4] parasites,[5] and bacteria.[6] It should be mentioned that the biological activities of the acridines are mainly due to the ability of the acridine moiety to intercalate between base pairs of double-stranded DNA through π–π interactions.[7] Acridine-1,8-dione dyes have been the focus of studies due to their special photophysical and photochemical properties. The structure of these dyes is similar to that of 1,4-dihydropyridine.

An extensive quantum chemical study has been carried out on 9,10-bis(4-fluorophenyl)-3,3,6,6-tetramethyl-3,4,6,7,9,10-hexahydroacridine-1,8(2H,5H)-dione (FTHD) and 10-(4-fluorophenyl)-3,3,6,6-tetramethyl-9-(3,4,5-trimethoxyphenyl)-3,4,6,7,9,10-hexahydroacridine-1,8(2H,5H)-dione (FTMPHD) with the main aim of exploring their geometrical, vibrational, and electronic properties. For the sake of a comparative study, theoretical work has also been done on 3,3,6,6-tetramethyl-9-(4-nitrophenyl)-3,4,5,6,7,9-hexahydro-1H-xanthene-1,8(2H)-dione (TNHXD) and 9-(benzo[d][1,3]dioxol-5-yl)-3,3,6,6-tetramethyl-3,4,5,6,7,9-hexahydro-1H-xanthene-1,8(2H)-dione (BDTHXD). To perform quantum chemical calculations and to calculate important properties, rigorous implementation of the DFT-based method has been carried out with a focus on those properties that are not easily obtainable via experimental methods. These properties include molecular geometry, vibrational frequencies, dipole moments and higher-order moments, thermochemical properties, etc. In the standard literature, several forms of DFT are available whose applications depend on the nature of the molecular systems under investigation and characteristics to be explored. In numerous publications, it has been mentioned that,[8–10] DFT offers a better trade-off between computational cost and accuracy for medium-sized molecules, and hence it has been justifiably implemented. A piece of inclusive information about the density functional theory and the methods based on it can be accessed from contemporary literature.[11]

This chapter involves a comprehensive comparison of different properties of 9,10-bis(4-fluorophenyl)-3,3,6,6-tetramethyl-3,4,6,7,9,10-hexahydroacridine-1,8(2H,5H)-dione (FTHD) and 10-(4-fluorophenyl)-3,3,6,6-tetramethyl-9-(3,4,5-trimethoxyphenyl)-3,4,6,7,9,10-hexahydroacridine-1,8(2H,5H)-dione (FTMPHD) with that of 3,3,6,6-tetramethyl-9-(4-nitrophenyl)-3,4,5,6,7,9-hexahydro-1H-xanthene-1,8(2H)-dione (TNHXD) and 9-(benzo[d][1,3]dioxol-5-yl)-3,3,6,6-tetramethyl-3,4,5,6,7,9-hexahydro-1H-xanthene-1,8(2H)-dione (BDTHXD) as suggested by density functional theory. In this chapter, the geometrical and some spectroscopic features of these compounds are discussed.

Theoretical molecular modeling is a powerful tool for studying the physical, chemical, and biological properties of chemical compounds.[12, 13] Molecular modeling has different levels of theories

DOI: 10.1201/9781003441328-2

utilized for studying chemical compounds and reactions' properties.[14, 15] One of the most accurate theoretical approaches is density function theory (DFT), which studies the properties of compounds and provides indications close to those of an experiment.[16, 17] Moreover, molecular structure can be studied and provided with significant information about compounds and their interactions by calculating the infrared (IR) spectrum theoretically, which confirms the expressed value IR result.[18, 19]

2.2 TOOLS OF STUDY: COMPUTATIONAL DETAILS

It has been observed that for a variety of systems of biological relevance, the combination of Becke's three-parameter hybrid exchange functional[20] with the Lee–Yang–Parr correlation functional[21] (B3LYP) functional is quite consistent. The B3LYP functional is a gradient-corrected hybrid functional and is immensely popular for studying a variety of systems of biological interest.[22–25] Apart from the popular functional, the standard split-valence basis set, combined with diffuse as well as polarization functions 6–311++G(d,p), has also been employed in the calculations. The computational work has been carried out by employing the Gaussian 09 suite of code[26] using different keywords for different types of jobs.

The objective of geometry optimization is to find an atomic arrangement that makes the molecule most stable. Molecules are most stable when their energy attains the lowest possible value. The most stable structures of molecules are obtained by geometry optimization, necessarily followed by frequency calculations. The sole purpose of geometry optimization is to predict a three-dimensional arrangement of atoms that makes the molecule most stable, i.e., having minimum energy. Positive real values of frequencies further ensure that the optimized structure is really the lowest possible energy structure. The optimized geometries of dione-derivative compounds are shown in Figure 2.1. The calculated geometrical parameters show a good proximity when compared with the experimental results. The average value for C-C bonds in rings is found to be in the range of 1.515–1.567

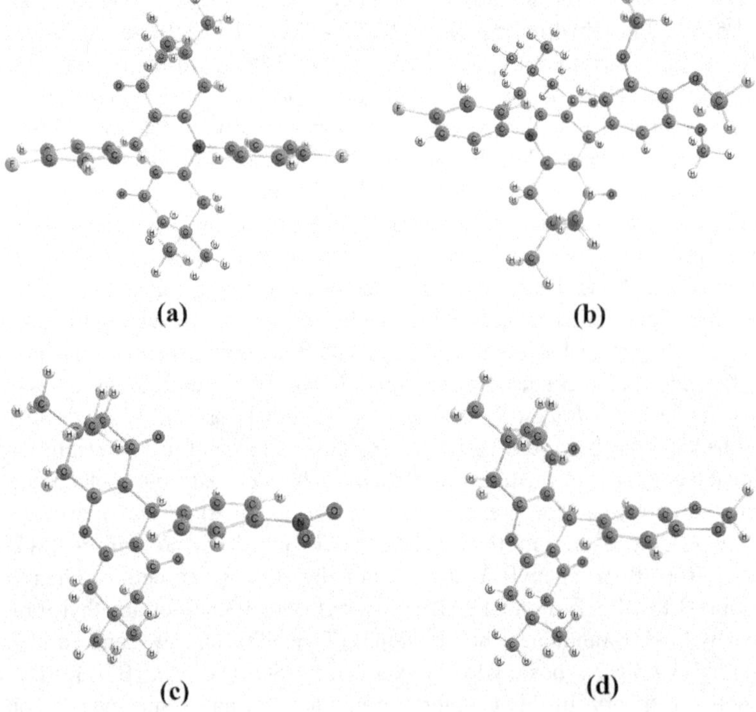

FIGURE 2.1 Optimized geometry of (a) FTHD, (b) FTMPHD, (c) TNHXD, and (d) BDTHXD.

Source: Regenerated from Ref. [8, 9].

Å and that for C-H bonds in the range of 1.090–1.100 Å, which are in good agreement with those reported for similar structures.[27, 28] The C-O bond length of 1.43 Å is also in good agreement with the literature data.[29, 30]

2.3 RESULTS AND DISCUSSION

2.3.1 Vibrational Analyses

Frequency calculations yield some of the most important information discussed in this thesis. IR and Raman spectra of molecules can be predicted for any optimized molecular structure. The position and relative intensity of vibrational bands can be gathered from the output of a frequency calculation. This information is independent of the experiment and can therefore be used as a tool to confirm peak positions in experimental spectra or to predict peak positions and intensities when experimental data is not available. Calculated frequencies are based on the harmonic model, while real vibrational frequencies are anharmonic. This partially explains discrepancies between calculated and experimental frequencies.

The total energy of a molecule comprising N atoms near its equilibrium structure may be written as:

$$\mathrm{E} = \mathrm{T} + V = \frac{1}{2}\sum_{i=1}^{3\mathrm{N}} q_i^2 + V_{eq} + \sum_{i=1}^{3\mathrm{N}}\sum_{j=1}^{3\mathrm{N}} \left(\frac{\partial^2 V}{\partial q_i \partial q_j}\right)_{eq} q_i q_j \tag{2.1}$$

Here the mass-weighted Cartesian displacements, q_i, are defined in terms of the locations X_i of the nuclei relative to their equilibrium positions $X_{i'eq}$ and their masses M_i:

$$q_i = \mathrm{M}_i^{1/2}\left(\mathrm{X}_i - \mathrm{X}_{ieq}\right) \tag{2.2}$$

V_{eq} is the potential energy at the equilibrium nuclear configuration.

For such a system, the classical mechanical equation of motion takes the form:

$$q_i = \sum_{i=1}^{3\mathrm{N}} f_{ij} q_i \qquad \mathrm{j} = 1, 2, 3 \ldots 3\mathrm{N} \tag{2.3}$$

The f_{ij} term quadratic force constants are the second derivatives of the potential energy with respect to mass-weighted Cartesian displacement, evaluated at the equilibrium nuclear configuration:

$$f_{ij} = \left(\frac{\partial^2 V}{\partial q_i \partial q_j}\right)_{eq} \tag{2.4}$$

The f_{ij} may be evaluated by numerical second differentiation:

$$\frac{\partial^2 V}{\partial q_i \partial q_j} = \frac{\Delta(\Delta V)^{£}}{\Delta q_i \Delta V q_j} \tag{2.5}$$

By numerical first differentiation of analytical first derivatives:

$$\frac{\partial^2 V}{\partial q_i \partial q_j} = \frac{\Delta\left(\partial V/\partial q_j\right)}{\Delta q_i} \tag{2.6}$$

or by direct analytical second differentiation, equation (2.6). The choice of procedure depends on the quantum mechanical model employed, that is, single-determinant or post-Hartree–Fock, and on practical matters such as the size of the system. Equation (2.3) may be solved by standard methods to yield a set of 3*N* normal-mode vibrational frequencies. Six of these (five for linear molecules) are zero as they correspond to translational and rotational (rather than vibrational) degrees of freedom. The vibrational frequencies are calculated for an isolated molecule in the gas phase, in contrast to experimental spectra determined in the liquid or solid phase of the sample with some impurities. It has been well established that $\nu_{gas} > \nu_{liquid} > \nu_{solid.}$[31] DFT calculations use the model considering elastic chemical bonds, which becomes inappropriate particularly at higher frequencies due to anharmonicity in vibrations.[32] This anharmonicity affects several vibrations, such as torsions and inversions, for instance, the umbrella modes of halomethyl radicals.[33] For lower frequencies usually having rotational modes, the coupling of vibrational modes and molecular interaction come into play, affecting the calculated frequencies. Therefore, it is required to obtain calculated frequencies in close agreement with their observed values. This can generally be achieved by proper scaling of calculated frequencies. Merrick et al.[34] suggested some scale factors on the basis of harmonic approximation, According to this, the scaling factor for B3LYP/6–311++G(d,p) is:

$$\nu_{scaled} = 0.9688\,\nu_{calculated}$$

The detailed descriptions of the vibrational assignment of dione derivatives have been carried out with the help of normal coordinate analyses. The FT-IR spectra of these molecules were recorded by the SHIMADZU (FT-IR 8400S) spectrometer in the region 4000–400 cm^{-1} using a KBr disc, as shown in Figures 2.2–2.5. The calculated vibrational frequencies with scaled values are tabulated

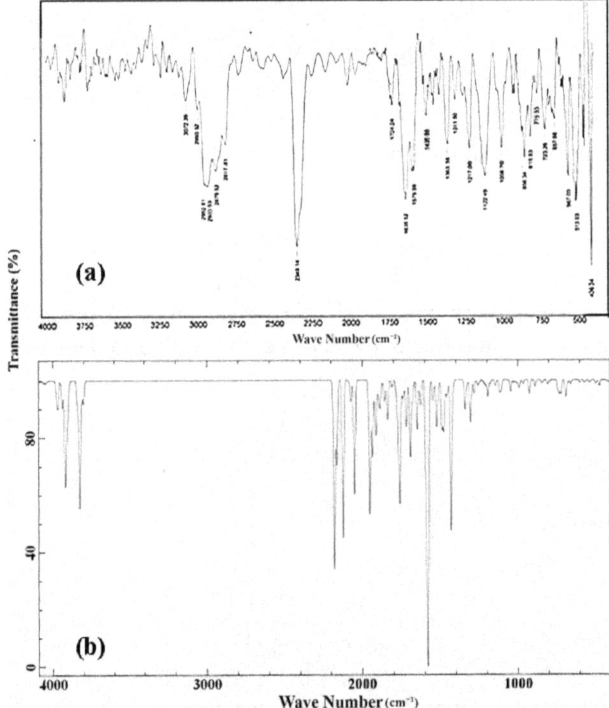

FIGURE 2.2 Experimental FTIR spectrum (a) of FTHD recorded in the region 4000–400 cm^{-1} with calculated IR spectrum (b) at B3LYP/6–311++G(d,p) level.

Source: From Ref. [8].

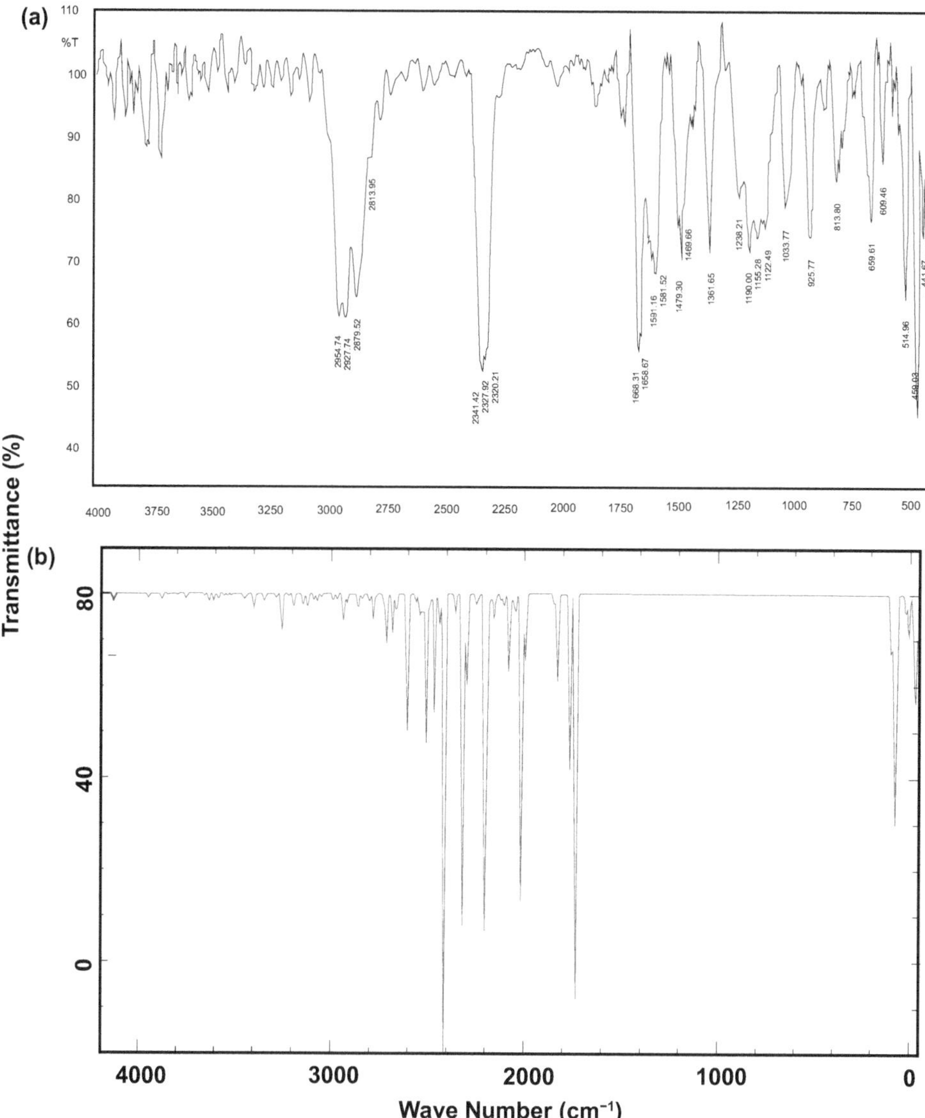

FIGURE 2.3 Experimental FTIR spectrum (a) of TNHXD recorded in the region 4000–400 cm^{-1} with calculated IR spectrum (b) at B3LYP/6–311++G(d,p) level.

Source: Panel (b) from Ref. [8].

in Tables 2.1–2.4. The vibrational assignments have been carried out by combining the result of the Gauss view 05 programs[18] symmetry considerations and the VEDA04 program.[35, 36] A detailed description of vibrational modes can be given by means of normal coordinate analysis.

2.3.1.1 Spectral Region above 2000 cm^{-1}

The characteristic CH vibration of the aromatic benzene ring is expected to appear between a range of 3200 cm^{-1} and 3000 cm^{-1}.[37, 38] In FTHD, the C-H stretching scaled frequency of 3040 cm^{-1} is very close to the experimental FT-IR frequency of 3024 cm^{-1}, and for FTMPHD, the C-H stretching scaled frequency of 3085 cm^{-1} is very close to the experimental FT-IR frequency of 3072 cm^{-1}. These values are calculated at 3195 cm^{-1}, 3117 cm^{-1}, 3102 cm^{-1}, 3070 cm^{-1}, 3015 cm^{-1} for the

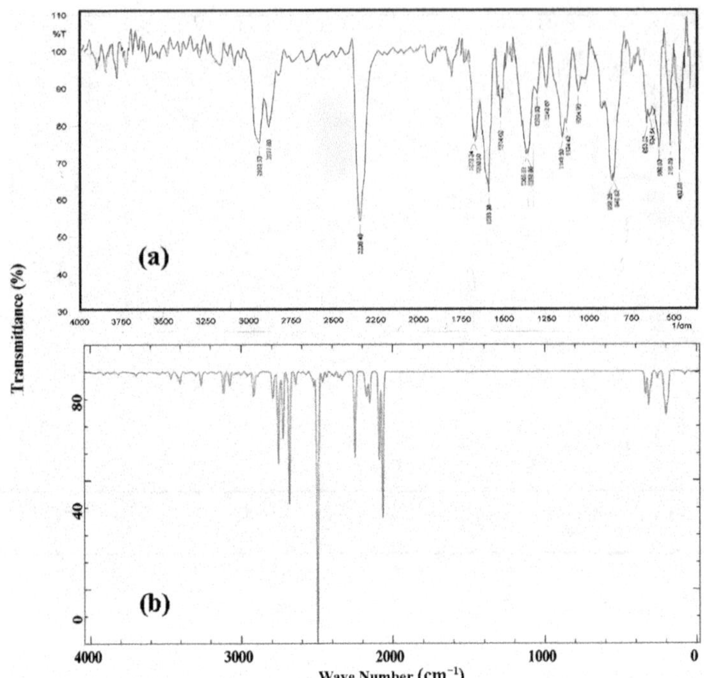

FIGURE 2.4 Experimental FTIR spectrum (a) of BDTHXD recorded in the region 4000–400 cm^{-1} with calculated IR spectrum (b) at B3LYP/6–311++G(d,p) level.

Source: From Ref. [9].

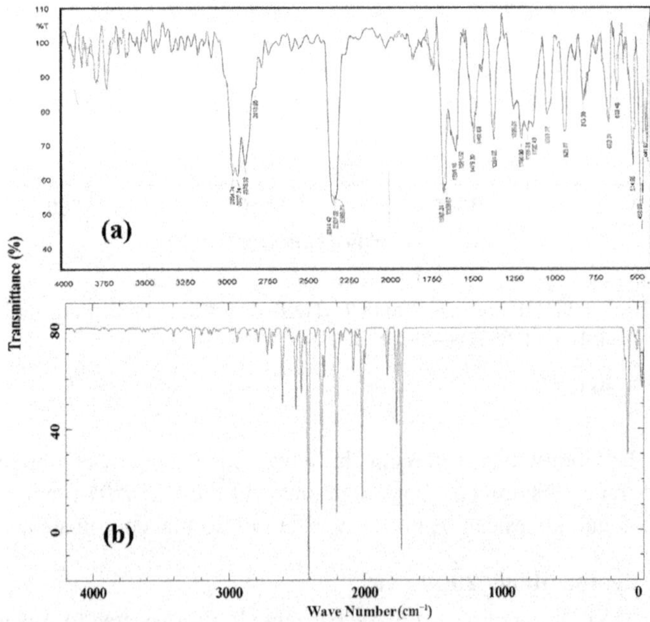

FIGURE 2.5 Experimental FTIR spectrum (a) of FTBMHD recorded in the region 4000–400 cm^{-1} with calculated IR spectrum (b) at B3LYP/6–311++G(d,p) level.

Source: From Ref. [9].

TABLE 2.1
Vibrational Analysis of Some Prominent Modes of FTHD at B3LYP/6–311++G (d,p) Level

Scaled Frequency (cm^{-1})	IR Intensity (cm^{-1})	FTIR Frequency (cm^{-1})	Mode of Vibration
3713	48.5	-	υ (H-O)adjR
3040	15.0	3024	υ (C-H)R
1639	163.4	1637	β(CCC)R + S(C-O-H)R
1587	4.1	1577	β(CC)R + β(CH)R
1490	10.9	1496	β(CC)R + β(CH)R
1449	23.6	1452	β(CC)R + S(CH_3)adjR
1359	1.3	1363	β(CC)R + β(CH_3)adjR
1298	1.7	1301	β(CCC)R + β(CH_2)R
1258	3.4	1259	R(H-C-C-H)R
1221	36.1	1220	β(CCC)R ++ β(CH)R+ω(CH_2)R
1167	35.6	1172	β(CC)R + ω(CH_2)inR
1132	10.9	1136	β(CC)R + S(H-C-C-H)R + ω(CH_3)adjR
1025	51.0	1027	β(CCC)R + β(CH)R + S(H-C-C-H)R
993	4.1	1004	S(H-C-C-H)R
931	0.2	929	β(CCC)R + t(H-C-C-H)R
837	34.5	840	ω(HCCH)R + ω(CH_2)R
759	4.3	771	R(H-C-C-H)R + β(C-N-C)R
706	1.1	700	t(H-C-C-H)R + ω(CH_2)R
661	5.8	663	ω(CH_2)R + ω(CH_3)adjR + β(HCCH)R + γ(CH)R
586	11.1	594	γ(C-H) + t(HCCH)R + ω(CH_3)adjR
563	5.4	565	ω(HCCH)R + β(O-H)R
525	2.5	524	ω(HCCH)R + R(CH_2)R + R(CH_3)adjR
515	2.2	511	ω(HCCH)R + BreathingR + R(CH_3)adjR
415	0.02	418	τ (HCCH)R

Source: From Ref. [8],

TABLE 2.2
Vibrational Analysis of Some Prominent Modes of FTMPHD at B3LYP/6–311++G (d,p) Level

Scaled Frequency (cm^{-1})	IR Intensity (cm^{-1})	FTIR Frequency (cm^{-1})	Mode of Vibration
3714	46.5	—	υ(H-O)adjR
3085	9.7	3072	$υ_{as}$(C-H)R
2993	23.8	2993	$υ_{as}$(C-H)adjR + $υ_{as}$ (C-H)R
2968	27.6	2952	$υ_{as}$(C-H)adjR
2920	34.9	2933	$υ_{as}$(C-H)adjR + υ(C-H)R
2897	16.7	2879	$υ_s$(C-H)adjR
1706	272.1	1724	υ(C-O)R + β(CCC)R
1640	180.1	1635	β(CCC)R + S(C-O-H)R
1567	134.2	1579	β(CC)R + β(C-H)adjR
1489	91.5	1496	β(CC)R + β(C-H)R
1359	6.1	1363	σ(CH_2)R + S(CH_3)adjR
1306	11.8	1311	R(CH_3)adjR + R(CH_2)

(*Continued*)

TABLE 2.2 (Continued)

Scaled Frequency (cm⁻¹)	IR Intensity (cm⁻¹)	FTIR Frequency (cm⁻¹)	Mode of Vibration
1220	11.4	1217	β(C-H)R + β(CH_2)R + β(CC)R
1122	10.8	1122	γ(CH_3)adjR
1007	52.7	1008.7	β(CC)R + V(C-O)adjR
853	3.5	856	ω(CH_3)adjR + ω(CH_2)R
827	6.5	815	ω(CH_2)R + γ(C-H)R+
783	16.02	775	β(CC)R+ ω(HCCH)R
729	12.5	723	ω(HCCH)R + ω(CH_2)R
655	4.9	657	ω(CH_2)R + ω(HCCH)R
566	14.5	567	ω(HCCH)R + ω(CH_2)R
515	11.4	513	ω(HCCH)R
442	3.8	426	τ(HCCH)R + ω(CH_3)adjR

Source: From Ref. [8].

TABLE 2.3
Vibrational Analysis of Some Prominent Modes of TNHXD at B3LYP/6–311++G (d,p) Level

Scaled Frequency (cm⁻¹)	IR Intensity (cm⁻¹)	FTIR Frequency (cm⁻¹)	Mode of Vibration
2933	21.1	2933	$υ_{as}$(C-H)R
2909	24.8	2878	$υ_{as}$(C-H)adjR + $υ_{as}$(C-H)R
1676	342.8	1670	$υ_{as}$(C-O)R + β(CCC)R
1658	203.5	1661	$υ_{as}$(C-C)R + β(C-H)adjR
1586	48.1	1585	$υ_s$(C-C)R + β(C-H)adjR
1513	64.1	1514	$υ_{as}$(O-N)
1360	15.7	1365	ω(H-C-H)
1355	9.7	1356	γ (HCCH)R + ω(CH_3)adjR
1302	8.8	1300	β(C-C)R
1143	155.2	1149	υ (C-O)adjR
1127	11.3	1124	ω(CH_2)R + ω(HCCH)R
851	11.9	858	ω(HCCH)R + ω(CH_2)R
840	8.9	849	σ(O-N-O)
650	7.5	654	σ(C-O-C)R

Source: From Ref. [9].

TABLE 2.4
Vibrational Analysis of Some Prominent Modes of BDTHXD at B3LYP/6–311++G (d,p) Level

Scaled Frequency (cm⁻¹)	IR Intensity (cm⁻¹)	FTIR Frequency (cm⁻¹)	Mode of Vibration
2960	16.6	2955	υ(C-H)R
2927	25.4	2928	$υ_{as}$(C-H)R
2905	3.1	2880	$υ_s$(C-H)R

(Continued)

TABLE 2.4 (Continued)

Scaled Frequency (cm⁻¹)		IR Intensity (cm⁻¹)	FTIR Frequency (cm⁻¹)	Mode of Vibration
1684		3.9	1668	υ(C-O)R + β(CCC)R
1662		163.3	1659	β(CC)R + β(CH_3)adjR
1601		8.5	1591	β(CC)R + β(CH_3)
1491		61.2	1479	σ(CH_2)R
1476		285.8	1470	σ(CH_2)R + σ(CH_3)adjR
1359		4.9	1362	β(CC)R
1223		16.5	1238	ω(HCCH)R + ω(CH_2)R
1181		428.5	1190	υ(C-O)R + β(CCC)R
1153		0.001	1155	β (HCCH)R
1125		10.7	1122	ω(HCCH)R
1040		127.0	1034	υ(C-O)R + ω(CH_2)R
917		1.2	926	ω(HCCH)R
806		8.1	814	υ(C-O)R
602		0.5	660	β(CCC)R
531		1.0	514	β(OCC)R
448	13.8	13.8	459	ω(CCCC)R

Source: From Ref. [9]. *Legend:* adj. = adjacent to, R = ring

Symbols: υ = stretching, υ_s = symmetric stretching, υ_{as} = asymmetric stretching, σ = scissoring, ρ = rocking, τ = twisting, ω = wagging, β = in plane bend, γ = out of plane

TNHXD and at 3195 cm⁻¹, 3097 cm⁻¹, 3048 cm⁻¹, 3044 cm⁻¹, 3025 cm⁻¹ for the BDTHXD. The experimental FT-IR frequency is exactly equal to the stretching scaled frequency, i.e., 2933 cm⁻¹ in the first molecule and second molecule, and the experimental FT-IR frequency 2928 cm⁻¹ is nearly equal to the stretching scaled frequency 2927 cm⁻¹.

2.3.1.2 Spectral Region below 2000 cm⁻¹

The ring C-C vibration known as the semicircular stretching mode is usually assigned in the range 1650–1400 cm⁻¹. These modes are sensitive to substitution but the actual positions are not found so much by the nature of the substitution by their position.[39] In the middle region, the FTHD C-C stretching scaled frequencies of 1639 cm⁻¹, 1587 cm⁻¹, and 1490 cm⁻¹ are very close to the experimental FT-IR frequency values of 1637 cm⁻¹, 1577 cm⁻¹ and 1496 cm⁻¹, and the FTMPHD C-C stretching scaled frequencies of 1640 cm⁻¹ and 1489 cm⁻¹ are very close to experimental FT-IR frequency values of 1635 cm⁻¹, 1496 cm⁻¹, and 2993 cm⁻¹. In the case of TNHXD and BDTHXD, the stretching vibration of the C=C band is observed at a scaled frequency of 1639 cm⁻¹, and the corresponding band is given at 1637 cm⁻¹ in the FT-IR spectrum. The theoretically scaled C-C stretching vibration of the first molecule is found in the range 1658–1326 cm⁻¹ and the FT-IR spectra at 1685 cm⁻¹ and 1585 cm⁻¹. The stretching vibrations of the C-H in-plane bending are also in good agreement with the corresponding FT-IR spectrum.

The C-O vibrational frequencies are the significant characteristic bands in vibrational spectra in ketones.[40] The asymmetric bands observed at 1676 cm⁻¹, 1143 cm⁻¹, 1179 cm⁻¹, and 991 cm⁻¹ are assigned to C-O stretching vibration and in FT-IR at 1670 cm⁻¹ and at 1149 cm⁻¹ for TNHXD. For BDTHXD, it's observed at 1253 cm⁻¹, 1181 cm⁻¹, 1040 cm⁻¹, 957 cm⁻¹, and in FT-IR at 1190 cm⁻¹ and 1034 cm⁻¹. The observation shows good agreement. Other lower frequencies found in the different modes of vibration.

2.3.2 NMR Spectroscopic Analysis

The NMR calculations can be done by using various methods such as the Gauge-Independent Atomic Orbital (GIAO),[41, 42] Continuous Set of Gauge Transformations (CSGT),[43] and Individual Gauges for Atoms in Molecules (IGAIM).[44] The vector potential representing the external magnetic field[45] and, consequently, the Hamiltonian and the NMR shielding tensor are not uniquely defined (see equation [2.7]). The GIAO eliminates the gauge dependence of properties and ensures a uniform description of the molecular system by making the basis functions explicitly dependent on the magnetic field with the inclusion of a complex phase factor that refers to the position of the basis function, which is centered on the nucleus.

The Hamiltonian (H) in an externally applied magnetic field (B_0) for a nuclear spin (I) is:

$$H = -\frac{\gamma h I \sigma B_0}{2\pi} \tag{2.7}$$

where γ is a gyromagnetic ratio, and σ is the NMR shielding tensor (chemical shift tensor), which can be written as:

$$\sigma = \begin{pmatrix} \sigma_{XX} & \sigma_{XY} & \sigma_{XZ} \\ \sigma_{YX} & \sigma_{YY} & \sigma_{YZ} \\ \sigma_{ZX} & \sigma_{ZY} & \sigma_{ZZ} \end{pmatrix} \tag{2.8}$$

The isotropic and anisotropic components of σ can then be obtained:

$$\sigma_{iso} = \frac{\sigma_{XX} + \sigma_{YY} + \sigma_{ZZ}}{3} \tag{2.9}$$

$$\sigma_{anis} = \frac{\sigma_{YY} - \sigma_{XX}}{2} \tag{2.10}$$

The chemical shift of any x proton (δ_X) is equal to the difference between isotropic magnetic shielding of tetramethylsilane (TMS) and proton (x), given by the equation: $\delta_X = \sigma_{iso} - \sigma_X$. ^{1}H and ^{13}C NMR spectra were obtained at 400 MHz and 100 MHz, respectively, using Bruker DRX- 400 spectrometer and $CDCl_3$ as the solvent. The NMR spectra of the FTHD and FTMPHD compound are calculated with the GIAO approach at B3LYP/6–311++G(d,p) method. The calculated and experimental chemical shifts (δ/ppm) of FTHD and FTMPHD, ^{1}H NMR are given in Tables 2.5 and 2.6.

In the ^{1}H NMR spectrum of FTHD, a quartet chemical shift in experimental value of 0.74 ppm and a doublet chemical shift in FTMPHD of 0.90 ppm in the range 0.74–0.88 ppm designate the presence of the -CH_2 and -CH_3 group, respectively. The OCH_3 group hydrogen attached to the

TABLE 2.5
Selected NMR Chemical Shifts (δ, in ppm) of FTHD with Their Assignments

	δ				δ		
Atom	Calculated	Expected	Assignment	Atom	Calculated	Expected	Assignment
H	1.69–2.02	1.74–2.05	[s, H(C_6H_6)]	C	160.0	160.0	[s, C-N]
H	0.76–0.85	0.74–0.88	[s, H(CH_3)]	C	128.4–149.5	129.2–149.6	[s, C-C]
H	7.12–7.25	7.15–7.29	[s, H(C_6H_4F)]	C	26.8–33.3	26.7–32.4	[s, CH_3]

Source: From Ref. [8].

TABLE 2.6
Selected NMR Chemical Shifts (δ, in ppm) of FTMPHD with Their Assignments

	δ				δ		
Atom	Calculated	Expected	Assignment	Atom	Calculated	Expected	Assignment
H	1.67–2.00	1.74–2.02	[s, H(C_6H_6)]	C	160.9	161.2	[s, C-N]
H	0.79–0.90	0.79–0.90	[s, H(CH_3)]	C	118.2–149.5	117.4–149.6	[s, C-C]
H	6.29	6.51	[s, H(C_6O_4H)]	C	26.7–32.9	26.7–32.8	[s, CH_3]
H	3.57–5.00	3.71–5.19	[s, H(-OCH_3)]	C	55.9–60.6	56.0–60.7	[s, OCH_3]

Source: From Ref. [8].

FTMPHD molecule at the C_6O_4H ring appears as a singlet at 3.71–5.19 ppm due to nitrogen inversion. In the FTHD and FTMPHD theoretical ^1H NMR spectrum, the chemical shift values lie in the range of 0.76–7.25 ppm (with respect to TMS), and the experimental signal at the range of 0.74–7.29 ppm. The chemical shift values of aromatic carbons in the spectrum lie between 100 and 150 ppm.[46]

The ^{13}C chemical shifts of carbonyl carbons vary from 150 to 220 ppm, and this depends on the decrease in electron donating of the attached atoms.[47] The C=O groups of carboxylic acids and derivatives are in the range of 150–185 ppm.[48] In the FTHD ^{13}C spectrum, a signal observed at 149.5 and 135.1 ppm is due to C9 and C39 carbons. For the FTMPHD ^{13}C spectrum, signals observed at 118.2 and 149.5 ppm re due to C39 and C9 carbons. The calculated chemical shifts of ^{13}C and ^1H are in good harmonization with the experimental chemical shift.

2.4 CONCLUSION

In this chapter, we discussed the results of DFT-based calculations on four dione derivatives, FTHD, FTMPHD, TNHXD, and BDTHXD, using the B3LYP/6–311++G(d,p) level of theory. The FT-IR spectrum of the studied compounds recorded using the KBr disc method has been analyzed and interpreted by vibrational mode assignments. The scaled theoretical wave numbers are in perfect agreement with the experimental FTIR values. According to the results obtained by the GIAO method of NMR calculations for FTHD and FTMPHD, the calculated chemical shifts are in perfect agreement with the experimental chemical shift values. These results suggest that the choice of an appropriate functional for the system under study, with a suitable basis set and an appropriate DFT method, is capable of reproducing the experimental data. We believe that this chapter will help to correlate the experimental and DFT-based theoretical spectroscopic results. Further, this chapter will help readers to grasp the discussion about spectroscopic results.

2.4.1 Acknowledgments

AKS is thankful to CSIR, SERB, and UGC for financial assistance in the form of a senior research fellowship (SRF), national postdoctoral fellowship (NPDF), and start-up grant, respectively. RK is also thankful to UGC, New Delhi, India, for senior research fellowship.

REFERENCES

[1]. Pulay, P., Zhou, X. and Fogarasi, G., 1993. Development of an ab initio based database of vibrational force fields for organic molecules. Recent Experimental and Computational Advances in Molecular Spectroscopy, 406, 99–111.

[2]. Josephrajan, T., Ramakrishnan, V.T., Kathiravan, G. and Muthumary, J., 2005. Synthesis and antimicrobial studies of some acridinediones and their thiourea derivatives. Arkivoc, 11, 124–136.

[3]. Frank, C.A., Forst, J.M., Grant, T., Harris, R.J., Kau, S.T., Li, J.H., Ohnmacht, C.J., Smith, R.W., Trainor, D.A. and Trivedi, S., 1993. Dihydropyridine KATP potassium channel openers. Bioorganic & Medicinal Chemistry Letters, 3(12), 2725–2726.
[4]. Jain, S.M., Khajuria, R.K., Dhar, K.L., Singh, S., Bani, S., Pahwa, G.S., Singh, G.B. and Sarin, A.N., 1992. Synthesis and pharmacological screening of 1, 8-dioxo-9-(substituted phenyl)-1, 2, 3, 4, 5, 6, 7, 8, 9, 10-decahydroacridines. ChemInform, 23(5), 191.
[5]. Wainwright, M., 2001. Acridine-a neglected antibacterial chromophore. Journal of Antimicrobial Chemotherapy, 47(1), 1–13.
[6]. Demeunynck, M., Charmantray, F. and Martell, A., 2004. Current Pharmaceutical Design, 7, 2001, 1703;(b) Demeunynck M. Expert Opin Ther Patents, 14(1).
[7]. Lee, H.H., Wilson, W.R., Ferry, D.M., van Zijl, P., Pullen, S.M. and Denny, W.A., 1996. Hypoxia-selective antitumor agents. 13. Effects of acridine substitution on the hypoxia-selective cytotoxicity and metabolic reduction of the bis-bioreductive agent nitracrine N-oxide. Journal of Medicinal Chemistry, 39(13), 2508–2517.
[8]. Shukla, D.V., Kumar, A., Srivastava, A.K., Misra, N. and Brahmachari, G., 2017. Spectral (FT-IR, NMR) analyses, molecular structures, and chemical bonding of two hexahydroacridine-1, 8 (2 H, 5 H)-dione derivatives: A comparative quantum chemical study. Polycyclic Aromatic Compounds, 37(5), 426–441.
[9]. Kumar, R., Kumar, A., Srivastava, A.K., Brahmachari, G., Tiwari, G. and Misra, N., 2020. Structural, spectroscopic analysis of two hexahydroacridine-1, 8 (2H, 5H)-dione derivatives and identification of drug like properties: Experimental and computational study. Materials Today: Proceedings, 29, 1050–1054.
[10]. Kumar, A., Srivastava, A.K., Gangwar, S., Misra, N., Mondal, A. and Brahmachari, G., 2015. Combined experimental (FT-IR, UV–visible spectra, NMR) and theoretical studies on the molecular structure, vibrational spectra, HOMO, LUMO, MESP surfaces, reactivity descriptor and molecular docking of Phomarin. Journal of Molecular Structure, 1096, 94–101.
[11]. Srivastava, A. K. and Misra, N., 2021. DFT Based Studies on Bioactive Molecules. Singapore: Bentham Science Publishers.
[12]. Tiwari, G., Kumar, A., Dwivedi, K.K. and Sharma, D., 2022. In silico investigation of electronic structure, binding patterns and molecular docking of nevirapine: An anti-HIV type-1 drug. Polycyclic Aromatic Compounds, 42(5), 2789–2804.
[13]. Srivastava, A.K., Kumar, A., Pandey, S.K. and Misra, N., 2016. Spectroscopic analyses, intra-molecular interaction, chemical reactivity and molecular docking of imerubrine into bradykinin receptor. Medicinal Chemistry Research, 25, 2832–2841.
[14]. Budyka, M.F., 2019. Semiempirical study on the absorption spectra of the coronene-like molecular models of graphene quantum dots. Spectrochimica Acta Part A: Molecular and Biomolecular Spectroscopy, 207, 1–5.
[15]. Abdelsalam, H., Teleb, N.H., Yahia, I.S., Zahran, H.Y., Elhaes, H. and Ibrahim, M.A., 2019. First principles study of the adsorption of hydrated heavy metals on graphene quantum dots. Journal of Physics and Chemistry of Solids, 130, 32–40.
[16]. Srivastava, A.K., Kumar, A. and Misra, N., 2016. Quantum chemical studies on cocsuline using density functional theory and its docking into dihydrofolate reductase receptor. Main Group Chemistry, 15(2), 97–106.
[17]. Kumar, A., Kumar, R., Misra, N., Srivastava, H., Tripathi, J.K. and Srivastava, A.K., 2022. Electronic structures and properties of small (BCN)_x (BCN) x (x = 1–5) clusters and $(BCN)_{12}$, $(BCN)_{12}$ nanotube. Pramana, 96, 1–11.
[18]. Kumar, R., Kumar, A., Srivastava, A.K., Brahmachari, G. and Misra, N., 2022. Spectroscopic and structural investigations on novel 6-amino-3-phenyl-4-(pyridin-4-yl)-2, 4-dihydropyrano [2, 3-c] pyrazole-5-carbonitrile by FT-IR, NMR, docking, and DFT methods. Polycyclic Aromatic Compounds, 42(5), 2288–2304.
[19]. Kumar, A., Dwivedi, A., Srivastava, A.K., Misra, N., Narayana, B., Samshuddin, S. and Sarojini, B.K., 2017. Molecular structures, vibrational spectra, electronic properties, and molecular docking of two pyrazoline derivatives containing 1-carboxamide and 1-carbothioamide: A comparative study. Polycyclic Aromatic Compounds, 37(4), pp. 267–279.
[20]. Becke, A.D., 1988. Density-functional exchange-energy approximation with correct asymptotic behavior. Physical Review A, 38(6), 3098.
[21]. Lee, C., Yang, W. and Parr, R.G., 1988. Development of the colle-salvetti correlation-energy formula into a functional of the electron density. Physical Review B, 37(2), 785.

[22]. Srivastava, A.K., Kumar, A., Misra, N., Manjula, P.S., Sarojini, B.K. and Narayana, B., 2016. Synthesis, spectral (FT-IR, UV-visible, NMR) features, biological activity prediction and theoretical studies of 4-amino-3-(4-hydroxybenzyl)-1H-1, 2, 4-triazole-5 (4H)-thione and its tautomer. Journal of Molecular Structure, 1107, 137–144.
[23]. Brahmachari, G., Mondal, A., Nayek, N., Kumar, A., Srivastava, A.K. and Misra, N., 2017. Experimental and quantum chemical studies on poriferasterol–A natural phytosterol isolated from Cassia sophera Linn.(Caesalpiniaceae). Journal of Molecular Structure, 1143, 84–191.
[24]. Dwivedi, A. and Kumar, A., 2021. Molecular docking and comparative vibrational spectroscopic analysis, HOMO-LUMO, polarizabilities, and hyperpolarizabilities of N-(4-bromophenyl)-4-nitrobenzamide by different DFT (B3LYP, B3PW91, and MPW1PW91) methods. Polycyclic Aromatic Compounds, 41(2), 387–399.
[25]. Brahmachari, G., Majhi, S., Mandal, B., Mandal, M., Kumar, A., Srivastava, A.K., Singh, R.B. and Misra, N., 2018. Stigmasterol from the flowers of peltophorum pterocarpum (DC) backer ex k. Heyne (Fabaceae)-isolation, spectral properties and quantum chemical studies. Journal of the Indian Chemical Society, 95(10), 1231–1244.
[26]. Frisch, M.J., Trucks, G.W., Schlegel, H.B., Scuseria, G.E., Robb, M.A., Cheeseman, J.R., Scalmani, G., Barone, V., Mennucci, B., Petersson, G.A. and Nakatsuji, H., 2009. Gaussian 09 Revision A. Wallingford, CT: Gaussian Inc., 139.
[27]. Schrader, B. ed., 2008. Infrared and Raman Spectroscopy: Methods and Applications. Essen, Germany: John Wiley & Sons.
[28]. Arnett, E.M. and Larsen, J.W., 1969. Stabilities of carbonium ions in solution. VI. Large Baker-Nathan effect for alkylbenzenonium ions. Journal of the American Chemical Society, 91(6), 1438–1442.
[29]. Kumar, A., Srivastava, A.K., Gangwar, S.K., Misra, N., Brahmachari, G., Mondal, A. and Mondal, S., 2020. FT-IR, UV–visible, and NMR spectral analyses, molecular structure, and properties of Nevadensin revealed by density functional theory and molecular docking. Polycyclic Aromatic Compounds, 40(2), 540–552.
[30]. Tiwari, G., Srivastava, A.K., Kumar, R. and Kumar, A., 2019. Quantum chemical and molecular docking studies on two potential hepatitis C virus inhibitors. Main Group Chemistry, 18(2), 107–121.
[31]. Silverstein, M., Basseler, G.C., and Morill, C., 1981. Spectrometric Identification of Organic Compounds. New York: Wiley.
[32]. Scott, A.P. and Radom, L., 1996. Harmonic vibrational frequencies: An evaluation of Hartree-Fock, Møller-Plesset, quadratic configuration interaction, density functional theory, and semiempirical scale factors. The Journal of Physical Chemistry, 100(41), 16502–16513.
[33]. Marshall, P., Srinivas, G.N. and Schwartz, M., 2005. A computational study of the thermochemistry of bromine-and iodine-containing methanes and methyl radicals. The Journal of Physical Chemistry A, 109(28), 6371–6379.
[34]. Merrick, J.P., Moran, D. and Radom, L., 2007. An evaluation of harmonic vibrational frequency scale factors. The Journal of Physical Chemistry A, 111(45), 11683–11700.
[35]. Jamróz, M.H., 2004. Vibrational Energy Distribution Analysis: VEDA 4, Program, Warsaw.
[36]. Jamróz, M.H., 2013. Vibrational energy distribution analysis (VEDA): Scopes and limitations. Spectrochimica Acta Part A: Molecular and Biomolecular Spectroscopy, 114, 220–230.
[37]. Kun Fun, H, Quah, C. K., Attia, M. I., El-Bechairy, M. F. and Al-Deeb, O. A., 2012. (E)-1-(1, 3-benzodioxol-5-yl)-4, 4-dimethylpent-1-en-3-one. Acta Crystallographica, E68, 634.
[38]. Brahmachari, G., Das, S., Kumar, A., Misra, N., Sharma, S. and Gupta, V.K., 2017. Structural confirmation, single X-ray crystallographic behavior, molecular docking and other physico-chemical properties of gerberinol, a natural dimethyl dicoumarol from Gerbera lanuginosa Benth.(Compositae). Journal of Molecular Structure, 1136, 214–221.
[39]. Karabacak, M., Cinar, Z. and Cinar, M., 2011. Structural and spectroscopic characterization of 2, 3-difluorobenzoic acid and 2, 4-difluorobenzoic acid with experimental techniques and quantum chemical calculations. Spectrochimica Acta Part A: Molecular and Biomolecular Spectroscopy, 79(5), 1511–1519.
[40]. Smith, B., 1998. Infrared Spectral Interpretation, A Systematic Approach. Washington DC: CRC Press.
[41]. Gauss, J., 1993. Effects of electron correlation in the calculation of nuclear magnetic resonance chemical shifts. The Journal of Chemical Physics, 99(5), 3629–3643.
[42]. London, F., 1937. Théorie quantique des courants interatomiques dans les combinaisons aromatiques. Journal de physique et le Radium, 8(10), 397–409.
[43]. Keith, T.A. and Bader, R.F., 1993. Calculation of magnetic response properties using a continuous set of gauge transformations. Chemical Physics Letters, 210(1–3), 223–231.

[44]. Keith, R., 1994. The ecosystem approach: Implications for the North. Northern Perspectives, 22(1), 3–6.
[45]. Griffiths, D.J., 2013. Introduction to Electrodynamics. New Delhi, India: Pearson.
[46]. Kalinowski, H.O., Berger, S. and Braun, S., 1988. Carbon-13 NMR Spectroscopy. Oak Ridge, USA: OSTI.
[47]. Kalsi, P.S., 2004. Spectroscopy of Organic Compounds. New Delhi, India: New Age International.
[48]. Vinod, K.S., Periandy, S. and Govindarajan, M., 2015. Spectroscopic analysis of cinnamic acid using quantum chemical calculations. Spectrochimica Acta Part A: Molecular and Biomolecular Spectroscopy, 136, 808–817.

3 Exploring the Properties of Vincosamide-N-Oxide, a Biologically Active Natural Compound by Density Functional Theory

Ashok Kumar Mishra and Satya Prakash Tewari

3.1 INTRODUCTION

Vincosamide-N-Oxide ($C_{26}H_{31}N_2O_9$) is one of the reported natural products possessing activity against lung cancer cell lines H1299.[1] The quantum chemical calculation through the Schrodinger approach is based on the wave function associated with the physical states that lead to inaccuracy due to the larger number of constituent atoms in this natural product. This shortcoming can be overcome via the route of electron density as per the theorem derived by Hohenberg and Kohn,[2] which established that the ground state properties of the system are a function of electron density. Electron density is easy enough to calculate as it depends on *x*-, *y*-, *z*- and spin coordinates only. Two densities for spin polarized systems may be possible: one for spin up and the other for spin down. The electron density can be described as a function with three variables in the Cartesian coordinates *x*, *y*, and *z* of the electrons. In this way, the ground state properties are a function of the function depending on the variables *x*, *y*, and *z*. This function of the function is called functional and indicates the reason behind the nomenclature of density functional theory (DFT).

DFT evolved from the Hohenberg–Kohn (H-K) theorem-I and theorem-II, pertaining to the ground state and higher energy state of the system, respectively, which explored the existence of the density functional component in order to estimate the molecular properties, if the form of the functional is known accurately. It developed an alternative approach to quantum chemistry.

In this view, a mapping exists between the eigenfunction (wave function) and density function, and the correlation needs approximation to solve the equation. The easiest form of approximation in solving the equations is the local density approximation (LDA), which leads to the Thomas–Fermi[3,4] term for kinetic energy and the Dirac[5] term for exchange energy. The corresponding functional is known as Thomas–Fermi–Dirac energy. Slater's Xα method, formulated in 1951 and a predecessor of the modern chemical approaches to DFT, has been developed as an approximate solution to the Hartree–Fock (H-K) equations. H-F exchange was approximated as:[6]

$$E_{x\alpha}\left[\rho\uparrow,\rho\downarrow\right]=-\frac{9}{4}\alpha\left(\frac{3}{4\pi}\right)^{1/3}\int\left[\rho_{\uparrow}^{\frac{4}{3}}(r)+\rho_{\downarrow}^{\frac{4}{3}}(r)\right]dr \tag{3.1}$$

The approximate functional in DFT methods partitions the electronic energy into various terms expressed as:

$$\mathrm{E}=\mathrm{E}^{\mathrm{T}}+\mathrm{E}^{\mathrm{V}}+\mathrm{E}^{\mathrm{J}}+\mathrm{E}^{\mathrm{XC}} \tag{3.2}$$

DOI: 10.1201/9781003441328-3

where the symbols have their usual meanings, All terms except the nuclear–nuclear repulsion are functions of electron density. Electron–electron density repulsion (E^J) can be shown as[7]:

$$E^J = \frac{1}{2}, \rho(\vec{r_1})(\Delta r_{12})^{-1}\rho(\vec{r_2})d\vec{r_1}\vec{r_2} \tag{3.3}$$

In practice, E^{XC} is frequently approximated as an integral involving only the spin densities and their gradients:

$$E^{XC}(\rho) = \int f(\rho_\alpha(\vec{r}), \rho_\beta(\vec{r}). \nabla\rho_\alpha(\vec{r})\nabla\rho_\beta(\vec{r}))d^3\vec{r} \tag{3.4}$$

Where $\rho_\alpha = \alpha$ spin density $\rho_\beta = \beta$ spin density, and ρ = total electron density ($\rho_\alpha + \rho_\beta$).

It was approximated by Hohenberg and Kohn in 1964 that every observable of a stationary system can be written as a functional of the ground state density and that the ground state energy can be computed using the variational method in terms of only the density. The concepts that emerged from these theorems laid the foundation of DFT wherein the energy functional is written as the sum of the two terms:

$$E[\rho(r)] = \int V_{ext}(r)\rho(r)dr + F[\rho(r)] \tag{3.5}$$

Interaction of the electrons with external potential $V_{ext(r)}$ due to coulomb interaction with nuclei gives rise to the first term, whereas the second term $F[\rho(r)]$ is the sum of the kinetic energy of the electrons and the contribution from electronic interactions:

$$N = \int \rho(r)dr => N - \int \rho(r)dr = 0 \tag{3.6}$$

Kohn and Sham suggested a different approach to derive the Hohenberg–Kohn (H-K) theorem for a set of interacting electrons[8] where the density functional $\mathrm{F}[\rho(\mathrm{r})]$ is approximated as the sum of three energies.

$$F[\rho(r)] = T_0[\rho(r)] + E_H[\rho(r)] + E_{xc}[\rho(r)] \tag{3.7}$$

where $T_0[\rho]$ is the kinetic energy of electrons in the systems, in which there is no electron–electron interaction, given by:

$$T_0[\rho] = \sum_{i=1}^{N} \int \psi_i^*(r)\left(-\frac{\nabla^2}{2}\right)\psi_i(r)dr \tag{3.8}$$

The second term $E_H[\rho(r)]$ is recognized as Hartree electrostatic energy, arising from the classical interaction between the two charge densities:

$$E_H[\rho(r)] = \frac{1}{2}\iint \frac{\rho(r_1)\rho(r_2)}{|r_1 - r_2|}dr_1dr_2 \tag{3.9}$$

The last functional, $E_{xc}[\rho(r)]$, is called exchange-correlation energy.

Consequently, the total expression for the energy of the system can be written:

$$E[\rho(r)] = \int V_{ext}(r)\rho(r)dr + T_0[\rho(r)] + E_H[\rho(r)] + E_{xc}[\rho(r)] \tag{3.10}$$

Kohn and Sham have shown the density) ρ(r) of the system as square modulus of a set of one electron orthogonal orbitals:

$$\rho(r)=\sum_{i=1}^{N}\left|\psi_1\ (r)\right|^2 \tag{3.11}$$

By initiating this expression for electron density and applying the approximate variational condition, the following one-electron K-S equation can be obtained:[9]

$$\left\{-\frac{\nabla_1^2}{2}-\left(\sum_{A=1}^{M}\frac{Z_A}{r_{12}}\right)\right\}+\int\frac{\rho(r_2)dr_2}{r_{12}}+V_{XC}\left[r_1\right]=\varepsilon_i\psi_i(r_1) \tag{3.12}$$

where $V_{XC}\left[r\right]=\frac{\delta E_{XC}\left[\rho(r)\right]}{\delta\rho(r)}$ is known as the functional derivative of exchange correlation energy, which is a self-consistent approach to solving Kohn–Sham equations.

3.2 INVESTIGATIONS BASED ON DFT

3.2.1 Computational Details and Molecular Structure

The optimized geometry is obtained through the quantum chemical calculations based on DFT executed through the available Gaussian 09 program package,[10] which permits running geometry optimization and frequency calculations simultaneously by adopting the (opt + freq) job option at a suitable Becke's third order, corrected on the basis of a Lee, Yang, and Parr correlation functional set combination of B3LYP/6–31+G(d,p) or B3LYP/6–31++G(d,p) depending upon the size of the macromolecular structure. The optimized geometry and corresponding 2D structure of Vincosamide-N-Oxide have been well reported[11] and are depicted in Figures 3.1 and 3.2, respectively.

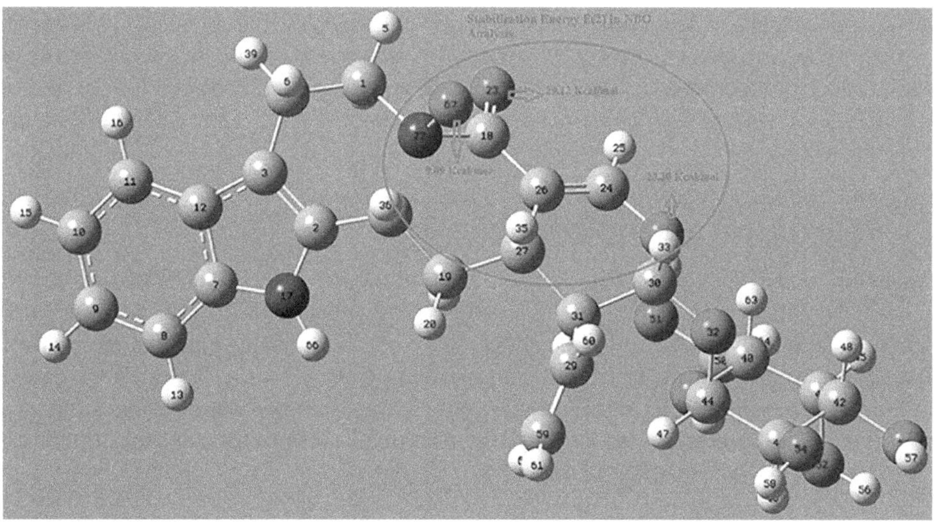

FIGURE 3.1 Optimized geometry of Vincosamide-N-Oxide.

Source: Ref. [11].

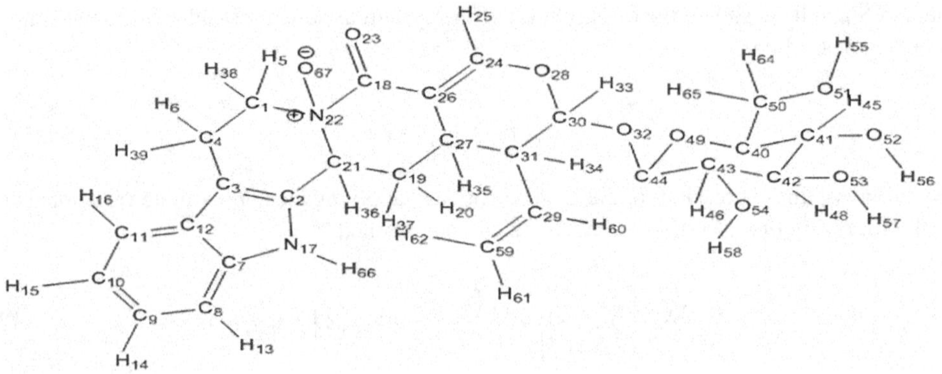

FIGURE 3.2 2D structure of Vincosamide-N-Oxide.
Source: Ref. [11].

TABLE 3.1
Thermodynamic Properties of Vincosamide-N-Oxide Bioactive Natural Compound

Thermodynamic Parameters					
ZPE (kcal/mol)	*E* (kcal/mol)	C_V (cal/mol- K)	*S* (cal/mol)	*H* (kcal/mol)	*G* (kcal/mol)
342.11	362.81	131.21	207.07	363.40	301.66

Source: Ref. [11].

By default, the program package performs thermodynamic calculations at 1 atmosphere P and 298.15 K, which results in heat capacity at a constant volume (*CV*), entropy (*S*), Gibb's free energy (*G*), enthalpy (*H*), zero-point energy (*ZPE*), and zero-point corrected vibrational energy (*E*), where zero-point energy correction accounts for the effect of persisting vibrations at 0 K. The final values of *E*, *H*, and *G* accounting for the thermal correction pertaining to the aforesaid natural compound are depicted in Table 3.1.

3.2.2 Molecular Electrostatic Potential (MESP)

The nucleophilic and electrophilic sites at the surface of the molecule during the reaction is understood to be the MESP surface,[12,13] and this potential is well suited for identifying one molecule among others during drug–receptor interactions.[14,15] The values of this potential are projected by color-code occurring in the order of red<orange<yellow<green<blue, wherein blue indicates the electron-poor region and red the electron-rich region. Recently, MESP has been described as a tool for interpretation and prediction pertaining to the macromolecules from molecular structure to the salvation effect.[16] MESP is expressed as:[17]

$$V(r) = \sum_{A} \frac{Z_A}{|R_A - r|} - \int \frac{Q(r)dr}{|r - r|}$$

The MESP surface of Vincosamide-N-Oxideis is depicted in Figure 3.3.

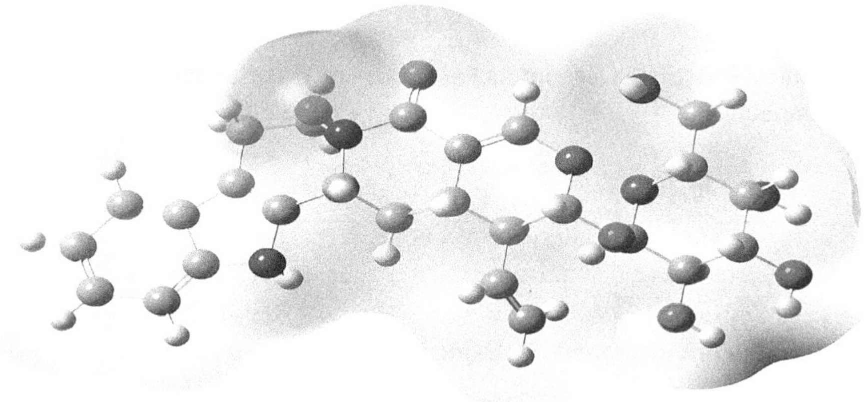

FIGURE 3.3 MESP surface of Vincosamide-N-Oxide.

Source: Ref. [11].

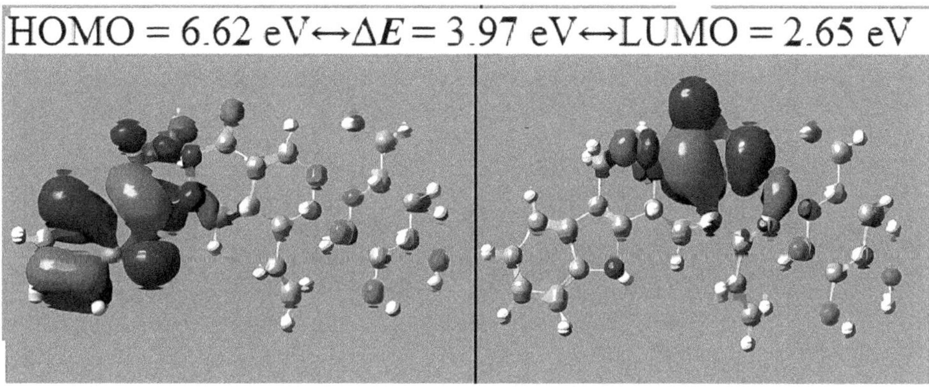

FIGURE 3.4 HOMO–LUMO energy level diagram: Vincosamide-N-Oxide.

Source: Ref. [11].

3.2.3 HOMO, LUMO, and Global Reactivity Descriptors

The softness of a molecule is measured in terms of the HOMO–LUMO energy band gap as per Pearson[18] and is applicable for describing the eventual charge transfer within the molecule. The global reactivity descriptors—ionization potential (I), electronic affinity (A), electro-negativity (χ), chemical potential (μ), global hardness (η), global softness (S), and global electrophilicity index (ω)—are expressed in terms of the HOMO–LUMO energy band gap.[19,20] The HOMO and LUMO qualitatively predicts the electron transportation and excitation[21,22] during reaction. HOMO and LUMO primarily act as an electron donor and acceptor, respectively, and its band gap measures the chemical reactivity,[23] as displayed in Figure 3.4, along with the global reactivity descriptors in Table 3.2.

3.2.4 Nonlinear Optical Properties

$$\xi = \xi^0 - \mu_1 E_l - \frac{1}{2}\alpha_{1m} E_l E_m - \frac{1}{6}\beta_{lmn} E_l E_m E_n + \ldots \quad (3.13)$$

TABLE 3.2
Global Chemical Reactivity Descriptors Parameters of Vincosamide-N-Oxide Bioactive Natural Compound

$I = -E_{HOMO}$ (eV)	$A = -E_{LUMO}$ (eV)	$\mu = -\chi$ (eV)	$\omega = \mu^2/2\eta$ (eV)	$\chi = (I + A)/2$ (eV)	$\eta = (I - A)/2$ (eV)
0.243	0.097	−0.17	0.197	0.170	0.072

TABLE 3.3
Optical Properties of Vincosamide-N-Oxide Bioactive Natural Compound

S. No.	Components of (α)	Values	Components of (β)	Values	Components of (μ)	Values
1	α_{xx}	421.1793	β_{xxx}	−551.21	μ_x	−5.5086
2	α_{yy}	−72.5888	β_{yyy}	247.745	μ_y	−6.8779
3	α_{zz}	346.2523	β_{zzz}	60.3753	μ_z	4.5148
4	α_{xy}	−4.8133	β_{xyy}	59.6891	μ_{total}	9.9012
5	α_{xz}	36.2401	β_{xxy}	54.5439	—	—
6	α_{yz}	308.9771	β_{xxz}	−78.407	—	—
7	α_0	231.6143	β_{xzz}	176.21	—	—
8	$\Delta\alpha$	465.1499	β_{yzz}	−117.39	—	—
9	α_0 (esu)	34.3252×10^{-24}	β_{yyz}	63.4955	—	—
10	$\Delta\alpha$ (esu)	68.9352×10^{-24}	β_{xyz}	106.346	—	—
11	—	—	β_0	368.344	—	—
12	—	—	β_0(esu)	3182.124×10^{-33}	—	—

Source: Ref. [11].

where (ξ) and ξ^0 are the energy of the perturbed and unperturbed biomolecules, respectively, E_l is the field at the origin, μ_l, α_{lm}, and β_{lmn} are the components of the dipole moment, polarizability, and first hyperpolarizability, respectively. The NLO parameters are defined as:

$$\mu = \left(\mu_x^2 + \mu_y^2 + \mu_z^2\right)^{1/2} \tag{3.14}$$

$$\alpha_0 = \frac{1}{3}\left(\alpha_{xx} + \alpha_{yy} + \alpha_{zz}\right) \tag{3.15}$$

$$\Delta\alpha = 2^{-1/2}\left[\left(\alpha_{xx} - \alpha_{yy}\right)^2 + \left(\alpha_{yy} - \alpha_{zz}\right)^2 + \left(\alpha_{zz} - \alpha_{xx}\right)^2 + 6\alpha^2 xz\right]^{1/2} \tag{3.16}$$

$$\beta_0 = \left(\beta_x^2 + \beta_y^2 + \beta_z^2\right)^{1/2} \tag{3.17}$$

where

$$\beta_x = \left(\beta_{xxx} + \beta_{yyy} + \beta_{zzz}\right), \beta_y = \left(\beta_{yyy} + \beta_{yzz} + \beta_{yxx}\right), \beta_z = \left(\beta_{zzz} + \beta_{zxx} + \beta_{zyy}\right)$$

$$\beta_0 = \left[\left(\beta_{xxx} + \beta_{yyy} + \beta_{zzz}\right)^2 + \left(\beta_{yyy} + \beta_{yzz} + \beta_{yxx}\right)^2 + \left(\beta_{zzz} + \beta_{zxx} + \beta_{zyy}\right)^2\right]^{1/2}$$

$\Delta\alpha$ = anisotropy of the polarizability, μ = total static dipole moment, β_0 = mean first-order hyperpolarizability, and α_0 = mean polarizability calculated according to the Kleinman symmetry.[24] The optical properties of some of the biologically active natural compounds studied by us have been depicted in Table 3.3.

3.2.5 Local Reactivity Descriptors

The ability to select the active sites relative to the higher layer within the natural compound is measured in terms of local softness $(s_k^{\pm})$, Fukui function $\left(f_k^{\pm}\right)$, and local electrophilicity$\left(\omega_k^{\pm}\right)$ calculated as per reported formula[25–28]:

$$\omega_k^{\pm} = \omega f_k^{\pm} \tag{3.18}$$

$$s_k^{\pm} = S f_k^{\pm} \tag{3.19}$$

$$f_k^{-} = [q(N) - q(N-1)] \tag{3.20}$$

$$f_k^{+} = [q(N+1) - q(N)] \tag{3.21}$$

$q(N+1)$, $q(N-1)$, and $q(N)$ denotes the atomic charge distribution for anionic, cationic and neutral chemical species respectively for nth atom. S and ω are the global softness and electrophilicity index, respectively. Morell et al.[29] reported that if a molecule is attacked by a hard reagent, then it tends to react at the site where the value of the local reactivity descriptor is smallest and that, if attacked by a soft agent, it has a tendency to react at the site where the value of the same is large. The local reactivity descriptors have been presented in Table 3.4.

TABLE 3.4
Local Reactivity Descriptors of Vincosamide-N-Oxide Bioactive Natural Compound

Position of Atoms	f_k^+	f_k^-	s_k^+	s_k^-	ω_k^+	ω_k^-
1C	−0.0426	−0.0184	−0.2958	−0.1278	−0.0084	−0.0036
2C	−0.1370	−0.0684	−0.9513	−0.475	−0.027	−0.0135
3C	−0.0399	0.0411	−0.2771	0.2854	−0.0079	0.0081
4C	0.0861	−0.1089	0.5979	−0.7562	0.0170	−0.0215
5H	0.0186	−0.0309	0.1292	−0.2146	0.0037	−0.0061
6H	0.0236	−0.0287	0.1639	−0.1993	0.0046	−0.0057
7C	0.0036	0.1279	0.0250	0.8881	0.0007	0.0252
8C	0.1646	−0.1252	1.1430	−0.8694	0.0324	−0.0247
9C	0.0436	−0.0279	0.3028	−0.1937	0.0086	−0.0055
10C	−0.0190	−0.0215	−0.1319	−0.1493	−0.0037	−0.0042
11C	0.1721	−0.0766	1.1951	−0.5319	0.0339	−0.0151
12C	0.0367	−0.0093	0.2548	−0.0646	0.0072	−0.0018
13H	0.0664	−0.0433	0.4611	−0.3007	0.0131	−0.0085
14H	0.0587	−0.0462	0.4076	−0.3208	0.0116	−0.0091
15H	0.0545	−0.0447	0.3784	−0.3104	0.0107	−0.0088
16H	0.0667	−0.0425	0.4632	−0.2951	0.0131	−0.0084
17N	0.0863	−0.0984	0.5993	−0.6833	0.017	−0.0194
18C	0.1372	−0.0473	0.9527	−0.3285	0.027	−0.0093
19C	0.0480	−0.0677	0.3333	−0.4701	0.0095	−0.0133
20H	0.0523	−0.0056	0.3632	−0.0389	0.0103	−0.0011

(Continued)

TABLE 3.4 (Continued)

Position of Atoms	f_k^+	f_k^-	s_k^+	s_k^-	ω_k^+	ω_k^-
21C	0.0789	0.0537	0.5479	0.3729	0.0155	0.0106
22N	0.0281	0.0442	0.1951	0.3069	0.0055	0.0087
23O	−0.0546	−0.0385	−0.3791	−0.2673	−0.0108	−0.0076
24C	−0.0670	−0.0133	−0.4652	−0.0924	−0.0132	−0.0026
25H	0.0107	−0.0142	0.0743	−0.0986	0.0021	−0.0028
26C	−0.0680	0.0474	−0.4722	0.3291	−0.0134	0.0093
27C	0.1049	0.0183	0.7284	0.1271	0.0207	0.0036
28O	0.0310	−0.0219	0.2153	−0.1521	0.0061	−0.0043
29C	−0.0237	0.032	−0.1646	0.2222	−0.0047	0.0063
30C	−0.0449	0.0093	−0.3118	0.0646	−0.0088	0.0018
31C	−0.1157	0.0364	−0.8034	0.2528	−0.0228	0.0072
32O	−0.0378	−0.0067	−0.2625	−0.0465	−0.0074	−0.0013
33H	0.0306	−0.0088	0.2125	−0.0611	0.0060	−0.0017
34H	0.0203	−0.0066	0.1410	−0.0458	0.0040	−0.0013
35H	0.0323	−0.0084	0.2243	−0.0583	0.0064	−0.0017
36H	0.0361	−0.0266	0.2507	−0.1847	0.0071	−0.0052
37H	0.0352	−0.0131	0.2444	−0.091	0.0069	−0.0026
38H	0.0462	−0.0232	0.3208	−0.1611	0.0091	−0.0046
39H	0.0514	−0.0325	0.3569	−0.2257	0.0101	−0.0064
40C	0.0927	0.0257	0.6437	0.1785	0.0183	0.0051
41C	−0.0334	−0.0167	−0.2319	−0.116	−0.0066	−0.0033
42C	0.0413	−0.0109	0.2868	−0.0757	0.0081	−0.0021
43C	−0.0046	0.0007	−0.0319	0.0049	−0.0009	0.0001
44C	0.0526	−0.0062	0.3653	−0.0431	0.0104	−0.0012
45H	0.0178	−0.0081	0.1236	−0.0562	0.0035	−0.0016
46H	0.0048	−0.0059	0.0333	−0.041	0.0009	−0.0012
47H	0.0115	0.0058	0.0799	0.0403	0.0023	0.0011
48H	0.0082	0.0002	0.0569	0.0014	0.0016	0.000
49O	−0.0702	0.0042	−0.4875	0.0292	−0.0138	0.0008
50C	−0.0328	−0.0144	−0.2278	−0.1	−0.0065	−0.0028
51O	−0.0772	0.0013	−0.5361	0.009	−0.0152	0.0003
52O	−0.0473	−0.0011	−0.3285	−0.0076	−0.0093	−0.0002
53O	−0.0395	−0.0081	−0.2743	−0.0562	−0.0078	−0.0016
54O	−0.0480	0.0063	−0.3333	0.0437	−0.0095	0.0012
55H	0.0315	0.0052	0.2187	0.0361	0.0062	0.0010
56H	0.0192	−0.0055	0.1333	−0.0382	0.0038	−0.0011
57H	0.0313	−0.0024	0.2173	−0.0167	0.0062	−0.0005
58H	0.0352	−0.0076	0.2444	−0.0528	0.0069	−0.0015
59C	−0.0789	−0.0327	−0.5479	−0.2271	−0.0155	−0.0064
60H	0.0260	0.0005	0.1805	0.0035	0.0051	0.0001
61H	0.0138	−0.0117	0.0958	−0.0812	0.0027	−0.0023
62H	0.0090	−0.0110	0.0625	−0.0764	0.0018	−0.0022
63H	0.0219	0.0026	0.1521	0.0181	0.0043	0.0005
64H	0.0223	−0.0096	0.1549	−0.0667	0.0044	−0.0019
65H	−0.0108	−0.0066	−0.0750	−0.0458	−0.0021	−0.0013
66H	0.1180	−0.0395	0.8194	−0.2743	0.0232	−0.0078
67O	−0.0891	−0.1392	−0.6187	−0.9666	−0.0176	−0.0274

Source: Ref. [11].

TABLE 3.5
NBO Data of Vincosamide-N-Oxide Bioactive Natural Compound

S. No.	Acceptor (*j*)	Donor (*i*)	*E*(*j*)-*E*(*i*) (au)	*F*(*i*,*j*) (au)	$E^{(2)}$ kcal/mol
1	σ* (C7-C8)	LP(σ)N17	0.37	0.048	7.34
2	π * (C18-N22)	LP(π)O23	0.54	0.112	29.12
3	π * (C18-C26)	LP(π)O23	0.66	0.091	15.25
4	π * (C24-C26)	LP(π)O28	0.36	0.082	23.30
5	π* (C30-H33)	LP(π)O28	0.82	0.062	5.50
6	π* (O28-C30)	LP(π)O32	0.58	0.073	11.47
7	π* (O32-C44)	LP(π)O49	0.56	0.071	11.11
8	π* (C40-C41)	LP(π)O49	0.66	0.053	5.16
9	π* (C40-H63)	LP(π)O49	0.77	0.058	5.23
10	π* (C40-C50)	LP(π)O51	0.67	0.056	5.80
11	π* (C41-C42)	LP(π)O52	0.68	0.059	6.37
12	π* (C42-C43)	LP(π)O53	0.68	0.059	6.27
13	π* (C42-C43)	LP(π)O54	0.69	0059	6.31
14	π* (C21-N22)	LP(π)O67	0.47	0.045	5.33
15	π* (C18-N22)	LP(π)O67	0.50	0.060	9.09

Source: Ref. [11].

3.2.6 NBO Analysis

The identification of individual bonds and energies corresponding to the loan-pair electrons in a chemical reaction is performed through NBO analysis,[30–32] which gives the second-order perturbation energy of the donor–acceptor. Empty orbital interaction leads to delocalization, and the stabilization energy $E^{(2)}$ associated with electron delocalization is computed.[33,34] The perturbation theory of the second order of the Fock matrix analysis in NBO basis pertaining to some molecules has been well reported in the literature.[35,36] For each acceptor (*j*) and donor (*i*), the stabilization energy $E^{(2)}$ associated with the delocalization $i \rightarrow j$ is assessed as:[37]

$$E^{(2)} = \Delta E_{ij} = q_{ij} \frac{F(i,j)^2}{\varepsilon i - \varepsilon j} \tag{3.22}$$

$F(i,j)$ = off diagonal Fock matrix element, q_i = donor orbital occupancy, ε_i and ε_j = orbital energies diagonal elements.

3.3 DRUG PROPERTIES

DFT optimization enables one to use the protein data bank (PDB) textual file format as the input file for the study of the protein–ligand interaction usually performed using the molecular docking approach executed via available computational tools.[38] The biological activity of isolated Vincosamide-N-Oxide against the severe acute respiratory syndrome corona virus (SARS-COV-2) identified as COVID-19, as already reported, and the binding affinity in protein–ligand interaction complex are presented in Table 3.6 and Figure 3.5, respectively.[39]

3.4 CONCLUSION AND FUTURE SCOPE

Biologically active natural products are the gift given by nature that play a pivotal role in the welfare of human beings as an alternative medicine with less cost and null toxicity due to their inherent

TABLE 3.6
Vincosamide-N-Oxide ($C_{26}H_{31}N_2O_9$) of Binding Affinity with the Target Proteins

Rank	6LU7		6CRV		6Y84	
(cluster)	ΔE (kcal/mol)	K	ΔE (kcal/mol)	K	ΔE (kcal/mol)	K
1	−5.81	55.52 μM	−4.59	434.05 μM	−8.81	348.65 nM
2	−4.78	312.98 μM	−4.47	532.26 μM	−8.77	370.30 nM
3	−4.49	511.38 μM	−4.28	730.11 μM	−8.41	686.48 nM
4	−3.93	1.31 mM	−4.05	1.07 mM	−8.08	1.19 μM
5	−3.89	1.40 mM	−4.00	1.17 mM	−7.92	1.58 μM
6	−3.75	1.79 mM	−3.49	2.78 M	−7.70	2.26 μM
7	−3.48	2.83 mM	−3.44	3.02 mM	−7.22	5.08 μM
8	−3.41	3.15 mM	−3.38	3.31 mM	−6.26	25.76 μM
9	−3.33	3.64 mM	−3.29	3.91 mM	−5.08	187.55 μM
10	−3.03	5.96 mM	−3.04	5.93 mM	—	—

Source: [Ref. 39].
Δ*E*: binding free energy, *K*: inhibition constant, 6LU7 and 6Y84: COVID 19 protease, 6CRV: spike glycoprotein.

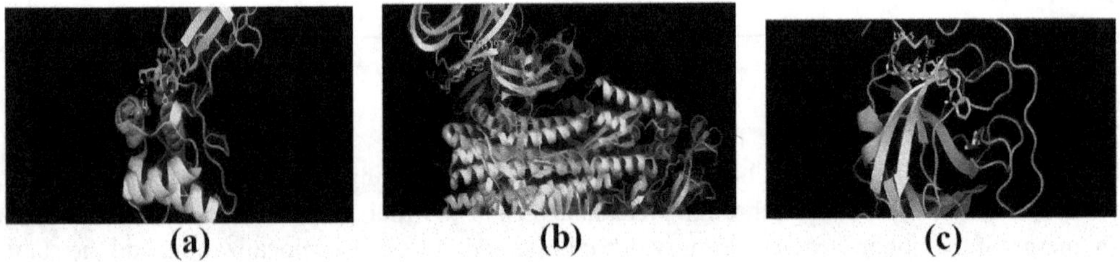

FIGURE 3.5 Ligand (Vincosamide-N-Oxide)—protein interaction: (a) 6 LU7, (b) 6 CRV, and (c) 6Y84.

Source: Ref. [39].

bioactivity. The causes behind the inherent bioactivity of this natural product are yet to be substantiated in terms of the properties associated with molecular structure. DFT-based calculations enable us to predict the thermodynamics, spectral, optical, and drug properties of the Vincosamide-N-Oxide, as well as a few of the natural products that are important for revealing the structure–property correlations. The prediction of the MESP surface is the significant outcome for the identification of reactive sites in Vincosamide-N-Oxide when the protein–drug interaction is studied. DFT-based quantum chemical calculations significantly evaluate the NBO, chemical reactivity descriptors, and drug activity of Vincosamide-N-Oxide.

The future scope for computational studies on natural products is full of the probability of developing a low-cost, user-friendly, very low-toxicity, and multifunctional drug agents for various fatal diseases by tracking the pathway of bioactivity at the molecular level. The present century is the right time to carry out the theoretical research for establishing the scientific basis of the inherent bioactivity of such naturally occurring biomolecules using existing physical principles at the molecular level due to the high-level quantum computational techniques that are now available[40] and the references that are also available for exploiting these computational techniques to carry out such research on biologically active molecules.[41–43]The present chapter presents a view of theoretical investigations using DFT methods pertaining to the structure and properties of some of these natural products in order to explore the chemistry of these macromolecular structures.

REFERENCES

1 Mishra, DP, Khan, MA, and Yadav, DK, et al., Monoterpene indole alkaloids from anthocephalus cadamba fruits exhibiting anticancer activity in human lung cancer cell line H1299. Chemistry Select. 3: 8468–8472, 2018. Available from: https://doi.org/10.1002/slct.201801475
2 Hohenberg, P, and Kohn, W, Inhomogeneous electron gas, Physical Rev. 136 (3B): 864–871, 1964. Available from: https://doi.org/10.1103/PhysRev.136.B864
3 Thomas, LH, The calculation of atomic fields proc. Cambridge Phil Soc. 23(5): 542–548, 1927. Available from: http://dx.doi.org/10.1017/S0305004100011683
4 Fermi, E, Un Metodo statistic per la determinazione di alcune prioprieta dell atomo. Rend Accad Naz Lincei. 6: 602–607, 1927.
5 Dirac, PAM, The Principles of Quantum Mechanics. UK: Oxford University Press, 257, 1930.
6 Slater, JC, A simplification of the Hartree-Fock method. Phys Rev. 81: 538–541, 1951.
7 Parr, RG, Gadre, SR, and Bartolotti, LJ, Local density functional theory of atoms and molecules. Proc Natl Acad Sci. 76(6): 2522–2526, 1976.
8 Kohn W, and Sham, LJ, Self-consistent equations including exchange and correlation effects. Physical Rev. 140 (4A): A1133, 1965. Available from: https://doi.org/10.1103/PhysRev.140.A1133
9 Haoyu, SY, Shaohong, LL, and Donald, GT, Perspective: Kohn-Sham density functional theory descending a staircase. J Chem Phys. 145: 130901, 2016.
10 Frisch, MJ, Trucks, GW, and Schlegel, HB, et al., Gaussian 09, Revision B.01. Wallingford CT: Gaussian Inc., 2010.
11 Mishra, AK, and Tewari, SP, Density functional theory calculations of spectral, NLO, reactivity, NBO properties and docking study of Vincosamide-N-Oxide active against lung cancer cell lines H1299. [Cited 2023] SN Appl Sci, 2:1021, 2020. Available from: https://doi.org/10.1007/s42452-020-2842-9
12 Scrocco, E, and Tomasi, J, Electronic molecular structure, reactivity and intermolecular forces: An euristic interpretation by means of electrostatic molecular potentials. Adv Quantum Chem. 11: 115–193, 1978. Available from: https://doi.org/10.1016/S0065-3276(08)60236-1
13 Luque, FJ, Lopez JM, and Orozco, M, Perspective on "electro-static interactions of a solute with a continuum: A direct utilization of ab initio molecular potentials for the prevision of solvent effects". Theor Chem Acc. 103:343–345, 2000. Available from: https://doi.org/10.1007/978-3-662-10421-7
14 Scrocco, E, and Tomasi, J, Top Curr Chem. 7: 95, 1973. Available from: www.springer.com/series/128
15 Li, Y, Liu, Y, and Wang, H, Synthesis, crystal structure, vibration spectral, and DFT studies of 4-aminoantipyrine and its derivatives. Molecules. 18:877–893, 2013. Available from: https://doi.org/10.3390/molecules18010877
16 Tomasi, J, Mennucci, B, and Cammi, R, MEP: A tool for interpretation and prediction. From molecular structure to solvation effects. Comput Theor Chem. 3:1–103, 1996. Available from: https://doi.org/10.1016/S1380–7323(96)80041–0
17 Kuan, JL, Ming, CC, and Wang, Y, Charge density distributions, interaction energies, and electrostatic potentials of hydrogen bonds. J Phys Chem. 98:11685–11693, 1994.
18 Pearson, RG, Absolute electronegativity and hardness correlated with molecular orbital theory. Proc. Natl. Acad. Sci. 83:8440–8441, 1986. Available from: https://doi.org/10.1073/pnas.83.22.8440
19 Pearson, RG, Absolute electro negativity and hardness: Applications to organic chemistry. J Org Chem. 54:1423–1430, 1989. Available from: https://doi.org/10.1021/jo00267a034
20 Padmanabhan, J, Parthasarathi, R, and Subramanian, V, et al., Electrophilicity-based charge transfer descriptor. J Phys Chem A. 111:1358, 2007.
21 Belletete, M, Morin, JF, and Leclerc, M, et al., A theoretical, spectroscopic, and photophysical study of 2-7-carbazolenevi-nylene-basedconjugatedderivatives. J Phys Chem. 109:6953–6959, 2005.
22 Zhenminga, D, Heping, S, and Yufang, L, Experimental and theoretical study of 10-methoxy-2-phenylbenzo[h]quinoline. Spectro Chim Acta Part A. 78:1143–1148, 2011.
23 Lewis, DF, Loannides, C, and Parke, DV, Interaction of a series of nitriles with the alcohol-inducible isoform of P450: Computer analysis of structure-activity relationships. Xenobiotica 24:401–408, 1994.
24 Kleinman, DA, Nonlinear dielectric polarization in optical media. Phys Rev. 126:1977, 1962.
25 Parr, RG, and Pearson, RG, Absolute hardness: Companion parameter to absolute electronegativity. J Am Chem Soc. 105:7512–7516, 1983.
26 Parr, RG, Szentpaly, LV, and Liu, S, Electrophilicity index. J Am Chem Soc. 121:1922–1924, 1999.
27 Chattaraj, PK, Lee, H, and Parr, RG, Principle of maximum hardness. J Am Chem Soc. 113:1854–1855, 1999.

28 Chattaraj, PK, and Giri, S, Stability, reactivity, and aromaticity of compounds of a multivalent superatom. J Phys Chem A. 111:11116–11121, 2007.
29 Morell, C, Grand, A, and Labbe, AT, New dual descriptor for chemical reactivity. J Phys Chem A. 109:205–212, 2005.
30 Pophristic, V, Goodman, L, and Guchhait, N, Role of lone-pairs in internal rotation barriers. J Phy Chem. A. 101:4290–4297, 1997.
31 Guo D, and Goodman, L, Nature of barrier forces in acetaldehyde. J Phys Chem. 100:12540–12545.
32 Weinhold, F, A new twist on molecular shape. Nature. 411:539–541, 1996, 2001.
33 Reed, AE, Curtiss, LA, and Weinhold, F, Inter molecular interactions from a natural bond orbital, donor-accept or view point. Chem Rev. 88:899–926, 1988.
34 Liu, CG, Su, ZM, and Guan, XH, Redox and photo iso-merization switching the second-order nonlinear optical properties of a tetrathiafulvalene derivative across six states: A DFT study. J Phys Chem C. 115:23946–23954, 2011.
35 Mishra, AK, and Gupta, V, Density functional theory calculations of electronic properties and bioactivity of natural macromolecule N-β-D-Glucopyranosyl Vincosamide showing activity against the diabetic mellitus and prediction of its derivative, Applied Innovative Research (AIR). [Cited-2022] 2: 61–79, 2020. Available from: http://nopr.niscpr.res.in/handle/123456789/54957
36 Brahmachari, G, Kumar, A, and Srivastava, AK, et al., Synthesis, spectroscopic characterization, X-ray analysis and theoretical studies on the spectral features (FT-IR, 1H-NMR), chemical reactivity, NBO analyses of 2-(4-fluorophenyl)-2-(4-fluorophenylamino) acetonitrile, and its docking into IDO enzyme. RSC Advances. 5:80967–80977, 2015.
37 Glendening, ED, Landis CR, and Weinhold, F, "Natural bond orbital methods" WIREs. Comput. Mol. Sci. 2: 1–42, 2012.
38 Morris, GM, Huey, R, and Sanner, MF, et al., Auto dock4 and auto dock tools4: Automated docking with selective receptor flexibility. J Comput Chem. 30:2785–2791, 2009.
39 Mishra AK, Gupta, V, and Tewari, SP, In silico screening of some naturally occurring bioactive compounds predicts potential inhibitors against SARS-COV-2 (COVID19) protease. Indian J. Biochem. Biophys. [Cited- 2023] 57: 416–425, 2021. Available from: http://nopr.niscpr.res.in/handle/123456789/58245
40 Young, DC, Computational Chemistry: A Practical Guide for Applying Techniques to Real-World Problems. New York: John Wiley and sons, 2001.
41 Srivastava, AK, and Misra, N, Computational studies of biologically active molecules and small clusters DFT and TDDF T approaches. PhD Thesis Lucknow University, 2018. Available from: http://hdl.handle.net/10603/202589 Accessed on 25/08/2018.
42 Srivastava, AK, Pandey, AK, and Jain S, et al., FT-IR spectroscopy, intra-molecular C–H–O interactions, HOMO, LUMO, MESP analysis and biological activity of two natural products, triclisine and rufescine: FT and QTAIM approaches. Spectrochima Acta. 136:682–689, 2015. Available from: https://doi.org/10.1016/j.saa.2014.09.082
43 Tewari, SP, and Mishra, AK, Computational study of some naturally occurring biomolecules. Ph.D. Thesis Dr. Shakuntala Misra National Rehabilitation University, Lucknow, 2020. Available from: http://hdl.handle.net/10603/338036

4 Drugs, Drug–Biomolecule Interactions, and Drugs Delivery Systems
Quantum Chemical Approaches

Soni Mishra and Abhishek Kumar Mishra

4.1 INTRODUCTION

4.1.1 Novel Drug Delivery Systems

Drug delivery system (DDS) is a device that allows a therapeutic substance to reach its correct site of action without affecting other organs or tissues. A drug delivery system is used to deliver the drug to the appropriate receptors. Novel drug delivery systems (NDDS) are innovative approaches and technologies that are designed to enhance the delivery and targeting of drugs to specific sites in the body. These systems aim to improve the therapeutic efficacy, safety, and patient compliance of drug treatments. Examples include liposomes, nanoparticles, microneedles, implantable drug delivery systems, gene delivery systems, etc.

4.1.2 Nanoparticles and Drug Delivery

Nanomedicine, an emerging field that integrates nanoscience knowledge and techniques with medical biology, is revolutionizing disease prevention and treatment. It encompasses the utilization of nanoscale compounds, including nanorobots, diagnostic nanosensors, and targeted drug delivery systems, within living cells. Typically, nanoparticles employed as drug carriers have dimensions below 100 nm on at least one side and consist of various materials such as natural or synthetic polymers, lipids, or metals, which confer their unique properties.[1–3] Compared to larger macromolecules, nanoparticles are more efficiently processed by cells, making them highly effective transport and delivery systems. Their appeal in therapeutic applications stems from their advantageous characteristics, such as a large surface-to-mass ratio, quantum properties, and the ability to encapsulate and deliver diverse compounds, which can overcome the limitations of conventional delivery systems.[4,5] Extensive research has been conducted on the development of nanoparticles for targeted drug delivery, and this delivery approach can be achieved through effective targeting or action recognition mechanisms.[6]

Nanostructures exhibit prolonged circulation within the bloodstream, allowing for the controlled release of drug compounds in accordance with prescribed dosages. As a result, they minimize fluctuations in plasma concentration and reduce undesirable side effects.[7] The selection of a nano drug delivery system is primarily determined by the biophysical and biochemical properties of the specific drug intended for treatment. Thus nanotechnology offers numerous advantages in the management of chronic diseases by enabling targeted drug delivery. However, the limited understanding of nanostructure toxicity remains a significant concern, emphasizing the need for further research to enhance the efficiency and safety of these high-performance systems for their secure utilization.[8] Careful design of these nanoparticles can help address the associated challenges, and quantum chemical calculations provide valuable insights at the atomistic level to elucidate these systems.

DOI: 10.1201/9781003441328-4

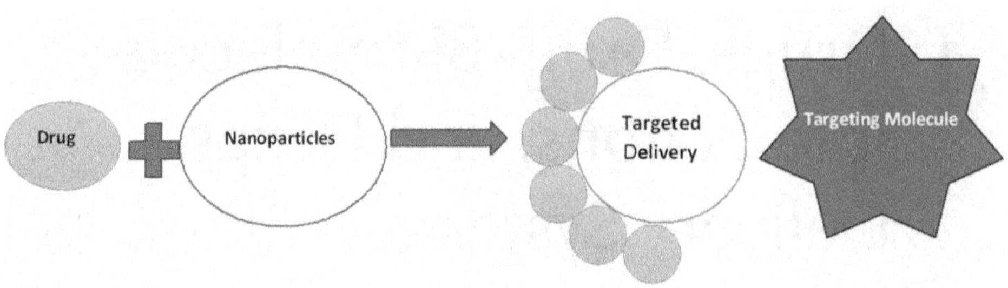

FIGURE 4.1 Nanotechnology in targeted drug delivery system.

Nanomedicines and nanomedicine delivery systems are rapidly advancing and involve the utilization of nanoscale materials for precise diagnostic purposes and controlled delivery of therapeutic agents to specific target areas.[8] Nanotechnology offers numerous benefits in the treatment of targeted human diseases and the localized delivery of specific drugs, as depicted in Figure 4.1. Notably, nanomedicine has demonstrated remarkable applications in various disease treatments, encompassing chemotherapeutic agents, biological agents, immunotherapeutic agents, and more.[8, 9] A nanoparticle-based approach has been developed that combines therapeutic and diagnostic strategies for cancer diagnosis.[8, 10] The initial generation of nanoparticle-based treatments is comprised of lipid systems like liposomes and micelles, which have obtained FDA approval.[11] These liposomes and micelles can incorporate inert nanoparticles such as gold or magnetic nanoparticles,[12] enabling enhanced drug delivery, imaging, and therapeutic functionalities. Moreover, nanostructures have been reported to assist in preventing drug degradation in the gastrointestinal tract and facilitating the controlled release of slow-melting drugs to the intended target area.

Nanoparticles are broadly divided into various categories, and, depending on their size and chemical properties, they exist in different ways such as spherical, tubular, and irregular structures (Figure 4.2). The most commonly nanomaterials used are:

- Dendrimers
- Carbon nanotubes
- Liposomes
- Fullerenes
- Quantum dots
- Ceramic nanoparticles
- Metal nanoparticles
- Semiconductor nanoparticles

Lipids are a group of molecules formed by the chemical bonding of fatty acids, which possess high solubility in organic solvents but low solubility in polar solvents such as water.[13] They possess diverse structures and functionalities, with one of the most abundant lipids being triacylglycerol (TAGs), which plants utilize for efficient energy storage; humans commonly consume lipids derived from vegetables in their daily diet.[14] Lipids play crucial roles in the functioning and structural organization of the nervous system. They can be transported within cells and enzymatically transferred from one location to the membrane bilayer.[13] When associated with proteins, lipids are transported through the bloodstream, allowing them to disperse in the aqueous plasma medium.[15] Ingested lipids are emulsified using bile acids, synthesized in hepatocytes from cholesterol, and transported via ATP, leading to the formation of micelles.[16] In the pharmaceutical industry, lipids find extensive use in drug delivery systems, facilitating rapid absorption into the bloodstream due to their characteristic property of selectively allowing passage of lipophilic (hydrophobic) substances. Research has shown the effectiveness of liposomes, spherical structures consisting of phospholipid

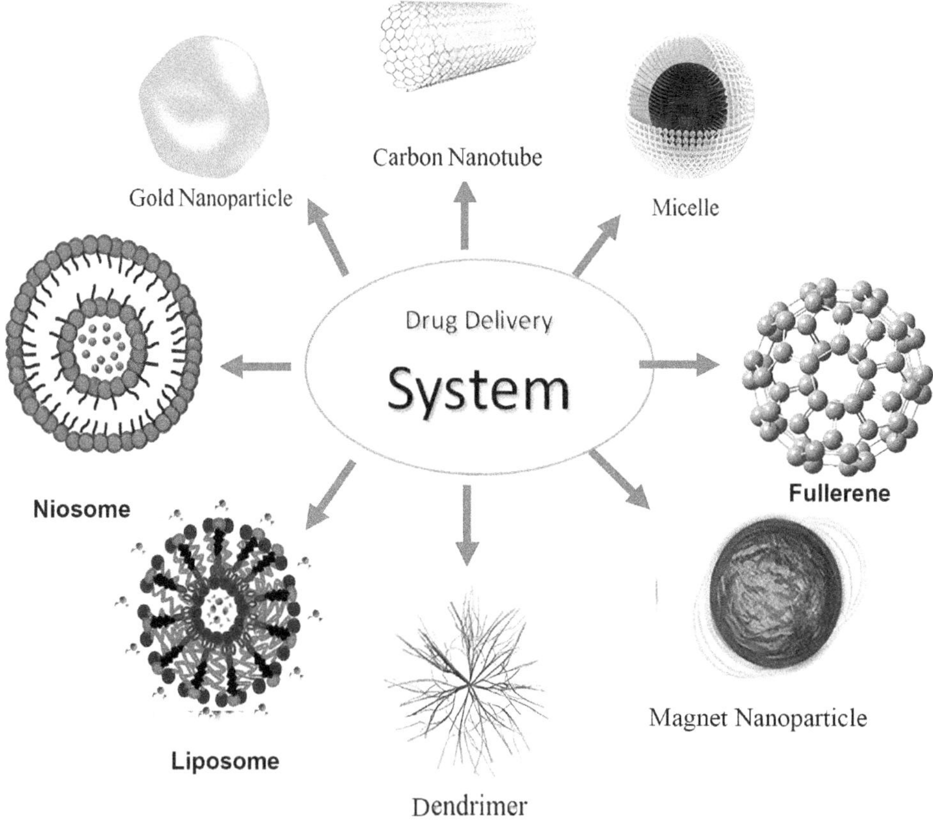

FIGURE 4.2 Drug delivery systems.

bilayers, for drug delivery, as they can be targeted to specific tissues or organs for generalized medication.[17]

Phospholipids are a form of lipid that consist of phosphorus and are ubiquitous in all organisms as the major component of biological membranes. These are greatly affected by pH; when disposed in water, they hydrate, forming hexagonal phases or lamellar, get easily oxidized in the presence of unsaturated fatty acids, and exhibit anti-oxidative properties. Being beneficial for human health is currently produced in the form of lecithin.[15]

In this chapter, the quantum chemical calculations of drug molecules are discussed along with the behavior of DPPC (1,2-dipalmitoyl-sn-glycero-3-phosphocholine) phospholipid, which leads to further advancement in providing the key for the architecture and activities of the membrane and has been an important aspect in providing a new perspective in the function of related biological systems. This has played a major role in quantifying and detecting the interaction occurring in the lipid bilayer.[18] Whenever the molecule DPPC interacts with different drugs, induced changes are observed in its structural organization, especially membranes, and sometimes dysfunctioning is also detected. Finally, we discuss the interactions of carbon nanotubes (CNTs) with DPPC through detailed quantum chemical calculations in drug delivery systems.

4.2 METHODOLOGY

The DFT method[19] was used as implemented in the Gaussian 03 program.[20] The electronic structure and optimized geometries of the molecules were calculated using Lee–Yang–Parr correlation functionals.[21–23] The GaussView program[24] was employed to facilitate the analysis of these results.

4.3 RESULT AND DISCUSSION

4.3.1 Quantum Chemical Calculations of Drug Molecules

4.3.1.1 Secnidazole Drug

Secnidazole (α,2-dimethyl-5-nitro-1H-imidazole-1-ethanol) is used as antiprotozoal, antiamebic, and antibacterial drug.[25] It is primarily used for the treatment of various infections caused by anaerobic bacteria and protozoa. The drug secnidazole has been evaluated for the treatment of bacterial vaginosis, and it was found that a single dose of secnidazole was as effective as the standard multiple-dose regimen, providing high cure rates and sustained improvement in symptoms.[26] Muzny et al. investigated the efficacy and safety of secnidazole in the treatment of trichomoniasis and showed that secnidazole was highly effective in eradicating trichomonas vaginalis infection, with minimal side effects.[27, 28] Secnidazole has been found to exhibit comparable efficacy and a more favorable tolerability profile, making it a viable treatment option for giardiasis.[29] Recently, we performed DFT-based simulations to explore the conformational stability of the secnidazole molecule (Figure 4.3).[30]

In this study, we conducted geometry optimizations to determine the most stable conformers of secnidazole, comparing their relative energies.[30] The electronic structure and optimized geometries of the molecule's stable conformers were calculated using DFT B3LYP/6–311++G(d,p) basis set.

Currently, racemic secnidazole is utilized in clinical trials, despite the presence of a chiral carbon in its molecular structure. As per the orientations of the ethanol component (Figure 4.4), three distinct configurations have been identified. Energy minimization was employed to determine the geometries of conformers. These calculations were conducted using the 4–31G, 6–31G, and 6–311++G(d,p) basis sets. The optimized geometries led to the determination of relative energies for these three conformers, which are presented in Table 4.1.

Our quantum chemical calculations employing DFT-based methods demonstrate that the energy differences among the observed conformers of secnidazole range from 0.5 to 3.0 kcal mol^{-1}. Nevertheless, the crystalline structure of secnidazole hemihydrate reveals the presence of only one conformer. The optimized structure of this particular conformer closely aligns with the experimentally observed structure, with minimal differences in bond lengths (less than 0.01 Å) and bond angles (around 1°, except for a few moieties exhibiting a difference of approximately 6°).[30]

4.3.1.2 Efavirenz Drug

Efavirenz is commonly referred to as (S)-6-chloro-4-(cyclopropylethynyl)-1, 4-dihydro-4-(trifluoromethyl)-2H-3, 1-ben-zoxazin-2-one (Figure 4.5) in chemical terms. It belongs to the

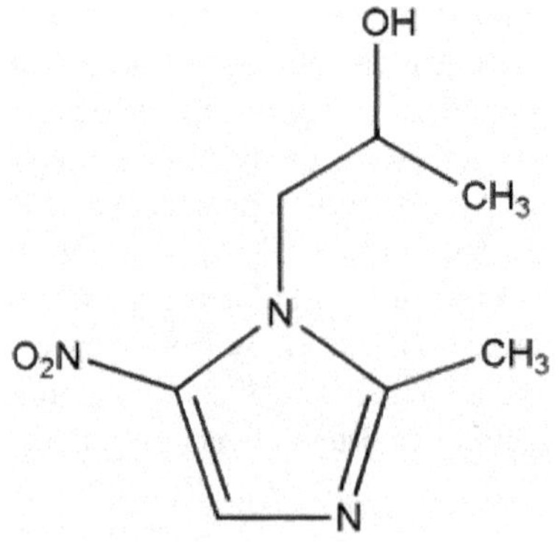

OH
CH_3
N
O_2N
CH_3
N

FIGURE 4.3 Schematics of the secnidazole molecule.[30]

Source: reprinted (adapted) with permission from *J. Phys. Chem. A* 2009, 113, 273–281. Copyright 2009 american chemical society.

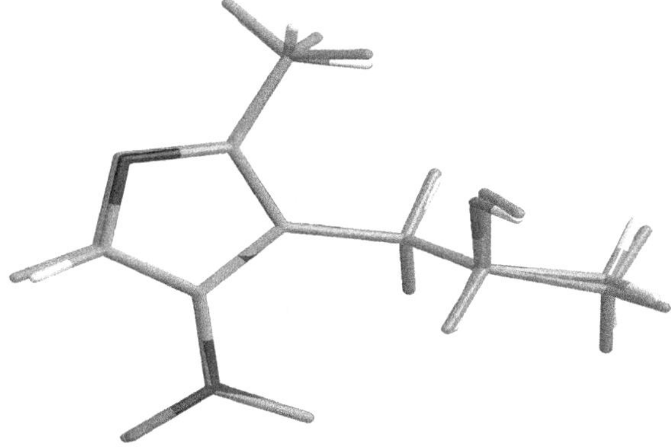

FIGURE 4.4 Single-crystal X-ray diffraction experimental structure and DFT-calculated conformations of secnidazole.[30]

Source: Reprinted (adapted) with permission from *J. Phys. Chem. A* 2009, 113, 273–281. Copyright 2009 American Chemical Society.

TABLE 4.1
Calculated Energies (in Hartree) of the Equilibrium Conformers of Secnidazole[30]

Conformer	4-31G	6-31G	6-311++G(d,p)
I	−662.2816	−662.9706	−663.3908
II	−662.2808	−662.9696	−663.3894
III	−662.2761	−662.9652	−663.3865

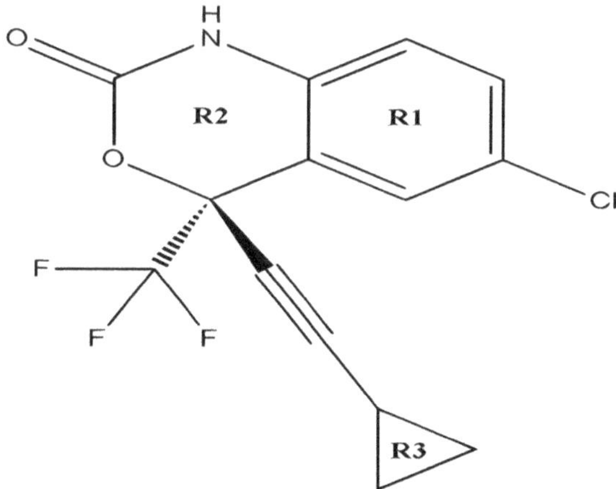

FIGURE 4.5 Schematics of the efavirenz molecule.[44]

Source: Reprinted (adapted) with permission from *Spectrochimica Acta A* 88, 2012, 116–123. Copyright 2012 Elsevier.

second-generation non-nucleoside inhibitors of HIV-1 reverse transcriptase (RT). The role of HIV-1 RT is to convert single-stranded viral RNA into double-stranded DNA, which is then integrated into the human host's genome.[31] Non-nucleoside inhibitors of HIV-1 RT (NNRTIs) have garnered significant attention recently.[32–34] However, efavirenz does not exhibit effectiveness against HIV-2 due to structural variations in the pocket of HIV-2 reverse transcriptase, leading to intrinsic resistance to the NNRTI class.[35] NNRTIs, including compounds like nevirapine, delavirdine, and efavirenz, have been approved for human use, contributing to their incorporation into multidrug combinations for HIV infection treatment.[36–40] Despite their distinct structural features, these allosteric inhibitors have shown similar modes of action.[33] They bind to the enzyme at the same site,[32, 33, 41] engaging in similar interactions with amino acids and displaying comparable patterns of resistance to specific mutations.[42]

We recently performed quantum chemical calculations of the efavirenz drug.[43,44] DFT calculations were employed to optimize the molecular geometries of efavirenz, as shown in Figure 4.5. The calculations utilized Becke's B3LYP hybrid exchange functionals with the 6–311++G(d,p) basis set,[25] along with the Lee–Yang–Parr correlational functional (B3LYP).[27–29]

Our study was aimed to analyze the conformational stability of efavirenz by systematically varying the dihedral angle C15C14C13C8. Figure 4.6 depicts the potential energy obtained from the calculations. Notably, two distinct minima are clearly evident at approximately 93º and 145º. These minima correspond to conformers 1 and 2, respectively, with the former representing the low-energy conformation and the latter the high-energy conformation. The energy gap between these conformers is approximately 1.5 kcal/mol. The main difference among these conformers arises from the orientation of the alkyne tail, which in turn impacts the positioning of the CF3 group due to the sp3 hybridization of C14. As a result, as shown in Figure 4.7, the oxygen atom of the oxazine ring can be positioned above or below the plane formed by the benzene ring.

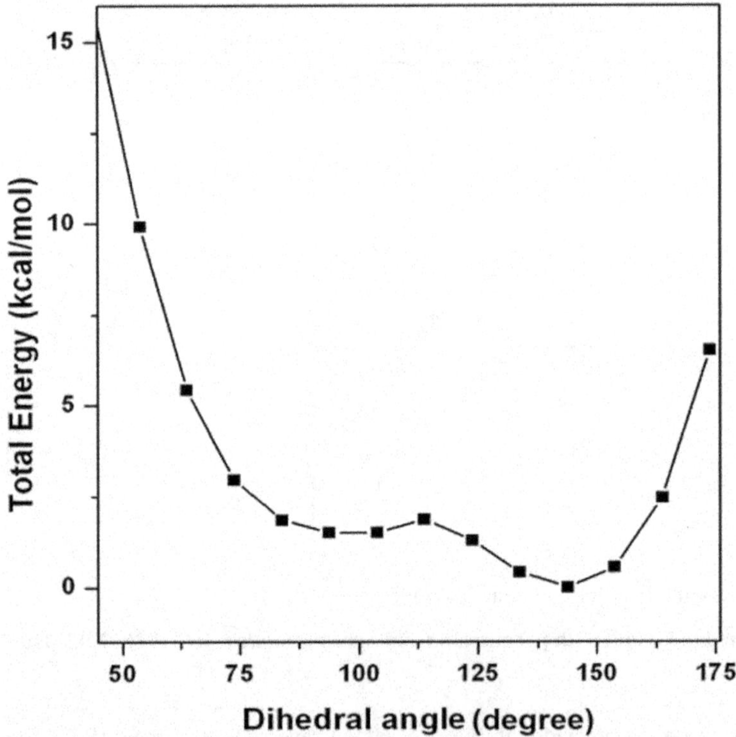

FIGURE 4.6 Potential energy curve of efavirenz as the C13-C14 bond undergoes rotation.[43]

Source: Reprinted (adapted) with permission from *Vib. Spectrosc.* 53, 2010, 112–116. Copyright 2010.

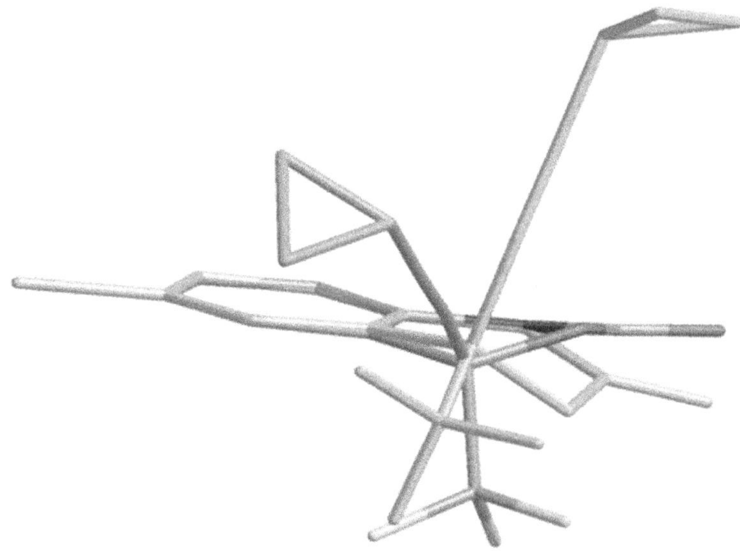

FIGURE 4.7 Comparison of conformers 1 and 2 (in orange) of efavirenz.[43]

Source: Reprinted (adapted) with permission from *Vib. Spectrosc.* 53, 2010, 112–116. Copyright 2010 Elsevier.

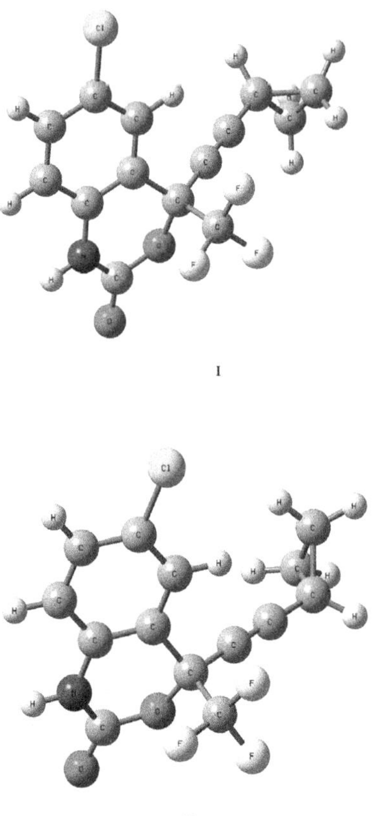

FIGURE 4.8 Optimized structure of efavirenz conformers I and II.[43]

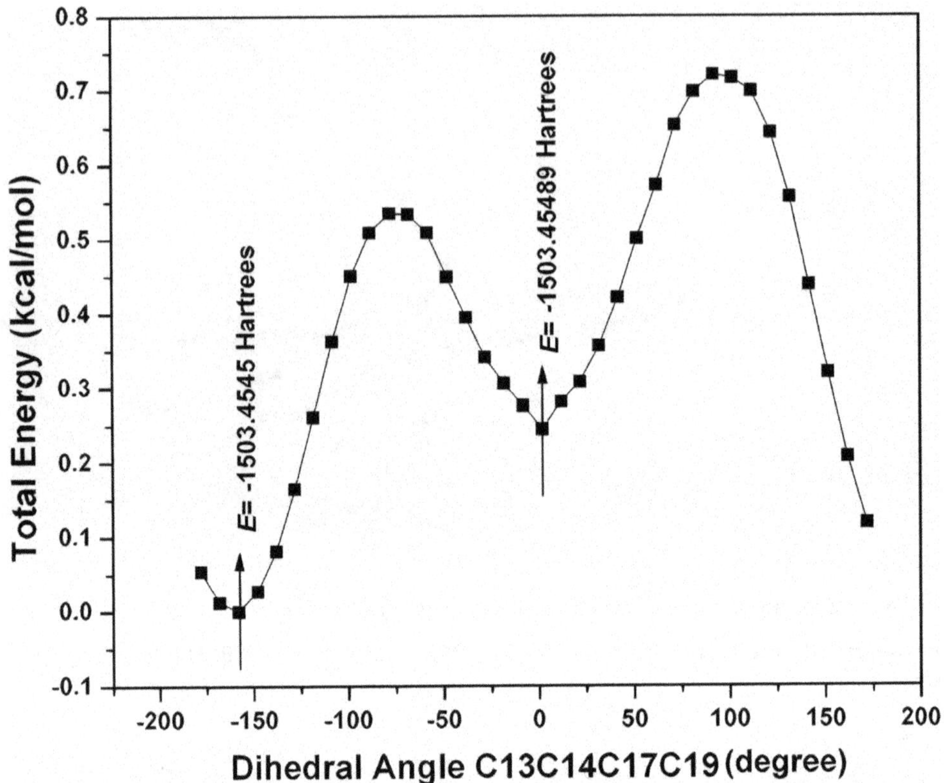

FIGURE 4.9 Potential energy curve of efavirenz as a function of the rotation around the C14-C17 bond.[44]

Source: Reprinted (adapted) with permission from *Spectrochimica Acta A* 88, 2012, 116–123. Copyright 2012 Elsevier.

The conformational landscape of efavirenz was explored by systematically examining the relative orientation of the benzoxazine and cyclopropyl rings. Figure 4.8 illustrates the different conformers observed in this investigation. The potential energy of the molecule was assessed by calculating the variation in total energy with respect to the dihedral angle τ(C13C14C17C19), employing the DFT/6–31G method (Figure 4.9). The potential energy curve displays two distinct minima, corresponding to conformers I and II. Conformer I is deeper and therefore represents the more stable conformation.

Our focus in this study was to investigate the conformational stability of efavirenz, which provides valuable information for quality control purposes, particularly in the field of medicine. To assess the conformational stability, we compared the optimized structures of conformer I and conformer II of efavirenz with the experimental conformation observed in the enzyme-inhibitor complex. This comparison was achieved by superimposing the structures using a least squares algorithm that minimizes the distances between corresponding nonhydrogen atoms. Notably, when comparing the optimized structure of conformer I with the X-ray crystallographic data of the enzyme-inhibitor complex at a resolution of 2.5 Å (Figure 4.10), the geometry optimization process accurately reproduced the experimental conformation. The main differences lie in the misorientation of the cyclopropyl ring. On the other hand, the main differences between the optimized structure of the conformer II and the one obtained from X-ray crystallographic data at 2.5 Å resolution of the enzyme-inhibitor complex structure are related to the cyclopropyl ring and CF3 group. This outcome suggests that, in the case of the enzyme-inhibitor complex, the intramolecular interactions are not sufficiently strong to induce significant conformational changes.

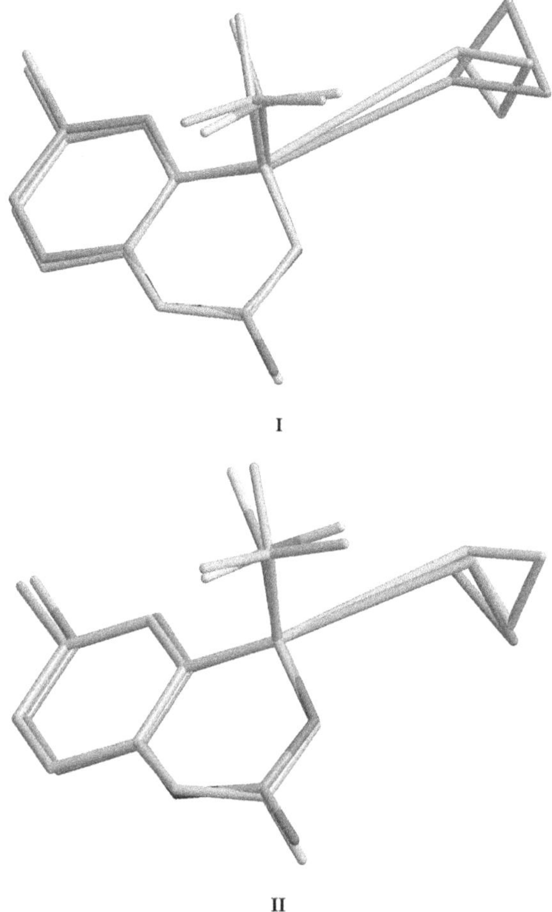

FIGURE 4.10 Comparison of optimized conformation conformers I and II with the ones from complex with HIV-1 RT.

4.4 QUANTUM CHEMICAL CALCULATION OF DPPC WITH DRUG MOLECULE

DPPC is a phospholipid that belongs to the class of glycerophospholipids, which are essential components of biological membranes. DPPC is commonly found in cell membranes, especially in lung surfactants, where it plays a crucial role in maintaining the stability and functionality of the pulmonary system. DPPC is composed of a glycerol backbone, two palmitoyl fatty acid chains, a phosphate group, and a choline headgroup. This molecular structure allows DPPC to form lipid bilayers, which are the primary structural components of cell membranes. The hydrophobic fatty acid chains face inward, while the hydrophilic headgroups face outward, creating a stable barrier that separates the intracellular and extracellular environments. DPPC and other phospholipids contribute to the fluidity, permeability, and stability of cell membranes. Additionally, DPPC has specific physiological functions, particularly in the lungs, where it reduces surface tension to prevent alveolar collapse during breathing. It is also widely used in research and industry as a model lipid for studying membrane properties and as a component in liposome formulations for drug delivery. Here, we discuss our recent work on quantum chemical studies of chlorpromazine drugs and their interaction with DPPC.[45] As a model compound, chlorpromazine (CPZ) is a widely used classical neuroleptic antipsychotic drug.[46] CPZ exhibits surfactant properties due to the presence of hydrophilic and

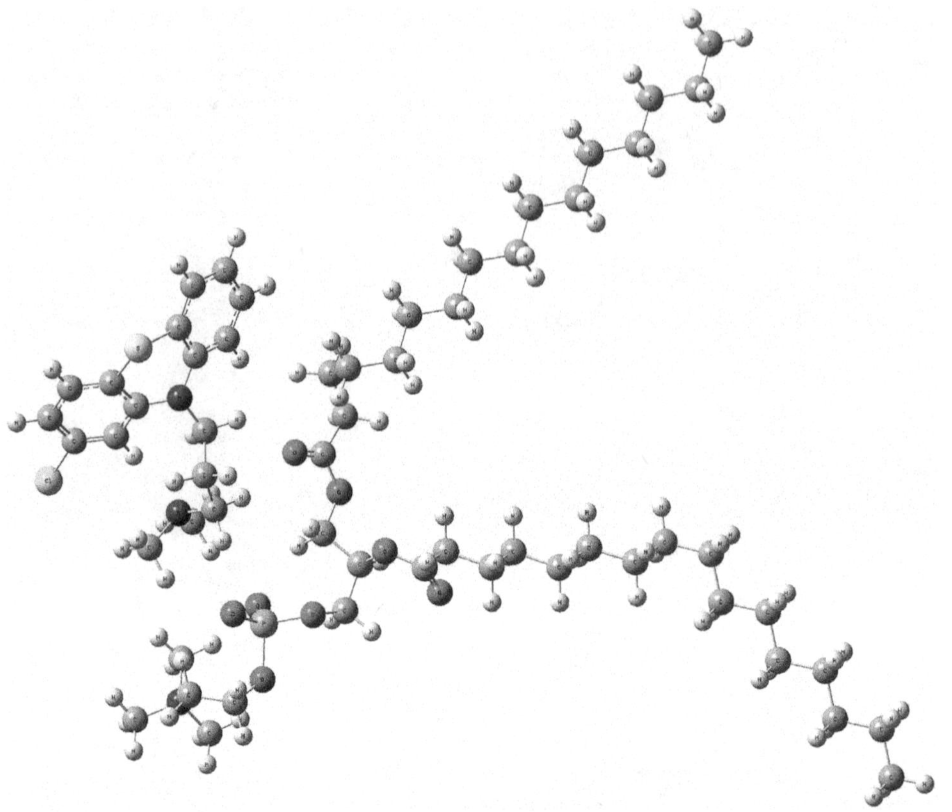

FIGURE 4.11 DFT optimized structure of DPPC and CPZ.[45]

hydrophobic groups in its molecular structure. Being an amphiphilic compound, CPZ has the ability to incorporate into lipid bilayers, thereby altering the physicochemical properties of biomembranes and influencing the activities of various membrane proteins. Understanding the molecular-level interactions of lipids and CPZ is crucial for comprehending the membrane-related biological effects of CPZ molecules. The optimized structure of DPPC and CPZ is given in Figure 4.11.

The molecular electrostatic potential (MEP) is a valuable means used to gain an understanding of the intermolecular interactions and molecular characteristics of small molecules and in interpreting the interaction of drug molecules with biomolecular systems.[47–50] The MEP at a specific point r within the molecular space is given by the following equation, expressed in atomic units:

$$V(r) = \sum_{A} \frac{Z_A}{|\vec{R}_A - \vec{r}|} - \int \frac{\rho(\vec{r}\,')dr'}{|\vec{r}\,' - \vec{r}|}$$

where Z_A represents the charge residing on nucleus A, situated at position R_A, while $\rho(r')$ denotes the electronic density function of the molecule. The first term in the equation describes the impact of the nuclei, and the second term represents the effect of electrons. These terms have opposite signs, resulting in contrasting effects. $V(r)$ represents the overall electrostatic effect at each point r, indicating the net impact produced by the total charge distribution (electrons + nuclei) of the molecule. The MEP helps to understand the reactivity of a molecule, highlighting the charged area available for reactions.[47] It offers a visual means to comprehend the relative polarity of the molecule. One crucial aspect that can be explored using potential maps is hydrogen bonding, which plays a significant role in biological recognition and interactions. The electrostatic potential surface (ESP), generated using

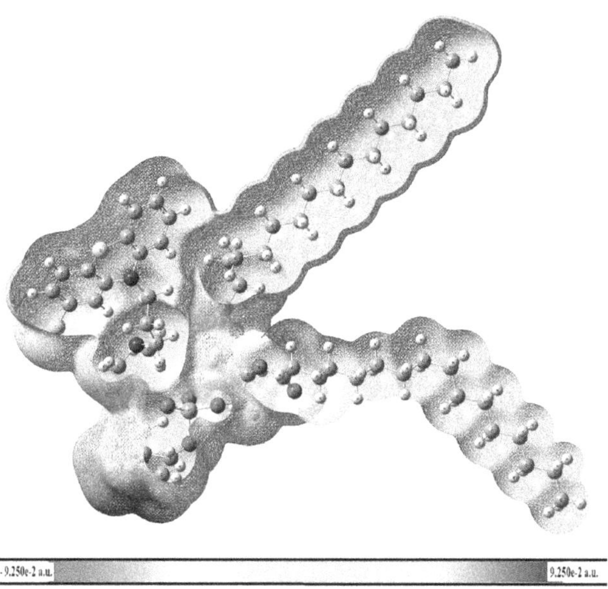

FIGURE 4.12 Isodensity surface for DPPC and CPZ, with the molecular electrostatic potential mapped onto it.[45]

the computer program GaussView,[30] aids in predicting sites and assessing relative reactivity toward electrophilic attack, as well as studying hydrogen bonding interactions. Mapping an electron density isosurface with the electrostatic potential surface provides insights into the size, shape, charge density, and sites of chemical reactivity within the molecules. In the plotted electrostatic potential surface for DPPC and CPZ using the 6–31G(d,p) basis set in GaussView ver.3, different colors represent various electrostatic potential values. The potential values are arranged in the following order: red < orange < yellow < green < blue. Regions near the oxygen atom in DPPC show the most negative potential, while the sulfur and chlorine atoms in the drug exhibit negative electrostatic potential.

In order to elucidate the hydrogen bonding phenomenon using MEP analysis, we generated MEP maps for the DPPC-CPZ system (Figure 4.12). The maps clearly demonstrated that stable complexes are formed when regions of red (representing negative potential) and blue (representing positive potential) colors come into close proximity. Conversely, when two regions of the same color (indicating similar electrostatic potential) approach each other, an unstable complex is formed. Since the DPPC molecule predominantly exhibits large positive and negative potentials at its periphery, small molecules tend to orient themselves in a manner that facilitates the formation of hydrogen bonds.

In the DPPC-CPZ complex, a significant hydrogen bonding occurs between the CO group of DPPC and the CH_3 group of CPZ. The negative charge on the oxygen atom of CO facilitates an attractive interaction with the positively charged hydrogen of CPZ. This interaction creates the potential for a stronger intermolecular hydrogen bond, specifically a CO. . .H bond, due to the close proximity of the H and O atoms at a distance of 1.618 Å. Our computational analysis indicates that the protonized amino groups show affinity with the negatively charged phosphate groups of the phospholipid molecules, while the CPZ rings are oriented toward the DPPC's hydrophobic part.

4.5 QUANTUM CHEMICAL CALCULATION OF DPPC WITH FUNCTIONALIZED CNTs

Very recently, employing DFT calculation initially, we conducted an investigation on the interaction between DPPC and CNT, which revealed a very weak interaction due to the absence of a direct

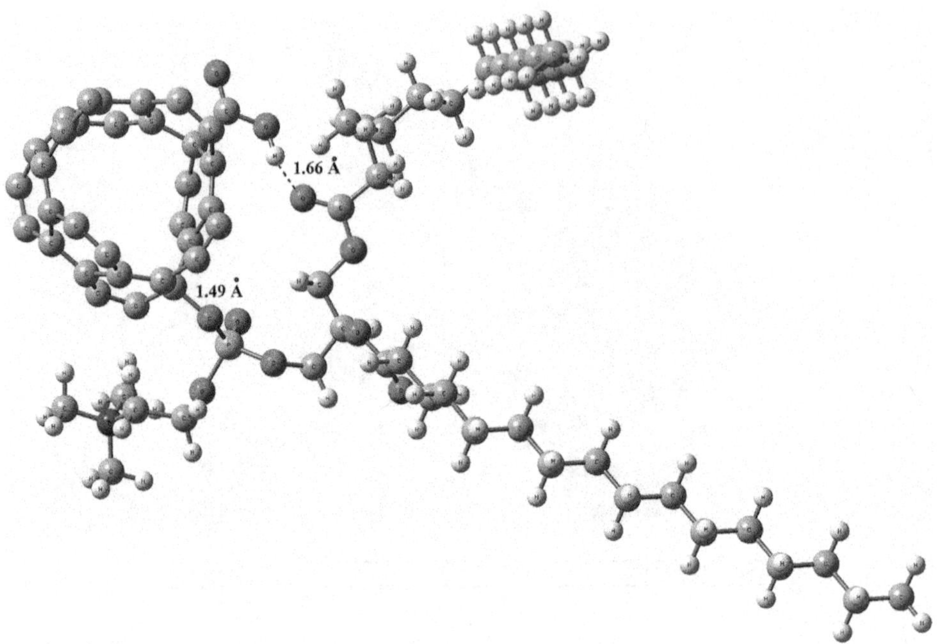

FIGURE 4.13 Interaction between DPPC and COOH-CNT structures.[51]

Source: Reprinted (adapted) with permission from *Physica Status Solidi (B) Basic Research*, 2023, 10.1002/pssb.202200574. Copyright 2023 Wiley.

chemical bond between the two.[51] Subsequently, we explored the interaction between COOH-functionalized CNT and DPPC by positioning them in close proximity in various configurations. In this instance, a robust interaction was detected, featuring a mild hydrogen bond between the H atom of the COOH group in CNT and one of the O atoms in DPPC, as illustrated in Figure 4.13, with a bond length measuring 1.66 Å. Moreover, a strong covalent bond was identified between the oxygen atom of the phosphate groups in DPPC and one of the carbon atoms in CNT, as depicted in Figure 4.13, with a bond length measuring 1.49 Å.

In essence, the objective of this study was to enhance our comprehension of the characteristics exhibited by drugs in their solid state. The detailed knowledge of the thermodynamics of an active pharmaceutical ingredient is an invaluable tool to avoid/solve problems in the production and formulation processes as well as to protect intellectual property against possible patent-related infringements. Thus polymorphism in pharmaceuticals has become an economic issue in several very expensive court cases. In general, the results of this work demonstrated that by means of quantum chemical calculations, it is possible to obtain a detailed description of properties that are relevant for the understanding of the polymorphism phenomenon and properties of lipids, providing additional support to the development of safe and efficient medicines.

4.6 CONCLUSIONS

We have seen throughout this chapter that quantum chemical calculations help in the choice of the most appropriate form of a drug, which is very important in its early stages of development, thus saving time and costs. Through these calculations, systematic isolation and early characterization of as many possible forms of a drug reduce the possibility of surprises in the early stages of production due to the identification of a new crystalline form or phase changes. Understanding the origins of multiple solid forms of a drug due to differences in packaging arrangement or conformation of molecules has become a fundamental step in the development of new active ingredients. The effectiveness of a drug can be related to its physical structure in a particular formulation. There is an emphasis on studying the physical properties of drugs in order to improve performance and delivery.

Overall, this study aimed to contribute to the understanding of the solid-state properties of drugs and drug delivery systems. Progress in research in this field will undoubtedly enhance the bioavailability of drugs, minimize toxicity, and promote adherence, among other factors that directly influence the quality of treatment for at-risk populations. Furthermore, this research will provide valuable support to existing programs aimed at combating the HIV/AIDS epidemic and addressing various neglected diseases such as malaria, schistosomiasis, African sleeping sickness, and leprosy, by contributing to the production of affordable, high-quality medicines. In a broader context, the pharmaceutical industry will reap significant benefits from the outcomes of this research.

REFERENCES

1. Suri S.S., Fenniri H., Singh B., Nanotechnology-based drug delivery systems. *J Occup Med Toxicol*, 16(2), 2007 doi: 10.1186/1745-6673-2-16
2. Brannon-Peppase L., Blanchette J.Q., Nanoparticle and targeted systems for cancer therapy. *Adv Drug Deliv Rev*, 56(11), 2004, 1649–1659. doi: 10.1016/j.addr.2004.02.014
3. Yokoyama M., Drug targeting with nano-sized carrier systems. *J Artif Organs*, 8(2), 2005, 77–84. doi: 10.1007/s10047-005-0285-0
4. Mitchell M.J., Billingsley M.M., Haley R.M., Wechsler M.E., Peppas N.A., Langer R., Engineering precision nanoparticles for drug delivery. *Nat Rev Drug Discov,* 20, 2021, 101–124. doi: 10.1038/s41573-020-0090-8
5. Blanco E., Shen H., Ferrari M., Principles of nanoparticle design for overcoming biological barriers to drug delivery. *Nat. Biotechnol*, 33, 2015, 941–951.
6. Boonme P., Patenting nanomedicines for nutraceutical ingredients. In: Souto E. (ed.), Patenting Nanomedicines. Heidelberg: Springer, 2012. doi: 10.1007/978-3-642-29265-1_12
7. Haba Y., Kojima C., Harada A., Ura T., Horinaka H., Kono K., Preparation of poly (ethylene glycol)-modified poly (amido amine) dendrimers encapsulating gold nanoparticles and their heat-generating ability. *Langmuir*, 23, 2007, 5243–6.
8. Cramer C.J., Essentials of Computational Chemistry: Theories and Models. New York: Wiley, 2002.
9. Kaur K., Mohan A., Girdhar M., Nanotechnology based approaches for carcinogenesis treatment: A review. *Trends Biomater. Artif. Organs*, 32(3), 2018, 121–127.
10. Dessale M., Mengistu G., Mengist H.M., *Int J Nanomedicine*, 17, 2022, 3735–3749
11. Shi X., Sun K., Baker J.R., Jr., Spontaneous formation of functionalized dendrimer-stabilized gold nanoparticles. *J Phys Chem C*, 112, 2008, 8251–8258. doi: 10.1021/jp801293a.
12. Park S.H., Oh S.G., Mun J.Y., Han S.S., Loading of gold nanoparticles inside the DPPC bilayers of liposome and their effects on membrane fluidities. *Coll Surf B*, 48, 2006, 112–8.
13. Abraham Domínguez-Avila J., González-Aguilar Gustavo A., Chapter 13—lipids. In: Postharvest Physiology and Biochemistry of Fruits and Vegetables (pp. 273–292). Queretaro, Mexico: Woodhead Publishing, 2019. doi:10.1016/B978-0-12-813278-4.00013–0.
14. Benjamins J.A., Murphy E.J., Seyfried T.N., Principles of molecular, cellular, and medical neurobiology, chapter 5 – lipids. Basic Neurochemistry (8th ed., pp. 81–100). doi: 10.1016/B978-0-12-374947-5.0000.5–5
15. Blanco A., Blanco G., Lipids. Medical Biochemistry (pp. 99–119). 2017. doi: 10.1016/B978-0-12-803550-4.00005–7
16. Litwack G., Chapter 9 – Lipids. Human Biochemistry (pp. 199–255). 2018. doi: 10.1016/B978-0-12-383864-3.00009–0
17. Aldred E.M., Buck C., Vall K., Lipids. Pharmacology: A Handbook for Complementary Healthcare Professionals (pp. 73–80). Churchill Livingstone, 2009. doi: 10.1016/B978-0-443-06898-0.00010–4
18. Schultz Z.D., Levin I.W., Vibrational spectroscopy of biomembranes. *Annu. Rev. Anal. Chem*, 4, 2011, 343–366, doi: 10.1146/annurev-anchem-061010–114048
19. Hohenberg P., Kohn W., Inhomogeneous electron gas. *Phys. Rev,* 136, B864, 1964, 136.
20. Frisch M.J., Trucks G.W., Schlegel H.B., Scuseria G.E., Robb M.A., Cheeseman J.R., Montgomery J.A., Vreven T., Kudin K.N., Burant J.C., Millam J.M., Iyengar S.S., Tomasi J., Barone V., Mennucci B., Cossi M., Scalmani G., Rega N., Petersson G.A., Nakatsuji H., Hada M., Ehara M., Toyota K., Fukuda R., Hasegawa J., Ishida M., Nakajima, Honda Y., Kitao O., Nakai H., Klene M., Li X., Knox J.E., Hratchian

H.P., Cross J.B., Adamo C., Jaramillo J., Gomperts R., Stratmann R.E., Yazyev O., Austin A.J., Cammi R., Pomelli C., Ochterski J.W., Ayala P.Y., Morokuma K., Voth G.A., Salvadorm P., Dannenberg J.J., Zakrzewski V.G., Dapprich S., Daniels A.D., Strain M.C., Farkas O., Malick D.K., Rabuck A.D., Raghavachari K., Foresman J.B., Ortiz J.V., Cui Q., Baboul A.G., Clifford S., Cioslowski J., Stefanov B.B., Liu G., Liashenko A., Piskorz P., Komaromi I., Martin R.L., Fox D.J., Keith T., Al-Laham M.A., Peng C.Y., Nanayakkara A., Challacombe M., Gill P.M.W., Johnson B., Chen W., Wong M.W., Gonzalez C., Pople J.A., Gaussian 03, Revision C.02, 2003.

21. Lee C.T., Yang W.T., Parr R.G., Development of the colle-salvetti correlation-energy formula into a functional of the electron density. *Phys. Rev. B*, 37, 1988, 785.
22. Parr R.G., Yang W., Density Functional Theory of Atoms and Molecules. New York: Oxford University Press, 1989.
23. Becke A.D., Density-functional thermochemistry. III. The role of exact exchange. *J. Chem. Phys*, 98, 1993, 5648.
24. Frisch A., Nielson A.B., Holder A.J., Gaussview User Manual. Pittsburgh, PA: Gaussian Inc, 2000.
25. Robinson A.E., Martindale: The extra pharmacopoeia 27th edition. *J. Pharm. Pharmacol*, 29(1), 1977, 647–648
26. Hillier S.L., Nyirjesy P., Waldbaum A.S., Schwebke J.R., Morgan F.G., Adetoro N.A., Braun C.J., Secnidazole treatment of bacterial vaginosis: A randomized controlled trial. *Obstet Gynecol*, 130(2), 2017, 379–386.
27. Kissinger P., Muzny C.A., Mena L.A., Lillis R.A., Schwebke J.R., Beauchamps L., Taylor S.N., Schmidt N., Myers L., Augostini P., Secor W.E., Bradic M., Carlton J.M., Martin D.H., Single-dose versus 7-day-dose metronidazole for the treatment of trichomoniasis in women: An open-label, randomised controlled trial. *Lancet Infect Dis*, 18(11), 2018, 1251–1259.
28. Muzny C.A., Van Gerwen O.T., Legendre D., Secnidazole: A treatment for trichomoniasis in adolescents and adults. *Expert Rev Anti Infect Ther*, 20(8), 2022, 1067–1076. doi: 10.1080/14787210.2022.2080656
29. Gardner T.B., Hill D.R., Treatment of giardiasis. *Clin Microbiol Rev*, 14(1), 2001, 114–28. doi: 10.1128/CMR.14.1.114-128.2001. PMID: 11148005; PMCID: PMC88965.
30. Mishra S., Chaturvedi D., Tandon P., Gupta V.P., Ayala A.P., Honorato S.B., Siesler H.W., Molecular structure and vibrational spectroscopic investigation of secnidazole using density functional theory. *J. Phys. Chem. A*, 113, 2009, 273–281
31. Parker W.B., White E.L., Shaddix S.C., Ross L.J., Buckheit R.W., Jr., Germany J.M., Secrist J.A., Vince R., Shannon W.M., Mechanism of inhibition of human immunodeficiency virus type 1 reverse transcriptase and human DNA polymerases alpha, beta, and gamma by the 5'-triphosphates of carbovir, 3'-azido-3'-deoxythymidine, 2',3'-dideoxyguanosine and 3'-deoxythymidine. A novel RNA template for the evaluation of antiretroviral drugs. *J Biol Chem*, 266, 1991, 1754.
32. Declerc E., Non-nucleoside reverse transcriptase inhibitors (NNRTIs) for the treatment of human immunodeficiency virus type 1 (HIV-1) infections: strategies to overcome drug resistance development. *Med Res Rev*, 16, 1996, 125.
33. Declerc E., Perspectives of non-nucleoside reverse transcriptase inhibitors (NNRTIs) in the therapy of HIV-1 infection. *Farmaco*, 54, 1999, 26.
34. Artico M., Non-nucleoside anti-HIV-1 reverse transcriptase inhibitors (NNRTIs): A chemical survey from lead compounds to selected drugs for clinical trials. *Farmaco*, 51, 1996, 305.
35. Ren J., Bird L.E., Chamberlain P.P., Stewart-Jones G.B., Stuart D.I., Stammers D.K., Structure of HIV-2 reverse transcriptase at 2.35-Å resolution and the mechanism of resistance to non-nucleoside inhibitors. *Proc Natl Acad Sci USA*, 99, 2002, 14410.
36. De Clercq E., The role of non-nucleoside reverse transcriptase inhibitors (NNRTIs) in the therapy of HIV-1 infection. *Antiviral Res*, 38, 1998, 153.
37. Gazzard B.G., Efavirenz in the management of HIV infection. *Int J Clin Pract*, 53, 1999, 60.
38. Merluzzi V.J., Hargrave K.D., Labadia M., Grozinger K., Skoog M., Wu J.C., Shin C.K., Eckener R.J.S., Hattox R.J., Adamms J., Rosenthal A.S., Faanes R., Eckner R.J., Koup R.A., Sullivan J.L., Inhibition of HIV-1 replication by a nonnucleoside reverse transcriptase inhibitor. *Science*, 250, 1990, 1411.
39. Romero D., Morge R.A., Genin M.J., Biles C., Busso M., Resnick L., Althaus I.W., Reuser F., Thomas R.C., Tarpley W.G., Bis(heteroaryl)piperazine (BHAP) reverse transcriptase inhibitors: structure-activity relationships of novel substituted indole analogs and the identification of 1-[(5-methanesulfonamido-1H-indol-2-yl)carbonyl]-4-[3-[(1-methylethyl)amino]pyridinyl]piperazinemonomethanesulfonate (U-90152S), a second-generation clinical candidate. *J Med Chem*, 36, 1993, 1505.

40. Young S.D., Britcher S.F., Tran L.O., Payne L.S., Lumma W.C., Lyle T.A., Huff J.R., Anderson P.S., Olsen D.B., Carroll S.S., Pettibone D.J., O'Brien J.A., Ball R.G., Ballani S.K., Lin J.H., Chen I.W., Schleif W.A., Long W.J., Byrnes V.W., Emine E.A., Sardana V.V., A novel, highly potent nonnucleoside inhibitor of the human immunodeficiency virus type 1 reverse transcriptase. *Antimicrob Agents Chemother*, 39, 1995, 2602.
41. Ren J.S., Esnouf R., Garman E., Somers D., Ross C., Kirby I., Keeling J., Darby G., Jones Y., Stuart D., Stammers D., High resolution structures of HIV-1 RT from four RT-inhibitor complexes. *Nat Struct Biol,* 2, 1995, 293.
42. Schinazi R.F., Larder B.A., Mellors J.W., *Int Antiviral News* 5, 1997, 129.
43. Mishra S., Chaturvedi D., Srivastava A., Tandon P., Ayala A.P., Siesler H.W., Quantum chemical and experimental studies on the structure and vibrational spectra of efavirenz. *Vib. Spectrosc,* 53, 2010, 112–116
44. Mishra S., Tandon P., Ayala A.P., Study on the structure and vibrational spectra of Efavirenz conformers using DFT: Comparison to experimental data, *Spectrochimica Acta A,* 88, 2012, 116–123.
45. Mishra S., Mishra A.K., Effects of chlorpromazine drug on DPPC lipid: Density functional theory study, *Int Environ Anal Chem*, 101, 2021, 1773–1784.
46. Agasosler A.V., Tungodden L.M., Cejka D., Bakstad E., Sydnes L.K., Holmsen H., *Biochem. Pharmacol,* 7, 2001, 817. doi:10.1016/S0006-2952(01)00542-1.
47. Tomasi J., In: Politzer P., Truhlar D. (eds.), Chemical Application of Atomic and Molecular Electrostatic Potentials: Reactivity, Structure, Scattering, and Energetics of Organic, Inorganic, and Biological Systems (p. 257). New York: Plenum Press 1981.
48. Moro S., Bacilieri M., Ferrari C., Spalluto G., Curr. *Drug Discov. Technol,* 2, 2005, 13. doi:10.2174/1570163053175439.
49. Murray J.S., Sen K., Molecular Electrostatic Potentials, Concepts and Applications. Amsterdam: Elsevier, 1996.
50. Weiner P.K., Langridge R., Blaney J.M., Schaefer R., and Kollman P.A., *Proc. Natl. Acad. Sci. USA: Biophysics*, 79, 1982, 3754. doi:10.1073/pnas.79.12.3754.
51. Mishra S., Mishra A.K., Understanding carbon nanotubes functionalization and their interactions with lipids toward drug delivery systems: A combined experimental and computational spectroscopic investigation. *Physica Status Solidi (B) Basic Research*, 2023. doi: 10.1002/pssb.202200574

5 Graphene-Based Nanomaterials (GBNs) and Their Biomedical Applications

Ruby Srivastava and Sravani Joshi

5.1 INTRODUCTION

Graphite and diamond are the basic three-dimensional crystal structures of carbon. The carbon atom can exist in different forms as zero to three dimensions (0–3D).[1–8] Carbon is the basis of life, is the most abundant material, and is widely used in organic chemistry. The research activity of carbon was accumulated with fullerenes (0D) in 1988 and the discovery of carbon nanotubes (1D) in 1991.[9–11] Various efficient methods has been developed to produce graphene in recent years. for example micromechanical exfoliation, chemical vapor deposition, epitaxial growth, and chemical synthesis.[12,13] Earlier in 2004, the monolayer of graphene was prepared by using a stick tape and a pencil. Graphene can also be prepared by reducing graphene oxide (GO). GO can be produced at a larger scale and can have high output at a low processing cost. GBNs have biomedical applications in biosensors, drug delivery systems, cell imaging, gene therapy, tissue engineering, photothermal therapy, and near-infrared (NIR) fluorescence imaging.[14–17] Advanced applications include biomedical equipment such as deep brain stimulators[18] and blood glucose sensors.

In the following sections, we discuss the history of GBNs, its structural details, biocompatibility, toxicity, and measures to reduce the toxic effects.

5.2 HISTORY OF GBNs

Geim and Novoselov (2004)[19] obtained graphene, a new 2D member of the carbon family via micromechanical exfoliation of a graphite sheet. Graphene is a six-membered benzene ring that is formed by sp^2 hybridization of monolayer carbon atoms.[20] The p orbital electrons of each carbon atom form a tightly stacked large π-bond to the 2D monolayer honeycomb lattice structure. Though the single layer of graphene has a thin layer (0.335 nm), graphene can be transformed into 0D fullerenes, 1D carbon nanotubes, or 3D graphite. Graphite oxide (GO) was synthesized earlier, and in 1962 reduced graphite oxide (rGO) was prepared through chemical reduction. Studies indicated that a monolayer of graphite on the surface of Pt (100) (1968–1969) and a monolayer of graphite from a carbon surface and silicon carbide was produced in 1970–1975. The name "graphene" was given to the monolayer of graphite in 1986 by Boehm.[21] The GBNs as graphene, graphene oxide (GO), reduced graphene oxide (rGO), graphene nanoribbons (GNBs), and graphene quantum dots (GQDs) show interesting features that can be used in material sciences as well as in the biological sciences due to their unique structures and geometry.[22, 23]

5.3 SYNTHESIS

Two main approaches, top-down and bottom-up, are used for the synthesis of graphene. The top-down approach includes reduction, mechanical cleavage (MC), and the Hummers method, whereas the bottom-up methods include arc discharge, epitaxial growth, organic synthesis, and chemical vapor deposition (CVD). The bottom-up method uses small-molecule carbon to form monolayers

DOI: 10.1201/9781003441328-5

and multilayers of graphene and GO. The details about these methods are given elsewhere.[2–8] In the following section, the structure and properties of the GBNs are discussed in detail.

5.4 STRUCTURES AND PROPERTIES

5.4.1 Graphene

Graphene is a 2D planar single carbon layer (hexagonal array) of the graphite structure. Each carbon has four bonds and sp^2 hybridization. Graphene is considered hydrophobic due to the absence of oxygen groups. Graphene is the thinnest 2D material that not only wraps into 0D fullerenes but also rolls into 1D carbon nanotubes and stacks into 3D graphite. Graphene is known as impractical material as the single layer of graphene is still unstable. According to the Landau theory, it is thermodynamically impossible to predict the 2D crystals of graphene.[24] Mermin–Wagner also predicted that 2D graphene crystals could not be stable at any temperature.[25] The deformation of graphene sheets increases its stability. As the surface of graphene does not consist of active groups, it is inert and relatively stable. Graphene has poor water solubility, and it possesses excellent electrical, thermal, and mechanical properties. Given its low-cost manufacturing, monolayer graphene produces attractive nanomaterials with a wide range of applications. Isolated nanographene crystals (NGCs) are reliable and exhibit good charge trapping capacities that can be used for superintegrated storage and tunable memory. The multiple dielectric layers of nanographenes are very stable in high-temperature annealing. Other advantages of these novel nanographene materials are their charge-trapping layer and high compatibility with CMOS technology.

5.4.2 Graphene Oxide (GO)

GO is a monolayer of graphene, which reflects superior physical and chemical properties. GO is produced by the reaction of strong acids and strong oxidants. GO owns both sp^2 and sp^3 hybrid orbitals representing both the aliphatic and the aromatic domains. Several functional groups such as -OH, -COOH, -O-, and -CH(O)CH- exist on the surface of GO.[26] The strong π–π interaction, hydrogen bonds, and hydrophobic forces enable GO to be an ideal candidate for the loading of drugs and drug delivery. As the various surface groups on GO make the surface modification easy, GO can be used for wider applications. A variety of oxygen-containing functional groups provide GO with a quantum Hall effect, optical transmittance, fluorescence quenching ability, and high electron mobility. This property of GO is used for the manufacturing of biosensors, design of fluorescent receptor, and a stable platform for covalent binding of biological entities.[27]

5.4.3 Reduced Graphene Oxide (rGO)

rGO is produced by the intergradation of graphene-to-graphite oxide by the chemical or thermal reduction of GO or graphite oxide.[25] The drug delivery system and photothermal effect of rGO are increased due to the sp^2 hybridization and restoration of the π–π structure. The modification in the functional groups improves hydrophilicity and biocompatibility, reduces the body opsonic damage, and prolongs the drug effect time. Being a drug carrier, rGO adsorbs the drugs rich in benzene rings or polyunsaturated bonds (doxorubicin [DOX], paclitaxel, and fluorouracil) on the plane through noncovalent interactions. The chemical structures of graphene, GO, and rGO are given in Figure 5.1.

5.4.4 Graphene Quantum Dots (GQDs)

GQDs has smaller diameter size (1–20 nm, fewer than 10 layers), sp^2–sp^2 carbon bonds, larger surface area, and an appropriate graft for surface functionalization. GQDs show stable photoluminescence and contain hydrophilic groups at their boundaries, such as -OH and -COOH, which can

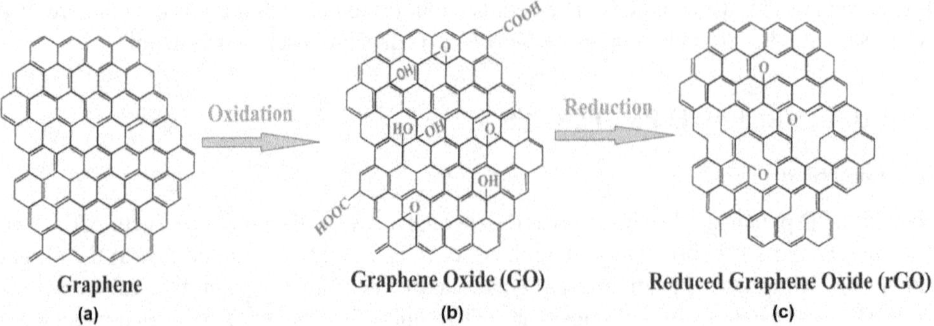

FIGURE 5.1 Schematic representation of (a) graphene, (b) graphene oxide (GO), and (c) reduced graphene oxide (rGO).

Source: Reproduced from *Chemistry* (2018) 8:123–137.

be modified as needed.[28] GQDs are nanoscopic fragments of graphene that show similar (physicochemical and optical) properties as traditional QDs. The toxic by-products produced by QDs restrict its usage for biological applications. Also, the inorganic (mostly metal sulfide) colloidal particles of QDs exhibit poor solubility and stability in aqueous media. These limitations of QDs are reduced by the use of GQDs, which can be oxidized to increase the polarity and peak emission wavelength in the presence of polar groups (oxygen) in the acquired structure. GQDs show excellent physicochemical properties. Oxidized GQDs (oGQDs) show better polarity and peak emission wavelength properties due to the presence of a polar group (oxygen) in the structure.

5.4.5 Graphene Nanoribbons (GNRs)

Graphene nanoribbons (GNRs) are quasi-one-dimensional carbon-based nanomaterials with graphene layer defects. GNRs consist of oxygen-containing functional groups on the surface and show conducive behavior. With many beneficial properties of graphene, GNRs have flexible and adjustable electrical properties due to their special edge-limiting effects, which can be used widely in quantum devices, transistors, and other similar devices.[29–32] The electrical properties of GNRs are affected by the applied stress, which can be used to adjust the bandgap width.[29] GNRs turn their original metallic properties into semiconducting properties.[30] With an increase in GNRs' width, the bandgap is decreased ultimately to zero. At zero bandgap, GNRs show metallic properties. Doping of a heteroatom also changes the bandgap. Boron- and nitrogen-doped GNRs have semiconducting properties.[33] Other properties included superspecific surface area, high tensile strength, magnetism, and good thermal conductivity. The physicomechanical properties of GNRs are affected by its length, width and defects.[34] For example, GNRs exhibit different absorption wavelengths in different directions.[35] The magnetic properties of GNRs differ due to a local edge state and vary according to various parameters such as width, defect, impurities, edges, and applied electric and magnetic fields. The effect of varied magnetic properties is due to the high sensitivity of the band structures to the edge configuration.

5.5 COMPUTATIONAL STUDIES ON GBNs

In recent years, with molecular modeling and data science, several *in silico* approaches have been developed to study the biomedical properties of GBNs. Regarding the computational predictions of graphene-/GO-based drug delivery system (GDDS), quantum mechanics (QM) and molecular dynamics (MD) simulations are used to accurately investigate the nonbonded interaction between the graphene/GO and ligand. The metadynamics (MTD) simulation method[36] and semiclassical MD simulation method are used for predicting molecular spectroscopy (MS).[37] Docking software such as the AutoDock program[38] and web-based servers are used to search a small molecule within the pocket of a graphene/GO model for biomedical applications. Huge *in vivo* testing expenses of nanomaterials prompted researchers to select these faster and lower-cost computational approaches to reduce the delay in *in vitro* biomedical applications. DFT calculations with machine learning tools were applied to build an interatomic potential for highly ordered shape memory materials (SMMs), graphene oxide nanoribbons. When an electric field is applied, the graphene oxide layer can shrink to a metastable phase with lower constant lattice, and it returns to the initial phase through an external mechanical force.[39] DFT results showed no magnetization for sufficiently narrow nanoribbons, while the data-driven approaches with machine learning interatomic potential predicted the suppression of the metastable phase for the same narrower nanoribbons. Finally, these data-driven approaches are used to see the realistic distribution of oxygen positions over the GO surface and the effect of vacancies and impurities (boron, nitrogen atoms).[40] Other computational studies have been carried out on polyaromatic hydrocarbon (PAH) containing 96 carbon atoms to investigate the application of graphene as a sensor for biomolecules. Adsorption energies for different molecules are compared to explore the scope of analytes of graphene-based sensors. Results indicated that larger biomolecules as phthalocyanine and tetrabenzoporphine exhibit stronger interactions than smaller systems, such as glycine, melamine, pyronin cation, and porphine.[41] The success of computational studies on graphene for nanopore DNA sequencing has made it possible to explore opportunities for other potential single- and few-atom-thick layers of 2D materials such as phosphorene and silicone. These materials exhibit technologically useful properties for DNA base detection. DFT calculations on the interaction of nanoribbon and nanopore devices with silicone showed that, due to the differences in structural and electronic properties of DNA bases, each base is expected to have a unique interaction with silicene, or the graphene monolayer material. The buckled honeycomb structure with tunable bandgap and high thermal stability of silicene make it an ideal material for designing active components of biosensing nanodevices.[42]

In silico studies on the GO surface on Aβ peptides showed that GO could decrease the β-strand propensity of the Aβ peptides, which is in agreement with the experimental results. Further, the comparison between GO and rGO revealed that the latter acts as a better inhibitor for fibril formation because it has extra van der Waals interactions than the former, leading to better adsorption of the Aβ peptide on the surface and enhanced conformational transitions.[43] Other studies also suggested that some carbon-based nanoparticles can promote and others can inhibit fibril formation of amyloidogenic peptides and proteins. These types of proteins can misfold and lead to the aggregate accumulation of insoluble fibril-like structures, which have been linked to diseases such as Alzheimer's, atherosclerosis, and type-II diabetes. The mechanisms of interactions that render these nanomaterials as inhibiting or promoting amyloid formation remain unclear. However, with the aid of advanced computational resources, it is now possible to explore the conformational features and dynamical interactions of proteins with complex nanomaterials at atomistic levels and time scales, previously inaccessible by means of experimentation. In order to explore the interactions of proteins with nanomaterials, it is important to first study the dynamic behavior and properties of the individual systems. Other studies on the computational model for a retinal prosthesis by the novel use of graphene-based microelectrodes with different topologies of the electrode arrangement were investigated to find the most suitable configuration for the arrangement that could serve as the ground in a hexagonal fashion around the central stimulating electrode. The results were used to optimize the implantable microelectrode design to find a cure to vision impairment diseases.[44] Due to the lower sensitivity of graphene toward the cisplatin (CPT), DFT studies were carried out on the absorption performance of graphene doping boron atoms (B/Gra), nitrogen atoms (N/Gra), and hexagonal boron nitride (BN/Gra) to form heteroquantum dots (QDs) as anticancer drug carriers. Result indicated that B/Gra QD can be used as drug carrier for CPT.[45]

In a recent study, a boron-nitrogen (BN)-doped graphene (G) was used for the drug delivery of hydroxyurea (HU) anticancer. DFT results showed that the graphene surface can be used as a drug delivery platform for HU anticancer. The BN-doped region increases the ionic state of the graphene surface for better participation in interactions.[46–49] So the surface can be used as a drug delivery platform by introducing a representative adsorbent for the carrier role by forming HUG complexes.[49] The role of the PNIPAM monomer in the adsorption of the nucleobase onto the GO showed the dynamic behavior of switchable interactive surfaces of DNA protein that is triggered by the PNIPAM-grafted GO.[50] DFT-based studies were performed on the electronic properties of the nanomaterials integrated on/off switchable aptasensor system. The PNIPAM molecule was represented by a single monomer (NIPAM), which is functionalized onto the GO flake as a GO/NIPAM complex. The aptamer was depicted by a single nucleobase adsorbed onto the surface of the GO/NIPAM complex, while nucleobases enable us to reflect the chemical properties of intricate biomolecules. Results indicated higher bandgap energy for the nucleobase and GO/NIPAM complex with stabilized hydrogen bonding within the GO/NIPAM complex. Conductivity was greatly enhanced by the presence of the GO surface, and the thermal responsive behavior of the MD-simulated PNIPAM-grafted GO surface induced an on/off surface state in a positive manner.[51]

A carbon nanocone-oxide (ONC) is used to conduct the drug delivery process of hydrea (Hyd) anticancer by performing DFT-based electronic and absorption studies. In an MD simulations study, the destruction of the cell membrane by graphene nanosheets (GN) was studied with *Escherichia coli* DH5a strain. Pristine graphene and graphene oxide nanosheets (GON) (C20O2(OH)2COOH) were interacted with 1-palmytoil-2-oleoyl-sn-glycero3-phosphatidyl-ethanolamine (POPE) and mixed POPE/POPG (POPE/1-palmytoil-2-oleoylsn-glycero-3-phosphatidyl-glycerol) membranes to cause lipid extraction and lead to bacterial membrane disruption.[52] Another MD simulation study showed the extraction of lipids from the (POPC/POPS) bilayer with GN insertion.[53] The advanced computational tools help to predict the cytotoxicity of GBNs and other nanomaterials of varied composition by using different chemical composition and molecular dynamics (MD) simulations studies. Tools such as the computational indicator of nanotoxicity (CIN2D) evaluate the biodistribution, pharmacokinetic parameters, and accumulation in a living organism or tissues.[54] Using nanographene oxide (NGO) modified by folic acid, computational studies were carried out to investigate the release rate of the drug (camptothecin) and its structural changes of the combined complex with nanographene oxide as a drug carrier.[55] Recently a computational platform e-Graphene was developed to predict the graphene-based drug delivery system by Quantum Genetic Algorithm and Cascade Protocol.[56–58] Density functional theory (DFT) and time-dependent density functional theory (TDDFT) were applied to study the metal adsorption on edge-chlorinated nanographenes (Cl NGRs) and graphene nanoribbons (GNRs).[59] With increased solubility and tunable band gaps, the complexes can be used for biosensors, electronics, and catalysis applications. The computational studies on noncovalent graphene–peptide based complexes exhibit excellent biomedical applications,[60] as short peptide probes act as biomarkers for the detection of diseases. See the noncovalent interactions in nanographene-peptide complexes in Figure 5.2.

5.6 FUNCTIONAL MODIFICATION

The structures of GBNs can be altered for wider applications. Graphene is highly hydrophobic, which affects its dispersion in water. The charge shielding effects of the functional groups bear the aggregation for the physiological environment of GO. GO causes inflammation as it has strong adsorption on proteins and is easily accepted and immersed by macrophages *in vivo*.[61] As GBNs do not have targeted, slow, and controlled release functions *in vivo*, they are not applied directly in biomedical applications. The surface modifications of GBNs help to increase their biomedical applications as they enhance stability, water solubility, and other functions as does a targeted, slow, and controlled release. Surface modifications are divided into (1) covalent and (2) noncovalent types. In covalent modifications, active double bonds and functional groups and polymers are added on the surface of the GBNs. Covalent bonds are produced with amides, free radicals, and other reactions under acidic conditions. These bonds chemically react with active groups on the surface of GBNs, which are modified through GO, as the large oxygen-containing functional groups can be used for

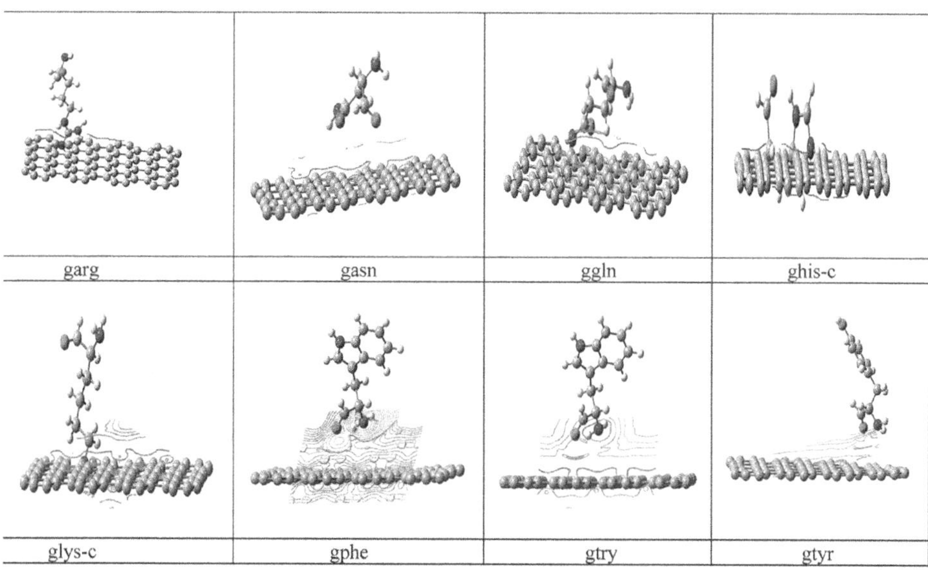

FIGURE 5.2 Noncovalent interaction (isosurface plots) of eight graphene–peptide complexes: graphene–arginine (garg), graphene–asparagine (gasn), graphene–glutamine (ggln), graphene–cationic histidine (ghis-c), graphene–cationic lysine (glys-c), graphene–phenylalanine (gphe), graphene–tryptophan (gtry), and graphene–tyrosine (gtyr).

Source: Reproduced from *RSC Adv.*, 2020, 10, 38654.

covalent modification. Since the covalent modifications cause a reduction in π electron networks, it enhances the properties of GBNs in diagnostics, imaging, and drug delivery. In a strong acidic medium, the structure of GO is modified, which increases its physicochemical properties. In a previous study, GO was functionalized by the formation of amide bond between polyethylene glycol (PEG) and GO lamellates (PEG-GO)[62] and was used to passively target the tumor sites in an effective manner. Grafting techniques are used to bind GBNs with polymers (poly[vinyl alcohol] [PVA], polyvinylpyrrolidone [PVP], chitosan [CS], sodium alginate [SA], and amphiphilic copolymer) to increase stability and biocompatibility and to reduce toxicity.[63] Covalent modifications help GBNs to produce a fast and effective drug release rate at the specific target site. GBNs are also used as stimulus-responsive GBN carriers. Many reactive object, such as -OH, -COOH, -CH(O)CH-, are added for the structural modification of GO. A molecular attachment mechanism functionalizes dual covalent modifications in GO. Benzoquinone, which is covalently formed on GO, expands its multiple applications in drug delivery, biosensing, imaging, tissue engineering, and photothermal therapy.[64–69]

Noncovalent modifications increase the dispersion, safety, reactive activation, and biosensor activities of GBNs in an effective manner. These forces included van der Waals forces, π–π interaction, ionic bonds, hydrogen bonds, electrostatic interaction, and coordination bonds between the modified groups. Noncovalent bonds are weak and are not affected by the internal and external environments. So noncovalently modified GBNs are less stable *in vitro* or *in vivo*, but the modification maintains the structures and properties of GBNs.[70] The alteration in noncovalent bonding in GBNs can be attained by polymer/biomacromolecule encapsulation or through the surface absorption of GBNs. The covalently changed GBNs are more appropriate for drug delivery as compared to the noncovalently modified GBNs as the conjugated sites prefer binding to a coating polymer. π–π interaction can adsorb both organic and inorganic components. Amantane-grafted porphyrin compounds (such as naphthalene, phenanthrene, pyrene, and gold nanoparticles) are bound to graphene and GO with π–π interactions.[71] The combination of GO with other functional groups easily form a stable GO colloidal suspension. Without hydrogen bond impurities, the combinations can act as a carrier for biomedical applications. Also, hydrophobic interactions between poloxam F127 and GO was used for high drug loading and a significant pH-responsive drug release behavior.[72] Recent studies showed that oxidized (GO) and reduced (rGO) graphene oxides are mostly used for biomedical applications like drug delivery and therapeutics.

5.7 BIOMEDICAL APPLICATIONS

GBNs have excellent electrical, photothermal, and mechanical properties that are widely applied in material sciences. Recently, GBNs found application in biomedical fields as biosensors, for antineoplastic drug delivery, in biological imaging, and in PTT. Over the past two decades, nanotechnology has produced a large number of novel materials that can be used in diagnosis and therapeutics. See Figure 5.3. In this section, we discuss the biomedical applications of GBNs.

5.7.1 Drug/Gene Delivery

Nanomaterial-based drug delivery systems have been used for the treatment of cancer in recent years. The purpose is to improve the therapeutic efficacy and reduce toxic side effects. GBNs are excellent drug/gene carriers. The 2D monolayer structure provides a larger specific surface area for the loading of drugs with higher load rate. GBNs have good mechanical strength and chemical stability for different environments and are easily functionalized except for graphene and GO (cytotoxicity). However, the cytotoxicity and other limitations can be reduced by surface modifications requiring organic materials such as linear polymers, nonlinear polymers, polysaccharides, amino acid proteins, and nonpolymers.[73] Linear polymerides (PLGA, PEG) can be used in drug and gene delivery at the same time.[73–74] In previous studies, a delivery material was prepared that is capable of adsorbing the hydrophobic aromatic molecule camptothecin analogue SN38 using PEG and GO. PLGA is biocompatible and biodegradable *in vivo*, so it is used for clinical applications in orthopedic fixation devices. PLGA/GBNs have wider applications in drug delivery systems. Another material, polyethylenimine/plasmid DNA-GO/PLGA scaffold, is used in tissue engineering.[67] PVP/GBNs are used as a biocompatible nontoxic surfactant stabilizer. PDA has excellent adhesion properties, which can

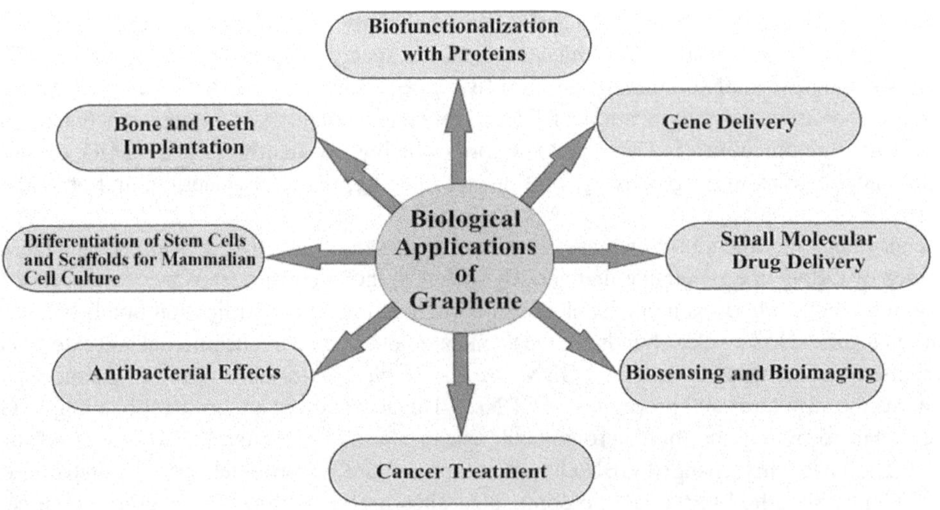

FIGURE 5.3 Biomedical applications of graphene-based nanomaterials (GBNs).

Source: Reproduced from *JNC* (2018) 8:123–137.

be used to modify the surface of GBNs or to stabilize nanocarriers. Nonlinear polymers, due to their spatial structures, are used as drug carriers: for example, polyethylenimine (PEI), polyamidoamine (PAMAM), hyperbranched polyglycerol (hPG), and cyclodextrin (CDS). Natural polysaccharides (chitosan, carboxymethyl cellulose, and hydroxyethyl cellulose) are used to implement GO drug loading modifications. Fe_3O_4 nanoparticles promote the temperature-dependent drug release of GO, and they can be used[70, 71] as a magnetically guided cell signal in several biomedical applications. TiO_2/GO based on TiO_2 nanoparticles (NPs) also hinder the growth of cancer cells by combining chemotherapy/PTT photocatalytic therapy.[73] GQDs exhibit excellent fluorescence properties with its biocompatibility, low toxicity *in vitro* and *in vivo*, stability, high permeability, ultrahigh specific surface area, and surface functionalization. The drug-loaded GO modified by mesoporous silica nanoparticles (MSN) can be used for the controlled release of drugs. Hydroxyapatite (HAP) has good bone conductivity, and it is widely used for bone repair, bone regeneration, and drug release. HAP can absorb drugs such as ibuprofen and 5-fluorouracil. Since GBNs can interact not only with the drugs but also with biomolecules, they are used as carriers and in the identification of nucleic acids due to their large sp^2-hybridized carbon area.[74] The gene delivery vector protects DNA from degradation and establishes high transfection capacity. Also, viral and nonviral vectors are used for gene delivery research. NIR irradiation of functionalized rGO can change the membrane integrity of endosomes, thus improving the intracellular lifetime of the drug or gene and their delivery efficacy.[75,76] See Figure 5.4.

5.7.2 Biosensor

GBNs are used for biosensor applications as they improve sensitivity. GBNs directly provide the electron transfer channel with the active sites of target receptors as they possess excellent optical and electrochemical properties. GBNs can act as electrochemical biosensors, field effect transistors (FETs) and optical biosensors (OBs), which involve Förster resonance energy transfer

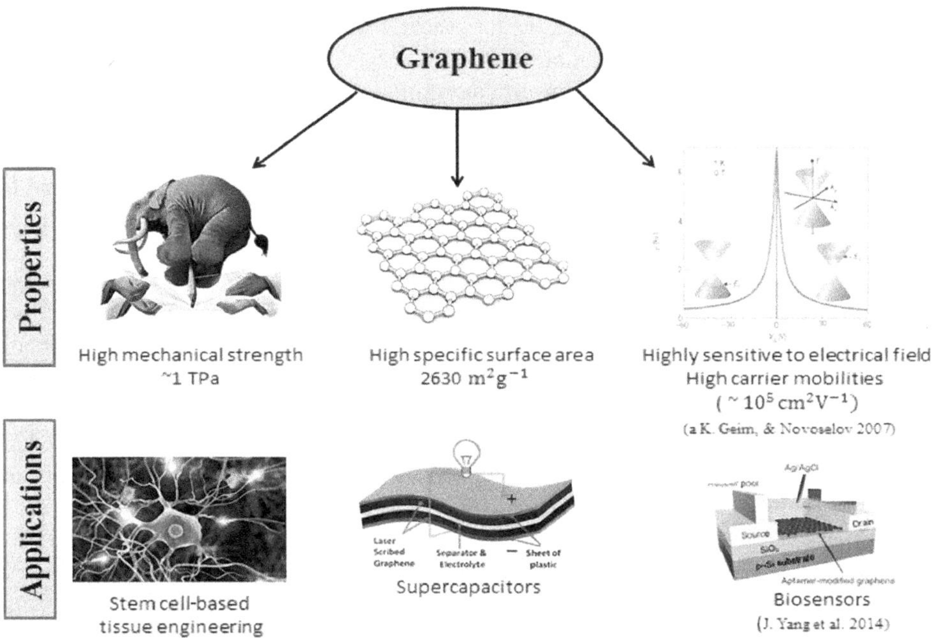

FIGURE 5.4 Schematic representation of graphene and its biomedical applications.

Source: ***Sains Malaysiana*** **46(7)(2017): 1125–1139.**

(FRET), chemiluminescence resonance energy transfer (CRET), electrochemiluminescence (ECL), surface-enhanced Raman scattering (SERS), and surface plasmon resonance (SPR) mechanisms.

GBN biosensors can detect biomolecules (RNA/DNA, protein) and other crucial biomarkers, which include hormone, glucose, nicotinamide dinucleotide adenine, adenosine triphosphate (ATP), and hydrogen peroxide. Yet these properties have not been fully utilized due to issues with the controlled, scaled, reproducible, and superficial preparation of GNR-based nanomaterials with defined structural properties. Electrochemical biosensors (ECB) use biomolecules to convert chemical signals into electrochemical signals through specific recognition. The large surface area of GBNs is conducive to the immobilization of the enzyme, whereas good conductivity ensures a highly sensitive electron transfer rate. However, surface defects between the surface charge area of graphene and analytes ensure a good electron transfer rate by the GBN sensor. The biosensors are simple in operation with low cost, flexible design, and high sensitivity.[77] Field effect transistor (FET) biosensors can easily detect various biomolecular analytes with high efficiency, precision, real-time detection, specificity, and label-free detection. rGO is used for FET biosensors and for EC or ECL assays[78,79] due to their high conductivity compared to GO. Biosensors (GO or rGO) exhibit fluorescence and fluorescence-quenching properties.[80,81] Optical biosensor (OB)-GBN, a fluorescent biosensor with quenching properties, has probe molecules complementary to the target, a fluorophore (signal), and a fluorophore carrier GBNs (quenchant).[82] Still, a few limitations need to be addressed as they hinders applications. GBN biosensors can be extended in clinical applications and should be developed to enable their reusability as these sensors have low cost and high efficacy.

5.7.3 Bioimaging

Bioimaging is used to identify the abnormal processes related to disease and in understanding the operating mechanisms of cells, tissues, and organs. The other applications of bioimaging include carrying contrast agents for photoacoustic imaging,[83] MR imaging,[84] and single-photon emission computed tomography (SPECT). GO and rGO are mainly used in *in vivo* and *in vitro* optical imaging as both have emission spectra with appropriate absorption wavelength. Moreover, GO is used for Raman imaging[85] and for subcellular imaging for cancer treatment.[86] GNRs are also used in superior optoelectronic and bioimaging applications.[87]

GO/rGO nanoparticles are used to investigate drug activation pathways that involve cellular activities. Fluorescent dye Cy5 with copper GNRs is also used for optical imaging. GQDs are also excellent materials for bioimaging. The fluorescent properties of GQDs can be tailored for biocompatibility with low *in vitro* and *in vivo* toxicity.[88] GQDs are physiologically stable and highly permeable, and they have ultrahigh specific surface area and ability for surface functionalization. The optical properties of GNRs are due to a special bandgap, which is utilized for bioimaging applications. The four main imaging techniques are fluorescence imaging (FRI), magnetic resonance imaging (MRI), Raman spectroscopy imaging (RSI), and photoacoustic imaging (PAI). QDs exhibit dual-modal imaging, two-photon fluorescence imaging, cellular imaging, diagnostic imaging, and targeted imaging. GQDs are used for the imaging and targeting of Golgi instruments in HeLa and MCF7 live cells,[89–92] human breast cancer cells, T4D cells, dermal fibroblast cells, A549 cells, MCF-7 cells, MDA-MB231 cells, human osteosarcoma MG-63 cells, stem cells, Chinese hamster ovary CHO-K1 cells, MC3T3 cells, human hepatic cancer cells (HuH-7 cells), L929 cells, and so on.[93,94] GQDs can be applied to the multiphoton fluorescence bioimaging. N-doped GQDs are fabricated with high sensing[94] and the surface modification further increase the efficiency of GQDs. Folic acid (FA) shows a high affinity to the folate receptor (FR), so FA-GQD complex was fabricated with carboxyl groups near the edge for diagnostic imaging and targeted cancer cell imaging and PTT both *in vivo* and *in vitro*.[95] FA-integrated GQDs (FA-GQDs) are used to accurately identify cancer cells from normal cells.[96] The strong interaction enhances the cellular absorption of FA-GQDs to specific cells for drug delivery. Combined PDT and PTT approaches are also proposed for imaging cancer cells. The *in vivo* bioimaging of mice have excitation light wavelengths varying from 465 to

500 nm with light emission from 580 to 620 nm,[97] with the best one having an excitation wavelength of 465 nm (emission wavelength of 620 nm). The NIR fluorescence of GQDs within 800–850 nm exhibit excellent penetration performance during the imaging of the heart, kidneys, liver, and spleen. Red-emitting GQD–europium complex (GQD/DBM)3/EuPhen/GQDs are also used for bioimaging of HeLa tumor-bearing nude mice.[98] Two-photon fluorescence imaging (TPFI) is observed in carbon dots (C-dots), GOs, and their derivatives, GQDs for deep-tissue imaging. The photofluorescence properties of GBNs are tuned by (1) doping hetero atoms, (2) fluorescence ability without passivation, and (3) increased TP absorption. Nitrogen-doped GQD (NGQD) are used for dynamic dual modal bioimaging in diagnostics.[99, 100] By integrating MRI and fluorescence imaging, the diagnostic efficiency is significantly improved: for example, Fe2O3@SiO2 and a GQD-integrated nanoprobe during cancer treatment targeting HeLa cells.[100]

With a single atomic layer and small lateral size, GQDs have an oxygen-rich surface, so utilizing them to load drug molecules increases physiological stability. The fluorescence properties of GQDs are used for drug delivery in cancer treatment.[101,102] Contrast agents (CAs) such as manganese (Mn) and gadolinium (Gd) are used for MRI as they exhibit nonspecific toxicity *in vivo*.[103,104] Paramagnetic metal ions induce toxicity in GBNs. NIR absorbance of GBN diffusion in PAI is used for tissue and organ therapy. rGO as a photoacoustic contrast agent absorb NIR more effectively for larger sp^2 domains. Raman signal imaging (RSI) has outstanding signal-to-noise ratio, inappreciable photobleaching, and multiplexing potential via a nondisturbing and nondestructive mechanism, which is used for bioimaging in GBNs. Gold/silver nanoparticles with GBNs improve the sensitivity of bio imaging. RSI is observed in silver/GO hybrid, Au-NGQD NPs,[90] and Ag/GO coupling with folic acid complexes.[104] GBNs is used in computed tomography imaging (CTPI), multimodal imaging (MMI), and radionuclide imaging (RNI).

5.7.4 Tissue Engineering

Biodegradable substitutes are used to repair and maintain the properties of tissues/organs or to change their functions. The materials in tissue engineering are used to repair tissues/organs that have been damaged due to wounds, infection, tumors, and deformities or to treat any toxicities; materials (calcium phosphate and calcium silicate) lacking in tissue-induced activities; materials (TiO_2 and HAP) lacking accumulation, mobility, and affability; or materials lacking the mechanical strength to attach the cells and conduct diffusion. Since GBNs have unique properties such as strength, flexibility, and strong adaptability of nonuniform surfaces, they are used as excellent materials in tissue engineering. GBNs can induce stem cells to identify tissue cells with specific functions and regulate the interactions between the cells. By chemical modifications, GBNs promote their interaction with various biomolecules.[105] The biodegradable biomaterials used in bone tissue engineering are (1) natural degradable polymers (collagen, chitosan, gelatin) or (2) synthetic degradable polymers (polylactic acid [PLA], polyglycolic acid [PGA], polycaprolactone [PCL], and polylactic acid-PGA copolymer [PLGA]). Natural degradable polymers are obtained from aquatic and land animal skin/bone. They require rigorous environmental and high safety standards. GO increases the mechanical strength of the degradable polymer scaffold to promote biomineralization, bone cell proliferation, and differentiation. GO-interacted materials improve the elastic modulus and tensile strength as GOg-PLA.[106] PCL/rGO is used to adsorb and deposit serum proteins, which promotes cell adhesion and proliferation. These materials are used to accelerate bone tissue repair and regeneration. GO-PLGA-Col matrices are used for bone tissue engineering. Graphene,[107] GO, and rGO sheet enhance the repair mechanism of MSC[108] and improve the defects of hydrogel.[109] The nervous system is constituted of electrically sensitive neurons. The ionic concentration gradients are generated by neuronal cells, which maintain the required voltage difference on the membrane through ion pumps. GBNs are also used in skin tissue engineering due to its large surface area and multipore size. Further modifications by polymers increase the adhesion and proliferation of fibroblasts.

5.7.5 Photothermal Therapy (PTT)

PTT reagent is used to treat local tumors by fast heating and long accession in the tumor area. It combines with other therapies to generate universal therapeutic effects and to reduce off-target toxicity. NIR-induced PTT promotes the uptake of drug cells by increasing the permeability of the cell membrane and blocking the outflow of drug and reversing multidrug resistance. Efficient candidates for PTT include high photothermal conversion efficiency capability, functional modifications, biocompatibility, and excellent tumor adhesion. GO and rGO both act as good photothermal agents due to their high NIR absorption and photothermal effect,[110] with rGO having a stronger photothermal conversion rate. The (PDD-PTT) combined GBNs can serve as a carrier of PDT reagent for excellent light absorption properties.

5.7.6 Antibacteria

The antibacterial activities of GBNs depend on their structural and physicochemical properties. The antibacterial properties depend on several factors such as lateral size, defects, charge, functional groups, hydrophilicity, and many more. GO exhibits the strongest antibacterial activity.[111] In other GBNs with low toxicity, antibacterial efficiency can be developed through surface modification. The antimicrobial (metal nanoparticles) activities of GBNs can be used for various biomedical applications.[112] The physicochemical action of GBNs deposited on the surface of bacteria is responsible for the antibacterial mechanism. The physical action includes cell wrapping and mechanical tearing of the cell membrane, and the chemical action is responsible for superoxide-induced oxidative stress response. The chemical damage is due to the light-induced electron-hole pairs, ROS generation, and/or charge transfer, which cause cell death by destroying glutathione or inducing DNA breaks. Suitable GBNs are selected based on the physicochemical characteristics of bacteria. GBNs are used in drug delivery, surface infection, dental fillers, membrane antibiotic fouling, water disinfection, and food packaging.[113] The antibacterial activities of GBNs and their nanocomposites include controlling microbial pathogens,[114] wound dressing,[102,103] tissue engineering,[115,116] packaging,[117] drug delivery,[118–120] and water purification.[121–122]

5.8 HEALTH AND ENVIRONMENTAL RISKS OF GBNs

With the diversified role of GBNs, there are several health and environmental risks. So it is very important to evaluate the toxicity of GBNs in biomedical applications.[123,124] Toxicity occurs due to the functional and environmental exposures through other products.[125] The level of GBNs' toxicity is measured by understanding the interactions at the cellular and molecular levels. GBNs are modified to overcome these limitations, so that the modified GBNs can be used in effective therapeutic delivery. The physicochemical properties (dose, shape, surface chemistry, exposure route, and purity) are responsible for exerting toxicity.[125] Surface chemistry is important for biofunctionalization, and a decrease in surface area increases the number of layers, which are vital to determine the specific surface area and bending stiffness. Though the lateral dimensions define only the material, the parameter is important for various biological phenomena.[126] GO is highly reactive as compared to other GBN family members. GBNs have unreacted and residual chemicals, resulting in inadequate washing, which inhibit their biological activities. GBNs enter into the body via blood circulation and affect different organs. Due to its effective nanosized properties, GBNs may enter organs by crossing the blood–air barrier, blood–testis barrier, blood–brain barrier, and blood–placental barrier. The toxicity of GBNs is due to cellular mechanisms such as oxidative stress, DNA damage, inflammatory response, apoptosis, autophagy, and necrosis.[127] So it is necessary to use GBNs with utmost caution in biomedical applications due to its toxicity and biosafety as graphene damage to mitochondria and the plasma membrane in PC12 cells and HepG2 cells was seen in previous studies.[127] rGO not only induces cytotoxicity, it produces genotoxic effects in HMSCs. The testing of toxic effects in animal models needs to strictly follow the animal ethics guidelines. All animal models such as zebrafish, mice, rats, and primates are used in *in vitro* screening to explore the *in vivo* accumulation and distribution of drugs in the organ structure. The neurotoxicity of GO in zebrafish was assessed in previous studies. As the zebrafish model has a homologous genome with humans, the animal model is usually selected for biomedical studies. Another study exhibited the pulmonary toxic potential of GOs in mice,[128] which produces the reactive oxygen, causes oxidative stress, and destroys the cellular structure of biomolecules. GO can promote sperm damage.[129] GBNs have strong protein adsorption capacities. As protein absorption in GBNs disappears after modification, they became unusable for drug delivery and treatment. Yet in a contradictory study, the hemolytic toxicity of chitosan is reduced with the accumulation of GO.

5.8.1 Impact on the Environment

The full application of GBNs will be realized only by reducing its cytotoxic effects on life systems and the environment. The toxic effects of GBNs on cells and animal models have been reported, with a few reports evaluating environmental effect. Sunlight reduces the stability of GO by changing its physicochemical properties and increases its environmental exposure. The migration effect of GO in water is affected by bivalent cations, natural organic complexes, and water forces. Exposure to GBNs causes serious effects in humans and aquatic species, which are exposed to the environment in many ways in the food chain, industrial wastes, pharmaceutical production waste, soil, and water.[130]

5.8.2 Methods to Reduce Toxicity

Though the cytotoxicity of GBNs to cancer cells and some bacteria/fungi proves its beneficial application, their toxic side effects in normal species and the environment limit their usage. The toxic effects of GBNs are reduced by surface modification, combination with macromolecules, synthesis of novel safe derivatives, exploration of the mechanism of toxicity, and advancing the preparation level. PEG approved by the U.S. Food and Drug Administration (FDA) is used to modify GBNs because of its nontoxic and nonimmunogenic behavior. Orally administered PEGylated GO tested on mice showed no appreciable tissue absorption. Intraperitoneally injected PEGylated GO in mice has not shown any toxic effects. Hydrophobic rGO accumulation on cell skeleton induces high ROS and apoptosis, whereas the toxicity is reduced. The toxicity of rGO could be reduced by increasing water solubility via surface modification.[131] The interaction of nanoproteins causes unpredictable side effects. FDA-approved albumin-based paclitaxel nanoparticles (Abraxane) is used as a promising drug for targeted delivery in oncology.[132] Graphene derivatives are used for various applications. For example, graphene is used to reduce physical damage to cells, and fluoroglycopene is biocompatible. Other GBN derivatives have shown excellent biofunctionality, water solubility, drug loading, and cell membrane permeability.

Advance preparation processing decreases the unreacted and residual chemicals, remarkably reducing toxicity, and ensures its safe use and applications. Safety and nontoxicity are vital as GBNs enter the body through blood circulation or biological barriers, and main organs are affected in these processess.[133,134]

5.9 CONCLUSIONS

With the so-called graphene revolution in 2004, GBNs have served as major potential candidate for application in translational medicine in recent years. The unique physiochemical, thermal, mechanical, and electrochemical properties are useful for many biomedical applications, which include drug/gene delivery, biosensors, biological imaging, tissue engineering, PTT, and antibacterial activities. The excellent optical properties of GBN-based hybrids are used for bioimaging in clinical diagnostics and PT cancer therapy. GO and rGO acts as the most potent antibacterial agents and are used in nanohybrids to synthesize novel antibacterial agents. GNRs have more flexible and adjustable properties due to their special edge-limiting effects, which is also effectively used for biomedical applications. Graphene quantum dots (GQDs) are less toxic, ultrasmall in a size, with excellent photo stability, tunable fluorescence properties, water solubility, and biocompatibility. GQDs are used as efficient candidates for bioimaging, drug delivery, therapeutics, and theranostics. It is hoped that the rapid development of GBNs with more attractive properties and functions can be used in a wide range for biosensing and nanomedicinal applications in the future.

5.9.1 Acknowledgments

RS acknowledges the financial assistance by DST WOSA project (SR/WOS-A/CS-69/2018). RS is thankful to her Mentor Dr. Shrish Tiwari, Bioinformatics, CSIR-Centre for Cellular and Molecular Biology, and Dr. G. Narahari Sastry, Director, CSIR-NEIST for the technical support.

5.9.2 Competing Interests

There is no conflict of interest for this manuscript.

REFERENCES

1. Jiang, X.; Ruan, G. Assembly and application advancement of organic-functionalized graphene-based materials: A review. J. Sep. Sci. 2020, 43 (8), 1544–1557.
2. Li, J,; Huamin, Z.; Zhaowu, Z.; Yiying, Z.; Tian X. Promising graphene-based nanomaterials and their biomedical applications and potential risks: A comprehensive review. ACS Biomater. Sci. Eng. 2021, 7 (12), 5363–5396. https://doi.org/10.1021/acsbiomaterials.1c00875.
3. Thabitha, P.; Dasari, S; McShan, D.; Dasmahapatra, A.K.; Tchounwou, P.B. A review on graphene-based nanomaterials in biomedical applications and risks in environment and health. Nano-Micro Lett. 2018, 10, 53. https://doi.org/10.1021/acsbiomaterials.1c00875.
4. Biswas M. C.; Islam, M. T.; Nandy, P. K.; Hossain, M. M. Graphene quantum dots (GQDs) for bio-imaging and drug delivery applications: A review. ACS Materials Lett. 2021, 3, 889–911. https://doi.org/10.1021/acsmaterialslett.0c00550.
5. Hamzah, A. A.; Selvarajan, R. S.; Majlis, B. Y. Graphene for biomedical applications: A review. Sains Malaysiana. 2017, 46 (7), 1125–1139.
6. Priyadarsini, S.; Mohanty, S.; Mukherjee, S.; Basu, S.; Mishra, M. Graphene and graphene oxide as nanomaterials for medicine and biology application. J. Nanostructure Chem. 2018, 8, 123–137. https://doi.org/10.1007/s40097-018-0265-6.
7. Younis, M. R.; He G.; Lin J.; Huang P. Recent advances on graphene quantum dots for bioimaging applications. Front. Chem. 2020, 8, 424. https://doi.org/10.3389/fchem.2020.00424.
8. Zakharova, O. V.; Mastalygina, E. E.; Golokhvast, K. S.; Gusev, A. A. Graphene nanoribbons: Prospects of application in biomedicine and toxicity. Nanomaterials. 2021, 11, 2425. https://doi.org/10.3390/nano11092425.
9. Hashimoto, A.; Takamura-Enya, T. Synthesis and in vitro biological evaluation of psoralen-linked fullerenes. Photochem. Photobiol. 2019, 95 (6), 1403–1411.
10. Chen, M.; Guan, R.; Yang, S. Hybrids of fullerenes and 2D nanomaterials. Adv. Sci. (Weinh). 2019, 6 (1), 1800941.
11. Negri, V.; Pacheco-Torres, J. Carbon nanotubes in biomedicine. Top Curr. Chem. (Cham). 2020, 378 (1), 15.
12. Madni, A.; Noreen, S.; Maqbool, I.; Rehman, F.; Batool, A.; Kashif, P. M.; Rehman, M.; Tahir, N.; Khan, M. I. Graphene-based nanocomposites: Synthesis and their theranostic applications. J. Drug Target. 2018, 26 (10), 858–883.
13. Zhang, Q.; Wu, Z.; Li, N.; Pu, Y. Q.; Wang, B.; Zhang, T.; Tao, J. S. Advanced review of graphene-based nanomaterials in drug delivery systems: Synthesis, modification, toxicity and application. Mater. Sci. Eng. C. 2017, 77, 1363–1375.

14. Xiang, C.; Zhang, Y.; Guo, W.; Liang, X. J. Biomimetic carbon nanotubes for neurological disease therapeutics as inherent medication. Acta Pharm. Sin. B. 2020, 10 (2), 239–248.
15. Johnson, A. P.; Gangadharappa, H. V.; Pramod, K. Graphene nanoribbons: A promising nanomaterial for biomedical applications. J. Control. Release. 2020, 325, 141–162.
16. Nanda, S. S.; Kaushal, S.; Shin, Y.; Yun, K.; Seong, S. A. A.; Hembram, K. P. S. S.; Georgia, C. P.; Yi, D. K. Cancer cell detection on the surface of top-gated monolayer graphene via raman spectroscopy. ACS Appl. Bio Mater. 2021, 4 (2), 1493–1498.
17. Wu, Y.; Deng, P.; Tian, Y.; Feng, J. X.; Xiao, J. Y.; Li, J. H.; Liu, J.; Li, G. L.; He, Q. G. Simultaneous and sensitive determination of ascorbic acid, dopamine and uric acid via an electrochemical sensor based on PVP-graphene composite. J. Nanobiotechnol. 2020, 18 (1), 112.
18. Zhao, S.; Li, G.; Tong, C.; Chen, W. J.; Wang, P. X.; Dai, J. K.; Fu, X. F.; Xu, Z.; Liu, X. J.; Lu, L. L.; Liang, Z. F.; Duan, X. J. Full activation pattern mapping by simultaneous deep brain stimulation and fMRI with graphene fiber electrodes. Nat. Commun. 2020, 11 (1), 1788.
19. Novoselov, K. S.; Geim, A. K.; Morozov, S. V.; Jiang, D.; Zhang, Y.; Dubonos, S. V.; Grigorieva, I. V.; Firsov, A. A. Electric field effect in atomically thin carbon films. Science. 2004, 306 (5696), 666–669.
20. Abbasi, E.; Akbarzadeh, A.; Kouhi, M.; Milani, M. Graphene: Synthesis, bio-applications, and properties. Artif. Cells, Nanomed., Biotechnol. 2016, 44 (1), 150–156.
21. Boehm, B. A Spiral model of software development and enhancement. Softw. Eng. Notes. 1986, 11, 14–24. https://doi.org/10.1145/12944.12948
22. Dasari, S. T. P.; McShan, D.; Dasmahapatra, A. K.; Tchounwou, P. B. A Review on graphene-based nanomaterials in biomedical applications and risks in environment and health. Nano-Micro Lett. 2018, 10 (3), 53.
23. Jana, A. Scheer, E. Polarz, S. Synthesis of graphene–transition metal oxide hybrid nanoparticles and their application in various fields. Beilstein J. Nanotechnol. 2017, 8, 688–714. https://doi. org/10.3762/bjnano.8.74
24. O'Hare, A.; Kusmartsev, F. V.; Kugel, K. I. A stable "flat" form of two-dimensional crystals: Could graphene, silicene, germanene be minigap semiconductors? Nano Lett. 2012, 12 (2), 1045–1052.
25. Fasolino, A.; Los, J. H.; Katsnelson, M. I. Intrinsic ripples in graphene. Nat. Mater. 2007, 6 (11), 858–861.
26. Zhao, H.; Ding, R.; Zhao, X.; Li, Y. W.; Qu, L. L.; Pei, H.; Yildirimer, L.; Wu, Z. W.; Zhang, W. X. Graphene-based nanomaterials for drug and/or gene delivery, bioimaging, and tissue engineering. Drug Discov. Today. 2017, 22 (9), 1302–1317.
27. Li, K. H.; Zhao, X. N.; Wei, G.; Su, Z. Q. Recent Advances in the Cancer Bioimaging with Graphene Quantum Dots. Curr. Med. Chem. 2018, 25 (25), 2876–2893.
28. Chung, S.; Revia, R. A.; Zhang, M. Graphene quantum dots and their applications in bioimaging, biosensing, and therapy. Adv. Mater. 2021, 33, e1904362.
29. Shemella, P.; Zhang, Y.; Mailman, M.; Ajayan, P. M.; Nayak, S. K. Electronic structure and band-gap modulation of graphene via substrate surface chemistry. Appl. Phys. Lett. 2007, 91, 042101.
30. Wassmann, T.; Seitsonen, A. P.; Saitta, A. M.; Lazzeri, M.; Mauri, F. Clar's theory, pi-electron distribution, and geometry of graphene nanoribbons. J. Am. Chem. Soc. 2010, 132, 3440.
31. Shen, H.; Shi, Y.; Wang, X. Synthesis, charge transport and device applications of graphene nanoribbons. Synth. Met. 2015, 210, 109.
32. Jaiswal, M.; Lim, C. H. Y. X.; Bao, Q.; Toh, C. T.; Loh, K. P.; Oezyilmaz, B. ACS Nano. 2011, 5, 888.
33. Huang, X.; Yin, Z.; Wu, S.; Qi, X.; He, Q.; Zhang, Q.; Yan, Q.; Boey, F.; Zhang, H. Graphene-based materials: Synthesis, characterization, properties, and applications. Small. 2011, 7, 1876.
34. Aparicio, E.; Tangarife, E.; Munoz, F.; Gonzalez, R. I.; Valencia, F. J.; Careglio, C.; Bringa, E. M. Nanotechnology. 2021, 32, 045709.
36. Laio, A., and Parrinello, M. Escaping free-energy minima. Proc. Natl. Acad. Sci. U.S.A. 2022, 99, 12562–12566. https://doi.org/10.1073/pnas.2024 27399.
37. Conte, R.; Ceotto, M. Semiclassical molecular dynamics for spectroscopic calculations. Quantum Chemistry and Dynamics of Excited States (pp. 595–628). 2020. https://doi.org/10.1002/9781119417774.ch19.
38. Morris, G. M.; Huey, R.; Lindstrom, W.; Sanner, M. F.; Belew, R. K.; Goodsell, D. S.; Olson, A. J. AutoDock4 and AutoDockTools4: Automated docking with selective receptor flexibility. J. Comput. Chem. 2009, 30, 2785–2791. https://doi.org/10.1002/jcc.21256.
39. León, C.; Melnik, R. Machine learning for shape memory graphene nanoribbons and applications in biomedical engineering. Bioengineering (Basel). 2022 Feb 23, 9 (3), 90. https://doi.org/10.3390/bioengineering9030090.

40. Harismah, K.; Sain, DK.; Al-Owaidi, MF.; Zandi, H.; Saimmai, A. Assessing a BN-doped graphene for the drug delivery of hydroxyurea anticancer. Biointerface Res. Appl. Chem. 2023, 13 (5), 485. https://doi.org/10.33263/BRIAC135.485.
41. Piras, A.; Ehlert, C.; Gryn'ova G. Sensing and sensitivity: Computational Chemistry of Graphene-Based Sensors. 2021, 11 (5), e1526. https://doi.org/10.1002/wcms.1526.
42. Dien, VK.; Li, WB.; Lin, KI.; Han, NT.; Lin, MF. Electronic and optical properties of graphene, silicene, germanene, and their semi-hydrogenated systems. RSC Adv. 2022, 12, 34851–34865. https://doi.org/10.1039/D2RA06722F.
43. Baweja, L.; Balamurugan, K.; Subramanian, V.; Dhawan, A. Effect of graphene oxide on the conformational transitions of amyloid beta peptide: A molecular dynamics simulation study. J. Mol. Graph. Model. 2015, 61 175–185. https://doi.org/10.1016/j.jmgm.2015.07.007.
44. Asghar, SA.; Pal, P.; Nazeer, K.; Mahadevappa, M. A Computational study of graphene as a prospective material for microelectrodes in retinal prosthesis and electric crosstalk analysis. 2020 42nd Annual International Conference of the IEEE Engineering in Medicine & Biology Society (EMBC), Montreal, QC, Canada, 2020, 2291–2294, https://doi.org/10.1109/EMBC44109.2020.9176388.
45. Munny, KN.; Ahmed, T.; Piya, AA.; Shamim, SUD. Exploring the adsorption performance of doped graphene quantum dots as anticancer drug carriers for cisplatin by DFT, PCM, and COSMO approaches. Struct. Chem. 2023. https://doi.org/10.1007/s11224-023-02150-y.
46. Ishii, T.; Philavanh, M.; Negishi, J.; Inukai, E.; Ozaki, JI. Preparation of chemically structure-controlled BN doped carbons for the molecular understanding of their surface active sites for oxygen reduction reaction. ACS Catal. 2022, 12, 1288–1297, https://doi.org/10.1021/acscatal.1c04806.
47. Darougari, H.; Rezaei-Sameti, M. The drug delivery appraisal of Cu and Ni decorated B12N12 nanocage for an 8-hydroxyquinoline drug: A DFT and TD-DFT computational study. Asian J. Nanosci. Mater. 2022, 5, 196–210, www.ajnanomat.com/article_154625.html.
48. Anafcheh, M.; A comparison between density functional theory calculations and the additive schemes of polarizabilities of the Li-F-decorated BN cages. J. Appl. Organomet. Chem. 2021, 1, 125–133. https://doi.org/10.22034/jaoc.2021.292384.1027
49. Herrera-Reinoza, N.; dos Santos, AC.; de Lima, LH.; Landers, R.; de Siervo, A. Atomically precise bottomup synthesis of h-BNC: Graphene doped with h-BN nanoclusters. Chem. Mater. 2021, 33, 2871–2882, https://doi.org/10.1021/acs.chemmater.1c00081.
50. Kim, J.; Serpe, MJ.; Lyon, LA. Hydrogel microparticles as dynamically tunable microlenses. J. Am. Chem. Soc. 2004, 126, 9512–9513.
51. Putri, AD.; Murti, BT.; Kanchi, S.; Sabela, M. I.; Bisetty, K.; Inamuddin, A. T.; Asiri, A. M. Computational studies on the molecular insights of aptamer induced poly(*N*-isopropylacrylamide)-graft-graphene oxide for on/off- switchable whole-cell cancer diagnostics. Sci Rep. 2019, 9, 7873. https://doi.org/10.1038/s41598-019-44378-x.
52. Tu, Y.; Lv, M.; Xiu, P.; Huynh, T.; Zhang, M.; Castelli, M.; Liu, Z.; Huang, Q.; Fan, C.; Fang, H.; et al. Destructive extraction of phospholipids from Escherichia coli membranes by graphene nanosheets. Nat. Nanotechnol. 2013, 8, 594–601.
53. Jo, B.C.; Yoon, H.J.; Ok, M.R.; Wu, S. Molecular dynamics simulation of cytotoxicity of graphene nanosheets to blood-coagulation protein. Biointerphases. 2017, 12, 01A403.
54. Tu, Y.; Lv, M.; Xiu, P.; Huynh, T.; Zhang, M.; Castelli, M.; Liu, Z.; Huang, Q.; Fan, C.; Fang, H.; et al. Destructive extraction of phospholipids from Escherichia coli membranes by graphene nanosheets. Nat. Nanotechnol. 2013, 8, 594–601.
55. Jo, B.C.; Yoon, H.J.; Ok, M.R.; Wu, S. Molecular dynamics simulation of cytotoxicity of graphene nanosheets to blood-coagulation protein. Biointerphases. 2017, 12, 01A403.
56. Tsukanov, A.A.; Turk, B.; Vasiljeva, O.; Psakhie, S.G. Computational indicator approach for assessment of nanotoxicity of two-dimensional nanomaterials. Nanomaterials. 2022, 12, 650. https://doi.org/10.3390/nano12040650.
57. Moradi, O.; Mahdavian, L. Simulation and computational study of graphene oxide nano-carriers, absorption, and release of the anticancer drug of camptothecin. J Mol Model. 2021 Aug 17, 27 (9):251.
58. Zheng, S.; Xiong, J.; Wang, L.; Zhai, D.; Xu, Y.; Lin, F. e-Graphene: A computational platform for the prediction of graphene-based drug delivery system by quantum genetic algorithm and cascade protocol. Front. Chem. 2021, 9:664355. https://doi.org/10.3389/fchem.2021.664355.
59. Srivastava, R. Interactions, electronic and optical properties of nanographene-peptide complexes: A theoretical study. RSC Adv. 2020 Oct 21, 10 (63):38654–38662. https://doi.org/10.1039/d0ra07961h.

60. Srivastava, R. Theoretical studies of optoelectronic, magnetization and heat transport properties of conductive metal adatoms adsorbed on edge chlorinated nanographenes. RSC Adv. 2018, 8, 17723–17731. https://doi.org/10.1039/C8RA02032A.
61. Kumari, S.; Sharma, P.; Ghosh, D.; Shandilya, M.; Rawat, P.; Hassan, M. I.; Moulick, R. G.; Bhattacharya, J.; Srivastava, C.; Majumder, S. Time-dependent study of graphene oxide-trypsin adsorption interface and visualization of nano-protein corona. Int. J. Biol. Macromol. 2020, 163, 2259–2269.
62. Kazempour, M.; Namazi, H.; Akbarzadeh, A.; Kabir, R. Synthesis and characterization of PEG-functionalized graphene oxide as an effective pH-sensitive drug carrier. Artif. Cells Nanomed. Biotechnol. 2019, 47 (1), 90–94.
63. Mullick, C. S.; Lalwani, G.; Zhang, K.; Yang, J. Y.; Neville, K.; Sitharaman, B. Cell specific cytotoxicity and uptake of graphene nanoribbons. Biomaterials. 2013, 34 (1), 283–293.
64. Chen, W.; Liu, P.; Min, L.; Zhou, Y. M.; Liu, Y.; Wang, Q.; Duan, W. F. Non-covalently functionalized graphene oxide-based coating to enhance thermal stability and flame retardancy of PVA film. Nano-Micro Lett. 2018, 10 (3), 39.
65. Sun, W.; Huang, S.; Zhang, S.; Luo, Q. Preparation, characterization and application of multi-mode imaging functional graphene Au-Fe(3)O(4) magnetic nanocomposites. Materials. 2019, 12 (12), 1978.
66. Li, H.; Jia, Y.; Liu, C. Pluronic® F127 stabilized reduced graphene oxide hydrogel for transdermal delivery of ondansetron: Ex vivo and animal studies. Colloids Surf. B. 2020, 195, 111259.
67. Han, X. M.; Zheng, K. W.; Wang, R. L.; Yue, S. F.; Chen, J.; Zhao, Z. W.; Song, F.; Su, Y.; Ma, Q. Functionalization and optimization-strategy of graphene oxide-based nanomaterials for gene and drug delivery. Am. J. Transl. Res. 2020, 12 (5), 1515–1534.
68. Wang, Z.; Shen, H.; Song, S.; Zhang, L. M.; Chen, W.; Dai, J. W.; Zhang, Z. J. Graphene oxide incorporated PLGA nanofibrous scaffold for solid phase gene delivery into mesenchymal stem cells. J. Nanosci. Nanotechnol. 2018, 18 (4), 2286–2293.
69. Chen, M.; Jiang, S.; Zhang, F.; Li, L. L.; Hu, H. L.; Wang, H. L. graphene oxide immobilized plga-polydopamine nanofibrous scaffolds for growth inhibition of colon cancer cells. Macromol. Biosci. 2018, 18 (12), No. e1800321.
70. Lan, M. Y.; Hsu, Y. B. Polyethylene glycol-coated graphene oxide loaded with erlotinib as an effective therapeutic agent for treating nasopharyngeal cancer cells. Int. J. Nanomed. 2020, 15, 7569–7582.
71. Mahdavi, M.; Fattahi, A.; Tajkhorshid, E. M.; Nouranian, S. Molecular insights into the loading and dynamics of doxorubicin on pegylated graphene oxide nanocarriers. ACS Appl. Bio Mater. 2020, 3 (3), 1354–1363.
72. Luo, Y.; Shen, H.; Fang, Y.; Cao, Y. H.; Huang, J.; Zhang, M. X.; Dai, J. W.; Shi, X. Y.; Zhang, Z. J. Enhanced proliferation and osteogenic differentiation of mesenchymal stem cells on graphene oxide-incorporated electrospun poly(lactic-co-glycolic acid) nanofibrous mats. ACS Appl. Mater. Interfaces. 2015, 7 (11), 6331–6339.
73. Lu, Y. J.; Lan, Y. H.; Chuang, C. C.; Lu, W. T.; Chan, L. Y.; Hsu, P. W.; Chen, J. P. Injectable thermo-sensitive chitosan hydrogel containing CPT-11-loaded EGFR-targeted graphene oxide and SLP2 shRNA for localized drug/gene delivery in glioblastoma therapy. Int. J. Mol. Sci. 2020, 21 (19), 7111.
74. Li, S.; Xiao, L.; Deng, H.; Shi, X. W.; Cao, Q. H. Remote controlled drug release from multi-functional Fe(3)O(4)/GO/chitosan microspheres fabricated by an electrospray method. Colloids Surf., B. 2017, 151, 354–362.
75. Ding, Y.; Zhou, L.; Chen, X.; Wu, Q.; Song, Z. Y.; Wei, S. H.; Zhou, J. H.; Shen, J. Mutual sensitization mechanism and selfdegradation property of drug delivery system for in vitro photodynamic therapy. Int. J. Pharm. 2016, 498 (1–2), 335–346.
76. Li, J.; Tan, S.; Kooger, R.; Zhang, C.; Zhang, Y. Micrornas as novel biological targets for detection and regulation. Chem. Soc. Rev. 2014, 43 (2), 506–517. https://doi.org/10.1039/c3cs60312a.
77. Kim, H.; Lee, D.; Kim, J.; Kim, T.-I.; Kim, W.J. Photothermally triggered cytosolic drug delivery via endosome disruption using a functionalized reduced graphene oxide. ACS Nano. 2013, 7 (8), 6735–6746. https://doi.org/10.1021/nn403096s.
78. Kim, H.; Kim, W.J. Photothermally controlled gene delivery by reduced graphene oxide-polyethylenimine nanocomposite. Small. 2013, 10 (1), 117–126. https://doi.org/10.1002/smll. 201202636.
79. Waehler, R.; Russell, S.J.; Curiel, D.T. Engineering targeted viral vectors for gene therapy. Nat. Rev. Genet. 2007, 8 (8), 573–587. https://doi.org/10.1038/nrg2141.
80. Thakur, S.; Karak, N. Alternative methods and nature-based reagents for the reduction of graphene oxide: A review. Carbon. 2015, 94, 224–242. https://doi.org/10.1016/j.carbon.2015.06.030.

81. Zhang, H.; Grüner, G.; Zhao, Y. Recent advancements of graphene in biomedicine. J. Mater. Chem. B. 2013, 2013 (1), 2542–2567. https://doi.org/10.1039/c3tb20405g.
82. Huang, X.; Boey, F.; Zhang, H.U.A. A brief review on graphenenanoparticle composites. Cosmos. 2010, 06 (02), 159–166. https://doi.org/10.1142/s0219607710000607.
83. Agharkar, M.; Kochrekar, S.; Hidouri, S.; Azeez, M.A. Trends in green reduction of graphene oxides, issues and challenges: A review. Mater. Res. Bull. 2014, 59, 323–328. https://doi.org/10. 1016/j.materresbull.2014.07.051.
84. Kim, Y. J.; Jeong, B. Graphene-based nanomaterials and their applications in biosensors. Adv. Exp. Med. Biol. 2018, 1064, 61–71.
85. Li, J.; Lyv, Z.; Li, Y.; Liu, H.; Wang, J.; Zhan, W.; Chen, H.; Chen, H.; Li, H. A theranostic prodrug delivery system based on Pt(IV) conjugated nano-graphene oxide with synergistic effect to enhance the therapeutic efficacy of Pt drug. Biomaterials. 2015, 51, 12–21. https://doi.org/10.1016/j.biomaterials.2015.01.074.
86. Zheng, XT.; Chen, P.; Li, C.M. Anticancer efficacy and subcellular site of action investigated by real-time monitoring of cellular responses to localized drug delivery in single cells. Small. 2012, 8 (17), 2670–2674. https://doi.org/10.1002/smll. 201102636.
87. Gurunathan, S.; Han, J.W.; Dayem, A.A.; Eppakayala, V.; Park, M.-R.; Kwon, D.-N.; Kim, J.-H. Antibacterial activity of dithiothreitol reduced graphene oxide. J. Ind. Eng. Chem. 2013, 19 (4), 1280–1288. https://doi.org/10.1016/j.jiec.2012.12.029.
88. Sahoo, N.G.; Bao, H.; Pan, Y.; Pal, M.; Kakran, M.; Cheng, H.K.F.; Li, L.; Tan, L.P. Functionalized carbon nanomaterials as nanocarriers for loading and delivery of a poorly water-soluble anticancer drug: A comparative study. Chem. Commun. 2011, 47 (18), 5235–5237. https://doi.org/10.1039/c1cc00075f.
89. Liu, K.; Wang, Y.; Li, H.; Duan, Y. A facile one-pot synthesis of starch functionalized graphene as nanocarrier for pH sensitive and starch-mediated drug delivery. Colloids Surf. B. 2015, 128, 86–93. https://doi.org/10.1016/j.colsurfb.2015.02.010.
90. Fan, X.; Jiao, G.; Zhao, W.; Jin, P.; Li, X. Magnetic Fe3O4–graphene composites as targeted drug nanocarriers for pH-activated release. Nanoscale. 2013, 5 (3), 1143–1152. https://doi.org/10. 1039/c2nr33158f.
91. Fan, H.; Yu, X.; Wang, K.; Yin, Y.; Tang, Y.; Tang, Y.; Liang, X. Graphene quantum dots (GQDs)-based nanomaterials for improving photodynamic therapy in cancer treatment. Eur. J. Med. Chem. 2019, 182, 111620.
92. Liu, H.; Kameoka, J.; Czaplewski, D. A.; Craighead, H. G. Polymeric nanowire chemical sensor. Nano Lett. 2004, 4, 671–675.
93. Chen, W.; Lv, G.; Hu, W.; Li, D.; Chen, S.; Dai, Z. Synthesis and applications of graphene quantum dots: A review. Nanotechnol. Rev. 2018, 7, 157–185.
94. Campbell, E.; Hasan, M. T.; Gonzalez Rodriguez, R.; Akkaraju, G. R.; Naumov, A. V. doped graphene quantum dots for intracellular multicolor imaging and cancer detection. ACS Biomater. Sci. Eng. 2019, 5, 4671–4682.
95. Younis, M. R.; He, G.; Lin, J.; Huang, P. Recent advances on graphene quantum dots for bioimaging applications. Front. Chem. 2020, 8, 424.
96. Fu, Y.; Gao, G.; Zhi, J. Electrochemical synthesis of multicolor fluorescent N-doped graphene quantum dots as a ferric ion sensor and their application in bioimaging. J. Mater. Chem. B. 2019, 7, 1494–1502.
97. Li, S.; Zhou, S.; Li, Y.; Li, X.; Zhu, J.; Fan, L.; Yang, S. Exceptionally high payload of the ir780 iodide on folic acid functionalized graphene quantum dots for targeted photothermal therapy. ACS Appl. Mater. Interfaces. 2017, 9, 22332–22341.
98. Hai, X.; Wang, Y.; Hao, X.; Chen, X.; Wang, J. Folic Acid encapsulated graphene quantum dots for ratio metric PH sensing and specific multicolor imaging in living cells. Sens. Actuators. B. 2018, 268, 61–69.
99. Zhu, S.; Song, Y.; Zhao, X.; Shao, J.; Zhang, J.; Yang, B. The photoluminescence mechanism in carbon dots (Graphene Quantum Dots, Carbon Nanodots, and Polymer Dots): Current state and future perspective. Nano Res. 2015, 8, 355–381.
100. Lu, H.; Li, W.; Dong, H.; Wei, M. Graphene quantum dots for optical bioimaging. Small. 2019, 15, 1902136.
101. Su, X.; Chan, C.; Shi, J.; Tsang, M.-K.; Pan, Y.; Cheng, C.; Gerile, O.; Yang, M. A Graphene quantum Dot@Fe3O4@SiO2 based nanoprobe for drug delivery sensing and dual-modal fluorescence and MRI imaging in cancer cells. Biosens. Bioelectron. 2017, 92, 489–495.
102. Fan, W.; Yung, B.; Huang, P.; Chen, X. Nanotechnology for multimodal synergistic cancer therapy. Chem. Rev. 2017, 117 (22), 13566–13638.

103. Cheng, D.; Yang, L.; Li, X.; Zhou, J.; Chen, Q.; Yan, S.; Li, N.; Chu, M.; Dong, Y.; Xie, Z.; Zhang, C. An electrochemical DNA sensing platform using carboxyl functionalized graphene as the electrode modified material. J. Electrochem. Soc. 2017, 164, H345– H351.
104. Pistone, A.; Iannazzo, D.; Ansari, S.; Milone, C.; Salamo, M.; Galvagno, S.; Cirmi, S.; Navarra, M. Tunable doxorubicin release from polymer-gated multiwalled carbon nanotubes. Int. J. Pharm. 2016, 515, 30–36.
105. Helmchen, F.; Denk, W. Deep tissue two-photon microscopy. Nat. Methods. 2005, 2, 932–940.
106. Singh, H.; Sreedharan, S.; Tiwari, K.; Green, N. H.; Smythe, C.; Pramanik, S. K.; Thomas, J. A.; Das, A. Two photon excitable graphene quantum dots for structured illumination microscopy and imaging applications: Lysosome specificity and tissue-dependent imaging. Chem. Commun. 2019, 55, 521–524.
107. Ghosh, S.; Chatterjee, K. Poly(Ethylene Glycol) functionalized graphene oxide in tissue engineering: A review on recent advances. Int. J. Nanomed. 2020, 15, 5991–6006.
108. Li, W. X.; Xu, J. W.; Chen, L.; Shan, M.; Shan, M. J.; Tian, X.; Yang, C. Y.; Lv, H. M.; Qian, X. M. A facile method to produce graphene oxide-g-poly(L-lactic acid) as an promising reinforcement for PLLA nanocomposites. Chem. Eng. J. 2014, 237, 291–299.
109. Talukdar, Y.; Rashkow, J.; Lalwani, G.; Kanakia, S.; Sitharaman, B. The effects of graphene nanostructures on mesenchymal stem cells. Biomaterials. 2014, 35 (18), 4863–4877.
110. Jin, L.; Zeng, Z. P.; Kuddannaya, S.; Yue, D.; Bao, J. N.; Wang, Z. L.; Zhang, Y. L. Synergistic effects of a novel free-standing reduced graphene oxide film and surface coating fibronectin on morphology, adhesion and proliferation of mesenchymal stem cells. J. Mater. Chem. B 2015, 3 (21), 4338–4344.
111. Shin, S. R.; Zihlmann, C.; Akbari, M.; Assawes, P.; Cheung, L.; Zhang, K. Z.; Manoharan, V.; Zhang, Y. S.; Yüksekkaya, M.; Wan, K. T.; Nikkhah, M.; Mehmet, R. D.; Tang, X. W. S.; Khademhosseini, A. Reduced graphene oxide-GelMA hybrid hydrogels as scaffolds for cardiac tissue engineering. Small. 2016, 12 (27), 3677–3689.
112. Sundaram, P.; Abrahamse, H. Phototherapy combined with carbon nanomaterials (1D and 2D) and their applications in cancer therapy. Materials. 2020, 13 (21), 4830.
113. Xia, M. Y.; Xie, Y.; Yu, C. H.; Chen, G. Y.; Li, Y. H.; Zhang, T.; Peng, Q. Graphene-based nanomaterials: The promising active agents for antibiotics-independent antibacterial applications. J. Control. Release. 2019, 307, 16–31.
114. Zhao, R. T.; Kong, W.; Sun, M. X.; Yang, Y.; Liu, W. Y.; Lv, M.; Song, S. P.; Wang, L. H.; Song, H. B.; Hao, R. Z. Highly stable graphene-based nanocomposite (GO-PEI-Ag) with broad-spectrum, long-term antimicrobial activity and antibiofilm effects. ACS Appl. Mater. Interfaces. 2018, 10 (21), 17617–17629.
115. Hegab, H.M.; ElMekawy, A.; Zou, A.; Mulcahy, D.; Saint, C.P.; Ginic-Markovic, M. The controversial antibacterial activity of graphene-based materials. Carbon. 2016, 105, 362–376. https://doi.org/10.1016/j.carbon.2016.04.046.
116. Tang, J.; Chen, Q.; Xu, L.; Zhang, S.; Feng, L.; Cheng, L.; Xu, Z.; Liu, H.; Peng, R. Graphene oxide–silver nanocomposite as a highly effective antibacterial agent with species-specific mechanisms. ACS Appl. Mater. Interfaces. 2013, 5 (9), 3867–3874. https://doi.org/10.1021/am4005495.
117. Fan, Z.; Liu, B.; Wang, J.; Zhang, S.; Lin, Q.; Gong, P.; Ma, L.; Yang, S. A novel wound dressing based on ag/graphene polymer hydrogel: Effectively kill bacteria and accelerate wound healing. Adv. Funct. Mater. 2014, 24 (25), 3933–3943. https://doi.org/10.1002/adfm.201304202.
118. Lu, B.; Li, T.; Zhao, H.; Li, X.; Gao, C.; Zhang, S.; Xie, E. Graphene-based composite materials beneficial to wound healing. Nanoscale. 2012, 4 (9), 2978–2982. https://doi.org/10.1039/c2nr11958g.
119. Shin, S.R.; Li, Y.-C.; Jang, H.L.; Khoshakhlagh, P.; Akbari, M.; Nasajpour, A.; Zhang, Y.S.; Tamayol, A.; Khademhosseini, A. Graphene-based materials for tissue engineering. Adv. Drug Deliver. Rev. 2016, 105, 255–274. https://doi.org/10.1016/j.addr. 2016.03.007.
120. Akhavan, O. Graphene scaffolds in progressive nanotechnology/stem cell-based tissue engineering of the nervous system. J. Mater. Chem. B. 2016, 4 (19), 3169–3190. https://doi.org/10. 1039/c6tb00152a.
121. Rutz, A.L.; Hyland, K.E.; Jakus, A.E.; Burghardt, W.R.; Shah, R.N. A multimaterial bioink method for 3d printing tunable, cell-compatible hydrogels. Adv. Mater. 2015, 27 (9), 1607–1614. https://doi.org/10.1002/adma.201405076.
122. Seethamraju, S.; Kumar, S.; Bharadwaj, K.B.; Madras, G.; Raghavan, S.; Ramamurthy, P.C. Million-fold decrease in polymer moisture permeability by a graphene monolayer. ACS Nano. 2016, 10 (7), 6501–6509. https://doi.org/10.1021/acsnano.6b02588.
123. Goenka, S.; Sant, V.; Sant, S. Graphene-based nanomaterials for drug delivery and tissue engineering. J. Control. Release. 2014, 173, 75–88. https://doi.org/10.1016/j.jconrel.2013.10.017.

124. Abraham, J.; Vasu, K.S.; Williams, C.D.; Gopinadhan, K.; Su, Y; Cherian, S.T.; Dix, J.; Prestat, E.; Haigh, S.J.; Grigorieva, I.V.; Carbone, P.; Geim, A.K.; Nair, R.R. Tunable sieving of ions using graphene oxide membranes. Nat. Nanotechnol. 2017, 12 (6), 546–550. https://doi.org/10. 1038/nnano.2017.21.
125. Bianco, A. Graphene: Safe or toxic? The two faces of the medal. Angew. Chem. Int. Edit. 2013, 52 (19), 4986–4997. https://doi. org/10.1002/anie.201209099.
126. McCallion, C.; Burthem, J.; Rees-Unwin, K.; Golovanov, A.; Pluen, A. Graphene in therapeutics delivery: Problems, solutions and future opportunities. Eur. J. Pharm. Biopharm. 2016, 104, 235–250. https://doi.org/10.1016/j.ejpb.2016.04.015.
127. Mahmoudi, E.; Ng, L.Y.; Ba-Abbad, M.M.; Mohammad, A.W. Novel nanohybrid polysulfone membrane embedded with silver nanoparticles on graphene oxide nanoplates. Chem. Eng. J. 2015, 277, 1–10. https://doi.org/10.1016/j.cej.2015.04.107.
128. Sanchez, V.C.; Jachak, A.; Hurt, R.H.; Kane, A.B. Biological interactions of graphene-family nanomaterials: An interdisciplinary review. Chem. Res. Toxicol. 2012, 25 (1), 15–34. https://doi.org/10.1021/tx200339h.
129. Ou, L.; Song, B.; Liang, H.; Liu, J.; Feng, X.; Deng, B.; Sun, T.; Shao, L. Toxicity of graphene-family nanoparticles: A general review of the origins and mechanisms. Part. Fibre Toxicol. 2016, 13 (1), 57. https://doi.org/10.1186/s12989-016-0168-y.
130. El-Yamany, N. A.; Mohamed, F. F.; Salaheldin, T. A.; Tohamy, A. A.; El-Mohsen, W. N. A.; Amin, A. S. Graphene oxide nanosheets induced genotoxicity and pulmonary injury in mice. Exp. Toxicol. Pathol. 2017, 69 (6), 383–392.
131. Nirmal, N. K.; Awasthi, K. K.; John, P. J. Effects of NanoGraphene Oxide on Testis, Epididymis and Fertility of Wistar Rats. Basic Clin. Pharmacol. Toxicol. 2017, 121 (3), 202–210.
132. Gao, Y.; Ren, X. M.; Zhang, X. D.; Chen, C. L. Environmental fate and risk of ultraviolet- and visible-light-transformed graphene oxide: A comparative study. Environ. Pollut. 2019, 251, 821–829.
133. Pal, N.; Banerjee, S.; Roy, P.; Pal, K. Reduced graphene oxide and PEG-grafted TEMPO-oxidized cellulose nanocrystal reinforced poly-lactic acid nanocomposite film for biomedical application. Mater. Sci. Eng. C. 2019, 104, 109956.
134. Gawde, K. A.; Sau, S.; Tatiparti, K.; Kashaw, S. K.; Mehrmohammadi, M.; Azmi, A. S.; Iyer, A. K. Paclitaxel and difluorinated curcumin loaded in albumin nanoparticles for targeted synergistic combination therapy of ovarian and cervical cancers. Colloids Surf., B. 2018, 167, 8–19.

6 Concept and Applications of Biomolecular Simulations

Vandana Kardam and Kshatresh Dutta Dubey

6.1 INTRODUCTION

Computer-based molecular modeling has gained a lot of attention from researchers in various fields during the past several years (Van Gunsteren & Berendsen, 1990). Biomolecules such as proteins, nucleic acids, and lipids require a truly multidisciplinary approach to modeling ("Introduction," 2008). Mathematicians analyze and develop suitable numerical models and algorithms, while computer scientists and engineers provide crucial implementational support for running large computer programs on high-speed and extended-communication platforms. Biologists describe the cellular picture, while chemists fill in the atomic and molecular details, and physicists extend these views to the electronic level and the underlying forces. The field is known by many different names, which emphasizes how interdisciplinary it is: computational biology, computational chemistry, *in silico* biology, computational structural biology, computational biophysics, theoretical biophysics, theoretical chemistry, and so on. Computational chemistry is the primary focus of this chapter (Schlick, 2002; Young, 2004).

Computational chemistry is a branch of chemistry that links experimental data with theoretical models by resolving chemical problems with computational methods (Straatsma & McCammon, 1992). Computational chemistry can be very effectively used to solve complex chemical and biological problems. One of the most commonly used constructs in computational chemistry is a model. A model is a simple way of describing and predicting scientific results. With the help of a model, the measurable and immeasurable properties are computed, and they are eventually compared to properties that have been found by experimentation. This comparison with experimental data verifies the accuracy of the computational model employed to investigate the chemical issues (Dutta Dubey et al., 2013). A very popular technique for generating molecular models for a variety of uses is molecular mechanics (Tsai, 2003). However, the molecules are dynamic, undergoing vibrations and rotations continuously. Flexibility and motion are clearly important to the biological functioning of biomolecules (Jakobsson, 2001).

Among biomolecules, proteins are the universal machines in nature. For instance, they serve as building blocks (like collagen in skin, bones, and teeth), transporters (like hemoglobin, which carries oxygen in the blood), reaction catalysts, and nanomachines (like myosin that is at the basis of muscle contraction) (Helfand, 1985; McCammon & Harvey, 1988). For some of these functions, the protein's structure or molecular architecture is sufficient (collagen, for instance), but for most of them, the function is closely related to internal dynamics. In these circumstances, evolution has optimized and fine-tuned the protein to precisely display the type of dynamics necessary for its function (Scheraga et al., 2007). Therefore, understanding protein dynamics is frequently necessary before understanding protein function (van Oss, 1979). Molecular dynamics is important for understanding protein dynamics, particularly because detailed experimental studies are quite difficult ("Introduction," 2002) (Dubey & Ojha, 2013). The molecular dynamics simulation technique (Berendsen & Van Gunsteren, 1984) is the most frequently used to examine the atomic level motion of biomolecules such as proteins, nucleic acid, etc. On the other hand, Quantum mechanics/molecular mechanics (QM/MM) calculations have become an essential tool for studying chemical reactions in complex systems. In QM/MM calculations, the system of interest is divided into two parts: the quantum mechanical region, which describes the chemical reactions and electronic structure

DOI: 10.1201/9781003441328-6

using *ab initio* or density functional theory methods, and the classical molecular mechanical region, which treats the rest of the system using classical force fields (Senn & Thiel, 2009) (Field et al., 1990). MD simulations and QM/MM calculations have proven to be invaluable tools in solving real-world problems in fields such as chemistry, biology, and materials science. For example, in drug discovery, MD simulations can be used to investigate the binding of small molecules to proteins, allowing researchers to design new drugs with improved efficacy and fewer side effects. QM/MM calculations are also used to study the mechanisms of enzymatic reactions, providing insights into how enzymes catalyze chemical reactions and how they can be modified to enhance their activity. In this chapter, we discuss the various protocols used in MD simulations and QM/MM calculations, followed by their applications (Boyd, 2004).

6.2 MOLECULAR DYNAMICS SIMULATIONS

The atoms in biological macromolecules are constantly in motion, and their molecular function and intermolecular interactions depend on the dynamics of the molecules involved. Molecular dynamics is the science of simulating the motion of a system of particles applied to biological macromolecules. The basic idea of molecular simulations is straightforward; a particle-based description of the system under investigation is constructed, and then the system is propagated by either deterministic or probabilistic rules to generate a trajectory describing its evolution over the course of the simulation. The first MD simulation of simple gasses was performed in the late 1950s (Alder & Wainwright, 1957). The first MD simulation of a protein was accomplished in late 1970 (McCammon et al., 1977), and the groundwork that enabled these simulations was recognized by the Nobel Prize in Chemistry in 2013 (Levitt, 2014). MD simulations are now widely employed for a variety of purposes, including dynamic surface exploration of proteins, the study of numerous protein macromolecular complexes (e.g., protein–protein, protein–DNA, and protein–ligand interactions), and lipid systems (Dubey et al., 2013).

MD simulation treats molecules like a ball on a spring. This model can simulate the dynamics of molecules. Molecular modeling aids in the generation of biomolecule structures by providing the geometrical coordinates of biomolecules available as NMR or X-ray data. X-ray crystallographic structures can be easily deduced using computational algorithms and then assigning the *x*-, *y*-, and *z*-coordinates to the molecules based on their geometry (Hollingsworth & Dror, 2018; Dubey & Ojha, 2013). The *ab initio* method, threading, and homology modeling are three major modeling methods.

6.2.1 Basic Principles of Molecular Dynamic Simulations

MD simulation requires the step-by-step numerical solution of the classical equations of the motion of atoms and molecules depending on the time evolution of the system, which can, in the simplest form, be written as (equations 6.1 and 6.2) (Leach, 2001; "Molecular Modeling: Molecular Mechanics," 2002):

$$F_i(t) = m_i r_i(t) \tag{6.1}$$

$$m_i r_i(t) = m_i d^2 r / dt^2 = -\Delta_i [U(r_1, r_2 \ldots r_N] \tag{6.2}$$

where F_i is the force acting on atom *i* at time *t*, m_i is the mass of atom *i*, and a_i is the acceleration of atom *i* at time *t*, *U* is the potential energy surface, which depends on the positions of the *N* particles in the system. Molecular dynamics simulations, in which Newton's equations of motion are solved for the atoms of the system and any surrounding solvent, offer the most accurate and comprehensive information. The key steps in numerically solving the classical equation of motion can be divided into two categories: evaluating energies and forces and propagating atomic positions and velocities

(McCammon & Harvey, 1988). Based on the initial atom coordinates of the system, new positions and velocities can be calculated at time t, and they will be moved to their new positions (Leach, 2001). As a result, a new conformation is formed. The cycle is then repeated for multiple time steps. A collection of energetically accessible conformations produced by this procedure is known as an ensemble.

For computing the potential energy, we must clearly understand its functional form, the force field. The basic principle behind the molecular dynamics simulations begins with the knowledge of force field (Hünenberger & van Gunsteren, 1999). Force field (FF) is the set of mathematical functions/parameters used to characterize the potential energy functions, including bonded and non-bonded potential terms of molecules and atoms (Mayo et al., 1990). These force field functions and parameters are the product of high-level quantum mechanical computations and experimentation. One functional form of such a force field is as follows (equation 6.3):

$$E_{total} = E_b + E_{nb} \quad (6.3)$$

where E_{total} represents the total potential energy and is the sum of bonded energy (E_b) and non-bonded energy (E_{nb}). Figure 6.1 demonstrates the simple visualization of different bonding and nonbonding terms acting on a molecule ("Introduction," 2002).

The bonded energy is associated with the internal degrees of freedom within molecules, and it consists of bond stretching (E_r) (Pearlman & Kollman, 1991), bond angle bending (E_θ), bond torsion (E_Φ), whereas nonbonded energy is an energy associated with the van der Waals interactions (E_{vdW}) and electrostatic interactions (E_{elec}). There are some differences in the forms of energies, depending upon the choice of FF and the parameters of the system of interest. For example, MM2/3 for organic compounds (Grančič et al., 2015) and AMBER (Weiner et al., 1984) or CHARMm (Brooks et al., 1983) for biological macromolecules. The bonded energy terms for most of the FF are the same (equation 6.4):

$$E_r = \sum K_r (r - r_0)^2$$

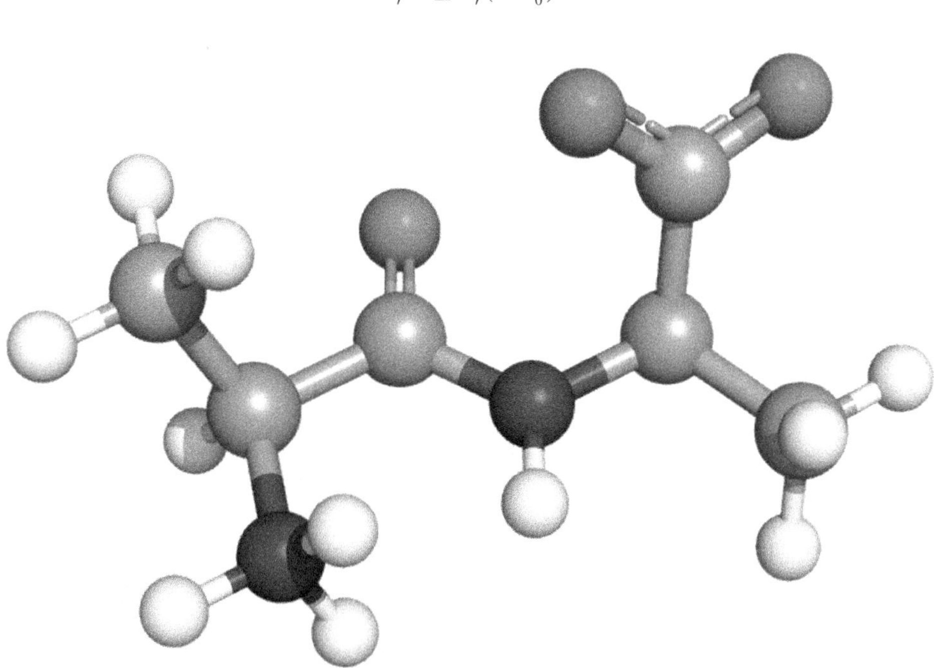

FIGURE 6.1 Different forces acting on a molecule present in a solvent like water.

$$E_\theta = \sum K_\theta(\theta - \theta_0)^2$$

$$E_\phi = \sum K_\Phi[1 + \cos(n\Phi - \Phi_0)] \tag{6.4}$$

where K_r, K_θ, and K_Φ are bond, bond angle, and dihedral angle force constants, respectively, and r_0, θ_0, and Φ_0 are the equilibrium values of bond distance, bond angle, and dihedral angle, respectively. n is the periodicity of the Fourier term. The nonbonded energy (E_{nb}) is composed of three parts (equation [6.5]): (1) a repulsive term preventing atoms from interpenetrating at very short distances; (2) an attractive term accounting for the London dispersion forces between atoms; and (3) an electrostatic term that is attractive or repulsive depending on whether the charges q_i and q_j are of opposite or the same sign (Karplus & Petsko, 1990).

$$E_{nb} = \sum\left\{4\varepsilon_{ij}[(\sigma_{ij}/r)^{12} - (\sigma_{ij}/r)^6 + (q_i q_j)/r]\right\} \tag{6.5}$$

The first two terms combine to give the Lennard–Jones potential (Verlet, 1967), where, ε_{ij} and σ_{ij} represent the atomic (i,j) pair Lennard–Jones well depth and diameter, r is the distance between the centers of the nonbonded atoms or group. The third term represents the electrostatic potential, where $q_i q_j$ is the product of charges present over a pair of atoms, D is the effective dielectric function for the medium, and r is the distance between the two charges. Use of atomic partial charges eliminates the need for a separate term to represent the hydrogen bond interaction. The reason is that, when the positive hydrogen attached to an electronegative atom comes within van der Waals distance of a negative acceptor atom, the coulomb attraction adds to the Lennard–Jones potential and results in a hydrogen bond (Karplus & Petsko, 1990).

The transferability of the functional form and parameters is a significant feature of a force field. Transferability means that the same set of parameters can be employed to model a series of similar molecules rather than having a separate set of parameters for individual molecules in the system (Leach, 2001).

The atomic positions (3N) of every atom in a biomolecular system are a function of potential energy. The motion equation must be solved numerically because the potential energy function is complex in nature, and the equations of motion have no analytical solution. Therefore, the equation of motion needs to be solved numerically (Despiau-Pujo, 2019). This has been accomplished using algorithms. The Verlet algorithm is the most basic and widely used algorithm used for integrating the equation of motion in molecular dynamics. The simplicity and self-starting nature of this algorithm and the ease with which the new positions can be determined from the old ones are its main advantages. It also uses less computer memory, which is also an advantage. To derive it, we start with a Taylor expansion of the coordinate of a particle around time t without using explicit velocities as given (equations 6. 6.6 and 7) (Badar et al., 2022):

$$r(t + \delta t) = r(t) + v(t)\delta(t) + \frac{1}{2}a(t)\delta t^2 + \cdots \tag{6.6}$$

$$r(t - \delta t) = r(t) - v(t)\delta(t) + \frac{1}{2}a(t)\delta t^2 + \cdots \tag{6.7}$$

Adding these two equations, we get (equations 6.8 and 6.9):

$$r(t + \delta t) = 2r(t) - r(t - \delta t) + \left(\frac{F}{m}\right)\delta t^2 + \cdots \tag{6.8}$$

or

$$r(t+\delta t)=2r(t)-r(t-\delta t)-\left(\frac{\Delta U}{m}\right)\delta t^2+\cdots \quad (6.9)$$

where r, v, and a are position, velocity (first derivative with respect to time), and acceleration (second derivative with respect to time). U is a function of the coordinates and can be obtained by force fields or potential-energy function. The Verlet algorithm associates the lack of an explicit velocity term and in reality, the velocities are not available until the positions have been calculated at the subsequent step.

However, during the summation of the two equations in the Verlet algorithm, the velocity is canceled. Therefore, the Velocity Verlet was developed to overcome the associated problem related to the Verlet algorithm (Martys & Mountain, 1999). It demonstrates better application of the fundamental Verlet algorithm just discussed. The positions, velocities, and accelerations in Velocity Verlet can be determined (equations 6.10 and 6.11):

$$r(t+\delta t)=r(t)+v(t)\delta(t)+\frac{1}{2}a(t)\delta t^2+\cdots \quad (6.10)$$

$$r(t+\delta t)=v(t)+\frac{1}{2}[a(t)+a(t+\delta t)]\delta \quad (6.11)$$

6.2.2 Periodic Boundary Conditions

Since the presence of atoms/molecules along with the water molecules makes the entire system physically fluid, there is a great possibility that a few particles will flow out of the box due to its dynamic nature. We can solve this problem with a simple trick: create a box replica that completely covers the original box on all sides. Now whenever a particle attempts to exit the central box, another enters from the adjacent replica at the same speed to maintain system balance. This phenomenon is known as periodic boundary condition (Berkowitz & McCammon, 1982). The appropriate treatment of boundaries and boundary effects is crucial to simulation methods as it enables "macroscopic" properties to be calculated using a relatively small number of particles (Leach, 2001). A 2D representation of periodic boundary conditions is shown in Figure 6.2.

6.2.3 Simulation Protocols

Given a potential energy function, there are various ways to approach the dynamics. Structural and dynamic properties for a simple homogeneous system, such as a box of water molecules with periodic boundary conditions, can be determined only in simulations of a few picoseconds (Stillinger & Rahman, 1974). Homogeneous systems such as proteins require longer simulations. Modern computers enable simulations of up to microseconds, long enough to fully characterize the librations of small protein groups, identifying the dominant contributions to atomic fluctuations and the folding/misfolding in protein systems.

Though each system under study will have its own difficulties and factors to take into account, these are the general steps involved in running a molecular dynamics simulation (Karplus & Petsko, 1990):

I. System preparation and parametrization
II. Energy minimization
III. Assigning the velocities
IV. Equilibration
V. Production run

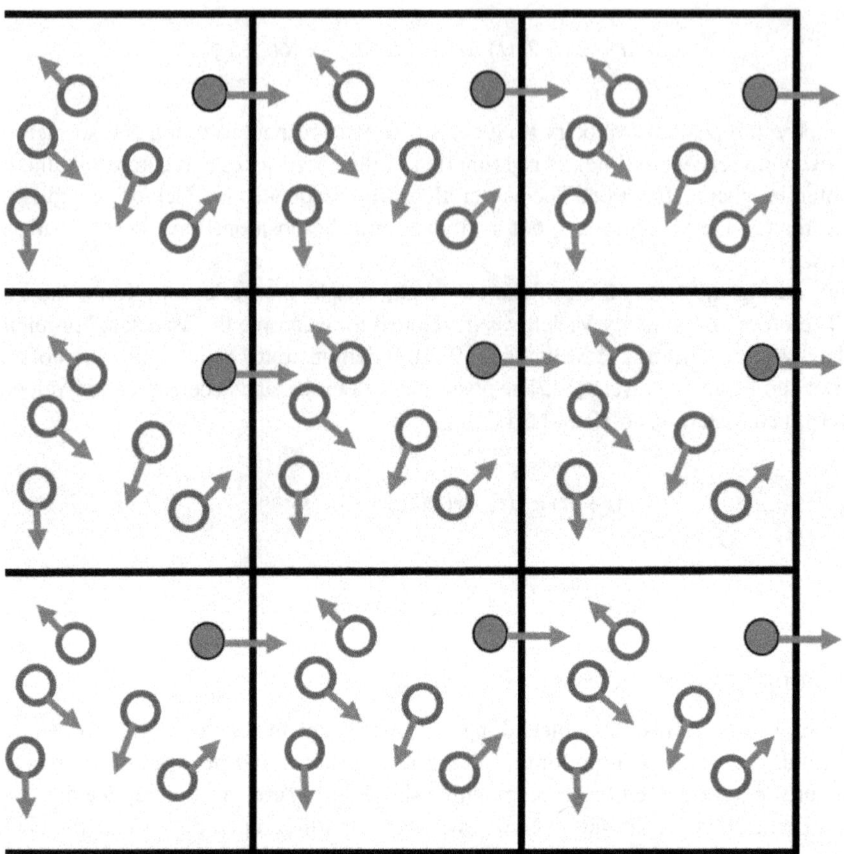

FIGURE 6.2 Periodic boundary conditions in 2D.

Prior to performing any of these steps, an appropriate amount of deliberation must be devoted to clearly defining the system and determining the appropriate simulation techniques. Here, we briefly discuss the steps involved in molecular dynamics simulations:

I. **System preparation and parametrization:** This step focuses on preparing the initial state required for the molecular simulations as an input, including building an initial structure, solvation, implementing the appropriate force field to the system, etc. An initial set of atomic coordinates and velocities is necessary to start a dynamic simulation. The coordinates can be determined from X-ray crystallographic or NMR structure data or by building a model. Since, in X-ray crystallography, it is not possible to determine the position of hydrogen atoms, it is essential to rectify this problem by adding missing hydrogen atoms into the initial structure obtained from X-ray crystallography. The subsequent step of parametrization involves the generation of input files used to simulate our system. For this, the choice of the appropriate force field is a crucial point. Standard amino acids and nucleic acids are frequently anticipated to be present in the libraries of molecular dynamics software (AMBER, CHARMM, GROMACS, NAMD, etc.). But occasionally one must manually create the library for the modeled drugs. Running *ab initio* calculations to accurately estimate the bonded and nonbonded parameters of force field equations is a good practice. The next step is to solvate the simulation box in an appropriate water box. Many explicit water models are available, but the most well liked are TIP3P, TIP4P, TIP5P, SPC, and SPC/E. These water models are established by quantum-mechanics-driven molecular mechanics and confirmed by experimental techniques. The precise nature and complexity of molecule hydration are mimicked by water models, including the orientation of solvent dipoles and efficient electrostatic shielding, subtle hydrogen bond network rearrangements, etc. To avoid polarization of the simulation ensemble, the total charge of the system should be kept zero by replacing certain solvent molecules with ions.

II. **Energy minimization:** This step entails determining the global minimum energy with respect to the placement of side chain atoms, which depicts the geometry of the specific atomic arrangements in which the net attractive force on each atom reaches its maximum. Different minimization algorithms are used to compute the minimum energy, but the most widely used methods are the steepest descent and conjugate gradient. The steepest descent minimizer approaches the energy minimum by using the first derivative of the energy function computed numerically. After one of the atoms has been slightly moved in one of the coordinate system's directions, the energy is calculated again for the initial geometry. Once all atoms have been relocated to new locations downhill on the energy surface, this process is repeated for each atom. The steepest descent method is frequently used for structures that are far from the minimum because the optimization process is slow near the minimum (Leach, 2001). The reversal of the advancement made in a previous iteration can be avoided in the conjugate gradient method. The gradient is calculated for each minimization step and is used as additional data to determine the new minimization direction vector. As a result, the direction toward the minimum is continuously refined by each subsequent step. The conjugate gradient is preferred for larger systems despite requiring more computational work and storage space than the steepest descent. Figure 6.3 displays the crucial steps just discussed in the MD simulations.

Solvation is a crucial process since numerous biological reactions take place in water-based solutions, and the impact of solvation on molecular conformation, electronic characteristics, binding energies, and other factors is crucial (Rosenthal et al., 1994). There are two main types of MD simulations: explicit and implicit (Scheraga, 1979; Stouten et al., 1993). All atoms in the system are explicitly modeled in explicit simulations, and

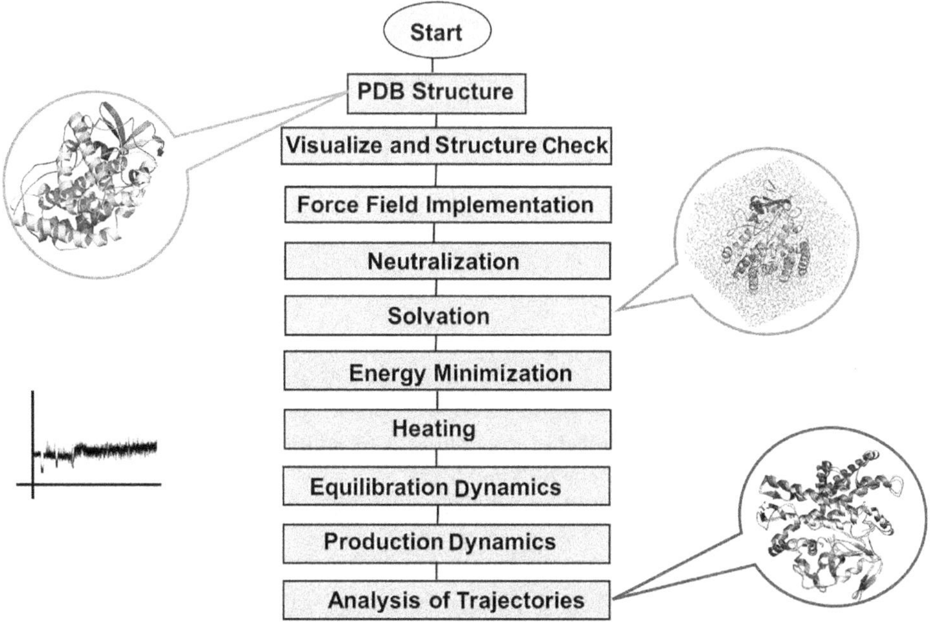

FIGURE 6.3 Displaying key steps of MD simulations of proteins.

their interactions are calculated using a force field. This approach can provide highly accurate results, but it requires a large amount of computational power and is limited by the size of the system that can be simulated. Implicit simulations, on the other hand, treat some or all of the solvent molecules as a continuum, simplifying the computational requirements(Roux & Simonson, 1999). This approach is often used to study large biomolecular systems, such as proteins and nucleic acids, which are difficult to simulate using explicit methods due to their size and complexity. Implicit simulations are also used to investigate properties such as protein–ligand binding and drug design. Both explicit and implicit simulations have their advantages and limitations, and the choice of which method to use depends on the specific research question being addressed. However, with advances in computational power and methodology, MD simulations continue to be an important tool for understanding the behavior of molecules and materials at the atomic level.

III. **Assigning the velocities:** To begin MD simulations, we need not just positions but also velocities. Therefore, starting velocities must be assigned, usually this is done by giving atoms random initial velocities in such a way that produces the correct Maxwell–Boltzmann distribution at the desired temperature (Braun et al., n.d.). A Maxwell–Boltzmann distribution is a Gaussian distribution, which can be obtained using a random number generator. Most random number generators are designed to produce random numbers that are uniform in the range of 0 to 1. However, the random assignment process may accidentally assign large velocities to a localized cluster of atoms, which makes the simulation unsteady. To resolve this problem, the standard protocol is to begin the simulation with the heat-up phase, where initial velocity assignments are made at a low temperature, which is then gradually raised to allow for dynamic relaxation. Until the simulation reaches the desired temperature, this gradual heating continues (Dubey & Ojha, 2013). The increment in velocities can be performed by:

a. **Reassigning the new velocities from the Maxwellian distribution (equation 6.12):**

$$P(v)\mathrm{d}v = \left(\frac{m}{2\pi k_B T}\right)^{\frac{1}{2}} \mathrm{e}\left[\frac{-mv^2}{2K_B T}\right]\mathrm{d}v \tag{6.12}$$

Moreover, the momentum of the center of mass of the simulation box is conserved by Newtonian dynamics, which is typically set to zero by removing the center-of-mass velocity from all particles after random assignment, preventing the simulation box from drifting (Dubey & Ojha, 2013).

b. **Scaling the velocities by uniform factor:** The associated temperature change can be calculated by multiplying the velocities by a factor and the temperature at time *t*, which is *T*(*t*) (equation 6.13).

$$\Delta T = \frac{1}{2}\sum_{i=1} 2\frac{m_i(\lambda v_i)^2}{N_{df}k_B} - \frac{1}{2}\sum_{i=1}\frac{2m_i v_i^2}{N_{df}k_B} \tag{6.13}$$

The simplest way to control the temperature is to multiply the velocities at each time step by the factor λ (equations 6.14 and 6.15):

$$\Delta T = \left(\lambda^2 - 1\right)T(t) \tag{6.14}$$

$$\lambda = \sqrt{\frac{T_0}{T(t)}} \tag{6.15}$$

where T is the current temperature as determined by the kinetic energy, and T_0 is the desired temperature. Having set up the system and assigning velocities, the simulations can commence (Dubey & Ojha, 2013).

IV. **Equilibration:** The first step of molecular dynamics simulation is the equilibration phase, which aims to bring the system to equilibrium from the starting configuration. The correct distribution of velocities typically necessitates adding or removing heat from the system by the thermostat as the system gets closer to the proper division of kinetic and potential energies. Therefore, it is suggested that a thermostated simulation is run before the desired production simulation. Langevin (Glatzel & Schilling, 2021; Grønbech-Jensen & Farago, 2014; Quigley & Probert, 2004) and Berendsen thermostats (Yang et al., 2018) are commonly used in molecular dynamics (MD) simulation studies to control the temperature of the system being simulated. The Langevin thermostat is a stochastic thermostat that simulates the effect of a heat bath on the system by introducing a frictional force and a random force to each particle. This results in the particles' experiencing both a deterministic and a random component to their motion, which helps to maintain the system at a constant temperature. On the other hand, the Berendsen thermostat is a deterministic thermostat that applies a scaling factor to the velocities of the system's particles to control the temperature. This scaling factor is calculated based on the difference between the current temperature and the desired temperature of the system. Regardless of which thermostat is used, it is important to carefully select the temperature control parameters and to ensure that the simulation results are validated against experimental data or other sources of information. One advantage of the Langevin thermostat is that it can be used to simulate systems at a range of temperatures, from very low to very high, without requiring the use of multiple thermostats. Additionally, the Langevin thermostat can be used to simulate systems with a wide range of time scales, from fast reactions to slow conformational changes.

However, one disadvantage of the Langevin thermostat is that it can introduce additional noise into the simulation, which can make it difficult to distinguish between true fluctuations in the system and noise from the thermostat. Additionally, the Langevin thermostat may not accurately capture certain types of dynamics, such as those involving long-range interactions.

On the other hand, the Berendsen thermostat is computationally efficient and can be used to maintain a constant temperature throughout a simulation, which can be particularly useful for long simulations or simulations of large systems. However, the deterministic nature of the Berendsen thermostat means that it may not accurately capture the true dynamics of the system, particularly at short time scales.

During this stage, various parameters, such as energy, density, pressure, temperature, etc., are being monitored together with the actual configurations (Braun et al., 2019). When these parameters achieve stable values, then the production phase can commence. One of the challenges in achieving equilibration is ensuring that the system's important properties are not changing systematically with simulation time. Even at the equilibration phase, slow fluctuations may still occur in the system, particularly if it possesses slow internal degrees of freedom. However, the essential properties of the system should no longer exhibit systematic trends from their starting structure. In order to preserve the initial configuration of the system during equilibration, constrained dynamics is a good practice. The procedure for the end of equilibration is somewhat different depending on the target ensemble for production (Hoover, 1985; Nosé, 1984a, 1984b). If the goal is a simulation in the NVT ensemble at a specific density, equilibration should be carried out in the NPT (Feller et al., 1995) ensemble. In this scenario, the system could be scaled to the desired average volume prior to running the production simulation.

V. **Production run:** Upon completion of the equilibration phase, the data acquisition process for subsequent analysis, commonly referred to as production, may commence. In this phase, the primary contrast with equilibration is the retention and analysis of the generated data. As a prerequisite for production, the appropriate equilibration for the intended production ensemble must always be carried out. It is during this stage that the system is simulated for a longer period, ranging from nanoseconds to microseconds, to capture the dynamics of the molecules accurately. In this step, the system is allowed to evolve under the influence of the force field and the specified ensemble conditions. The production run involves a series of technical steps that must be carefully executed to obtain meaningful results. First, the simulation box dimensions and initial conditions are set up using the output from the previous equilibration steps. Then the simulation parameters, such as the integration time step, temperature, pressure, and force field, are selected based on the specific system and the scientific questions being addressed. Next, the simulation is initiated, and the system is propagated through time using numerical integration algorithms such as the Verlet algorithm. During the simulation, the system's coordinates, velocities, and energies are recorded at regular intervals to generate a trajectory of the system's evolution. Several techniques are employed during the production run to ensure the simulation's accuracy and stability. These include periodic boundary conditions to simulate an infinite system, long-range electrostatics and van der Waals interactions, and energy conservation algorithms to maintain the system's total energy. The production run stage addresses the scientific questions, and the data obtained can be analyzed to gain insight into the system's properties and behavior. The simulation results can be analyzed using various techniques, such as trajectory visualization, thermodynamic properties calculation, and experimental data comparison. In summary, the production run step of molecular dynamics simulations is a critical stage that requires careful execution of technical steps to obtain accurate and meaningful results. It allows researchers to capture the dynamic behavior of molecules over a longer time scale and gain insight into the system's properties and behavior.

6.2.4 Applications of MD Simulations

I. **Protein folding:** Protein folding is a complex process that involves the transition of a protein from its unfolded state to its native, biologically active conformation. MD simulations can be used to study the folding pathways of proteins and the stability of protein structures (Daggett & Levitt, 1994). By simulating the motion of protein molecules, researchers can observe how the protein folds and refolds over time (Lazaridis & Karplus, 1997). This information can be used to identify the most stable protein conformations and to understand the role of different protein domains and interactions in the folding process (Eisenberg & McLachlan, 1986). One of the main advantages of MD simulations in protein folding studies is that they can provide atomic-level detail about the folding process (Karplus & Šali, 1995). This allows researchers to observe the formation of specific secondary structure elements and of key interactions between amino acid residues (Lindorff-Larsen et al., 2011). In addition, MD simulations can be used to study the effect of environmental factors, such as pH (Dubey et al., 2011) and temperature, on the folding process.

II. **Conformational studies:** MD simulations can be used to study the conformational changes of proteins and other biomolecules. By simulating the motion of molecules over time, researchers can observe how the molecule transitions through different conformational states. This information can be used to identify the most stable conformations of the molecule and to understand the molecular mechanisms underlying biological processes such as protein–protein interactions and enzyme catalysis. MD simulations can also be used to study the effects of mutations and post-translational modifications on protein structure and function. By simulating the motion of mutant or modified protein molecules, researchers can observe how the protein structure and stability are affected. This information can be used to understand the molecular basis of diseases and to design new drugs that can target specific mutant proteins.

III. **Ligand binding:** MD simulations can be used to study the binding of ligands to protein targets. By simulating the motion of the ligand and protein molecules, researchers can predict the binding affinity and specificity of the ligand for the protein target. This information can be used to design new drug molecules that can bind more effectively to the protein target (Ljungberg et al., 2001). MD simulations can also be used to study the molecular mechanisms underlying ligand binding. For example, researchers can observe the formation of specific interactions between the ligand and protein residues and the effect of ligand binding on protein conformation and stability. This information can be used to design new drugs that can target specific protein–ligand interactions and to understand the molecular basis of drug resistance. Overall, MD simulations are a powerful tool for studying the behavior of molecular systems in computational chemistry. They provide atomic-level detail about the structure and function of proteins and other biomolecules, and they can be used to design new drugs, understand disease mechanisms, and predict the behavior of molecules in different environments.

IV. **Drug design:** Molecular dynamics (MD) simulations have become an essential tool in the drug design process (Kerrigan, 2013). MD simulations can provide information about the dynamic behavior of a drug molecule and its interactions with its target (Wu et al., 2022). This information can be used to optimize drug design and improve its efficacy and safety. In drug design, MD simulations can be used to study the interaction of a drug molecule with its target (Yue et al., 2023). This can include the binding of the drug to a protein target or the interaction of the drug with a lipid bilayer in a cell membrane. MD simulations can provide insights into the binding affinity of the drug, the stability of the drug-target complex, and the dynamics of the drug–target interaction. One of the challenges in drug design is the optimization of drug potency and selectivity. MD simulations can be used to study the binding of a drug molecule to its target at the atomic level, allowing researchers to identify key interactions between the drug and the target. This information can be used to modify the structure of the drug molecule to optimize its binding affinity and selectivity. MD simulations can also be used to study the effects of mutations in the target protein on drug binding. This can be important in understanding the mechanism of drug resistance and in designing drugs that are effective against resistant strains. MD simulations can also be used in conjunction with other computational methods, such as quantum mechanics/molecular mechanics. In conclusion, MD simulations are a powerful tool in drug design, providing insights into the interaction of a drug molecule with its target at the atomic level. MD simulations can be used to optimize the design of drug molecules, improve their efficacy and safety, and understand the mechanisms of drug resistance.

V. **Free energy calculations:** MD simulations can be used to calculate the free energy of molecular systems (Åqvist et al., 2002; Murcko, 1995). Free energy calculations are important for understanding the thermodynamics of biological processes and for predicting the behavior of molecules in different environments (Florián et al., 2000; Gilson et al., 1997). By simulating the motion of molecules in different environments and under different conditions, researchers can calculate the energy required to transition between different states of the system. This information can be used to predict the thermodynamic properties of the system, such as its stability and reactivity, and to identify the most favorable pathways for chemical reactions and other processes (Gilson & Zhou, 2007). There are several different methods for performing free energy calculations using

MD simulations (Lazaridis et al., 2002). These methods allow researchers to calculate the free energy differences between different states of the system, such as the binding of a ligand to a protein or the transition of a molecule through different conformational states. Two commonly used methods for calculating free energies are the linear interaction energy (LIE) method and the molecular mechanics/Poisson–Boltzmann surface area (MMPBSA) method.

a. **Linear interaction energy (LIE):** The LIE method is a popular approach for calculating the free energy of protein–ligand binding (Aqvist & Marelius, 2001). It is a linearized form of the energy function and is based on the assumption that the nonbonded interaction energy between the protein and ligand can be divided into coulombic and a van der Waals contributions (Alam & Naik, 2009). The LIE method calculates the free energy of binding (equation 6.16):

$$\Delta G\_\text{bind} = -\alpha \Delta E\_\text{vdw} - \beta \Delta E\_\text{coul} - \gamma \Delta G\_\text{solv} \tag{6.16}$$

where α, β, and γ are empirical coefficients, ΔE_vdw is the change in the van der Waals energy upon binding, ΔE_coul is the change in the coulombic energy upon binding, and ΔG_solv is the solvation free energy of the ligand. The LIE method is computationally efficient and has been used to study a wide range of protein–ligand systems (Hansson et al., 1998; Singh et al., 2005). However, it relies on empirical parameters and may not be accurate for all systems.

b. **Molecular mechanics/Poisson–Boltzmann surface area (MMPBSA):** The MMPBSA method is a more complex approach that combines molecular mechanics calculations with continuum electrostatics and solvation models to calculate the free energy of protein–ligand binding (Dominy & Brooks, 1999; Edinger et al., 1997). The method is based on the decomposition of the total free energy of binding into several terms:

$$\Delta G\_\text{bind} = \Delta E\_\text{MM} + \Delta G\_\text{sol} - T\Delta S \tag{6.17}$$

where ΔE_MM is the change in the molecular mechanics energy upon binding, ΔG_sol is the solvation free energy of the system, and $T\Delta S$ is the entropic contribution to the free energy. The MMPBSA method has been shown to be accurate for a wide range of protein–ligand systems (Pearlman, 2005; Perdih et al., 2009), but it is computationally demanding and requires careful parameterization of the models used to calculate the solvation and entropy terms (Gohlke & Case, 2004; Jayaram et al., 1998). In conclusion, both the LIE and MMPBSA methods are useful for calculating free energies in various systems. The choice of method depends on the specific system being studied, the available computational resources, and the accuracy required. Other methods to calculate free energy in MD simulation methods are as follows:

c. **Thermodynamic integration (TI):** The TI method is a rigorous approach for calculating free energy changes between two states, such as the bound and unbound states of a protein–ligand complex. The method involves perturbing the system from the initial state to the final state via a series of intermediate states and calculating the free energy change associated with each step (Lee et al., 2017). The total free energy change is then obtained by integrating over all the intermediate states. TI is computationally expensive, but it is considered one of the most accurate methods for calculating free energies.

d. **Free energy perturbation (FEP):** The FEP method is a variant of the TI method that is used to calculate the free energy change associated with a small perturbation in the system, such as the introduction of a mutation or a small ligand modification (Brandsdal & Smalås, 2000). The method involves perturbing the system gradually from the initial state to the final state and then calculating the free energy change at each step. The total free energy change is then obtained by integrating over all the intermediate states. FEP is also computationally expensive, but it is a powerful tool for studying the effects of small changes in the system.

e. **Alchemical free energy simulations:** Alchemical free energy simulations involve the transformation of one molecule into another molecule by gradually changing the nonbonded interactions between the atoms in the system. The free energy change associated with the transformation is calculated using statistical mechanics and thermodynamic integration methods. This approach is computationally expensive but can be used to study a wide range of transformations, such as ligand binding, conformational changes, and protein–protein interactions. These are just a few examples of the many methods available for calculating binding free energies in MD simulations. The choice of method depends on the specific system being studied, the accuracy required, and the available computational resources.

6.2.5 Challenges and Limitations in Molecular Dynamics Simulations

The classical MD simulations possess several challenges (Durrant & McCammon, 2011), which are addressed next.

6.2.5.1 Force Field Refinement and Computational Demands

The accuracy of force fields used in molecular dynamics simulations requires further refinement. Additionally, the high computational demands of these simulations limit routine simulations to durations of a microsecond or less, resulting in inadequate sampling of conformational states.

6.2.5.2 Quantum Effects and Approximations

Classical force fields employed in molecular dynamics simulations often approximate quantum effects, especially in systems involving transition metal atoms. Addressing this limitation involves incorporating quantum mechanical calculations to improve the accuracy of the simulations.

6.2.5.3 Incorporating Quantum Mechanics

To overcome the challenge of quantum effects, researchers have developed a hybrid approach where specific regions of interest are simulated using quantum mechanics, while the rest of the system is approximated using molecular dynamics. This technique has been successfully applied in studying catalytic mechanisms and other important molecular motions.

6.2.5.4 Neglecting Electronic Polarization

In classical molecular dynamics simulations, the dynamic nature of electronic polarization, which influences the flow of electrons among chemically bonded atoms, is often ignored. This neglect fails to capture the true behavior of atoms in their environments. Developing widely accepted polarizable force fields to address this limitation has proven challenging.

6.2.5.5 Time Scale Limitations

Molecular dynamics studies are typically limited to short time scales. Consequently, long time scale processes and conformational changes relevant to drug binding and other biochemical processes are difficult to explore adequately. Current simulations are often limited to millionths or billionths of a second.

6.2.5.6 Overcoming Time Scale Limitations

Researchers have proposed several methods to overcome the time scale limitations of molecular dynamics simulations. Accelerated molecular dynamics (aMD) reduces energy barriers artificially, allowing proteins to transition between conformations that are inaccessible with conventional molecular dynamics. Utilizing specialized hardware, such as graphics processing units (GPUs) and dedicated processors, has also shown promise in extending simulation time scales and capturing important events such as protein folding, unfolding, and drug binding.

So far in this chapter, we have explained that molecular dynamics (MD) simulations are powerful tools used to study the behavior of molecules and materials at the atomic level. However, in order to accurately capture the behavior of systems containing certain chemical reactions, electronic processes, or catalytic mechanisms, traditional MD simulations may not be sufficient. In such cases, quantum mechanical/molecular mechanical (QM/MM) calculations are employed.

6.3 QM/MM CALCULATIONS

The pioneering work of Warshel and Levitt in 1976 was the genesis of the QM/MM (quantum mechanics/molecular mechanics) era. They introduced the concept of QM/MM and presented a method that included features that are now considered essential in this field (Ryde, 2016; Senn & Thiel, 2007b). They applied this method to an enzymatic reaction, which laid the foundation for subsequent developments.

The QM/MM approach has been combined with other computational methods for biomolecular systems with varying emphasis, as well as application surveys (Lin & Truhlar, 2007; Senn & Thiel, 2007a, 2007b; Vreven & Morokuma, 2006). The QM/MM approach is now an established and valuable tool not only for modeling biomolecular systems but also for investigating inorganic/organometallic and solid-state systems, as well as studying processes in explicit solvent. In subsequent sections, we introduce the basic methodological features of the QM/MM approach, covering topics such as the QM/MM partitioning, the selection of appropriate QM and MM methods, the treatment of the QM–MM interactions and the QM/MM boundary region and finally concluding with the various types of applications where the QM/MM approach has been particularly useful.

6.3.1 General Overview of QM/MM

6.3.1.1 QM/MM Partitioning

Quantum mechanics/molecular mechanics is a computational method used in the simulation of chemical reactions that involve large systems, such as enzymes or biomolecules. The method divides the system into two regions: the inner region (I) and the outer region (O) (Watanabe et al., 2014). The inner region is treated quantum-mechanically, while the outer region is described using a force field. Due to the strong interactions between the QM and MM regions, the total energy of the system cannot be simply written as the sum of the energies of the subsystems. Therefore, coupling terms have to be considered, and special precautions need to be taken at the boundary between the subsystems, especially if it cuts through covalent bonds. This region is referred to as the boundary region, which may contain additional atoms (link atoms) that cap the QM subsystem and are not part of the entire system. Depending on the type of QM/MM scheme, this region may also consist of atoms with special features that appear in both the QM and MM calculations. In fixed QM–MM partitioning, the boundary between the QM and MM regions is defined once and for all at the start of the calculation. For QM/MM schemes that allow the boundary to change during the course of a simulation, suitable for processes with shifting active region(s) such as defect propagation in materials or solvated ions, adaptive partitioning or "hot spot" methods are used.

6.3.1.2 Choice of QM and MM Methods

The QM/MM formalism can accommodate almost any combination of QM and MM methods. The choice of the QM method follows the same criteria as in pure QM (Lowe & Peterson, 2011) studies and can vary from semiempirical to high-level *ab initio* electron correlation methods. For electronic or polarized embedding, the QM code must be able to perform the self-consistent field (SCF) treatment in the presence of the external point-charge field that represents the MM charge model. In practice, many current biomolecular QM/MM applications use density-functional theory (DFT) (Bagayoko, 2014; Hohenberg & Kohn, 1964; Obot et al., 2015; Orio et al., 2009) as the QM method due to its favorable computational-effort/accuracy ratio. Semiempirical QM methods (Heimdal & Ryde, 2012; Luque et al., 2000; Mlynsky et al., 2014) have been most popular traditionally, and they remain important for QM/MM molecular dynamics. The SCC-DFTB (self-consistent charge density–functional tight-binding) method (König et al., 2005), which is DFT based, is increasingly being applied in biomolecular QM/MM studies. High-level *ab initio* methods, such as those based on Moller–Plesset perturbation theory or coupled-cluster theory, are at the high end in terms of accuracy and effort. The recent development of linear scaling local correlation methods has significantly extended the size of systems that can be treated with such methods, up to several tens of atoms. Although not an electronic structure QM method in the usual sense, the empirical valence bond (EVB) method has been applied with much success, notably by Warshel and collaborators (Kamerlin & Warshel, 2011; Warshel & Florián, 2002; Warshel & Weiss, 1980), to model the influence of the solvent and protein environment on reactions. For the choice of MM method, several biomolecular force fields are available, including AMBER (Götz et al., 2014), CHARMM, GROMOS, and OPLS-AA.

6.3.1.2.1 *Classical MM Energy Expression*

The classical potential energy function, known as the force field, for the MM region (molecular mechanics) contains bonded terms (such as bond stretching, angle bending, torsions, and improper torsions), Lennard–Jones-type van der Waals terms, and coulomb interaction terms between rigid point charges, unless otherwise stated. A simple example of this type of MM energy expression is presented in equation (6.1), in which different symbols and constants are used to represent the bond lengths, angles, torsions, force constants, and Lennard–Jones parameters.

6.3.1.2.2 *QM/MM Energy Expression*

6.3.1.2.2.1 *Subtractive and Additive QM/MM Schemes*

Two types of QM/MM (quantum mechanics/molecular mechanics) schemes are used in computational chemistry: subtractive and additive. In subtractive QM/MM schemes (see Figure 6.4a), the entire system undergoes an MM calculation, followed by a QM calculation on the inner subsystem and another MM calculation on the same inner subsystem. Finally, the QM/MM energy

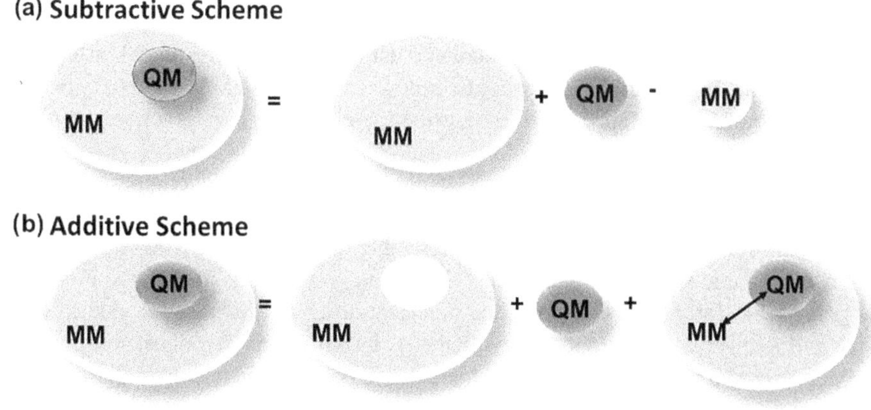

FIGURE 6.4 (a) Subtractive scheme, (b) additive scheme.

of the entire system is calculated by summing up the results from the MM and QM calculations and subtracting the results from the second MM calculation (see equation [6.18]). This approach corrects for the artifacts caused by link atoms, which connect the QM and MM regions. However, it requires a complete set of MM parameters for the inner subsystem, which can be difficult to obtain. Moreover,

$$E_{QM/MM}^{sub} = E_{MM} + E_{QM} - E_{MM} \tag{6.18}$$

the coupling between the subsystems is handled entirely at the MM level, which can cause problems, especially with the electrostatic interaction.

Additive QM/MM schemes (see Figure 6.4b), on the other hand, calculate the MM subsystem and QM subsystem separately. Then they add the contributions of both systems, along with an explicit coupling term between them. The QM subsystem is treated at the QM level, while the MM subsystem is treated at the MM level. Most QM/MM schemes used in current research are of the additive type (Brandt & Jacob, 2022). The coupling term between the QM and MM subsystems includes bonded, van der Waals, and electrostatic interactions between QM and MM atoms (see equation [6.19]).

$$E_{QM/MM}^{add} = E_{MM} + E_{QM} + E_{QM/MM} \tag{6.19}$$

- **Electrostatic QM–MM interactions:** The electrostatic QM–MM interaction refers to the interaction between the charge density of the quantum mechanical (QM) region and the charge model used in the molecular mechanical (MM) region (Sneskov et al., 2011). The interaction can be handled using different approaches, characterized by the extent of mutual polarization, and classified as mechanical embedding, electrostatic embedding, and polarized embedding (Roßbach & Ochsenfeld, 2017). In mechanical embedding schemes (Zhang, 2013), the QM and MM regions are mechanically coupled, and the forces between the QM and MM regions are treated using classical mechanics. The MM region is treated using a force field, while the QM region is treated using quantum mechanics. The mechanical coupling between the QM and MM regions is achieved by adding additional force terms to the MM force field, which represent the interaction between the QM and MM regions.

  In the electrostatic embedding scheme, the QM and MM regions are separated by a boundary, and the electrostatic interactions between the QM and MM regions are treated using a classical electrostatic potential (Kästner et al., 2007; Zhang, 2013). The MM region is treated using a force field, while the QM region is treated using quantum mechanics. The electrostatic potential is usually calculated using a point charge model or a multipole expansion. This embedding scheme is also known as the QM/MM boundary approach, where a coulombic potential is used to describe the electrostatic interactions between the QM and MM regions.

  In the polarized embedding scheme (Reinholdt et al., 2021; Thellamurege & Hirao, 2014), the QM and MM regions are coupled, and the electrostatic interactions between the QM and MM regions are treated using a polarizable force field. The MM region is treated using a polarizable force field, while the QM region is treated using quantum mechanics. The polarizable force field considers the polarization of the MM region due to the QM region and vice versa. This embedding scheme is also known as the polarizable embedding approach, where a polarizable force field is used to describe the electrostatic interactions between the QM and MM regions.

  The accurate description of electrostatic forces on the QM subsystem arising from the environment is essential for the realistic modeling of biomolecules. Electrostatic embedding is the most popular embedding scheme for biomolecular applications due to the readily available MM atomic partial charges, and the MM charge model's inclusion in the QM Hamiltonian is efficient. However, this approach is computationally demanding, and special care is required at the QM–MM boundary to avoid overpolarization. There is ongoing research to develop polarizable biomolecular force fields that can be used in polarized embedding QM–MM calculations.

- **Other nonbonded and bonded QM/MM interactions:** In the realm of quantum mechanics/molecular mechanics (QM/MM) interactions, there are more than just electrostatic interactions to consider. Nonbonded and bonded contributions also play a role in the QM/MM coupling term. The treatment of nonbonded interactions, such as van der Waals interactions, is relatively straightforward at the MM level (Sneskov et al., 2011). The commonly used Lennard–Jones potential is used to describe van der Waals interactions, and suitable parameters are required for the QM atoms in the inner region. However, finding the correct Lennard–Jones parameters can be challenging, as certain QM atoms may not be covered by the force field's atom types and assignment rules. Additionally, the parameters may need to be changed during a reaction, which can further complicate matters.

  Fortunately, the short-range nature of van der Waals interactions means that only QM atoms closest to the boundary significantly contribute to the interaction, reducing the potential for error. To further minimize error, researchers have re-optimized the QM Lennard–Jones parameters against QM data for small amino acid models. This approach yielded Lennard–Jones radii that were slightly larger than those of the underlying force field, which compensated for the overpolarization at the boundary.

  In terms of bonded interactions, such as bond stretching, angle bending, and torsional interactions, standard MM parameter sets are typically used and complemented with additional bonded terms as necessary. While there are some challenges in finding the appropriate parameters for nonbonded and bonded interactions, these complications can be overcome with pragmatic solutions. Despite the challenges, the importance of accurately modeling nonbonded and bonded QM/MM interactions cannot be overstated, as they are critical in predicting the behavior of complex molecular systems.

  This section discusses the treatment of covalent bonds that cross the boundary between the quantum mechanical (QM) and molecular mechanical (MM) subsystems. This is an important issue in QM/MM calculations because it is often necessary to cut through a covalent bond in order to separate the QM and MM regions. In this section, we introduce some labeling conventions that are used to designate the QM and MM atoms that are directly connected and discuss three key issues that arise when cutting covalent bonds: capping the dangling bond of the QM atom, preventing overpolarization of the QM density, and avoiding double-counting of interactions. We also discuss the three main classes of boundary schemes used to address these issues.

- **Labeling conventions:** To facilitate discussion, we use some labeling conventions to identify the QM and MM atoms that are directly connected by the covalent bond that crosses the QM–MM boundary. The QM atom directly connected to the MM atom is labeled QM^a, and the MM atom directly connected to the QM^a atom is labeled MM^a. These atoms are sometimes referred to as boundary, frontier, or junction atoms. The MM atoms that are directly bonded to MM^a are labeled MM^b, and the next shell of MM atoms is labeled MM^c, and so on, following the molecular graph outward from MM^a. Similarly, the corresponding naming convention applies to the QM side, with atoms QM^b being one bond away from QM^a, QM^c being two bonds away, and so on. If a link-atom scheme is used, the dangling bond of QM^a is saturated by the link atom L (see Figure 6.5).

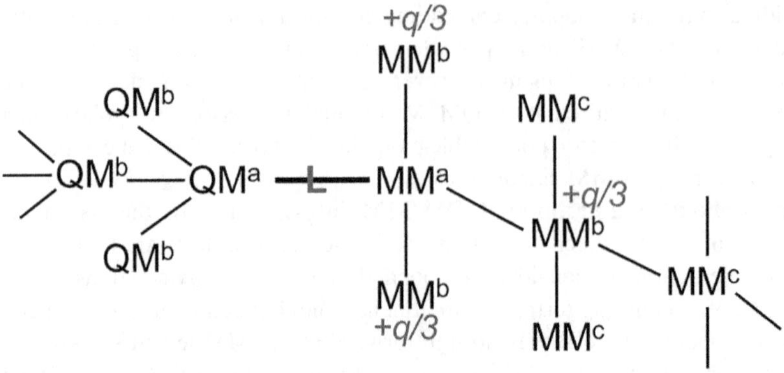

FIGURE 6.5 Labeling of atoms at the edge of the inner and outer regions. The partial charges demonstrate a charge-shift system. The MM^b atoms receive an equal distribution of the original MM charge q of MM^a, which has been removed.

6.3.1.2.2.2 Key Issues When Cutting Covalent Bonds

When a covalent bond is cut by the QM–MM boundary, three key issues need to be addressed:

1. **Capping the dangling bond of the QM atom QM^a:** Truncating the QM region by treating the bond as being homolytically or heterolytically cleaved is unrealistic. The dangling bond of the QM atom must be capped to avoid unphysical effects.
2. **Preventing overpolarization of the QM density:** When using link atoms, the MM charges close to the cut can overpolarize the QM density. This can be prevented by using polarized embedding or electrostatic embedding.
3. **Avoiding double-counting of interactions:** The bonded MM terms involving atoms from both subsystems must be selected carefully to avoid double-counting of interactions.

6.3.1.2.2.3 Boundary Schemes

Three main classes of boundary schemes are used to address the issues of cutting covalent bonds: link-atom schemes, boundary-atom schemes, and dual-space schemes (Boulanger & Thiel, 2012; Zhang & Lin, 2010).

- **Link-atom schemes:** Link-atom schemes introduce an additional atomic center L (usually a hydrogen atom) that is not part of the real system. It is covalently bound to QM^a and saturates its free valency. The L atom is treated as an MM atom and is connected to MM^a by a harmonic potential. The QM/MM interaction is then represented by a pairwise potential between the QM^a atom and the L atom. The use of a link atom allows the QM region to be fully described and avoids any unphysical effects associated with truncated bonds. However, the introduction of an extra atom can affect the energetics of the system and can lead to problems with charge transfer.
- **Boundary-atom schemes:** Boundary-atom techniques offer a novel way of dealing with the frontier (boundary) atom in molecular mechanics simulations (Benighaus & Thiel, 2008). They introduce a special boundary atom that serves as a mediator between the classical and quantum mechanical regions of the system. This boundary atom mimics the electronic properties of the MM moiety and saturates the free valency of the QM region. The schemes avoid the complexity of introducing additional atoms and are based on a monovalent pseudopotential that is positioned at the site of the frontier atom. This pseudopotential is parameterized to reproduce desired features, such as the electronic behavior of a methyl group in a C-C single bond. Different schemes have been proposed,

including adjusted connection atoms for semiempirical QM methods, the pseudobond approach for *ab initio* and DFT methods, and the use of tailored pseudopotentials within plane-wave QM methods. Other recent developments in this area show promise but have not yet been widely adopted.

- **Frozen localized orbitals:** The use of frozen localized orbitals to saturate the dangling bond at the QM/MM boundary has been a popular approach since Warshel and Levitt introduced it. Various methods have been developed based on the idea of placing a set of suitably oriented localized orbitals on one of the frontier atoms and freezing some of these orbitals during the SCF iterations.

  The local self-consistent field (LSCF) (Hégely et al., 2016; Loos et al., 2007) method involves constructing a strictly localized bond orbital (SLBO) for the frontier bond using a localization scheme. This SLBO, which is assumed to be transferable, is excluded from the SCF optimization in the QM/MM calculation and is oriented along the QM^a-MM^a vector. Recent modifications include the use of extremely localized molecular orbitals (ELMOs) or frozen core orbitals at the frontier atom, and the optimized-LSCF method, which allows the SLBOs to mix with their antibonding counterparts.

  The frozen orbitals method, developed by Friesner and colleagues, is a variant of the LSCF procedure that differs in some technical details and heavily parameterizes the QM–MM interactions at the boundary (Ferré et al., 2002; Kairys & Jensen, 2000; Murphy et al., 2000).

  The generalized hybrid orbitals (GHO) method (Gao et al., 1998) constructs localized hybrid orbitals and freezes some of them but places the set of localized hybrid orbitals on MM^a rather than QM^a. The orbital pointing toward QM^a is active and participates in the SCF iterations, while the remaining "auxiliary" hybrids are kept frozen and are not allowed to mix with the other orbitals.

  The effective fragment potentials (EFP) approach uses separate calculations on model systems to derive a set of one-electron terms that account for the environment's electrostatic, inductive, and repulsive interactions (Gordon et al., 2001; Gurunathan et al., 2016; Nemukhin et al., 2003). This EFP is then incorporated into the Hamiltonian of a QM calculation, where it describes the effects of the environment on the QM part. Initially designed for solvent environment modeling in QM calculations, the EFP method was extended to biomolecular systems using an LSCF-type procedure to treat covalent bonds across the boundary. Recent developments have lifted the limitation that the fragments be fixed in space.

  The boundary methods used in quantum mechanics/molecular mechanics (QM/MM) calculations have been extensively studied. Among them, the link-atom approach is the most popular and has been assessed in numerous studies. The variations in the handling of charges at the boundary have been the focus of most of these evaluations. Comparisons between link-atom and localized-orbital schemes have also been conducted.

While hybrid orbital-based approaches provide a more fundamental boundary description at the QM level, they are considerably more complex from a technical perspective. This is due to the orthogonality constraints required to prevent the mixing of frozen and active orbitals in the self-consistent field treatment. Additionally, localized orbitals and the parameter sets related to them must be determined beforehand through calculations on model compounds. These parameters are often not transferable and may need modification when the MM force field, the QM method, or basis set is changed.

Overall, the performance of link-atom and localized-orbital schemes is comparable, and both can provide reasonable accuracy when used with care. However, the link-atom approach is more widely used in practice due to its simplicity and efficiency, while localized-orbital schemes are more theoretically rigorous but require more computational resources.

6.4 CASE STUDIES OF MD AND QM/MM METHODS

To illustrate the capabilities of MD and QM/MM calculations, we present a selection of intriguing examples that showcase their diverse applications. Each example highlights the unique advantages of these methodologies and demonstrates their impact on different areas of research. These computational methods provide valuable insights into molecular dynamics, binding mechanisms, catalytic reactions, and revolutionizing research in various scientific fields.

6.4.1 Investigation of Role of a Crucial Dyad and Mechanistic Elucidation of Hydroxylation Mechanism in CYP450 from Mint Family

The cytochrome P450 (CYP) superfamily of enzymes is involved in a variety of physiological and xenobiotic metabolism reactions, including the biosynthesis and metabolism of endogenous compounds, as well as the metabolism of exogenous drugs and toxins. The CYPs from the supermint family play a crucial role in the biosynthesis of secondary metabolites, such as terpenoids and alkaloids (Isin & Guengerich, 2007; Shaik et al., 2010). In a recent study, the authors studied two wild type (WT) enzymes, CYP71D18 (L6H) and CYP71D13 (L3H) of the supermint family and their corresponding mutants (MT). They carried out MD simulations of wild type as well as mutant enzymes. It is noteworthy that in all four simulations—WT-L6H, WT-L3H, F363I mutant of L6H, and I366F mutant of L3H—a significant role was observed for a pair of aromatic molecules located near the heme center. This observation is supported by representative snapshots obtained from molecular dynamics (MD) studies. Figure 6.6a illustrates that the aromatic pair composed of F120 and F363 acts as two gates, resembling a doorway. In the case of WT-L6H, F363 is nearly

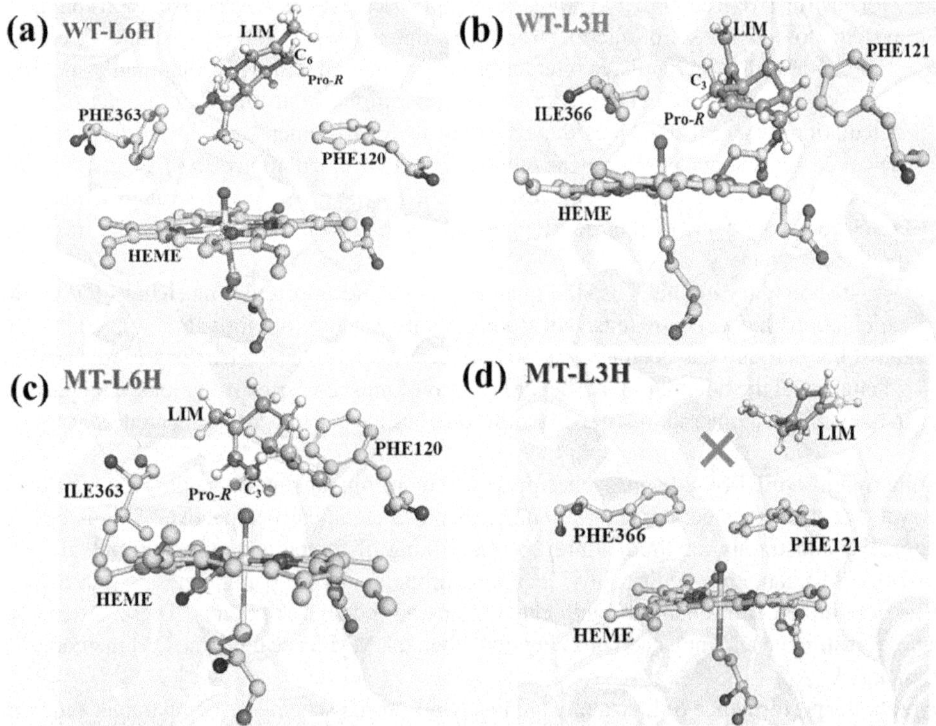

FIGURE 6.6 Interactions of substrate in proximity with residues 120/121 and 363/366: (a) WT-L6H, (b) WT-L3H, (c) MT-F363I-L6H, (d) MT-I366F-L3H.

perpendicular to the heme, creating a narrow passage for the substrate to reach the heme site. This limited accessibility explains the relatively lower activity observed in the WT-L6H enzyme. In both WT-L3H and the F363I mutant of L6H, the substitution of F363 with I leads to the removal of one gate, facilitating easier substrate access to the heme site. This could explain the enhanced activity observed in WT-L3H and the F363I mutant of L6H enzymes (Figure 6.6b and 6.6c). Interestingly, in the I366F mutant of the L3H enzyme, both aromatic pairs maintain a parallel position with respect to the heme plane, effectively blocking the substrate access channel (Figure 6.6d). Consequently, the I366F mutant of the L3H enzyme exhibits no activity due to the restricted substrate access to the heme site.

The authors used MD simulation technique to investigate the catalytic mechanism of the mint CYP71D18 (L6H) and CYP71D13 (L3H). They proposed a mechanism for the catalytic reaction, which involves the formation of a reactive intermediate, followed by single electron transfer and subsequent insertion of oxygen into the substrate molecule, leading to the formation of a product. The proposed mechanism is consistent with experimental observations of other CYPs and provides insights into the catalytic function of the mint CYPs (see Figure 6.7).

Overall, this study provides a detailed computational analysis of the catalytic mechanism of the mint CYPs and highlights the crucial role of the aromatic dyad in their function. The findings of this study could have important implications for the design of new drugs targeting CYP enzymes, particularly in the context of plant secondary metabolite biosynthesis.

6.4.2 Assessing the Impact of Various Water Models on the Structure and Function of Three Enzymes in CYP5450

The paper focuses on investigating the impact of different water models on the structure and function of cytochrome P450 enzymes using molecular dynamics simulations (Yadav et al., 2022). As we know, water is a ubiquitous solvent in biological systems, and the choice of a water model can have a significant impact on the results of molecular dynamics simulations. Many water models are commonly used in molecular dynamics simulations of biomolecules, but it is not clear which of

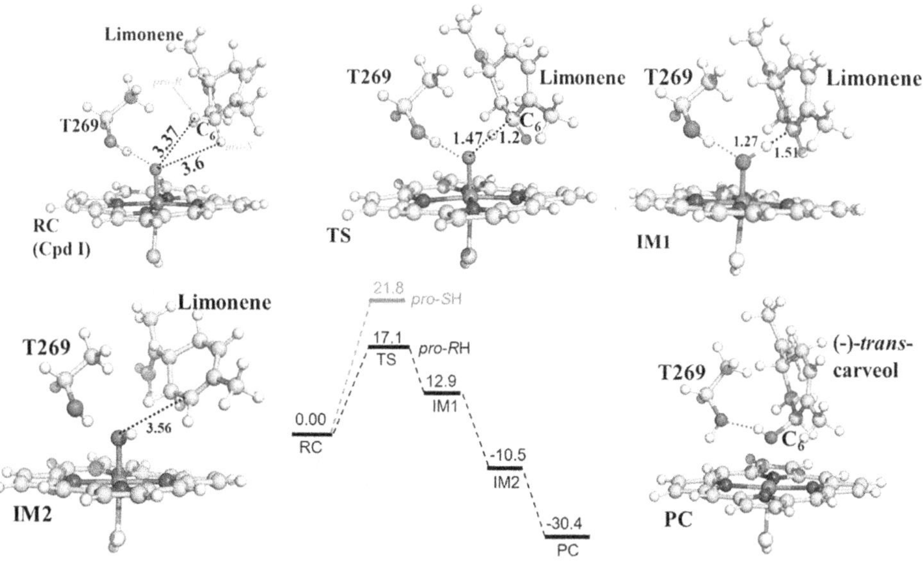

FIGURE 6.7 Predicted energy profile diagram via QM/MM calculations and key intermediates for WT-L6H enzyme.

these models is best suited for simulating the behavior of cytochrome P450 enzymes. The authors compared the performance of three of the most used water models—TIP3P, SPC/E, and OPC—in predicting the structural and dynamical properties of cytochrome P450 enzymes.

In this study, the authors performed molecular dynamics simulations of three cytochrome P450 enzymes—$CYP450_{BM3}$, $CYP450_{BS\beta}$ and $CYP450_{OleT}$—using the TIP3P, SPC/E, and OPC water models. They then analyzed the resulting simulations to determine the effects of the water models on the structure and function of the enzymes. The authors used a variety of methods to assess the accuracy of the simulations, including the root mean square deviation (RMSD) of the enzyme-substrate complex, the number of hydrogen bonds formed between the enzyme and water, and the catalytic activity of the enzyme.

The authors found that the choice of water model had a significant impact on the predicted structure and function of the cytochrome P450 enzymes. The study shows that the enzyme complexes remained stable in all three water models. Nevertheless, the OPC water model is more effective in maintaining stability for the polar active sites, including CYP450OleT and CYP450BSβ, while the TIP3P water model is better suited for the hydrophobic site found in CYP450BM3 (see Figure 6.8).

To analyze the substrate oxidation process, we utilized QM/MM calculations on selected snapshots from the MD trajectories. Given that the transfer of a hydrogen atom from the substrate to Cpd I is crucial to this process, we conducted a potential energy surface (PES) scan for the hydrogen atom transfer (HAT) step using various water models for each enzyme. The authors found that the HAT barrier results suggested that the TIP3P water model is more effective in peroxygenase while the OPC water model is more efficient in monooxygenase. However, the SPC/E model is the least efficient in peroxygenase, and TIP3P is the least efficient in monooxygenase among all three models.

6.4.3 Effect of Allostery on the Capping Loop and Its Role in Catalysis in Dipeptide Epimerases of Enolase Family

The first functionally characterized epimerase, Ala-Glu-Epimerase (AEE), specifically catalyzes the 1,1-proton transfer reaction. Likewise, O-succinyl benzoate synthases (OSBS), another member of the enalose family, specifically catalyzes dehydration reactions. In spite of similar active sites of both these enzymes, a glycine residue of OSBS enzyme close to the active site is replaced by an aspartic acid in AEE. As a result, it is possible to believe that nature replaced an aspartate residue

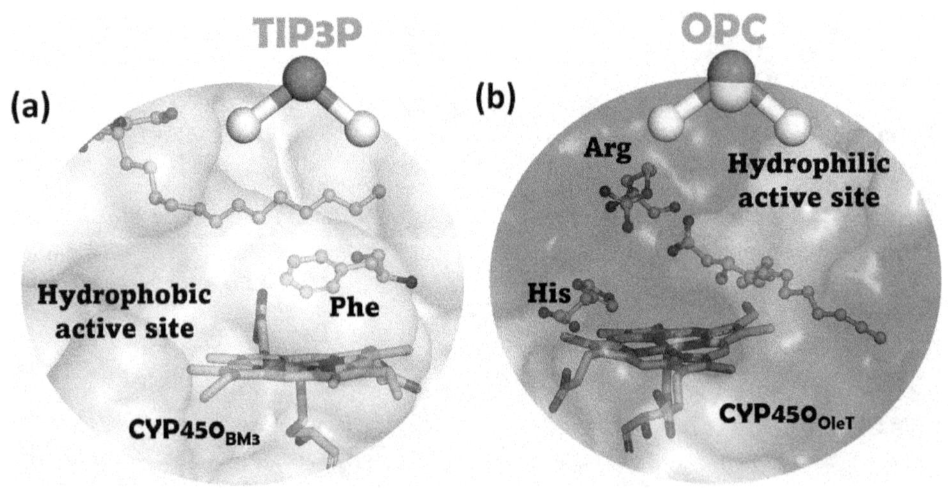

FIGURE 6.8 (a) Hydrophobic active site of CYP450BM3, (b) hydrophilic active site of CYP450OleT.

in AEE to evolve it for 1,1-proton transfer reactions during the evolution from a common progenitor of Ala-Glu-Epimerase (AEE) and O-succinyl benzoate synthases (OSBS) enzyme. In this work, using comprehensive MD simulations, the authors have shown how a mutation of Asp321 to glycine triggers an allosteric conformational change in the capping loop that benefits the substrate recognition for OSBS activity. In AEE enzymes, the capping loop undergoes a significant conformational change upon substrate entry, even though it is not directly involved in the reaction. This capping loop acts as a gatekeeper, and its role was investigated in the AEE enzyme. The active site architecture of the AEE enzyme revealed a well organized structure. The dipeptide substrate's α-ammonium group forms a hydrogen bond with Asp321 and Asp323 residues, while the substrate's carboxylate group forms a hydrogen bond with the guanidium group of Arg24. A perturbation near D321–D333 in the active site can induce a significant conformational change in the capping loop (see Figure 6.9a). To validate this allosteric effect, the D321 residue was mutated to glycine, resulting in the capping loop losing its interaction with the substrate and reverting to an open-channel conformation (see Figure 6.9c). The corresponding distance plots are given in Figures 6.9b and d. This mutant showed reduced catalytic efficiency for the native substrate. The finding aligns with previous experimental results, where the D321G mutation led to a complete loss of native activity in the epimerase enzyme.

Further, the authors have extended their study by evaluating the mechanism of 1,1-proton transfer using hybrid QM/MM calculations. The reaction proceeds with the first proton transfer from the substrate's Cα carbon to deprotonated Lys268, forming an unstable intermediate (IM). In the subsequent step, IM readily abstracts a proton from Lys162 to yield the thermodynamically stable product (PC), L-Ala-D-Glu dipeptide. The rate-determining step is the initial proton transfer, and the presence of the coordinated Mg^{2+} ion lowers the transition state barrier, facilitating an efficient reaction (see Figure 6.10).

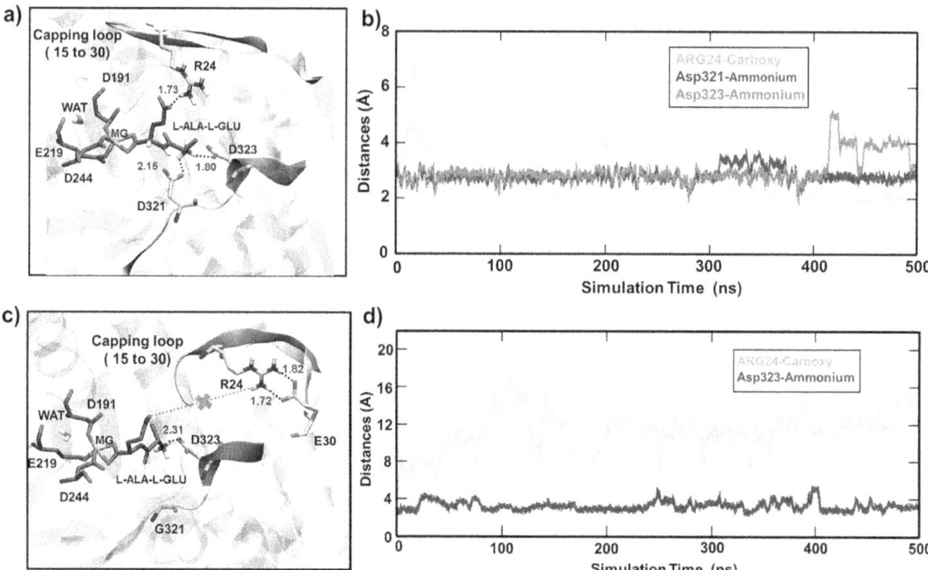

FIGURE 6.9 (a) Snapshot showing the position of residues R24, D321, and D323, the active site and substrate in AEE; (b) corresponding distance vs. simulation time plot, Arg24 to carboxylate of the substrate (grey), Asp321 to ammonium of substrate (black), and Asp323 to ammonium of substrate (dark grey); (c) position of residues R24, E30, G321, and D323; active site and substrate in D321G mutant of AEE; (b) corresponding distance vs. simulation time plot; Arg24 to carboxylate of the substrate (grey), Asp323 to ammonium of substrate (black).

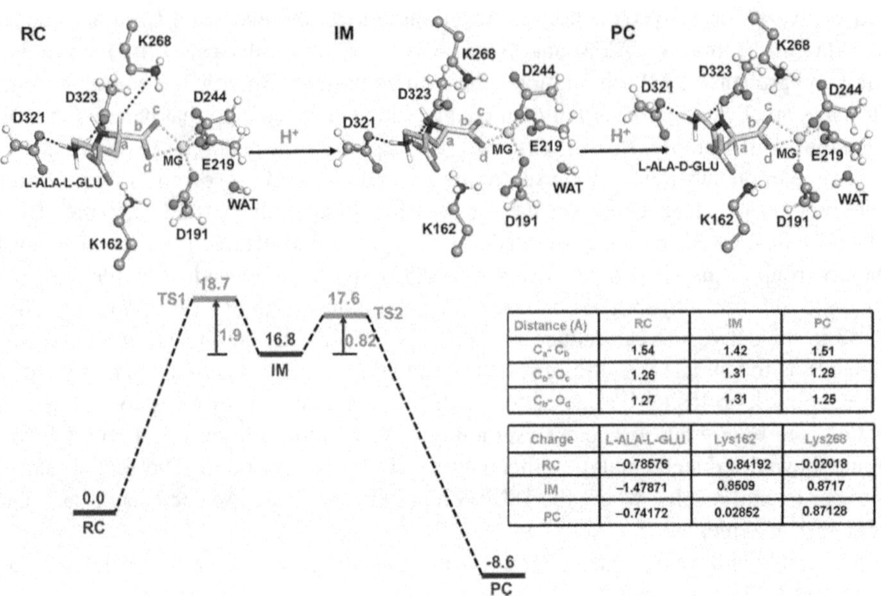

Distance (Å)	RC	IM	PC
C_a- C_b	1.54	1.42	1.51
C_b- O_c	1.26	1.31	1.29
C_b- O_d	1.27	1.31	1.25

Charge	L-ALA-L-GLU	Lys162	Lys268
RC	–0.78576	0.84192	–0.02018
IM	–1.47871	0.8509	0.8717
PC	–0.74172	0.02852	0.87128

FIGURE 6.10 Energy profile diagram for 1,1-proton transfer reaction. Key residues of the QM region are shown by the ball and stick model. A corresponding energy profile and barriers for the process are shown in the schematic diagram at bottom of the figure. The reported energy values are in kcal/mol.

6.5 CONCLUSION

The realm of molecular dynamics (MD) simulations has experienced a remarkable transformation with the advent of enhanced computational efficiency and robotic techniques. These powerful tools offer a unique opportunity to delve into the intricate details of molecular phenomena, bridging the gap between theoretical models and experimental investigations. With two decades of valuable insights and methodological advancements, MD simulations have become a trusted approach, empowering researchers to confidently tackle a diverse range of chemical and biological challenges. Through the judicious use of MD simulations, we can unlock the secrets of protein folding, elucidating the three-dimensional structure of the target protein or complex. In cases where experimental structures are unavailable, these simulations can provide reliable alternatives, generating homology models of comparable accuracy. Moreover, the integration of free energy methods into MD simulations promises to revolutionize screening pipelines and facilitate the discrimination of candidate molecules. As quantum mechanical (QM) methods find their place in free energy calculations, we can anticipate a significant boost in accuracy when determining binding free energies. This breakthrough enables us to unravel the intricacies of protein-ligand binding, delving into the effects of polarization and electronic interactions. The coming years hold tremendous potential for gaining a deeper understanding of protein–protein interactions, thanks to these advancements.

In parallel, QM/MM calculations have emerged as a state-of-the-art technique to investigate the electronic properties and chemical reactions within the context of a realistic molecular environment. The hybrid quantum mechanics/molecular mechanics (QM/MM) calculations have emerged as the leading computational approach for investigating reactive and electronic processes in biomolecular systems. Over the past few decades, the increasing number of publications utilizing QM/MM techniques demonstrates the maturity and effectiveness of this methodology. The QM/MM formalism enables the partitioning of the total energy, including the QM–MM interaction energy, into various components, allowing for the analysis of the protein environment's impact on the species involved in the catalytic cycle, even at the level of individual residues.

In this chapter, through illustrative case studies, we have highlighted the extensive use of MD simulation to investigate the crucial conformation changes in the protein systems responsible for the enzymatic activities, along with the ability of QM/MM calculations to offer detailed mechanistic insights into enzymatic reactions.

In conclusion, the combination of MD simulations and QM/MM calculations has revolutionized our understanding of biomolecular systems. These computational techniques, each with its own strengths and limitations, complement one another and offer a holistic view of molecular behavior.

REFERENCES

Alam, M. A., & Naik, P. K. (2009). Applying linear interaction energy method for binding affinity calculations of podophyllotoxin analogues with tubulin using continuum solvent model and prediction of cytotoxic activity. *Journal of Molecular Graphics and Modelling*, *27*(8), 930–943.

Alder, B. J., & Wainwright, T. E. (1957). Phase transition for a hard sphere system. *The Journal of Chemical Physics*, *27*(5), 1208–1209.

Åqvist, J., Luzhkov, V. B., & Brandsdal, B. O. (2002). Ligand binding affinities from MD simulations. *Accounts of Chemical Research*, *35*(6), 358–365.

Aqvist, J., & Marelius, J. (2001). The linear interaction energy method for predicting ligand binding free energies. *Combinatorial Chemistry & High Throughput Screening*, *4*(8), 613–626.

Badar, M. S., Shamsi, S., Ahmed, J., & Alam, M. A. (2022). Molecular Dynamics Simulations: Concept, Methods, and Applications. In N. Rezaei (Ed.), *Transdisciplinarity* (pp. 131–151). Springer International Publishing. https://doi.org/10.1007/978-3-030-94651-7_7

Bagayoko, D. (2014). Understanding density functional theory (DFT) and completing it in practice. *AIP Advances*, *4*(12), 127104.

Benighaus, T., & Thiel, W. (2008). Efficiency and accuracy of the generalized solvent boundary potential for hybrid QM/MM simulations: Implementation for semiempirical Hamiltonians. *Journal of Chemical Theory and Computation*, *4*(10), 1600–1609.

Berendsen, H. J. C., & Van Gunsteren, W. F. (1984). Molecular dynamics simulations: Techniques and approaches. *Molecular Liquids: Dynamics and Interactions*, 475–500.

Berkowitz, M., & McCammon, J. A. (1982). Molecular dynamics with stochastic boundary conditions. *Chemical Physics Letters*, *90*(3), 215–217.

Boulanger, E., & Thiel, W. (2012). Solvent boundary potentials for hybrid QM/MM computations using classical drude oscillators: A fully polarizable model. *Journal of Chemical Theory and Computation*, *8*(11), 4527–4538.

Boyd, D. B. (2004). Computational Medicinal Chemistry for Drug Discovery Edited by Patrick Bultinck, Hans De Winter, Wilfried Langenaeker, and Jan P. Tollenaere. Marcel Dekker, New York. 2004. xiv + 794 pp. 18 × 26 cm. ISBN 0-8247-4774-7. $195.00. *Journal of Medicinal Chemistry*, *47*(25), 6433. https://doi.org/10.1021/jm040170c

Brandsdal, B. O., & Smalås, A. O. (2000). Evaluation of protein–protein association energies by free energy perturbation calculations. *Protein Engineering*, *13*(4), 239–245.

Brandt, F., & Jacob, C. R. (2022). Systematic QM region construction in QM/MM calculations based on uncertainty quantification. *Journal of Chemical Theory and Computation*, *18*(4), 2584–2596.

Braun, E., Gilmer, J., Mayes, H. B., Mobley, D. L., Monroe, J. I., Prasad, S., & Zuckerman, D. M. (2019). Best practices for foundations in molecular simulations [Article v1. 0]. *Living Journal of Computational Molecular Science*, *1*(1), 5957.

Braun, E., Gilmer, J., Mayes, H. B., Mobley, D. L., Monroe, J. I., Prasad, S., & Zuckerman, D. M. (2019). Best practices for foundations in molecular simulations [Article v1.0]. *Living Journal of Computational Molecular Science*, *1*(1). https://doi.org/10.33011/livecoms.1.1.5957

Brooks, B. R., Bruccoleri, R. E., Olafson, B. D., States, D. J., Swaminathan, S. a, & Karplus, M. (1983). CHARMM: A program for macromolecular energy, minimization, and dynamics calculations. *Journal of Computational Chemistry*, *4*(2), 187–217.

Daggett, V., & Levitt, M. (1994). Protein folding↔ unfolding dynamics. *Current Opinion in Structural Biology*, *4*(2), 291–295.

Despiau-Pujo, E. (2019). A brief overview on Molecular Dynamics simulations of plasma-surface interaction in reactive ion etching. *66th American Vacuum Society (AVS) International Symposium.*

Dominy, B. N., & Brooks, C. L. (1999). Development of a generalized Born model parametrization for proteins and nucleic acids. *The Journal of Physical Chemistry B*, *103*(18), 3765–3773.

Dubey, K. D., Chaubey, A. K., & Ojha, R. P. (2011). Role of pH on dimeric interactions for DENV envelope protein: An insight from molecular dynamics study. *Biochimica et Biophysica Acta (BBA)-Proteins and Proteomics, 1814*(12), 1796–1801.

Dubey, K. D., & Ojha, R. P. (2013). Molecular dynamics simulations: Applicability and scopes in computational biochemistry. *Nanoscience and Computational Chemistry Research Progress, 371.*

Dubey, K. D., Tiwari, R. K., & Ojha, R. P. (2013). Recent advances in protein-ligand interactions: Molecular dynamics simulations and binding free energy. *Current Computer-Aided Drug Design, 9*(4), 518–531. https://doi.org/10.2174/15734099113096660036

Durrant, J. D., & McCammon, J. A. (2011). Molecular dynamics simulations and drug discovery. *BMC Biology, 9*(1), 71. https://doi.org/10.1186/1741-7007-9-71

Dutta Dubey, K., Kumar Tiwari, R., & Prasad Ojha, R. (2013). Recent advances in protein– ligand interactions: Molecular dynamics simulations and binding free energy. *Current Computer-Aided Drug Design, 9*(4), 518–531.

Edinger, S. R., Cortis, C., Shenkin, P. S., & Friesner, R. A. (1997). Solvation free energies of peptides: Comparison of approximate continuum solvation models with accurate solution of the Poisson–Boltzmann equation. *The Journal of Physical Chemistry B, 101*(7), 1190–1197.

Eisenberg, D., & McLachlan, A. D. (1986). Solvation energy in protein folding and binding. *Nature, 319*, 199–203.

Feller, S. E., Zhang, Y., Pastor, R. W., & Brooks, B. R. (1995). Constant pressure molecular dynamics simulation: The Langevin piston method. *The Journal of Chemical Physics, 103*(11), 4613–4621.

Ferré, N., Assfeld, X., & Rivail, J. (2002). Specific force field parameters determination for the hybrid ab initio QM/MM LSCF method. *Journal of Computational Chemistry, 23*(6), 610–624.

Field, M. J., Bash, P. A., & Karplus, M. (1990). A combined quantum mechanical and molecular mechanical potential for molecular dynamics simulations. *Journal of Computational Chemistry, 11*(6), 700–733.

Florián, J., Goodman, M. F., & Warshel, A. (2000). Free-Energy Perturbation Calculations of DNA Destabilization by Base Substitutions: The Effect of Neutral Guanine⊙ Thymine, Adenine⊙ Cytosine and Adenine⊙ Difluorotoluene Mismatches. *The Journal of Physical Chemistry B, 104*(43), 10092–10099.

Gao, J., Amara, P., Alhambra, C., & Field, M. J. (1998). A generalized hybrid orbital (GHO) method for the treatment of boundary atoms in combined QM/MM calculations. *The Journal of Physical Chemistry A, 102*(24), 4714–4721.

Gilson, M. K., Given, J. A., Bush, B. L., & McCammon, J. A. (1997). The statistical-thermodynamic basis for computation of binding affinities: A critical review. *Biophysical Journal, 72*(3), 1047–1069.

Gilson, M. K., & Zhou, H.-X. (2007). Calculation of protein-ligand binding affinities. *Annual Review of Biophysics and Biomolecular Structure, 36*, 21–42.

Glatzel, F., & Schilling, T. (2021). Comments on the validity of the non-stationary generalized Langevin equation as a coarse-grained evolution equation for microscopic stochastic dynamics. *The Journal of Chemical Physics, 154*(17), 174107.

Gohlke, H., & Case, D. A. (2004). Converging free energy estimates: MM-PB (GB) SA studies on the protein–protein complex Ras–Raf. *Journal of Computational Chemistry, 25*(2), 238–250.

Gordon, M. S., Freitag, M. A., Bandyopadhyay, P., Jensen, J. H., Kairys, V., & Stevens, W. J. (2001). The effective fragment potential method: A QM-based MM approach to modeling environmental effects in chemistry. *The Journal of Physical Chemistry A, 105*(2), 293–307.

Götz, A. W., Clark, M. A., & Walker, R. C. (2014). An extensible interface for QM/MM molecular dynamics simulations with AMBER. *Journal of Computational Chemistry, 35*(2), 95–108.

Grančič, P., Bylsma, R., Meekes, H., & Cuppen, H. M. (2015). Evaluation of All-Atom Force Fields for Anthracene Crystal Growth. *Crystal Growth & Design, 15*(4), 1625–1633.

Grønbech-Jensen, N., & Farago, O. (2014). Constant pressure and temperature discrete-time Langevin molecular dynamics. *The Journal of Chemical Physics, 141*(19), 194108.

Gurunathan, P. K., Acharya, A., Ghosh, D., Kosenkov, D., Kaliman, I., Shao, Y., Krylov, A. I., & Slipchenko, L. V. (2016). Extension of the effective fragment potential method to macromolecules. *The Journal of Physical Chemistry B, 120*(27), 6562–6574.

Hansson, T., Marelius, J., & Åqvist, J. (1998). Ligand binding affinity prediction by linear interaction energy methods. *Journal of Computer-Aided Molecular Design, 12*, 27–35.

Hégely, B., Bogár, F., Ferenczy, G.G., Kállay, M. (2016). A QM/MM program using frozen localized orbitals and the Huzinaga equation. In: Szabados, Á., Kállay, M., Szalay, P. (eds) *Péter R. Surján. Highlights in Theoretical Chemistry, 12*. Springer, Berlin, Heidelberg.

Heimdal, J., & Ryde, U. (2012). Convergence of QM/MM free-energy perturbations based on molecular-mechanics or semiempirical simulations. *Physical Chemistry Chemical Physics, 14*(36), 12592–12604.

Helfand, E. (1985). Dynamics of conformational transitions in polymers. *Journal of Polymer Science: Polymer Symposia*, *73*(1), 39–54.

Hohenberg, P., & Kohn, W. (1964). Density functional theory (DFT). *Physical Review, 136*(1964), B864.

Hollingsworth, S. A., & Dror, R. O. (2018). Molecular Dynamics Simulation for All. *Neuron*, *99*(6), 1129–1143. https://doi.org/10.1016/j.neuron.2018.08.011

Hoover, W. G. (1985). Canonical dynamics: Equilibrium phase-space distributions. *Physical Review A*, *31*(3), 1695.

Hünenberger, P. H., & van Gunsteren, W. F. (1999). Empirical classical force fields for molecular systems. *Potential Energy Surfaces: Proceedings of the Mariapfarr Workshop in Theoretical Chemistry*, 177–214.

Introduction. (2002). In *An Introduction to Computational Biochemistry* (pp. 1–10). John Wiley & Sons, Ltd. https://doi.org/https://doi.org/10.1002/0471223840.ch1

Introduction. (2008). In *Computational Chemistry and Molecular Modeling: Principles and Applications* (pp. 1–15). Springer Berlin Heidelberg. https://doi.org/10.1007/978-3-540-77304-7_1

Isin, E. M., & Guengerich, F. P. (2007). Complex reactions catalyzed by cytochrome P450 enzymes. *Biochimica et Biophysica Acta*, *1770*(3), 314–329. https://doi.org/10.1016/j.bbagen.2006.07.003

Jakobsson, E. (2001). Computational Biochemistry and Biophysics Edited by Oren M. Becker (Tel Aviv University), Alexander D. MacKerell, Jr. (Uni-versity of Maryland), Benoît Roux (Cornell University), and Masa-katsu Watanabe (Wavefunction, Inc.). Marcel Dekker: New York and Ba. *Journal of the American Chemical Society*, *123*(50), 12745. https://doi.org/10.1021/ja015256z

Jayaram, B., Liu, Y., & Beveridge, D. L. (1998). A modification of the generalized Born theory for improved estimates of solvation energies and p K shifts. *The Journal of Chemical Physics*, *109*(4), 1465–1471.

Kairys, V., & Jensen, J. H. (2000). QM/MM boundaries across covalent bonds: A frozen localized molecular orbital-based approach for the effective fragment potential method. *The Journal of Physical Chemistry A*, *104*(28), 6656–6665.

Kamerlin, S. C. L., & Warshel, A. (2011). The empirical valence bond model: Theory and applications. *Wiley Interdisciplinary Reviews: Computational Molecular Science*, *1*(1), 30–45.

Karplus, M., & Petsko, G. A. (1990). Molecular dynamics simulations in biology. *Nature*, *347*(6294), 631–639. https://doi.org/10.1038/347631a0

Karplus, M., & Šali, A. (1995). Theoretical studies of protein folding and unfolding. *Current Opinion in Structural Biology*, *5*(1), 58–73.

Kästner, J., Thiel, S., Senn, H. M., Sherwood, P., & Thiel, W. (2007). Exploiting QM/MM capabilities in geometry optimization: A microiterative approach using electrostatic embedding. *Journal of Chemical Theory and Computation*, *3*(3), 1064–1072.

Kerrigan, J.E. (2013). Molecular Dynamics Simulations in Drug Design. In *In Silico Models for Drug Discovery. Methods in Molecular Biology*, *993*. Humana Press, Totowa, NJ.

König, P. H., Hoffmann, M., Frauenheim, T., & Cui, Q. (2005). A critical evaluation of different QM/MM frontier treatments with SCC-DFTB as the QM method. *The Journal of Physical Chemistry B*, *109*(18), 9082–9095.

Lazaridis, T., & Karplus, M. (1997). "New view" of protein folding reconciled with the old through multiple unfolding simulations. *Science*, *278*(5345), 1928–1931.

Lazaridis, T., Masunov, A., & Gandolfo, F. (2002). Contributions to the binding free energy of ligands to avidin and streptavidin. *Proteins: Structure, Function, and Bioinformatics*, *47*(2), 194–208.

Leach, A. R. (2001). Molecular modelling : Principles and applications. In *TA—TT*—(2nd ed). Prentice Hall. https://doi.org/LK—https://worldcat.org/title/45008511

Lee, T.-S., Hu, Y., Sherborne, B., Guo, Z., & York, D. M. (2017). Toward fast and accurate binding affinity prediction with pmemdGTI: An efficient implementation of GPU-accelerated thermodynamic integration. *Journal of Chemical Theory and Computation*, *13*(7), 3077–3084.

Levitt, M. (2014). Birth and future of multiscale modeling for macromolecular systems (Nobel Lecture). *Angewandte Chemie International Edition*, *53*(38), 10006–10018.

Lin, H., & Truhlar, D. G. (2007). QM/MM: What have we learned, where are we, and where do we go from here? *Theoretical Chemistry Accounts*, *117*, 185–199.

Lindorff-Larsen, K., Piana, S., Dror, R. O., & Shaw, D. E. (2011). How fast-folding proteins fold. *Science*, *334*(6055), 517–520.

Ljungberg, K. B., Marelius, J., Musil, D., Svensson, P., Norden, B., & Åqvist, J. (2001). Computational modelling of inhibitor binding to human thrombin. *European Journal of Pharmaceutical Sciences*, *12*(4), 441–446.

Loos, P. F., Fornili, A., Sironi, M., & Assfeld, X. (2007). Removing extra frontier parameters in QM/MM methods: A tentative with the Local Self-Consistent Field approach. *Computer Letter*, *4*, 473–486.

Lowe, J. P., & Peterson, K. (2011). *Quantum Chemistry*. Elsevier.

Luque, F. J., Reuter, N., Cartier, A., & Ruiz-López, M. F. (2000). Calibration of the quantum/classical Hamiltonian in semiempirical QM/MM AM1 and PM3 methods. *The Journal of Physical Chemistry A, 104*(46), 10923–10931.

Martys, N. S., & Mountain, R. D. (1999). Velocity Verlet algorithm for dissipative-particle-dynamics-based models of suspensions. *Physical Review E, 59*(3), 3733.

Mayo, S. L., Olafson, B. D., & Goddard, W. A. (1990). DREIDING: A generic force field for molecular simulations. *Journal of Physical Chemistry, 94*(26), 8897–8909.

McCammon, J. A., Gelin, B. R., & Karplus, M. (1977). Dynamics of folded proteins. *Nature, 267*(5612), 585–590. https://doi.org/10.1038/267585a0

McCammon, J. A., & Harvey, S. C. (1988). *Dynamics of Proteins and Nucleic Acids*. Cambridge University Press.

Mlynsky, V., Banas, P., Sponer, J., van der Kamp, M. W., Mulholland, A. J., & Otyepka, M. (2014). Comparison of ab initio, DFT, and semiempirical QM/MM approaches for description of catalytic mechanism of hairpin ribozyme. *Journal of Chemical Theory and Computation, 10*(4), 1608–1622.

Molecular Modeling: Molecular Mechanics. (2002). *An Introduction to Computational Biochemistry* (pp. 285–314). John Wiley & Sons, Ltd. https://doi.org/https://doi.org/10.1002/0471223840.ch14

Murcko, M. A. (1995). Computational methods to predict binding free energy in ligand-receptor complexes. *Journal of Medicinal Chemistry, 38*(26), 4953–4967.

Murphy, R. B., Philipp, D. M., & Friesner, R. A. (2000). Frozen orbital QM/MM methods for density functional theory. *Chemical Physics Letters, 321*(1–2), 113–120.

Nemukhin, A. V., Grigorenko, B. L., Topol, I. A., & Burt, S. K. (2003). Flexible effective fragment QM/MM method: Validation through the challenging tests. *Journal of Computational Chemistry, 24*(12), 1410–1420.

Nosé, S. (1984a). A molecular dynamics method for simulations in the canonical ensemble. *Molecular Physics, 52*(2), 255–268.

Nosé, S. (1984b). A unified formulation of the constant temperature molecular dynamics methods. *The Journal of Chemical Physics, 81*(1), 511–519.

Obot, I. B., Macdonald, D. D., & Gasem, Z. M. (2015). Density functional theory (DFT) as a powerful tool for designing new organic corrosion inhibitors. Part 1: An overview. *Corrosion Science, 99*, 1–30.

Orio, M., Pantazis, D. A., & Neese, F. (2009). Density functional theory. *Photosynthesis Research, 102*, 443–453.

Pearlman, D. A. (2005). Evaluating the molecular mechanics Poisson– Boltzmann surface area free energy method using a congeneric series of ligands to p38 MAP kinase. *Journal of Medicinal Chemistry, 48*(24), 7796–7807.

Pearlman, D. A., & Kollman, P. A. (1991). The overlooked bond-stretching contribution in free energy perturbation calculations. *The Journal of Chemical Physics, 94*(6), 4532–4545.

Perdih, A., Bren, U., & Solmajer, T. (2009). Binding free energy calculations of N-sulphonyl-glutamic acid inhibitors of MurD ligase. *Journal of Molecular Modeling, 15*, 983–996.

Quigley, D., & Probert, M. I. J. (2004). Langevin dynamics in constant pressure extended systems. *The Journal of Chemical Physics, 120*(24), 11432–11441.

Reinholdt, P., Kongsted, J., & Lipparini, F. (2021). Fast approximate but accurate QM/MM interactions for polarizable embedding. *Journal of Chemical Theory and Computation, 18*(1), 344–356.

Rosenthal, S. J., Jimenez, R., Fleming, G. R., Kumar, P. V., & Maroncelli, M. (1994). Solvation dynamics in methanol: Experimental and molecular dynamics simulation studies. *Journal of Molecular Liquids, 60*(1–3), 25–56.

Roßbach, S., & Ochsenfeld, C. (2017). Influence of coupling and embedding schemes on QM size convergence in QM/MM approaches for the example of a proton transfer in DNA. *Journal of Chemical Theory and Computation, 13*(3), 1102–1107.

Roux, B., & Simonson, T. (1999). Implicit solvent models. *Biophysical Chemistry, 78*(1–2), 1–20.

Ryde, U. (2016). QM/MM calculations on proteins. *Methods in Enzymology, 577*, 119–158.

Scheraga, H. A. (1979). Interactions in aqueous solution. *Accounts of Chemical Research, 12*(1), 7–14.

Scheraga, H. A., Khalili, M., & Liwo, A. (2007). Protein-folding dynamics: Overview of molecular simulation techniques. *Annual Review of Physical Chemistry, 58*, 57–83.

Schlick, T. (2002). Biomolecular Structure and Modeling: Historical Perspective. In *Molecular Modeling and Simulation: An Interdisciplinary Guide* (pp. 1–32). Springer New York. https://doi.org/10.1007/978-0-387-22464-0_1

Senn, H. M., & Thiel, W. (2007a). QM/MM methods for biological systems. *Atomistic Approaches in Modern Biology: From Quantum Chemistry to Molecular Simulations*, 173–290.

Senn, H. M., & Thiel, W. (2007b). QM/MM studies of enzymes. *Current Opinion in Chemical Biology, 11*(2), 182–187.

Senn, H. M., & Thiel, W. (2009). QM/MM methods for biomolecular systems. *Angewandte Chemie International Edition, 48*(7), 1198–1229.

Shaik, S., Cohen, S., Wang, Y., Chen, H., Kumar, D., & Thiel, W. (2010). P450 Enzymes: Their Structure, Reactivity, and Selectivity—Modeled by QM/MM Calculations. *Chemical Reviews, 110*(2), 949–1017. https://doi.org/10.1021/cr900121s

Singh, P., Mhaka, A. M., Christensen, S. B., Gray, J. J., Denmeade, S. R., & Isaacs, J. T. (2005). Applying linear interaction energy method for rational design of noncompetitive allosteric inhibitors of the sarco- and endoplasmic reticulum calcium-ATPase. *Journal of Medicinal Chemistry, 48*(8), 3005–3014.

Sneskov, K., Schwabe, T., Christiansen, O., & Kongsted, J. (2011). Scrutinizing the effects of polarization in QM/MM excited state calculations. *Physical Chemistry Chemical Physics, 13*(41), 18551–18560.

Stillinger, F. H., & Rahman, A. (1974). Improved simulation of liquid water by molecular dynamics. *The Journal of Chemical Physics, 60*(4), 1545–1557. https://doi.org/10.1063/1.1681229

Stouten, P. F. W., Frömmel, C., Nakamura, H., & Sander, C. (1993). An effective solvation term based on atomic occupancies for use in protein simulations. *Molecular Simulation, 10*(2–6), 97–120.

Straatsma, T. P., & McCammon, J. A. (1992). Computational alchemy. *Annual Review of Physical Chemistry, 43*(1), 407–435.

Thellamurege, N. M., & Hirao, H. (2014). Effect of protein environment within cytochrome P450cam evaluated using a polarizable-embedding QM/MM method. *The Journal of Physical Chemistry B, 118*(8), 2084–2092.

Tsai, C. S. (2003). *An Introduction to Computational Biochemistry.* Germany: John Wiley & Sons.

Van Gunsteren, W. F., & Berendsen, H. J. C. (1990). Computer simulation of molecular dynamics: Methodology, applications, and perspectives in chemistry. *Angewandte Chemie International Edition in English, 29*(9), 992–1023.

van Oss, C. J. (1979). *A Review of: Principles of Protein Structure*, G. E. Schulz and R. H. Schirme, Springer-Verlag; New York, Heiddberg, Berlin.

Verlet, L. (1967). Computer" experiments" on classical fluids. I. Thermodynamical properties of Lennard-Jones molecules. *Physical Review, 159*(1), 98.

Vreven, T., & Morokuma, K. (2006). Hybrid methods: Oniom (qm: mm) and qm/mm. *Annual Reports in Computational Chemistry, 2*, 35–51.

Warshel, A., & Florián, J. (2002). The empirical valence bond (EVB) method. *Encyclopedia of Computational Chemistry.* P. Ragué Schleyer, N.L. Allinger, T. Clark, J. Gasteiger, P.A. Kollman, H.F. Schaefer, P.R. Schreiner, W. Thiel, W.L. Jorgensen and R.C. Glen, *John Wiley and Sons.* Hoboken, New Jersey.

Warshel, A., & Weiss, R. M. (1980). An empirical valence bond approach for comparing reactions in solutions and in enzymes. *Journal of the American Chemical Society, 102*(20), 6218–6226.

Watanabe, H. C., Kubař, T., & Elstner, M. (2014). Size-consistent multipartitioning QM/MM: A stable and efficient adaptive QM/MM method. *Journal of Chemical Theory and Computation, 10*(10), 4242–4252.

Weiner, S. J., Kollman, P. A., Case, D. A., Singh, U. C., Ghio, C., Alagona, G., Profeta, S., & Weiner, P. (1984). A new force field for molecular mechanical simulation of nucleic acids and proteins. *Journal of the American Chemical Society, 106*(3), 765–784.

Wu, X., Xu, L., Li, E., & Dong, G. (2022). Application of molecular dynamics simulation in biomedicine. *Chemical Biology & Drug Design, 99*(5), 789–800.

Yadav, S., Kardam, V., Tripathi, A., Shruti, T. G., & Dubey, K. D. (2022). The performance of different water models on the structure and function of cytochrome P450 enzymes. *Journal of Chemical Information and Modeling, 62*(24), 6679–6690. https://doi.org/10.1021/acs.jcim.2c00505

Yang, Z., Chen, J., Zhou, Y., Huang, H., Xu, D., & Zhang, C. (2018). Understanding the hydrogen transfer mechanism for the biodegradation of 2, 4, 6-trinitrotoluene catalyzed by pentaerythritol tetranitrate reductase: Molecular dynamics simulations. *Physical Chemistry Chemical Physics, 20*(17), 12157–12165.

Young, D. (2004). *Computational Chemistry: A Practical Guide for Applying Techniques to Real World Problems.* Germany: John Wiley & Sons.

Yue, J., Li, Y., Li, F., Zhang, P., Li, Y., Xu, J., Zhang, Q., Zhang, C., He, X., & Wang, Y. (2023). Discovery of Mcl-1 inhibitors through virtual screening, molecular dynamics simulations and in vitro experiments. *Computers in Biology and Medicine, 152*, 106350.

Zhang, Y. (2013). Electrostatic Interaction of the electrostatic-embedding and mechanical-embedding schemes for QM/MM calculations. *Communications in Computational Chemistry, 1*(2), 109–117.

Zhang, Y., & Lin, H. (2010). Flexible-boundary QM/MM calculations: II. Partial charge transfer across the QM/MM boundary that passes through a covalent bond. *Theoretical Chemistry Accounts, 126*, 315–322.

7 Soft Computing Technique towards the Geometry Optimization of Atomic Clusters

Ranita Pal, Bhrigu Chakraborty, and Pratim Kumar Chattaraj

7.1 INTRODUCTION

In recent years, computation has grown to be an essential component of scientific study, particularly when tackling issues that are challenging to address analytically. Even though the classical approaches, involving computation based only on the mathematical formulation of the problem, are still widely used, they frequently encounter insurmountable challenges such as a lack of system knowledge, nonlinearity, and instability. The soft computing (SC) domain of computer science involves dealing with imprecise and uncertain information.[1, 2] SC techniques aim to solve complex problems that traditional methods, such as classical logic and probability theory, cannot adequately address. They are based on the idea of mimicking humanlike decision making, which is inherently uncertain, imprecise, and incomplete. It is a combination of strategies and techniques that aims to offer an approximate solution to a challenging problem rather than an accurate one. In this way, SC methods can be used to solve problems in various fields, including engineering, finance, medicine, and many more.

SC techniques compose of metaheuristic algorithms and are thus independent of the type of the problem, in contrast to classical techniques of solving numerical problems, which are dependent on the same (e.g., differential, integro-differential, eigenvalue, linear/nonlinear algebraic equations) and involve various mathematical formulations. They operate on the idea that the problem to be solved may be reduced to a minimizing or maximizing problem with a suitable objective or fitness function, respectively, without bothering about the derivatives of that function. It should be mentioned that, although SC approaches use metaheuristics, the selection of suitable control parameters is a crucial step and may be problem dependent, thus requiring some adaptive alterations. They can either be inspired by nature or by biological/natural processes, to put it broadly. The majority of them fall under the second category, where computing techniques imitate various natural processes,[3] including the behavior of animals (like fireflies, bees, ants, or birds), brain functioning[4] (especially, the brain's reaction to neural stimulus), or the cooling and annealing of metals. The majority of SC approaches employ random numbers at some point in their algorithm, making them stochastic in nature. They do, however, incorporate some bias to direct the search toward the more pertinent regions of the terrain.

Several techniques fall under the umbrella of SC, including Bayesian networks, neural networks, evolutionary algorithms, fuzzy logic, etc. Owing to their strengths and weaknesses, each of these techniques is suitable for different types of problems in the fields of pattern recognition, data mining, control systems, image and natural language processing, to name a few. Their basic objective is to construct computational systems that can successfully address ill-defined and complex problems using conventional, hard computing techniques. Evolutionary algorithms, fuzzy logic, and neural networks have been previously implemented in screening large compound databases to identify

DOI: 10.1201/9781003441328-7

potential drug candidates, predicting the binding affinities of certain compounds to specific targets and property optimization of various new materials including alloys, polymers, and ceramics. In chemical informatics, decision trees and neural networks are commonly employed to predict chemical and physical molecular properties and to optimize reaction conditions. Recently, combining multiple SC methods to address intricate problems, called hybrid methods, have become popular as they often produce superior outcomes than utilizing a single technique. However, it is noteworthy that since the outcomes obtained from these methods are approximate, their accuracy and dependability are highly influenced by the quality of the input data, thus necessitating thorough preprocessing and validation procedures.

7.2 GLOBAL OPTIMIZATION (GO)

The PES for chemical clusters is a complex hypersurface in a $3N$-dimensional space, which is computationally difficult to work with. To make it easier, a transformed PES is generally preferred. One common transformation method includes smoothing out the ruggedness[5, 6] of the PES by locally optimizing the structures predicted in GO. The topology of a PES is heavily influenced by the type of interactions present between the structural units in a cluster, which can be roughly categorized into short-range and long-range interactions. Examples of the former include covalent and hydrogen bonds and dispersion interactions, while the latter include coulombic interaction (charge–charge and charge–dipole). It is noted in the GO of both models[7–9] and real clusters that it is generally easier to achieve GO for clusters dominated by long-range interactions compared to those dominated by short-range interactions. The rationale behind this observation is that for clusters dominated by the latter, regions with large distances have relatively very small effects on one another, leading to a high number of local minima with similar energy and structure. This results in a funnel-like structure on the PES, which introduces barriers that make the PES rugged and difficult to navigate. Consequently, many GO methods struggle to escape from these funnels. On the other hand, the sensitivity of the energy to local structural differences is higher for clusters with predominantly long-range interactions, leading to local minima that are well separated from one another on the PES.

The problem of global optimization can be properly explained by considering Figure 7.1, where the potential energy (PE) curve of a diatomic system is shown. P_0 is the point at which it attains the

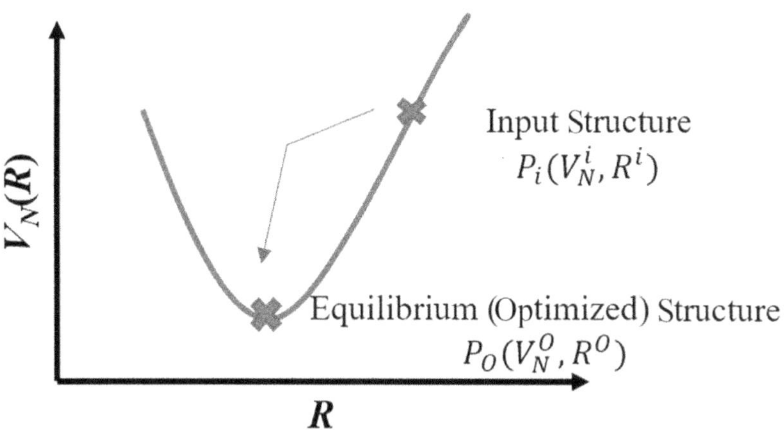

FIGURE 7.1 Representative potential energy curve for a diatomic system.

Source: Reprinted from Ref. [10] with permission from American Chemical Society.

lowest energy, and V_N^0 corresponds to its equilibrium potential energy with an interatomic distance of R_0. Starting from an initial guess geometry with an interatomic distance of R_i, at point P_i and a PE of V_N^i, performing a harmonic motion, we get:

$$V_N - V_N^O = \frac{k}{2}\left(R - R_O\right)^2 \tag{7.1}$$

At point P_i:

$$\left(\frac{\partial V_N}{\partial R}\right)_i = k\left(R_i - R_O\right) \tag{7.2}$$

In general, at all points:

$$\left(\frac{\partial^2 V_N}{\partial R^2}\right) = k \tag{7.3}$$

Substituting equation (7.3) on equation (7.2), we have:

$$\left(\frac{\partial V_N}{\partial R}\right)_i = \left(\frac{\partial^2 V_N}{\partial R^2}\right)\left(R_i - R_O\right) \tag{7.4}$$

$$\therefore R_O = R_i - \left(\frac{\partial V_N}{\partial R}\right)_i / \left(\frac{\partial^2 V_N}{\partial R^2}\right) \tag{7.5}$$

Employing the preceding equations in a multidimensional potential energy surface (PES), where $V_N \equiv E$ and $R \equiv q$, the initial matrices for the derivative of energy with respect to geometry (gradient matrix, g_i) and force constant (Hessian matrix, H_i) can be represented as:

$$\boldsymbol{g}_i = \begin{pmatrix} \left(\partial E / \partial q_1\right)_i \\ \left(\partial E / \partial q_2\right)_i \\ \vdots \\ \left(\partial E / \partial q_n\right)_i \end{pmatrix} \tag{7.6}$$

$$\boldsymbol{H}_i = \begin{pmatrix} \partial^2 E / \partial q_1 \partial q_1 & \partial^2 E / \partial q_1 \partial q_2 & \cdots & \partial^2 E / \partial q_1 \partial q_n \\ \partial^2 E / \partial q_2 \partial q_1 & \partial^2 E / \partial q_2 \partial q_2 & \cdots & \partial^2 E / \partial q_2 \partial q_n \\ \vdots & \ddots & \ddots & \vdots \\ \partial^2 E / \partial q_n \partial q_1 & \cdots & \cdots & \partial^2 E / \partial q_n \partial q_n \end{pmatrix} \tag{7.7}$$

Thus:

$$\boldsymbol{q_O} = \boldsymbol{q}_i - \boldsymbol{H}^{-1}\boldsymbol{g} \tag{7.8}$$

where $\boldsymbol{q}_i = \begin{pmatrix} q_{i1} \\ q_{i2} \\ \vdots \\ q_{in} \end{pmatrix}$ and $\boldsymbol{q}_O = \begin{pmatrix} q_{O1} \\ q_{O2} \\ \vdots \\ q_{On} \end{pmatrix}$

are the matrices for the input and equilibrium structures, respectively.

In quantum chemistry, only a handful of problems are exactly solvable analytically, whereas for the rest we rely upon numerical solutions. In an optimization problem, considering a nonlinear function $f(x)$ with several minima and maxima, we are required to solve for all x_i^* for which $f'(x_i^*) = 0$ and then obtain the sign of $f''(x_i)$ to determine the nature of the extrema.

- **Newton–Raphson approach:**
  Let $f'(x) = g(x) = 0$.
  Starting from a point x_i, where $g(x_i) \neq 0$ but is close to $g(x) = 0$, say, $g(x_i + h) = 0$.
  Taylor expansion provides us with:

$$g(x_i + h) \simeq g(x_i) + hg'(x_i) = 0 \tag{7.9}$$

$$\Rightarrow h = -\frac{g(x_i)}{g'(x_i)} \quad \Rightarrow \quad x_i + h = x_i - \frac{g(x_i)}{g'(x_i)} \tag{7.10}$$

  The new point is calculated as:

$$x_{i+1} = x_i - \frac{g(x_i)}{g'(x_i)}; \; I = 0, 1, 2, \cdots \tag{7.11}$$

  Usually, it converges to one of the solutions of $g(x) = f'(x) = 0$, although finding the global minimum (GM) of $f(x)$ is not inherently certain.
- **Steepest descent/ascent (SD/SA):**
  In the neighborhood of a minimum,

$$f(x_i + h) = f(x_i) + hf'(x_i) + \frac{h^2}{2!} f''(x_i) \tag{7.12}$$

  if the Taylor expansion is convergent,

$$\frac{h^2}{2!} f''(x_i) < hf'(x_i) \tag{7.13}$$

  Then

$$f(x_i + h) < f(x_i) \text{ if } hf'(x_i) < 0 \tag{7.14}$$

  a. **Steepest descent:** Moving in the opposite direction of the local gradient at x_i by λ (a small step length) results in a decrease in the function value.

$$h = -\lambda f'(x_i) \tag{7.15}$$

It continues until we reach a local minimum, i.e., a point x^* for which $f'(x^*) = 0$.

b. **Steepest ascent:** Similarly, a local maximum is reached when

$$h = \lambda f'(x_i) \tag{7.16}$$

This procedure has certain limitations, such as:

- o Getting stuck in a local optimal solution is possible, since $f'(x^*) = 0$ would imply $h = 0$ and thus, no further change in the function $f(x)$.
- o The value of λ can be too small (descent will be very slow) or too large (may surpass the minimum).

- **Newton's approach:**

This method takes care of such problems, i.e., the descent being too slow or the overshooting, but it can also get stuck at a local optimum.
Minimizing $f(x_i + h)$ *w.r.t.* h, we have:

$$f(x_i + h) = f(x_i) + hf'(x_i) + \frac{h^2}{2!} f''(x_i) \tag{7.17}$$

$$f'(x_i) + hf''(x_i) = 0 \Rightarrow h = -\frac{f'(x_i)}{f''(x_i)} \tag{7.18}$$

$$x_{i+1} = x_i - \frac{f'(x_i)}{f''(x_i)} \tag{7.19}$$

The movement direction is still opposite that of the local gradient, but each step length is adjusted based on the local curvature.

7.2.1 Particle Swarm Optimization (PSO)

Kennedy and Eberhart introduced PSO, a nature-inspired optimization algorithm, in 1995.[11, 12] It is a population-based metaheuristic that takes inspiration from the collective behavior of birds or fish. The algorithm operates by moving a swarm of particles throughout the search space, with each particle representing a possible solution to the optimization problem at hand. The movement of the particles is determined by their current position in the search space, their velocity, and the best solutions they have found so far (personal best) and those found by their immediate neighbors (global best). By iteratively updating the position and velocity of each particle, the algorithm aims to converge the swarm to the optimal solution.

Let us consider a function of multivariate nature $f(\vec{X}) = f(x_1, x_2, \ldots, x_N)$ where, $x_1, x_2, \ldots, x_N$ are the continuous variables lying in the range $(-L, L)$. In an N-dimensional search space containing n number of particles, let the ith particle's position at time t be $\vec{X}_i(t)$, and let $\vec{V}_i(t)$ be the corresponding velocity vector. Then:

$$\vec{X}_i(t) \equiv \left(x_{i1}(t), x_{i2}(t), \cdots, x_{iN}(t)\right) \tag{7.20}$$

$$\vec{V}_i(t) \equiv \left(v_{i1}(t), v_{i2}(t), \cdots, v_{iN}(t)\right) \tag{7.21}$$

The initial values of the position $\vec{X}_i(t=0)$ and velocity $\vec{V}_i(t=0)$ vectors are assigned randomly within the ranges $-a$ to a and $-v$ to v, respectively. The fitness of each particle I is calculated while searching for the global optimum solution, which can be represented as $f_i(t)=f(\vec{X}_i(t))$. At any particular time, t, $f_{ibest}(t)$, and $f_{i,lbest}(t)$ are the minima of the function attained by the ith particle and that attained by its immediate neighbors in the swarm, respectively, and $\vec{X}_{ibest}(t)$ and $\vec{X}_{i,lbest}(t)$ are their respective position vectors. Now, the ith particle improves its position via cognitive as well as social components. The cognitive part comes from subtracting its current best ($\vec{X}_i(t)$) from its personal best ($\vec{X}_{ibest}(t)$) position vectors, while the social component comes from the difference between its current best ($\vec{X}_i(t)$) and its neighborhood best ($\vec{X}_{i,lbest}(t)$) position vectors.

The updated position and velocity vectors are:

$$\vec{V}_i(t+1)\equiv\eta\vec{V}_i(t)+c_1r_{1i}(t)\left(\vec{X}_{ibest}(t)-\vec{X}_i(t)\right)+c_2r_{2i}(t)\left(\vec{X}_{i,lbest}(t)-\vec{X}_i(t)\right) \tag{7.22}$$

$$\vec{X}_i(t+1)\equiv\vec{X}_i(t)+\chi\vec{V}_i(t) \tag{7.23}$$

The degree to which the ith particle is inclined to maintain its velocity state is determined by the inertial weight η, while c_1 and c_2 represent the cognitive and social acceleration coefficients, respectively. To allow for oscillations between the two attractors $\vec{X}_{ibest}(t)$and $\vec{X}_{i,lbest}(t)$, the acceleration coefficients are modulated by two random numbers, $r_1(t)$ and $r_2(t)$, that range between 0 and 1. The correction size is limited by the constriction coefficient χ to prevent the swarm from disintegrating. As the algorithm approaches the optimal solution, the magnitude of the random oscillations is decreased, and the two attractors converge to form one. $\vec{X}_{ibest}(t)$and $\vec{X}_{i,lbest}(t)$ get updated:

$$\begin{aligned}\vec{X}_{ibest}(t+1)-\vec{X}_i(t),&\quad \textit{if } f\left(\vec{X}_i(t+1)\right)\geq f\left(\vec{X}_i(t)\right)\\ =\vec{X}_{i+1}(t),&\quad \textit{if } f\left(\vec{X}_i(t+1)\right)< f\left(\vec{X}_i(t)\right)\end{aligned} \tag{7.24}$$

$$\vec{X}_{i,lbest}(t)=min\left\{f\left(\vec{X}_i(t)\right)\right\},\qquad \forall\, X_i \epsilon\, M_i \tag{7.25}$$

where $M_i=\left\{\vec{X}_{i-l}(t),\vec{X}_{i-l+1}(t),\cdots,\vec{X}_i(t),\vec{X}_{i+1}(t),\cdots,\vec{X}_{i+l}(t)\right\}$ defines the neighborhood topology. In this case, the particles interact locally and explore multiple local minima to finally reach the global minimum. The convergence rate is slow in this case, but the probability of finding the global minima is much higher.

There is another topology (known as the *gbest* swarm (Figure 7.2b), where the particles imitate the *gbest* individual, resulting in faster convergence of the problem. However, in this process, scope of exploration is limited to the neighborhood of the *gbest*. Thus the solution fails to reach the GM unless the *gbest* is very close to it. The velocity and position vectors evolve here as follows:

$$\vec{V}_i(t+1)\equiv\eta\vec{V}_i(t)+c_1r_{1i}(t)\left(\vec{X}_{ibest}(t)-\vec{X}_i(t)\right)+c_2r_{2i}(t)\left(\vec{X}_{gbest}(t)-\vec{X}_i(t)\right) \tag{7.26}$$

$$\vec{X}_i(t+1)\equiv\vec{X}_i(t)+\chi\vec{V}_i(t+1) \tag{7.27}$$

7.2.2 Firefly Algorithm (FA)

Proposed in the year 2008 by Xin-She Yang,[13, 14] FA is also a metaheuristic optimization algorithm inspired by the flashing behavior of fireflies. In FA, each firefly represents a potential solution to an optimization problem. A fitness function determines the attractiveness of a firefly (solution).

The fireflies move toward one another in the search space, and their movements are guided by the brightness of their flashing light. Brighter fireflies attract other fireflies toward them, and weaker fireflies move toward the brighter ones. This process is repeated iteratively, with the brightness of the fireflies' lights being updated based on their position and fitness.

Let us consider an n-dimensional function, $f(\vec{X})$; $\vec{X} = x_1, x_2, \ldots, x_N$, with a possible solution vector, $\vec{X}^i = x_1^i, x_2^i, \cdots, x_N^i$. Here, the position vectors represent each firefly I in the search space. The objective function, $O(\vec{X}^i)$, gives the fitness or brightness of the ith firefly at position $\vec{X}^i$. The quality of the solution is directly proportional to the brightness of the firefly. Hence the purpose of the algorithm is to find the vector X^* corresponding to which the related objective function achieves the global maximum, making this a maximization problem. The movement of a firefly toward another firefly is determined by the brightness of the other firefly and the distance between them. The global minimum is the final solution, which has the highest intensity. The intensity of the glow of the ith firefly (I_i) is given by:

$$I_i = \beta_0 e^{-\gamma r_i^2},\ r \geq 0 \tag{7.28}$$

where γ is responsible for controlling the reducing rate of I_i, and β_0 is the intensity at the position of the ith firefly (i.e., at $r_i = 0$). The fireflies I and j attract each other based on the difference in their brightness, and the ith firefly moves towards the jth (assuming $I_j > I_i$) guided by:

$$\vec{X}(t+1) = \vec{X}^i(t) + \beta_0 e^{-\gamma r_{ij}^2}\left(\vec{X}^j(t) - \vec{X}^i(t)\right) + \alpha_t \vec{R}^i(t) \tag{7.29}$$

At time t, the position of the ith firefly is denoted by $\vec{X}^i(t)$, while the position of the jth firefly is $\vec{X}^j(t)$. The distance (Euclidean) between the ith and jth fireflies is represented by r_{ij}. A vector, $\vec{R}^i(t)$ is defined as its components having random numbers between the values 0 and 1. α_t is known as the randomization coefficient, and its value is $10^{-2}L\theta^t$ ($L = \gamma^{-2}$ and $0 < \theta < 1$). Equation (7.29), can be further represented in component form as:

$$x_{ki}(t+1) = x_{ki}(t) + \beta_0 e^{-\gamma r_{ij}^2}\left(x_{kj}(t) - x_{ki}(t)\right) + \alpha_i r_{ki}(t) \tag{7.30}$$

$$I = 1, 2, \cdots, n;\ k = 1, 2, \cdots, N$$

Now a few special cases may arise:

- If $\beta_0 = 0$, a random search is initiated on the objective function space.
- $\alpha_t = 0$, $\beta_0 > 0$, $\gamma = 0$, equation (7.29) will be the same as an *lbest* PSO.
- $\alpha_t = 0$, $\beta_0 > 0$, $\gamma = 0$ and $I_j > I_i$ at time t indicates that equation (7.29) will be the same as a *gbest* PSO
- A general $\beta(r)$ is possible and can be represented as $\beta(r) = \beta_0 e^{-\gamma r^m}$ $(m \geq 1)$.

The algorithm iteratively updates the position of the fireflies based on their attractiveness to other fireflies, intending to find the global minimum of the objective function. The algorithm terminates when the number of iterations reaches its maximum value or when a satisfactory level of fitness is attained.

7.2.3 Artificial Bee Colony (ABC) Algorithm

Proposed in 2005 by Karaboga, the ABC algorithm[15] is a population-based optimization algorithm that has gained widespread use in solving various optimization problems. In ABC, the population comprises three different types of bees. Each employed bee is a representation of a potential solution to the problem and shares information about its fitness with the other bees in the colony. The quality of the food supply, measured by the quantity of nectar, determines the fitness level. The onlooker bees then select a potential solution based on the information passed on from the employed bees and update their position accordingly. Finally, the scout bees randomly search for new potential solutions. The algorithm uses an approach that combines local as well as global search methods to find the optimal solution. The employed bees perform a local search around their current position, while the onlooker bees perform a global search by choosing a potential solution based on the information provided by the employed bees. The scout bees introduce random exploration into the search process by searching for new potential solutions outside the current search space. Each of the four stages that make up the ABC algorithm (the initiation phase, the employed bees phase, the observer bees phase, and the scout bee phase) is discussed next.

During the first phase (initialization), the positions of the food sources are randomly selected from within the defined boundaries:

$$x_{ij} = x_j^{min} + rand(0,1)\left(x_j^{max} - x_j^{min}\right) \tag{7.31}$$

The notations $i = 1,\cdots,SN$ and $j = 1,\cdots,D$ are used, where SN is the number of solutions in the problem, where each solution is a food source and is a D-dimensional vector. x_{ij} is a parameter for the ith employed bee on the jth dimension, whereas x_j^{min} and x_j^{max} represent the lower and upper boundaries, respectively.

During the employed bee phase, each bee assesses the quality of its current solution and explores a new solution in its local area. The bee replaces its current solution with the new one if it has a higher fitness level. To update the position of the ith candidate in the jth dimension during this phase, the following equation is used:

$$v_{ij} = x_{ij} + \phi_{ij}\left(x_{ij} - x_{kj}\right) \tag{7.32}$$

where $\phi_{ij}\left(x_{ij} - x_{kj}\right)$ is the step size, and $k = 1,2,\cdots,SN$, and $j = 1,2,\cdots,D$ are two indices selected randomly.

During the onlooker bee phase in the ABC algorithm, the fitness of each employed bee's solution is evaluated, and the onlooker bees choose a solution based on a probability that is proportional to its fitness value. The probability for selecting the ith solution is given as:

$$p_i = \frac{fit_i}{\sum_{i=1}^{SN} fit_i} \tag{7.33}$$

where fit_i is the fitness of the ith solution. Once a solution is chosen, the onlooker bees generate a new solution by incorporating small modifications to the selected solution within its neighborhood. The new solution is then evaluated for its fitness, and if it is better than the previous solution, it is replaced. If not, the old solution is kept.

The scout bees then generate new solutions by selecting random positions within the search space. The new solutions generated by the scout bees are evaluated for fitness. If a new solution has better fitness than the solution it replaces, it is kept; otherwise, the old solution is retained. The number of employed bees that are replaced by the onlooker bees is controlled by a parameter called abandonment limit. It is a user-defined parameter that sets the maximum number of times an employed bee's solution can be replaced without improving its fitness before it is replaced by a new solution generated by a scout bee. This phase can help the algorithm to converge to a better solution (global minimum) by avoiding premature convergences (local minima). If the abandoned solution is x_i, the scout bee replaces it with a fresh solution in the following way:

$$x_{ij} = x_j^{min} + r \quad \text{and} \quad (0,1)\left(x_j^{max} - x_j^{min}\right) \tag{7.34}$$

7.2.4 Bonobo Optimizer (BO)

The BO is a new metaheuristic optimization algorithm where the social and reproductive behavior of the bonobos are modeled.[16] The algorithm simulates the collaboration and communication among these animals to find optimal solutions to optimization problems. BO is a population-based optimization technique where the initial solutions are generated randomly and are updated iteratively throughout the execution of the code. Each of the bonobos represents a solution and there is a fitness value evaluated for each of these solutions. The entire population is divided into different subgroups with random sizes. The solution having the best fitness value in a bonobo community is termed the alpha bonobo, and it is the "current" best solution. After mating, the bonobos belonging to different subgroups recombine. The solution gets updated through these processes until the stopping criterion is met.

The most important aspect of this algorithm is that the parameters controlling the optimization process adapt themselves to changing circumstances during the evolution of the code. The initial parameter values are as follows: phase count, $pc = 0$; change of phase, $cp = 0$; negative phase count, $npc = 0$; positive phase count, $ppc = 0$; phase probability, $p_p = 0.5$; directional probability, $p_d = 0.5$; probability of extra-group mating, $p_{xgm} = p_{xgm_{initial}}$; temporary subgroup size, $tsgs_{factor} = tsgs_{factor\_initial}$. These parameters evolve as the algorithm proceeds.

A phase is positive (PP) if there are sufficient resources available such as food, high success of mating, protection, etc. In the opposite case, the phase is negative (NP). These are determined by the fitness of the α^{bonobo}. The improvement of the fitness suggests a PP, whereas, in either case, the phase is considered negative. Parameters *npc* and *ppc* count the number of iterations the situation is in the negative and positive phases respectively, throughout the execution of the code.

Initially, taking inspiration from the fission-fusion type of social behavior of bonobos, the algorithm determines the maximum size of *tsgs* ($tsgs_{max}$) for a population size of N:

$$tsgs_{max} = \max(2, tsgs_{factor} * N) \tag{7.35}$$

Next, the algorithm randomly determines the value of *tsgs*, between 2 and $tsgs_{max}$ and selects *tsgs* bonobos from the total population of $N-1$ bonobos (excluding the *i*th bonobo). This is followed by a comparison of the fitness of the best bonobo in this group with that of the *i*th bonobo. The superiority of the former makes it the *p*th bonobo and is selected for mating. If, however, the latter is superior, a random bonobo becomes the *p*th bonobo. Additionally, there is a possibility that the α^{bonobo} is selected for mating through this process.

Now for creating new bonobos, they have a total of four mating strategies. These are promiscuous, restrictive, consortship, and extra-group mating, where the first two occur in the PP and the last two in NP. To decide the mating strategy, a random number, $r_1 \in (0,1)$ is generated. If $r_1 \le p_p$, the offspring are generated through promiscuous or restrictive mating strategy, whereas either of the other two mating strategies can be followed when $r_1 > p_p$.

For the promiscuous and restrictive types of mating (when $r_1 \le p_p$), offspring is created as:

$$\begin{aligned} new\_bonobo_j &= bonobo_j^i + r_1 \times scab \times \left(\alpha_j^{bonobo} - bonobo_j^i\right) \\ &+ (1 - r_1) \times scsb \times flag \times \left(bonobo_j^i - bonobo_j^p\right) \end{aligned} \tag{7.36}$$

where flag = 1 and −1 are for promiscuous and restrictive types of mating, respectively. *scab* and *scsb* are the sharing coefficients for the α^{bonobo} and the chosen *p*th bonobo.

Now for consortship or extra-group mating strategies ($r_1 > p_p$) another random number (r_2) in the range (0,1) is generated. If $r_2 \le p_{xgm}$, extra-group mating strategy is incorporated, else, consortship type of mating is followed.

For extra-group mating, yet another random number(r_3) is generated and it is compared with the directional probability (p_d). If $r_3 \leq p_d$ and the jth variable of the alpha bonobo (α_j^{bonobo}) is greater than or equal to that of ith bonobo ($bonobo_j^i$), the new_bonobo_j is generated using equation (7.37).

$$new\_bonobo_j = bonobo_j^i + \beta_1 \times \left(Var\_max_j - bonobo_j^i\right) \tag{7.37}$$

In cases when $r_3 > p_d$ and $\alpha_j^{bonobo} \geq bonobo_j^i$, equation (7.38) is used.

$$new\_bonobo_j = bonobo_j^i - \beta_2 \times \left(bonobo_j^i - Var\_min_j\right) \tag{7.38}$$

For $\alpha_j^{bonobo} < bonobo_j^i$ and $r_3 \leq p_d$,

$$new\_bonobo_j = bonobo_j^i - \beta_1 \times \left(bonobo_j^i - Var\_min_j\right) \tag{7.39}$$

For $\alpha_j^{bonobo} < bonobo_j^i$ and $r_3 > p_d$,

$$new\_bonobo_j = bonobo_j^i + \beta_2 \times \left(Var\_max_j - bonobo_j^i\right) \tag{7.40}$$

where

$$\beta_1 = e^{\left(r_4^2 + r_4 - \frac{2}{r_4}\right)} \tag{7.41}$$

$$\beta_2 = e^{\left(-r_4^2 + 2r_4 - \frac{2}{r_4}\right)} \tag{7.42}$$

Here, β_1 and β_2 are the two intermediate parameters, r_4 is a random number ($0 < r_4 < 1$); Var_max_j is the upper boundary, and Var_min_j is the lower boundary value for the jth variable.

Again, for consortship type of mating $r_2 > p_{xgm}$, equation (7.43) is used.

$$new\_bonobo_j = \begin{cases} bonobo_j^i + flag \times e^{-r_5} \times \left(bonobo_j^i - bonobo_j^p\right) & if\left(flag = 1 \,||\, r_6 \leq p_d\right) \\ bonobo_j^p, \; otherwise \end{cases} \tag{7.43}$$

where r_5 and r_6 are random numbers. It is suggested that during the PP, the movement probabilities of the bonobos toward the α^{bonobo} are more (exploitation), while during the NP, the tendency of random movement is higher (exploration).

The algorithm uses different mechanisms, such as memory, communication, and migration, to balance the exploration with exploitation in the search space. The memory mechanism helps the algorithm to remember the best solutions found so far, while the communication mechanism enables individuals to share their knowledge and experiences to improve the diversity of the population. The migration mechanism allows individuals to explore new regions of the search space and avoid getting trapped in local optima.

7.2.5 Artificial Neural Network (ANN)

Inspired by the working of the human brain, ANN developed as a machine learning model.[17, 18] It consists of layers of interconnected nodes, also called neurons, through which information is processed to generate output. The whole network is comprised of several components, including an input set ($x_1, x_2, \ldots, x_k, \ldots, x_n$), a set of weights ($w_1, w_2, \ldots, w_k, \ldots, w_n$), an activation function, a biasing term *(b)*, and the output. A weight (w_i) is associated with each of the inputs (x_i) and the products of the inputs with their respective weights are fed into each neuron, which gives the net input as:

$$net = \sum_{i=1}^{n} x_i w_i = \vec{X} \cdot \vec{W} \tag{7.44}$$

For *j*th neuron:

$$net_j = \sum_{i=1}^{n} x_i w_i^j \tag{7.45}$$

The next step in an artificial neural network (ANN) involves applying an activation function. This function determines the output of a neuron based on its respective weights. By making use of the activation function (Φ), the net input received by a neuron is compared with a threshold (θ_j). If Φ (net_j) exceeds the threshold, we get the output (y_j):

$$y_j = \Phi\left(net_j\right) \tag{7.46}$$

The most widely used activation functions in artificial neural networks are the sigmoid function, signum function, step function, rectified linear unit (ReLU), and tanh functions.

7.2.5.1 Single Layer Perceptron (SLP)

An SLP is a basic type of ANN that is commonly used for binary classification tasks. It is comprised of a single layer of neurons that are connected to the input layer of the network (Figure 7.2a). The perceptron's output is calculated by taking a linear combination of the input values, which is then passed through a threshold function to generate the final output.

Let the set of input variables be $x_1, x_2, \ldots, x_k, \ldots, x_n$, which are fed to the input layer neurons 1, 2, …, *n*, the net input to the *j*th neuron is given by:

$$net_j = \sum\nolimits_{i=1}^{n} x_i w_{ij}; \quad j = 1, 2, \ldots, m \tag{7.47}$$

Now, this net_j is passed through the threshold function *f*(*z*), which gives the output (y_j):

$$y_j = f(net_j) \tag{7.48}$$

In single layer perceptron, the threshold function used is mostly a step function where we get binary values (0 or 1).

$$F(net_j) = \begin{cases} 1, & \text{if } net_j > 0 \\ 0, & \text{otherwise} \end{cases} \tag{7.49}$$

Also, linear or sigmoid functions can be used, and the corresponding outputs will be:

- Linear: $y_j = \Phi\left(net_j\right) = m \cdot net_j$ (7.50)
- Sigmoid: $y_j = \Phi(net_j) = \dfrac{1}{1 + e^{-nst_j}}$ (7.51)

7.2.5.2 Backpropagation Network (BPN)

A BPN is a type of ANN that consists of multiple layers and is often used for supervised learning tasks such as classification and regression. The information in BPN flows in a feedforward manner, starting from the input layer and passing through one or more hidden layers before reaching the output layer (Figure 7.2b). When multiple hidden layers are present, the neurons of one hidden layer are connected to those of the next hidden layer, and so on up to the output layer. This creates a hierarchy of layers where the neurons in each layer take inputs from its previous layer and create outputs for the next layer. The net input for each hidden layer is given by:

$$net_j = \sum_{i=1}^{n_j} x_i w_{ij}; \qquad j = 1, 2, \cdots, n_H \tag{7.52}$$

Now considering a sigmoid activation function and a threshold θ, the net output for the *j*th hidden layer is given by:

$$h_j = f\left(net_j\right) = \frac{1}{1 + e^{-\left(net_j + \theta_j\right)}} \tag{7.53}$$

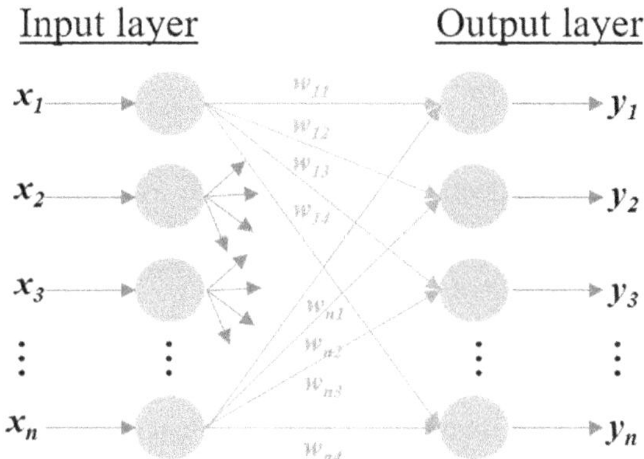

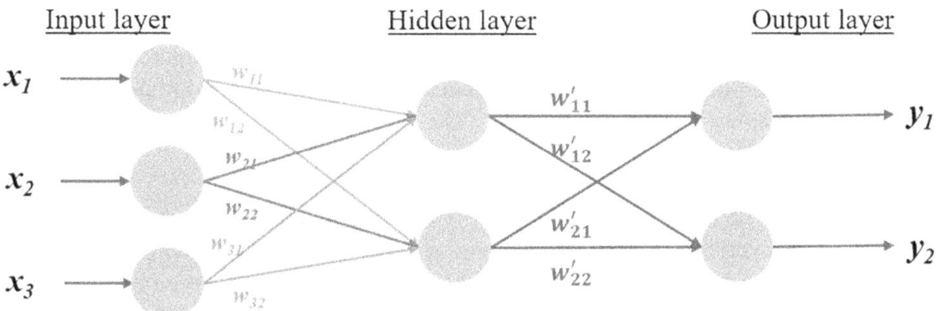

FIGURE 7.2 (a) SLP and (b) BPN as utilized in ANN.

Source: Reprinted from Ref. [10] with permission from American Chemical Society.

To reduce the error between the real and the predicted output to a minimum, the network employs this technique of backpropagation, which adjusts the weights of the connections between neurons. The network's performance can be enhanced by adjusting the number of hidden layers and the number of neurons contained in each of them. However, increasing those numbers can also increase the complexity of the network, which may lead to overfitting or slower training times.

7.2.6 Convolutional Neural Network (CNN)

A special type of neural network (artificial) that is highly effective in recognition tasks, especially video and image recognition is CNN.[19–22] Inspired by the neuron of the human nervous system and utilizing a process called convolution to identify patterns in the input data. From input data, CNN adaptively learns from spatial hierarchies, is designed in such a way that allows them to identify a complex vision, and is now widely used in high-tech automated applications such as self-driving vehicles, medical imaging, etc. Known for automatically learning and identifying features from input data, CNNs are a class of neural network that is achieved with convolutional layers that are connected to one or more pooling layers, which are then followed by connected layers. In contrast to an ANN, the weights and biases of neurons across all hidden layers in a CNN are identical. The unique features of input data are detected by the convolutional layers that contain kernels or filters. The pooling layers reduce the size of the data, reducing computational cost and preventing overfitting. The output of the convolutional and pooling layers is processed by the fully connected layer, and the final output is produced. An activation function is often applied to the final output to determine the probability of each class.

7.2.6.1 Forward Propagation

To describe the process of forward propagation in a convolutional neural network (CNN), we can refer to Figure 7.3a. The convolution operation is denoted by an asterisk (*). Let us assume that the input image is X and the dimension is (n, n). The filter is assumed to be f with dimensions (f, f) and the output Z_1 having dimensions of $((n - f + 1), (n - f + 1))$ and is obtained as:

$$\boldsymbol{Z}_1 = \boldsymbol{X} * \boldsymbol{f} \tag{7.54}$$

Next, the sigmoid activation function is applied to Z_1, resulting in $\mathbf{A} = sigmoid\ (\boldsymbol{Z}_1)$. The fully connected layer receives the features extracted from the convolutional layer and produces the final result, which is the same as conventional AN. A linear transformation is then performed by the fully connected layer, $\boldsymbol{Z}_2 = \boldsymbol{W}^T \cdot \boldsymbol{A} + b$, where W is a (m, n) matrix with m being the input and n the number of hidden layers, followed by a nonlinear transformation, $f(x) = \dfrac{1}{1+e^{-x}} \Rightarrow (0,1)$, which uses an activation function. The sigmoid function $\mathbf{Z}_2$ is then applied to obtain the output. The parameters $\boldsymbol{f}$, $\boldsymbol{W}$, and b are initialized randomly by the CNN model, trained, and updated during backpropagation to achieve the prediction.

7.2.6.2 Backward Propagation

Figures 7.3b and 7.3c depict the overall workflow of the method. In the backward propagation stage of the CNN, the network calculates the error between the targeted and obtained output, which is also known as the loss function. The network then uses a gradient descent method to update the parameters of the network. Initially, the parameters are randomly initialized, and the error is not at its minimum value. The gradient descent method works by adjusting the parameters based on the slope of the error function until it reaches the point where the error is at its minimum value.

$$a_{new} = a_{old} - \left(lr \cdot \frac{\partial E}{\partial a} \right) \tag{7.55}$$

Input → Convolution → Sigmoid → Linear Transformation → Sigmoid → Output

X $\quad Z_1 = X * f$ $\quad A = Sigmoid\,(Z_1)$ $\quad Z_2 = W^T \cdot A + b$ $\quad Sigmoid\,(Z_2)$

(a) CNN Forward Propagation

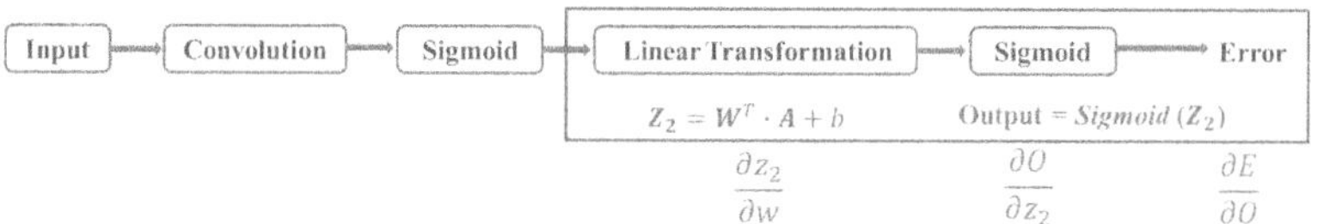

(b) CNN Backward Propagation: Fully Connected Layer

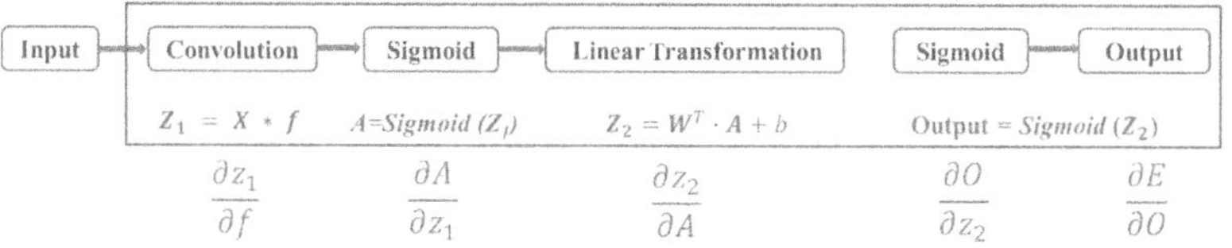

(c) CNN Backward Propagation: Convolution Layer

FIGURE 7.3 Workflow of (a) forward and (b) backward propagations in (c) fully connected and (c) convolution layers.

Source: Reprinted from Ref. [10] with permission from American Chemical Society.

where a_{new} and a_{old} are the new and the old parameters, respectively, lr is the learning rate constant, and $\frac{\partial E}{\partial a}$ is the gradient.

$\frac{\partial E}{\partial w}$ and $\frac{\partial b}{\partial w}$ are calculated in the fully connected layer:

- Change in error *w.r.t.* output:

$$E = \frac{(y-O)^2}{2} \quad \Rightarrow \quad \frac{\partial E}{\partial O} = -(y-O) \tag{7.56}$$

- Change in output *w.r.t.* z_2 (linear transformation of the output):

$$f(x) = \frac{1}{(1+e^{-x})}$$

$$\Rightarrow f'(x) = (1+e^{-x})^{-1}\left[1-(1+e^{-x})^{-1}\right]$$

$$\Rightarrow f'(x) = sigmoid(x)\left[1 - sigmoid(x)\right]$$

Derivative of the sigmoid:

$$\frac{\partial O}{\partial z_2} = (O)(1-O) \tag{7.57}$$

- Change in z_2 *w.r.t.* w (weights):

$$\boldsymbol{Z}_2 = \boldsymbol{W}^T \cdot \boldsymbol{A} + b$$

$$\frac{\partial z_2}{\partial w} = A \tag{7.58}$$

Hence:

$$\frac{\partial E}{\partial w} = \frac{\partial E}{\partial O} \cdot \frac{\partial O}{\partial z_2} \cdot \frac{\partial z_2}{\partial w} = -(y-O)\cdot(O)\cdot(1-O)\cdot A \tag{7.59}$$

$$W_{new} = W_{old} - \left(lr \cdot \frac{\partial E}{\partial w} \right) \tag{7.60}$$

A similar process is performed on the bias matrix.

The convolution layer:

$$\frac{\partial E}{\partial f} = \left(\frac{\partial E}{\partial O} \cdot \frac{\partial O}{\partial z_2} \cdot \frac{\partial z_2}{\partial A} \cdot \frac{\partial A}{\partial z_1} \right) * \frac{\partial z_1}{\partial f} \tag{7.61}$$

where

$$\boldsymbol{Z}_1 = \boldsymbol{X} * \boldsymbol{f} \Rightarrow \frac{\partial z_1}{\partial f} = X \text{ ;}$$

$$\mathbf{A} = sigmoid(\boldsymbol{Z}_1) \Rightarrow \frac{\partial A}{\partial z_1} = A(1-A)$$

$$\boldsymbol{Z}_2 = \boldsymbol{W}^T \cdot \boldsymbol{A} + b \Rightarrow \frac{\partial z_2}{\partial A} = W^T$$

$$f_{new} = f_{old} - \left(lr \cdot \frac{\partial E}{\partial f} \right) \tag{7.62}$$

In the training phase, the weights of the filters in the convolutional layers are modified using backpropagation, an algorithm used to compute the gradient of the loss function with respect to the weights. The loss function computes the discrepancy between the network's predicted and the actual outputs, and backpropagation fine-tunes the weights to decrease this discrepancy.

7.2.7 Basin Hopping (BH)

BH is a stochastic global optimization technique that combines local optimization and random perturbations to explore the energy landscape of the system.[23] The algorithm transforms the PES of the system, denoted by ($E(\vec{X})$), into a series of interconnected staircases, while maintaining the GM and the relative energies of the local minima (Figure 7.4).

The basic idea of basin hopping is to start with an initial configuration of the system and perform a local optimization to find the nearest local minimum of the energy. Then the system is randomly perturbed to explore the neighboring regions of the energy landscape. If the energy (E_{new}) of the new configuration is lower energy than that of the previous one (E_{old}), it is accepted as the new minimum. If not, the perturbed configuration is accepted with a probability determined by the Metropolis criterion. According to the Metropolis criterion, $\exp\left[(E_{old} - E_{new})/kT\right]$ is evaluated, and if it is greater than a random number drawn in the range [0, 1], E_{new} is accepted; otherwise, the previous one is accepted. Here the temperature is an adjustable parameter though it is not used for annealing.

The BH algorithm repeats this cycle of local optimization and random perturbation, gradually lowering the temperature of the system to explore deeper energy basins, until the energy reaches a

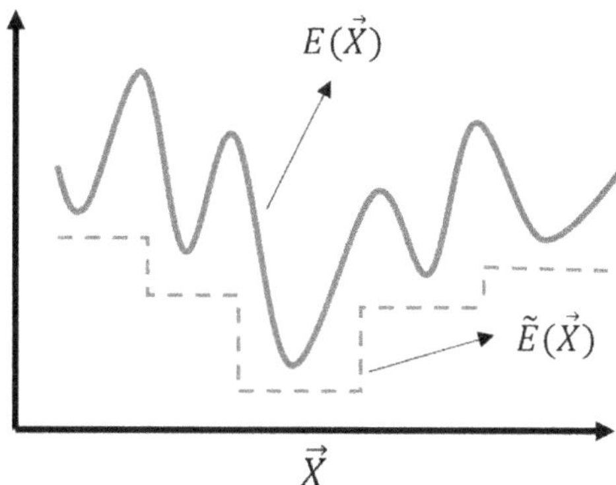

FIGURE 7.4 Reduction of the energy landscape into an energy staircase using the BH algorithm.

Source: Reprinted from Ref. [10] with permission from American Chemical Society.

global minimum or a stopping criterion is met. Basin hopping is an efficient method to obtain the global minimum even when multiple local minima are present.

7.2.8 Simulated Annealing (SA)

SA is a GO technique inspired by the metallurgy technique of annealing. This technique involves heating a material and then slowly cooling it in order to improve its crystalline structure.[24, 25] In SA, the objective is to find the global minimum of a cost or energy function by exploring the search space using a randomized approach.

The algorithm begins with a random solution to the multidimensional function $f(X)$, ($X = x_1, x_2, \ldots, x_n$) at an initial temperature (T_0). It then proceeds iteratively by introducing small random perturbations to the current solution, simulating the effects of thermal noise. If the kth variable undergoes a fluctuation, it generates a new value according to the following equation:

$$x_k^{'} = x_k + \Delta x_k\left(1-2r\right) = x_k + \Delta s r_1\left(1-2r\right) \tag{7.63}$$

The equation describes how a new value is generated for the kth variable, where r is a random number within the range 0–1, and Δs is the step length (small) specified by the user. The size of fluctuation, Δx_k, is defined as Δs times r_1. The term $\left(1-2r\right)$ determines the sign (+ or −) of the fluctuation with an equal probability.

Suppose a new solution is found, and it has a higher cost or energy than the current solution. In that case, it may be accepted with a probability that depends on two factors: the temperature (T) and the energy difference between the new (F') and the current solutions (F): $\Delta F = F' - F$.

The probability is given by:

$$P\left(\Delta F\right) = e^{-\Delta F/k_B T} \tag{7.64}$$

If $P\left(\Delta F\right) > 1$, the move is accepted; if not, the Metropolis criterion is incorporated to generate a random number r. If $r < P\left(\Delta F\right)$, the move gets accepted (uphill movement); otherwise, the move is rejected.

During each iteration, the temperature is systematically decreased based on a cooling schedule, which is usually represented by an exponential or logarithmic function. The cooling schedule plays a crucial role in determining the probability of accepting suboptimal solutions as the search

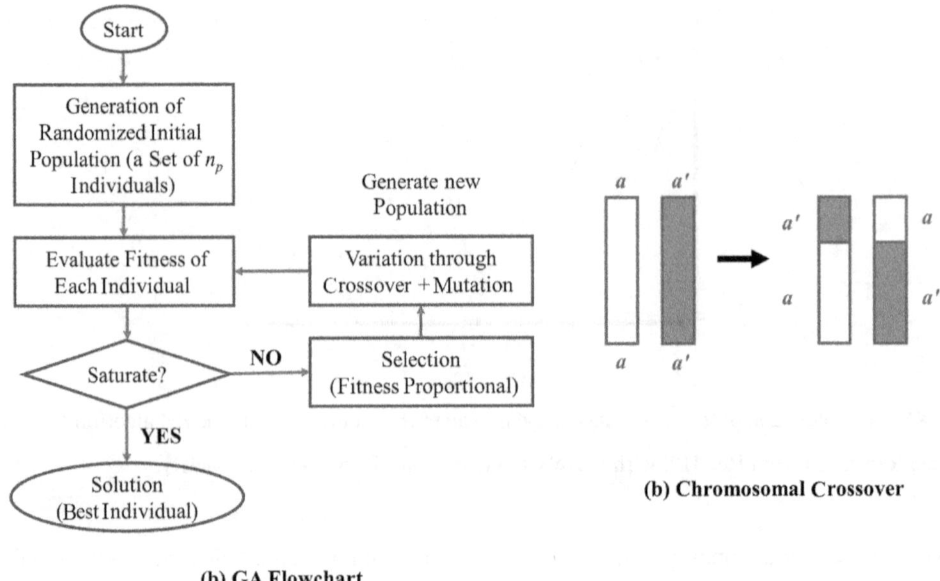

FIGURE 7.5 Flowchart of the GA method and a representative chromosomal crossover.

Source: Reprinted from Ref. [10] with permission from American Chemical Society.

proceeds. This enables the algorithm to explore the search space more extensively at the beginning and gradually converge toward an optimal solution as the temperature decreases.

7.2.9 Genetic Algorithm (GA)

A heuristic population-based search and optimization algorithm called genetic algorithm (GA) takes inspiration from the principles of natural selection and genetics in biological evolution.[26] The algorithm creates a population of candidate solutions represented as chromosomes or strings of genetic material, with each chromosome representing a possible solution to the optimization problem (Figure 7.5).

The algorithm initially generates a population of chromosomes represented as $\{X_i\}$ ($i = 1, 2, \ldots, n_p$), which is done either randomly or by means of some heuristic methods. It then iteratively evolves the population of chromosomes over several generations. In each generation, the chromosomes are evaluated based on a fitness function $f(X)$, ($X = x_1, x_2, \ldots, x_n$) that measures how well they perform on the optimization problem. The fitness function maps the input variables, X_i, to a scalar value, f_i, and evaluates the solutions which are better suited for mating. The average fitness value of the population, f_{av}, can be calculated using the fitness value, f_i, of each chromosome.

The algorithm applies genetic operators such as selection, crossover, and mutation to generate new offspring chromosomes. The selection process identifies the most robust chromosomes (based on f_{av}) in the current generation, which are then chosen as parents for the next generation. Crossover, on the other hand, combines genetic material from two parents to generate new offspring chromosomes. With a probability of p_c, crossover typically occurs at a randomly selected location within the chromosome, resulting in two distinct offspring chromosomes that differ from both parents. While mutation introduces a small probability of random changes, p_m, of the genetic material in the offspring chromosomes. The mutation operator incorporates new genetic information into the population and hopes not to get stuck at the local minima. The offspring chromosomes then form the next generation, and the process repeats until either a maximum number of generations is completed or a satisfactory level of fitness is reached. GA is a powerful optimization technique as it can explore large search spaces and handle nonlinear and nonconvex optimization problems.

7.3 CASE STUDIES

7.3.1 CNN and PSO in the Determination of GM Structures

The energy of N_4^{2-}, N_6^{4-}, C_5, Au_n ($n=2-8$), and Au_nAg_m ($2 \leq n+m \leq 8$) clusters[27–29] is learned and predicted using a CNN model. To begin with, molecular dynamics simulation performed using ADMP[30, 31] generates random geometries within a specified range in the 3D space, and their energy is calculated. This initial set is used to set up the CNN model, which is then used to generate numerous systems needed while searching for the GM. The architecture of the model includes a convolutional layer and two fully connected layers. The final layer uses an activation function of a linear nature, whereas sigmoid functions are used in the other layers. The objective function in the PSO is the final energy that is derived from the CNN model. Figure 7.6 shows the flowchart of the complete model. Once convergence is achieved in the PSO, a postprocessing step is carried out using Gaussian 09[32] to obtain the final optimized geometry. Figures 7.7a and 7.7b depict the optimized geometries.

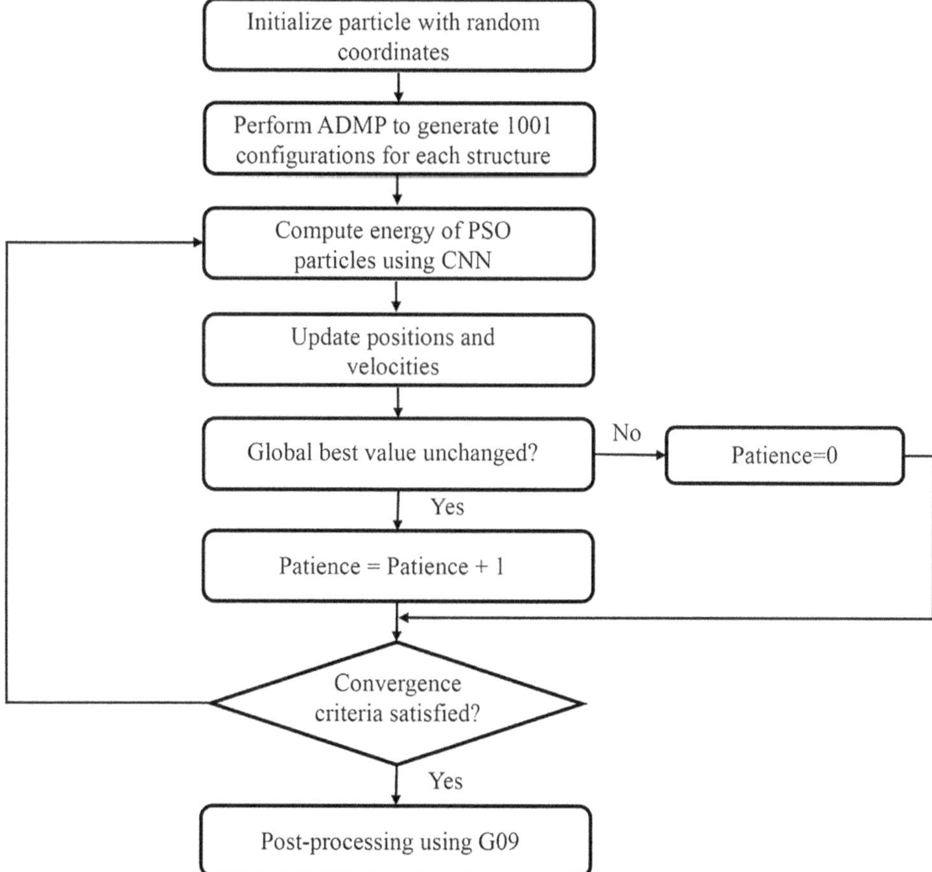

FIGURE 7.6 Workflow of the SC model made from the combination of CNN, PSO, and G09.

Source: Reprinted from Ref. [27] with permission from Springer Nature.

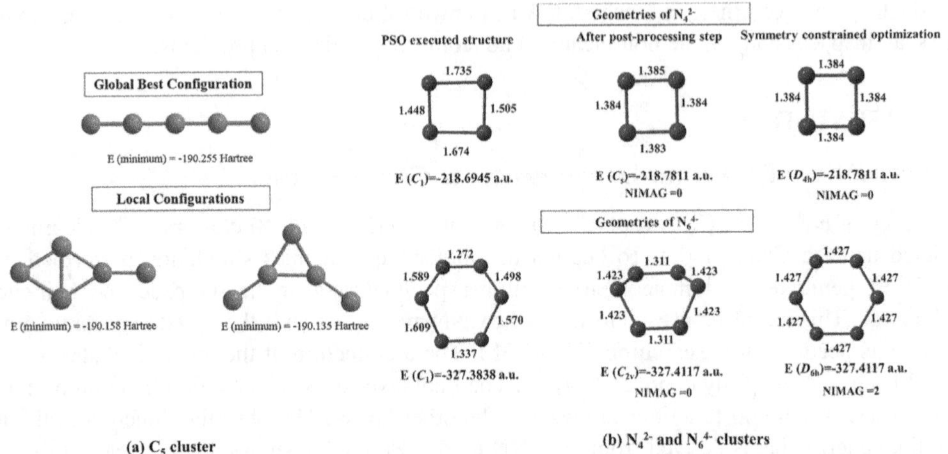

FIGURE 7.7 Global and local minimum energy structures for C5, N_4^{2-}, and N_6^{4-} clusters.

Source: Reprinted from Ref. [27] with permission from Springer Nature.

The linear geometry of the C_5 GM structure was identified through the use of this highly efficient optimization method that surpasses the Gaussian 09 program's normal process of optimizing random guess input structures. The success rate of this method was 77–90%, making it highly effective. The efficiency of the optimization method increases as the training data size grows and a greater percentage of configurations converge toward the *gbest* position, which is typical of any SC algorithm. The current technique outperforms SA and BH with respect to execution time, the number of iterations required to reach convergence, and the energy of the final optimized geometry. The success rates for N_4^{2-} and N_6^{4-} were 87.5% and 85%, respectively. The GMs of Au_n and Au_nAg_m were the same as those reported in previous studies.

7.3.2 FA Integrated with DFT for the GO of Al_4^{2-} Clusters

A hybrid approach that combines FA and DFT has been employed to achieve global optimization of the Al_4^{2-} cluster.[33] Figure 7.8 represents a firefly "particle" in the search space. The performance of this approach is compared to that of PSO on the same system. The modified FA approach is found to perform better than PSO in terms of both success rate (~98–99% for FA, ~91–92% for PSO) and efficiency. The GM structure of Al_4^{2-} is known to be planar, and incorporating planarity enhances the convergence rate by reducing the number of variables. On the other hand, the nonplanar approach often leads to getting trapped in a local minimum and immature convergence. In the case of Al_4^{2-}, this modified FA is a better method for locating the GM. Moreover, a correlation between the aromaticity and energy of the cluster indicates that the former increases with the stability of the cluster.

7.3.3 ABC Algorithm in the Determination of GM Structures of Hypercoordinate Clusters

The ABC algorithm, introduced by Zhang and Dolg in 2015,[34] contains several innovative features and can use deterministic local search procedures when required without affecting the swarm's ability to do a global search or its self-organization. We have worked on some small clusters with unusual properties using this modified ABC approach to find their respective GM structures. For example, studying the PES of $CSi_nGe_{4-n}^{2+/+/0}$ (n = 1–3) complexes indicated that the GM structures in their di-cationic states contain a planar tetracoordinate carbon (ptC) (Figures 7.9a, 7.9b, and 7.9c).[35] In

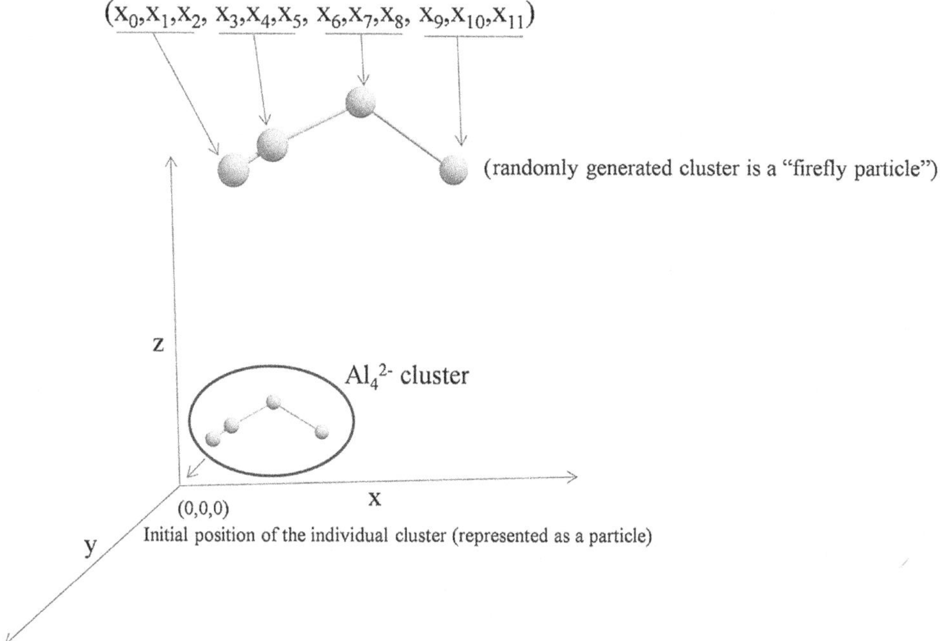

FIGURE 7.8 Representative firefly particle represented in the multidimensional search space.

Source: Reprinted from Ref. [33] with permission from Springer Nature.

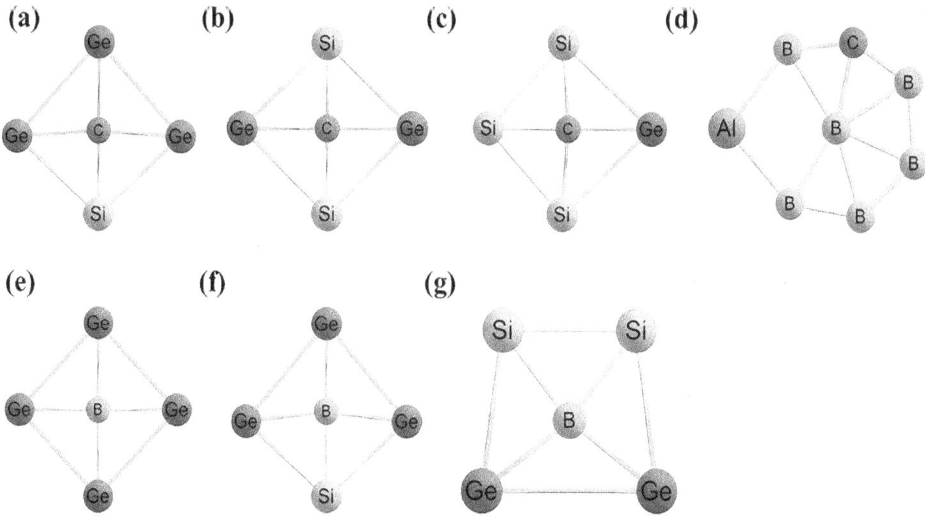

FIGURE 7.9 GM structures of (a) $CSiGe_3^{2+}$, (b) $CSi_2Ge_2^{2+}$, (c) CSi_3Ge^{2+}, (d) $CB_6Al^{0/+}$, (e) BGe^{4+}, (f) $BSiGe^{3+}$, and (g) CSi_2Ge^{2+} clusters.

Source: Reprinted from Ref. [10] with permission from American Chemical Society.

another study, we discovered that the PESs of $CB_6Al^{0/+}$ complexes contain a planar hexacoordinate boron (phB) in their GM structures (Figure 7.9d).[36] We have also reported a planar tetracoordinate boron (ptB) within the GM structures of $BSi_nGe_{4-n}^+$ (n = 0–2) clusters (Figures 7.9e, 7.9f, and 7.9g).[37] Other molecular groups with exotic properties, such as the reactivity and cooperativity of small boron clusters,[38] helical molecular orbitals of conjugated linear molecules,[39] helical-shape spin densities,[40] alkalides,[41, 42] molecular electrides,[43, 44] hypercoordinated carbon[45, 46] and boron clusters,[47] frustrated Lewis pairs (FLPs),[48, 49] magic clusters,[50, 51] fluxional molecules,[52–54] and so on, could be investigated using soft computing techniques.

At present, our team is utilizing the ABC algorithm to investigate the potential energy surfaces of small clusters that contain hypercoordinated elements from the main group in their global minimum structures. Additionally, we are exploring the application of the BO approach to determine the GO of different carbon clusters. In addition, we have created a deep generative model using a variational autoencoder for molecular graphs, which we have named NEVAE.[55] By conducting a comparative analysis of the effectiveness of different soft computing techniques on specific systems, we can gain better insights into the suitability of each method as a global optimizer.

7.4 SUMMARY

Computational chemistry has significantly benefitted from the application of soft computing techniques, which provide efficient, rapid, and reliable solutions to complex problems that are hard to solve analytically. Once we identify the most suitable SC technique for a particular problem, solving chemical and physical problems becomes straightforward. Metaheuristic techniques are particularly helpful as they can be customized to the requirements of the problem. Essential steps in these approaches involve formulating a fitness function that takes relevant constraints into account, initiating the population with seeds, enabling individuals to explore the search space, and establishing a suitable termination criterion. If required, an additional SC technique can be utilized, where the result of the first technique can act as input for the second technique to achieve the global optimum, or a postprocessing step may be adequate. With the increasing availability of computational resources, there is no limit to the type and amount of problems that can be tackled with SC. We have conducted studies on the global optimization of numerous chemical systems, utilizing CNN, PSO, FA, ABC, BO, and hybrid techniques, which provide better-quality outcomes through the use of two or more SC techniques (and/or one technique augmented by an electronic structure calculation method). The chemistry community has widely adopted these techniques, which have resulted in notable progress in various fields such as chemical informatics, material design, and drug discovery.

7.4.1 Acknowledgments

PKC would like to thank Professor Ambrish K. Srivastava for kindly inviting him to contribute a chapter to the book entitled *Computational Studies: From Molecules to Materials*. He thanks DST, New Delhi for the J. C. Bose National Fellowship, grant number SR/S2/JCB-09/2009. He is also thankful to all his students and collaborators whose work is discussed here. RP and BC thank CSIR and MHRD, respectively, for their fellowships.

7.4.2 Conflict of Interest

The authors declare that they have no conflict of interest, financial and/or otherwise.

REFERENCES

(1) Kecman, V. *Learning and Soft Computing: Support Vector Machines, Neural Networks, and Fuzzy Logic Models*; MIT Press: Cambridge, MA, **2001**.
(2) *Fundamentals of the New Artificial Intelligence*; Munakata, T., (Ed.), Texts in Computer Science; Springer: London, **2007**.
(3) Russell, S. J.; Norvig, P. *Artificial Intelligence: A Modern Approach*; Prentice Hall Series in Artificial Intelligence; Prentice Hall: Upper Saddle River, NJ, **2010**.

(4) Luger, G. F. *Artificial Intelligence: Structures and Strategies for Complex Problem Solving* (6th ed.). Addison-Wesley: Boston, MA, **2008**.
(5) Wales, D. J.; Scheraga, H. A. Global optimization of clusters, crystals, and biomolecules. *Science*, **1999**, *285*, 1368–1372.
(6) Doye, J. P.; Wales, D. J. Thermodynamics of global optimization. *Phys. Rev. Lett.*, **1998**, *80*, 1357.
(7) Doye, J. P.; Wales, D. J. Structural transitions and global minima of sodium chloride clusters. *Phys. Rev. B*, **1999**, *59*, 2292.
(8) Doye, J. P.; Miller, M. A.; Wales, D. J. Evolution of the potential energy surface with size for Lennard-Jones clusters. *J. Chem. Phys.*, **1999**, *111*, 8417–8428.
(9) Doye, J. P.; Wales, D. J.; Zetterling, F. H.; Dzugutov, M. The favored cluster structures of model glass formers. *J. Chem. Phys.*, **2003**, *118*, 2792–2799.
(10) Pal, R.; Das, P.; Chattaraj, P. K. Global optimization: A soft computing perspective. *J. Phys. Chem. Lett.* **2023**, *14*, 3468–3482.
(11) Kennedy, J.; Eberhart, R. Particle swarm optimization. In *Proceedings of ICNN'95_ International Conference on Neural Networks*, Perth, WA, Australia, November 27–December 1, 1995; IEEE. **1995**; *4*, 1942–1948.
(12) Kennedy, J.; Eberhart, R. C. *Swarm Intelligence*. Morgan Kaufmann Publishers: San Francisco, CA, **2001**.
(13) Yang, X. S. *Nature-Inspired Metaheuristic Algorithms* (1st ed.). Luniver Press: Frome, **2008**.
(14) Yang, X. S. Firefly algorithms for multimodal optimization. In *Proceedings of the Stochastic Algorithms: Foundations and Applications: 5th International Symposium (SAGA 2009)*, Sapporo, Japan, October 26–28, 2009; Watanabe, O., Zeugmann, T., (Eds.). Springer: Berlin. **2009**; 169–178.
(15) Karaboga, D. *An Idea based on Honey Bee Swarm for Numerical Optimization*; Technical Report TR06.; Erciyes University Press: Erciyes **2005**.
(16) Das, A. K.; Pratihar, D. K. A new bonobo optimizer (BO) for real-parameter optimization. In *2019 IEEE Region 10 Symposium (TENSYMP)*; IEEE. **2019**; 108–113.
(17) MacGregor, R. J. *Neural and Brain Modeling*; Academic Press: San Diego, CA, **1987**.
(18) Haykin, S. *Neural Networks: A Comprehensive Foundation* (2nd ed.). Prentice Hall PTR: Upper Saddle River, NJ, **1998**.
(19) Hubel, D. H.; Wiesel, T. N. Receptive fields and functional architecture of monkey striate cortex. *J. Physiol.* **1968**, *195*, 215–243.
(20) Fukushima, K.; Miyake, S. Neocognitron: A self-organizing neural network model for a mechanism of visual pattern recognition. In *Competition and Cooperation in Neural Nets*; *Lecture Notes in Biomathematics* (Vol. 45, pp. 267–285). Springer: Berlin, Heidelberg, **1982**.
(21) Le Cun, Y.; Boser, B.; Denker, J. S.; Henderson, D.; Howard, R. E.; Hubbard, W.; Jackel, L. D. Handwritten digit recognition with a back-propagation network. In *Advances in Neural Information Processing Systems 2 (NIPS 1989)* (pp 396–404). Morgan Kaufmann Publishers, **1989**.
(22) LeCun, Y.; Bottou, L.; Bengio, Y.; Haffner, P. Gradient-based learning applied to document recognition. *Proc. IEEE.* **1998**, *86*, 2278–2324.
(23) Wales, D. J.; Doye, J. P. K. Global optimization by basin-hopping and the lowest energy structures of Lennard-Jones clusters containing up to 110 atoms. *J. Phys. Chem. A.* **1997**, *101*, 5111–5116.
(24) Kirkpatrick, S.; Gelatt, C. D.; Vecchi, M. P. Optimization by simulated annealing. *Science.* **1983**, *220*, 671–680.
(25) Černý, V. Thermodynamical approach to the traveling salesman problem: An efficient simulation algorithm. *J. Optim. Theory Appl.* **1985**, *45*, 41–51.
(26) Holland, J. H. *Adaptation in Natural and Artificial Systems: An Introductory Analysis with Applications to Biology, Control, and Artificial Intelligence*; University of Michigan Press: Ann Arbor, MI, **1975**.
(27) Mitra, A.; Jana, G.; Pal, R.; Gaikwad, P.; Sural, S.; Chattaraj, P. K. Determination of stable structure of a cluster using convolutional neural network and particle swarm optimization. *Theor. Chem. Acc.* **2021**, *140*, 30.
(28) Mitikiri, P.; Jana, G.; Sural, S.; Chattaraj, P. K. A machine learning technique toward generating minimum energy structures of small boron clusters. *Int. J. Quantum Chem.* **2018**, *118*, No. e25672.
(29) Jana, G.; Mitra, A.; Pan, S.; Sural, S.; Chattaraj, P. K. Modified particle swarm optimization algorithms for the generation of stable structures of carbon clusters, C_n ($n = 3-6, 10$). *Front. Chem.* **2019**, *7*, 485.
(30) Schlegel, H. B.; Millam, J. M.; Iyengar, S. S.; Voth, G. A.; Daniels, A. D.; Scuseria, G. E.; Frisch, M. J. Ab initio molecular dynamics: Propagating the density matrix with Gaussian orbitals. *J. Chem. Phys.* **2001**, *114*, 9758–9763.

(31) Iyengar, S. S.; Schlegel, H. B.; Voth, G. A. Atom-centered density matrix propagation (ADMP): Generalizations using Bohmian mechanics. *J. Phys. Chem. A.* **2003**, *107*, 7269–7277.
(32) Frisch, M. J.; Trucks, G. W.; Schlegel, H. B.; Scuseria, G. E.; Robb, M. A.; Cheeseman, J. R.; Scalmani, G.; Barone, V.; Mennucci, B.; Petersson, G. A.; Nakatsuji, H.; Caricato, M.; Li, X.; Hratchian, H. P.; Izmaylov, A.F.; Bloino, J.; Zheng, G.; Sonnenberg, J.L.; Hada, M.; Ehara, M.; Toyota, K.; Fukuda, R.; Hasegawa, J.; Ishida, M.; Nakajima, T.; Honda, Y.; Kitao, O.; Nakai, H.; Vreven, T.; Montgomery Jr.; J.A.; Peralta, J.E.; Ogliaro, F.; Bearpark, M.; Heyd, J. J.; Brothers, E.; Kudin, K.N.; Staroverov, V. N.; Kobayashi, R.; Normand, J.; Raghavachari, K.; Rendell, A.; Burant, J. C.; Iyengar, S. S.; Tomasi, J.; Cossi, M.; Rega, N.; Millam, J. M.; Klene, M.; Knox, J. E.; Cross, J. B.; Bakken, V.; Adamo, C.; Jaramillo, J.; Gomperts, R.; Stratmann, R. E.; Yazyev, O.; Austin, A. J.; Cammi, R.; Pomelli, C.; Ochterski, J.W.; Martin, R. L.; Morokuma, K.; Zakrzewski, V. G.; Voth, G. A.; Salvador, P.; Dannenberg, J. J.; Dapprich, S.; Daniels, A.D.; Farkas, O.; Foresman, J. B.; Ortiz, J. V.; Cioslowski, J; Fox, D.J. *Gaussian 09*. Gaussian, Inc.: Wallingford, CT, **2009**.
(33) Mitra, A.; Jana, G.; Agrawal, P.; Sural, S.; Chattaraj, P. K. Integrating firefly algorithm with density functional theory for global optimization of Al_4^{2-} clusters. *Theor. Chem. Acc.* **2020**, *139*, 32.
(34) Zhang, J.; Dolg, M. ABCluster: The artificial bee colony algorithm for cluster global optimization. *Phys. Chem. Chem. Phys.* **2015**, *17*, 24173–24181.
(35) Das, P.; Khatun, M.; Anoop, A.; Chattaraj, P. K. $CSi_nGe_{4-n}^{2+}$(n = 1–3): Prospective systems containing planar tetracoordinate carbon (ptC). *Phys. Chem. Chem. Phys.* **2022**, *24*, 16701–16711.
(36) Das, P.; Patra, S. G.; Chattaraj, P. K. $CB6Al^{0/+}$: Planar hexacoordinate boron (phB) in the global minimum structure. *Phys. Chem. Chem. Phys.* **2022**, *24*, 22634–22644.
(37) Das, P.; Chattaraj, P. K. $BSi_nGe_{4-n}^{+}$(n = 0–2): Prospective systems containing planar tetracoordinate boron (ptB). *J. Chem. Sci.* **2023**, *135*, 1.
(38) Zhao, D.; He, X.; Li, M.; Wang, B.; Guo, C.; Rong, C.; Chattaraj, P. K.; Liu, S. Density functional theory studies of boron clusters with exotic properties in bonding, aromaticity and reactivity. *Phys. Chem. Chem. Phys.* **2021**, *23*, 24118–24124.
(39) Hendon, C. H.; Tiana, D.; Murray, A. T.; Carbery, D. R.; Walsh, A. Helical frontier orbitals of conjugated linear molecules. *Chem. Sci.* **2013**, *4*, 4278–4284.
(40) Zhang, W.; Zhao, Y.; An, X.; Fu, J.; Zhang, J.; Zhao, D.; Liu, S.; Rong, C. Cooperativity and reactivity properties of medium-sized boron clusters: A combined density functional theory and information-theoretic approach study. *Mol. Phys.* **2022**, doi: 10.1080/00268976.2022.2157774.
(41) Dye, J. L.; Ceraso, J. M.; Lok, M.; Barnett, B. L.; Tehan, F. J. Crystalline salt of the sodium anion (Na^-). *J. Am. Chem. Soc.* **1974**, *96*, 608–609.
(42) Liu, G.; Fedik, N.; Martinez-Martinez, C.; Ciborowski, S. M.; Zhang, X.; Boldyrev, A. I.; Bowen, K. H. Realization of Lewis basic sodium anion in the $NaBH_3^-$ cluster. *Angew. Chem., Int. Ed.* **2019**, *58*, 13789–13793.
(43) Postils, V.; Garcia-Borràs, M.; Solà, M.; Luis, J. M.; Matito, E. On the existence and characterization of molecular electrides. *Chem. Commun.* **2015**, *51*, 4865 – 4868.
(44) Saha, R.; Das, P.; Chattaraj, P. K. Molecular electrides: An in silico perspective. *Chem. Phys. Chem.* **2022**, *23*, No. e202200329.
(45) Pei, Y.; An, W.; Ito, K.; Schleyer, P. v. R.; Zeng, X. C. Planar pentacoordinate carbon in CAl5+: A global minimum. *J. Am. Chem. Soc.* **2008**, *130*, 10394–10400.
(46) Leyva-Parra, L.; Diego, L.; Yañez, O.; Inostroza, D.; Barroso, J.; Vasquez-Espinal, A.; Merino, G.; Tiznado, W. Planar hexacoordinate carbons: Half covalent, half ionic. *Angew. Chem., Int. Ed.* **2021**, *60*, 8700–8704.
(47) Yu, H. L.; Sang, R. L.; Wu, Y. Y. Structure and aromaticity of $B_6H_5^+$ cation: A novel borhydride system containing planar pentacoordinated boron. *J. Phys. Chem. A.* **2009**, *113*, 3382–3386.
(48) Welch, G. C.; Juan, R. R. S.; Masuda, J. D.; Stephan, D. W. Reversible metal-free hydrogen activation. *Science.* **2006**, *314*, 1124.
(49) Welch, G. C.; Stephan, D. W. Facile heterolytic cleavage of dihydrogen by phosphines and boranes. *J. Am. Chem. Soc.* **2007**, *129*, 1880–1881.
(50) Hakkinen, H.; Manninen, M. How "Magic" is a magic metal cluster? Phys. *Rev. Lett.* **1996**, *76*, 1599 – 1602.
(51) Reimann, S. M.; Koskinen, M.; Hakkinen, H.; Lindelof, P. E.; Manninen, M. Magic triangular and tetrahedral clusters. *Phys. Rev. B: Condens. Matter.* **1997**, *56*, 12147–12150.
(52) Martinez-Guajardo, G.; Cabellos, J. L.; Diaz-Celaya, A.; Pan, S.; Islas, R.; Chattaraj, P. K.; Heine, T.; Merino, G. Dynamical behavior of borospherene: A nanobubble. *Sci. Rep.* **2015**, *5*, 11287.

(53) Pan, S.; Ghara, M.; Kar, S.; Zarate, X.; Merino, G.; Chattaraj, P. K. Noble gas encapsulated B_{40} cage. *Phys. Chem. Chem. Phys.* **2018**, *20*, 1953–1963.
(54) Saha, R.; Kar, S.; Pan, S.; Martinez-Guajardo, G.; Merino, G.; Chattaraj, P. K. A Spinning umbrella: Carbon monoxide and dinitrogen bound MB_{12}^- clusters (M = Co, Rh, Ir). *J. Phys. Chem. A.* **2017**, *121*, 2971–2979.
(55) Samanta, B.; De, A.; Jana, G.; Gomez, V.; Chattaraj, P. K.; Ganguly, N.; Gomez-Rodriguez, M. NeVAE: A deep generative model for molecular graphs. *J. Mach. Learn. Res.* **2020**, *21*, 4556–4588.
(56) Sarkar, K.; Bhattacharyya, S. P. *Soft-Computing in Physical and Chemical Sciences: A Shift in Computing Paradigm*; CRC Press: Boca Raton, FL, **2017**.

8 17 Atoms Magnesium Nanoclusters for Purification of Air-Forming Gases

Sara Ahmadi and Mahmood Reza Dehghan

8.1 INTRODUCTION

One of the main concerns of the modern world today, given the variety of environmental pollutants, is finding new sources and candidates for clean energy consumption. Hydrogen gas is one of the main candidates in this field with its potential capabilities for consumption. It is one of the main gases in rocket fuel, so its preparation and purification from a cheap source such as air is being considered by military industries. Recently, the main activities and research reported focused on the production of new materials for hydrogen storage. On the other hand, the storage capacity of hydrogen per unit weight of conventional metal alloys is very low (about 2.0% by mass) and not enough for use in a fuel cell vehicle. Therefore, studies on the application of alloys containing light elements have focused on high-performance storage materials. Magnesium is one of the most important candidates for reversible hydrogen storage materials due to its high storage capacity (7.6% by mass). Magnesium hydrides are thermodynamically stable, and the dehydrogenation of magnesium hydrides requires high temperatures (>550 K). Recently, many studies have been conducted on magnesium nanoclusters[1–3]

As we know, air is a combination of different gases: oxygen, hydrogen, nitrogen, carbon monoxide, and a small number of other gases. It is actually the most available and cheapest source for the preparation of each of these gases. If we can find compounds that have the capacity to store or transmit any of these gases, they can be used for potential applications. Among the many studies conducted to find materials with high storage capacity, the focus is on reverse engineering to determine whether compounds can be found to purify and separate different gases from the air.

Recently, magnesium 17 atom nanoclusters (Mg_{17}) with the special arrangement (two four-member rings at the top and bottom and an octagon ring between them) have received special attention, as shown in Figure 8.1.

8.2 NANOCLUSTERS CHEMISTRY

Clusters are collections of atoms or molecules that may contain particles from three to tens or hundreds of millions of members.[4, 5] When such assemblies reach a diameter of nanometers, we call them nanoclusters. Nanoclusters have special properties and potential applications in microelectronics, magnetic storage, optical data storage, sensors, transducers, biological markers, chemical reactors, and catalysts.[6–9] Catalytic properties in nanoclusters are the result of the interaction between a set of dependent parameters, such as nanocluster shape and size, oxidation states, and quantum size.[5, 10] The constituent particles of nanoclusters may be the same, which leads to homo-atom (or homo-molecule) nanoclusters, A_n, or they can be of two or more different species, which leads to hetero-atom or (hetero-molecule) nanoclusters, A_aB_b. The sizes of nanoclusters are very diverse and vary from several atoms up to a mass (bulk). The physical and chemical properties of nanoclusters depend on their size, which is very different from the properties of individual atoms, bulk, and similar molecules.[5, 11] In fact, the geometric shape and stability of nanoclusters change with the size of

 DOI: 10.1201/9781003441328-8

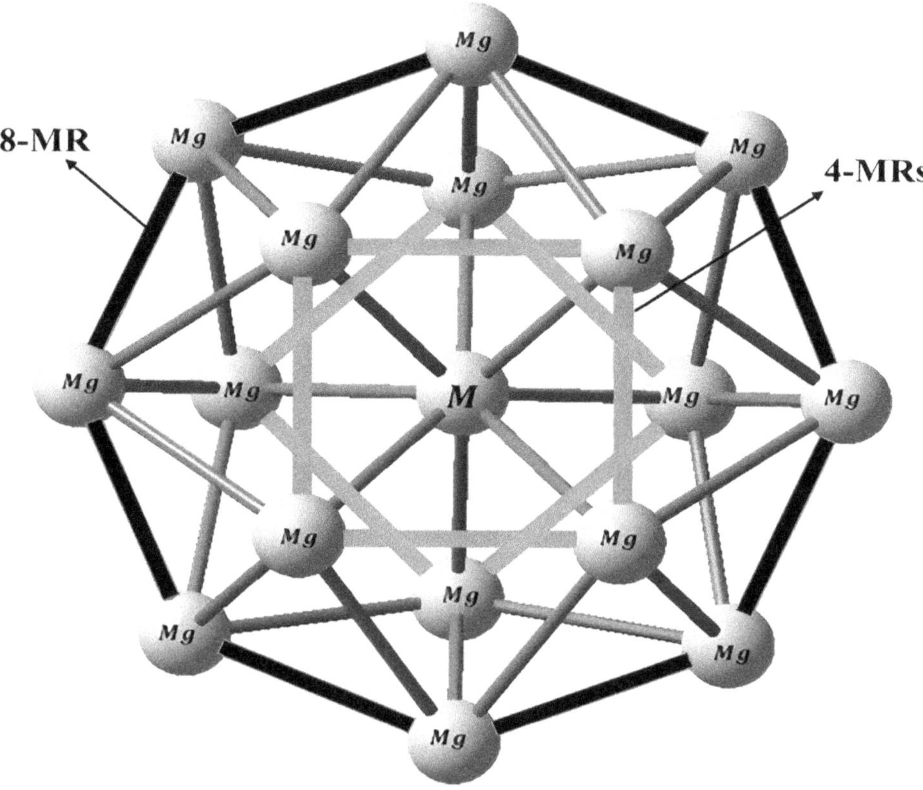

FIGURE 8.1 Schematic view of the Mg_{17} nanoclusters. The gray bonds represent the two 4-membered rings (4-MRs) and the thick black bonds represent one 8-membered ring (8-MR), M is the central atom.

the nanocluster changes. Structurally, nanoclusters take many different shapes, in some cases, they have specific geometric shapes and some are amorphous and liquid. Some of the properties of nanoclusters have been noticed and used since ancient times. The colored glass that was already used by the ancient Romans or the stained glass of the church window from the Middle Ages, owe their color to the tiny particles of nanoclusters of gold, copper, and silver embedded in the glass.

One of the early achievements of the study of clusters was the discovery of the C_{60} cluster, known as fullerene, with its impressive geometric shape. The scientific and valuable results that were obtained in the field of studying metal, carbon, and, increasingly, other materials caused the chemistry of clusters to be placed as an independent and legal discipline among the basic branches of physics and chemistry.

Cluster, molecule, and bulk are three completely different concepts that should be paid attention to.[4, 5, 7, 11] Molecules often have a well-defined structure, composition, and a limited number of isomers, while clusters often have a large number of isomers with close energies. The number of isomers increases sharply as the cluster size increases. For example, the Ar_{13} cluster has hundreds of isomers, while, the Ar_8 cluster has fewer than ten isomers. Therefore, nanoclusters can show different isomers in which the atoms are connected with different geometric patterns. The actual number of the cluster's isomers depends to some extent on the interatomic potential used in the calculations and simulations.

The structure of molecules shows one of the important characteristics that distinguish nanoclusters from molecules. Molecules such as N_2 and H_2 exist in nature under ambient temperature and pressure, while nanoclusters are created in the laboratory under vacuum and cold flow conditions, and their stability varies widely with the size of the nanocluster. Unlike molecules that have weak interactions with one another, nanoclusters have stronger interactions and often connect to form larger nanoclusters.[5, 7]

The study of nanoclusters, both theoretically and experimentally, has its own methods. It is noteworthy that, despite the advances in the field of chemistry at the atomic scale, it is still difficult to determine the correct atomic arrangement of small clusters in a laboratory. First principles calculation can provide accurate and reliable information about the geometry and electronic structure of small clusters. It is important to state that the chemistry of clusters does not fall within the scope of any of the chemistry of molecular, solid, or nuclear states. In fact, solid-state chemistry works with symmetric samples that are assumed to be unlimited, and molecular and nuclear state chemistry is about samples that do not exceed a few hundred constituents in any way.[4]

8.3 TYPES OF NANOCLUSTERS

In terms of bond type, nanoclusters can be divided into four types: covalent, van der Waals, ionic, and metallic bonds.

1. **Covalent nanoclusters:** Elements that have a very high ionization energy are not able to transfer electrons; nor are elements that have a low affinity for electrons able to absorb electrons. Atoms of such elements tend to share their electrons with atoms of other elements or with other atoms of the same element, in such a way that both atoms in their respective valence layers reach the octet configuration and thus become stable. Such communication through the subscription pairing of electrons between different or the same types is known as a covalent bond. The most famous one is fullerene with its impressive geometric shape. Figure 8.2 shows different examples of covalent bonds in nanoclusters.

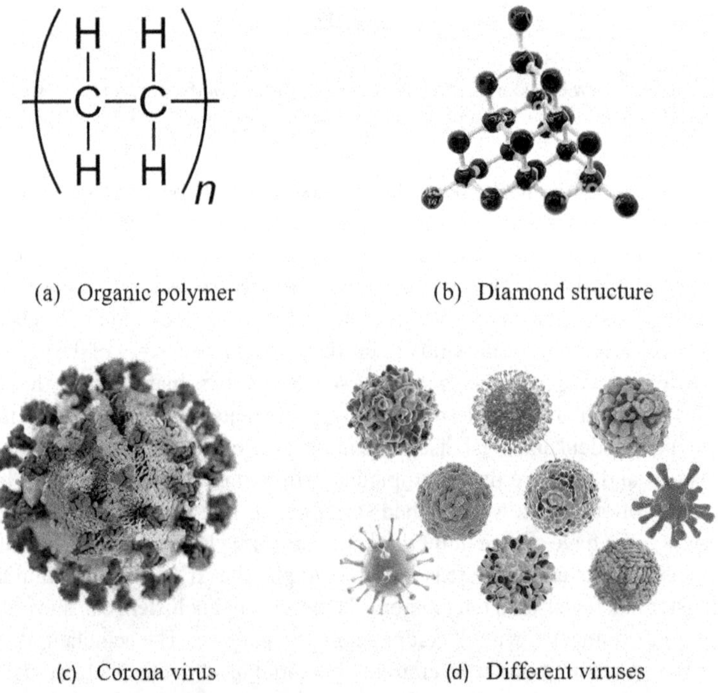

(a) Organic polymer (b) Diamond structure

(c) Corona virus (d) Different viruses

FIGURE 8.2 Sample structures of the covalent nanoclusters.

Sources: (b) adopted from https://encryptedtbn0.gstatic.com/images?q=tbn:ANd9GcT8EIKfrSsRGXSg 05a82rUBq-35CNSvI2maQp6N-2eZe2E35Y7JMmYTFkylOpNdDXxIIM&usqp=CAU, (c) adopted from www.un.org/en/coronavirus, (d) adopted from: www.dreamstime.com/stock-photo-viruses-different-shapes-highly-detailed-d-illustration-image81329412

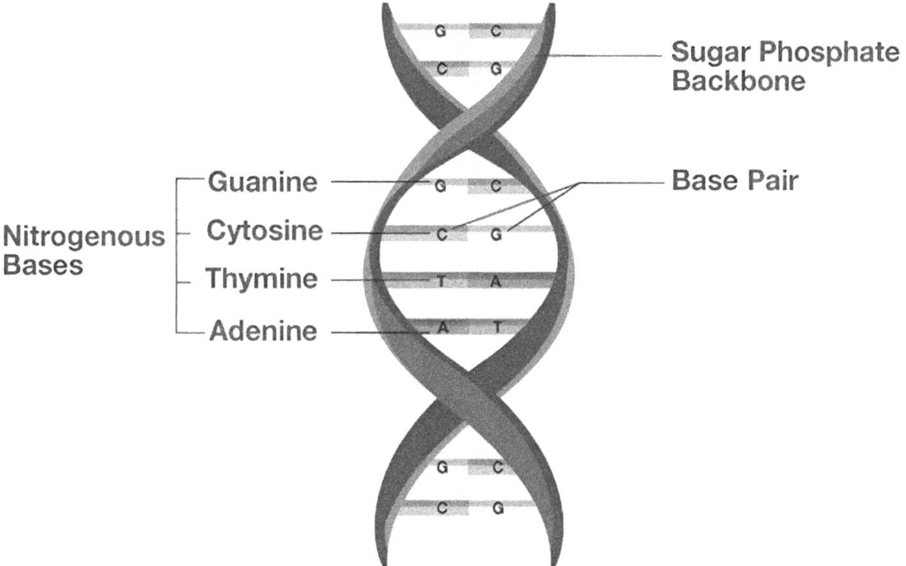

FIGURE 8.3 DNA strand as sample structure of van der Waals nanoclusters. Since hydrogen bonding is a type of strong van der Waals interaction, two DNA strands together can be considered a van der Waals cluster.

Source: Adopted from https://byjus.com/biology/dna-structure/

2. **Van der Waals nanoclusters:** In van der Waals nanoclusters, the atoms are connected by short-range forces that depend only on the distance between two atoms. This group of nanoclusters, when they have structures with high symmetries, are extremely stable. Van der Waals–bonded rare gas nanoclusters are held together by weak forces rooted in electrical interactions. Figure 8.3. illustrates a DNA strand as a good sample structure of the Van der Waals nanoclusters.
3. **Ionic nanoclusters:** These types of nanoclusters are held together by strong attractive forces between oppositely charged ions. Ionic nanoclusters have recently been the center of attention of many researchers, especially due to their use in photography, and AgBr nanoclusters are an example of this category, as illustrated in Figure 8.4.
4. **Metallic clusters:** The last category of nanoclusters consists of atoms that show the properties of metallic bonds. For small atomic nanoclusters such as Na, Al, Mg, Au, etc., the atoms are held together by the forces exerted by the free electron (Figure 8.5). Metal nanoclusters are the most suitable and least expensive option for studying systems with limited size, both from the experimental and theoretical points of view. Metal nanoclusters have many isomers due to weak bonding. The issue that is important here is to determine the most stable structure among a large number of isomers, which is a very difficult task. Therefore, the study of nanoclusters plays a fundamental role from the point of view of basic chemistry.

Studying nanoclusters has become one of the most active areas of chemistry with various achievements. Studying the structure and properties of small metal clusters provides useful information that shows different properties between individual and bulk atoms. Considering the dependence of the physical and chemical properties of nanoclusters on their size, such studies have been seriously placed on the agenda of the research areas of chemistry in the current century. Today, extensive research has been done on the electronic and structural properties of neutral and ionic metal clusters

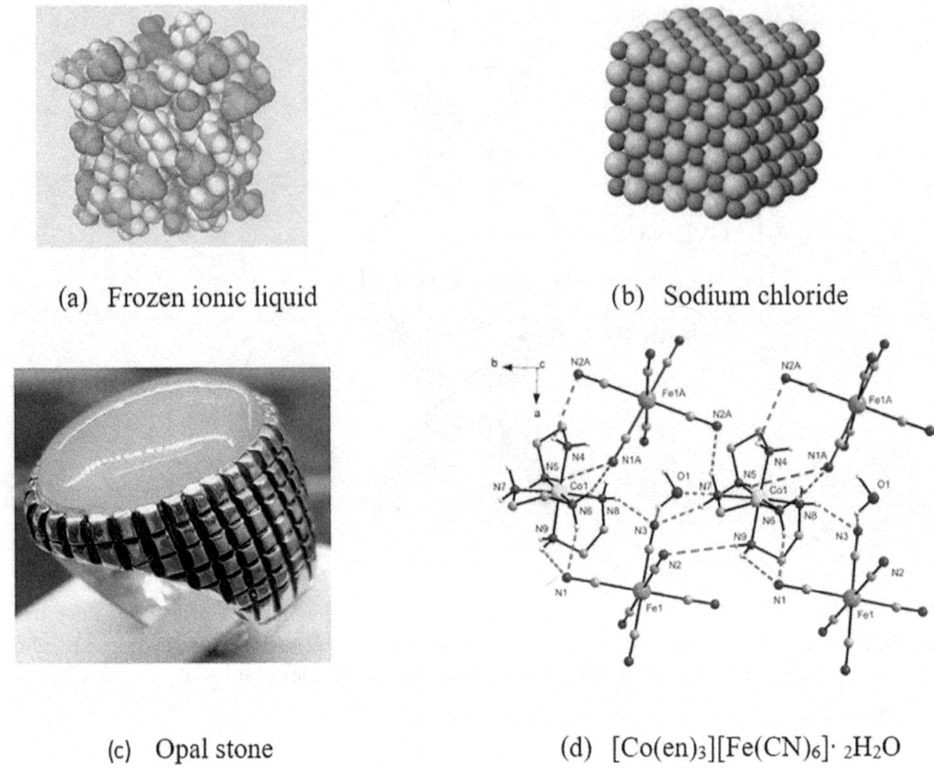

FIGURE 8.4 Sample structure of the ionic nanoclusters.

Sources: (a) Adopted from https://en.wikipedia.org/wiki/Ionic_liquid, (b) adopted from https://en.wikipedia.org/wiki/File:Sodium-chloride-3D-ionic.png, (c) www.ratnabhandar.com/gemstones-%28sold-out%29/opal-%28etho%29/opal-398-ct-%28-86925-%29-201812-15488/15488, (d) adopted from *Chemistry Central Journal*, 2013. 7: pp. 1–18.

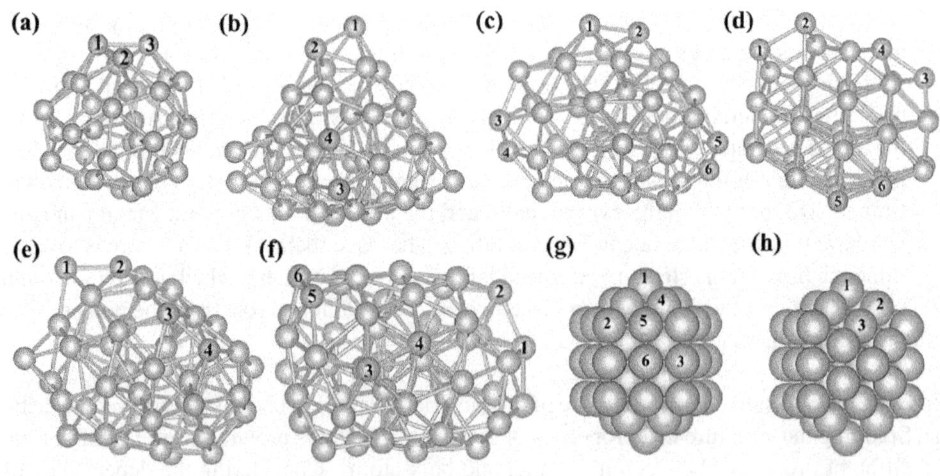

FIGURE 8.5 Sample structure of the metal nanoclusters.

Source: Adopted from *ACS Catalysis*, 2018. 8(10): pp. 9702–9710.

such as magnesium, gold, etc., using different computational methods. A wide range of theoretical studies have been conducted on the electronic and structural properties of neutral and ionic magnesium clusters.[12–15] Recently, Xia and coworkers investigated the electronic and structural properties of neutral and ionic Mg_n nanoclusters (n = 3–20) using the density functional theory and showed that the neutral Mg nanocluster with 17 atoms (Mg_{17}) with point group D_{4d} exhibits extraordinary stability due to its aromaticity and complete adherence to Huckel $4N + 2$ rule.[16]

8.4 SURFACE ABSORPTION AND VARIOUS ABSORPTION METHODS

Surface adsorption, or just *adsorption* is the process of atoms, ions, biomolecules, gases, liquid, or dissolved solid molecules sticking to a surface. In this process, a thin layer of adsorbed material is created on the surface of the adsorbent. This process is present in many physical, biological, and chemical systems. On the other hand, *absorption* is the penetration of the collected materials into the absorbent. Since these two phenomena often occur simultaneously, the word "sorption" is used for these phenomena. Due to the fact that nanomaterials have a high surface-to-volume ratio, in recent years, with the expansion of studies in the nano field, the phenomenon of absorption has been investigated and welcomed more than in the past. Figure 8.6 shows a simple view of the absorption phenomenon on the absorbent surface.

Surface absorption is caused by two forces. The first effective force in surface adsorption is caused by solvent-disliking (lyophobic) properties of the soluble component to the solvent, and the second is the specific tendency of the dissolved component toward the surface of the solid adsorbent. Based on these forces, there are three types of surface absorption, discussed next.

8.5 PHYSICAL ADSORPTION

In physical surface adsorption, molecules are attracted to the surface of the adsorbent material by van der Waals forces. The van der Waals force is proportional to the square of the distance between the two particles. In this type of absorption, there is no electron transfer due to the van der Waals forces. In physical adsorption, the strength of absorption depends on the visible physical properties in the adsorptive. As a result, the physical properties of the adsorbent have no effect on the absorption process. The characteristic of this type of absorption is the low absorption energy and reversibility of the absorption process. The adsorption energy is usually less than 50 kJ/mol. In this type of absorption, the activation energy is close to zero, and the resulting heat is very low. It

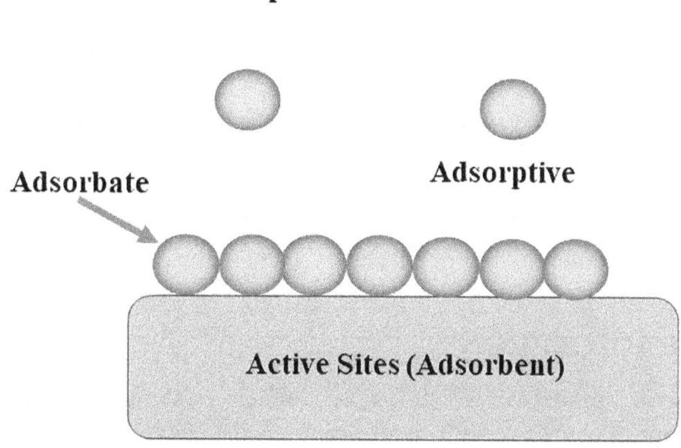

FIGURE 8.6 Schematic of absorption process.

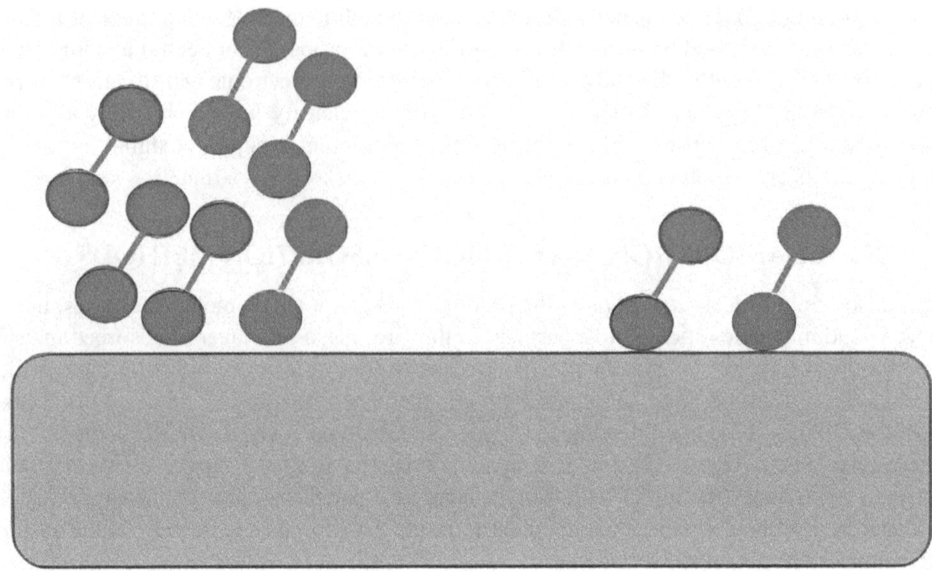

FIGURE 8.7 Schematic of physical adsorption.

happens quickly. This process is mainly visible at low temperatures; therefore, the absorption efficiency increases with decreasing temperature. In addition, by increasing the pressure, absorption can be increased to the extent that absorption takes place in several layers. In physical absorption, all substances can act as absorbent and absorbed. As a result of all these features, physical absorption is widely used in industry. An example of physical adsorption is surface adsorption by activated carbon. Figure 8.7 shows a schematic of physical adsorption.

8.6 CHEMICAL SURFACE ADSORPTION

In chemical surface adsorption, the absorbed molecules are kept on the surface of the material with chemical bonds. In the process of forming a bond with the adsorbent, the molecules that are chemically absorbed undergo a change in the internal electron configuration. The bonds inside some molecules are stretched and weakened, and some are even broken. As a results, in this absorption, electron transfer is done between the particle and the surface. Chemical surface adsorption is an irreversible process because a lot of energy is required to carry out the reverse reaction. Unlike physical surface absorption, in this type of absorption, with increasing temperature, the rate of absorption or reaction increases. Resultantly, the efficiency of surface absorption increases. In general, in surface chemical adsorption, the activation energy is high, the adsorption speed is low, and, due to the formation of chemical bonds, the heat of adsorption is high. Also, the nature of absorption depends on the properties of the absorbent and absorbed. Increasing the pressure also slightly changes the amount of absorption. Chemical adsorption is rarely used in environmental engineering. But physical surface absorption is common. Figure 8.8 shows a schematic of chemical adsorption.

8.7 EXCHANGE ADSORPTION

Exchange adsorption is a process in which ions accumulate on the surface and charge the surface. The determining factor of exchange surface absorption is the charge of ions. The higher the charge of the absorbed ion, the greater the intensity of absorption. For example, ions with 3 charges have a faster adsorption speed to the opposite surface than ions with 1 charge. If the charge of the ions is the

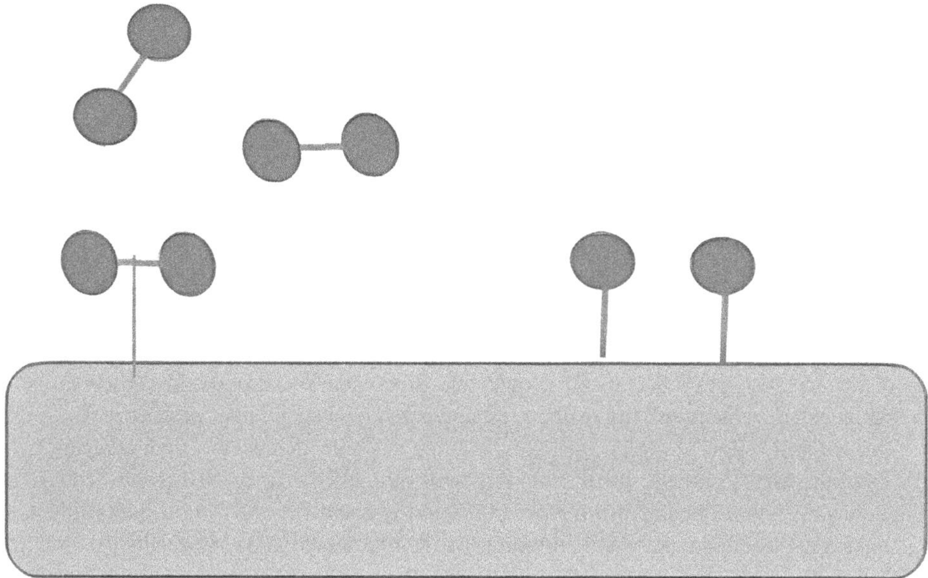

FIGURE 8.8 Schematic of chemical adsorption.

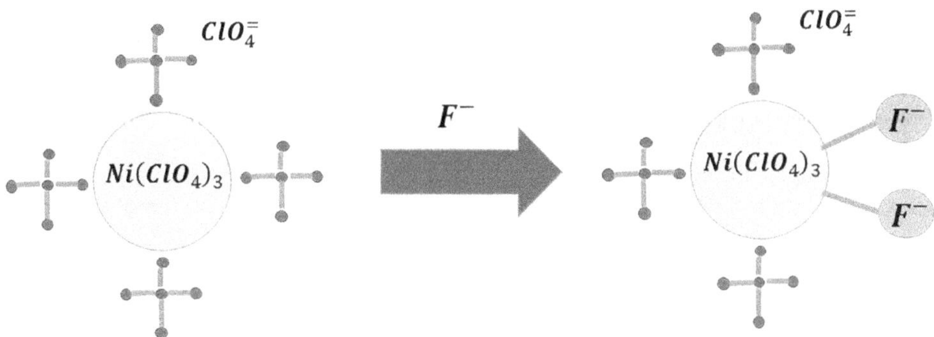

FIGURE 8.9 Schematic view of exchange adsorption.

same, the absorption intensity depends on the size of the molecule. The smaller the molecule size, the easier it is to penetrate deeper past the absorbent surface. This type of absorption, like chemical surface absorption, requires a• lot of energy to carry out the reverse reaction; therefore, it is irreversible. A schematic view of exchange adsorption is illustrated in Figure 8.9.

Most surface absorption phenomena are caused by a combination of these three types of absorption, and usually it is not possible to distinguish them easily.

In the description of surface absorption, it can be said that, among the distinguishing points of surface studies from the volume, the surface is very quickly covered with absorbent material. We justify this sentence with the following relationship:

$$Z = \frac{N.C}{4} = \frac{P}{4KT}\sqrt{\frac{8KT}{NM}} = \frac{P}{\sqrt{2\pi MKT}} \tag{8.1}$$

where P is the pressure, T is the temperature, M is the mass, and N is the number of particles. Also, k is Boltzmann constant, and Z is the number of collisions per unit area per unit time.

8.8 FACTORS AFFECTING SURFACE ABSORPTION

Some factors affecting absorption are as follows:

- **Absorbent surface area:** Because surface absorption is a surface phenomenon, the amount of absorption depends on the specific surface area ratio of the absorbent body. The specific surface area is the total surface area of the absorbent that participates in the absorption process. The smaller the absorbent particles are, the higher will be the absorption rate due to the increase in the contact surface.
- **Type of absorbent:** the size and diameter of the molecules of the absorbent component must be such that they can be absorbed and enter the adsorptive.
- **pH:** The pH of the solution in which the absorption process takes place affects the amount of surface absorption because hydrogen and hydroxide ions are quickly absorbed by the surface of the absorbent, thus affecting the amount of absorption of other ions. In general, surface absorption of organic pollutant soluble in water increases with decreasing pH.
- **Temperature:** Surface absorption reactions are exothermic, so the absorption rate increases with a decrease in temperature, and if chemical absorption is dominant, the absorption rate increases with an increase in temperature. It is noteworthy that the main factor of surface absorption is physical absorption, and in physical absorption, the amount of absorption decreases with increasing temperature.
- **Mixing:** The more mixing, the faster the absorption will be.
- **Absorbent concentration:** The higher the concentration of the absorbent component in the solution, the higher the absorption rate will be.

8.9 EXAMINING ADSORPTION BEHAVIORS THROUGH THEORETICAL CALCULATIONS

In order to investigate the adsorption behavior between the adsorbent and adsorbed species through theoretical calculations, it is necessary to evaluate various parameters, which are introduced next.

- **Absorption energy:** The adsorption energy(E_{ad}), binding energy($E_{(bind)}$), and deformation energy ($E_{(def)}$) are calculated based on following equations:

$$E_{ad} = E_{M...X} - (E_M + E_X) \tag{8.2}$$

$$E_{bind} = E_{M...X} - (E_{M\ in\ complex} + E_{X\ in\ complex}) \tag{8.3}$$

$$E_{def} = E_{ad} - E_{bind} \tag{8.4}$$

  where $E_{M...X}$, E_M, and E_X are the optimized energies of the complex (nanocluster/X), nanocluster, and X molecule, respectively.

  High absorption energy indicates strong absorption, which is evidence of chemisorption. On the other hand, the positive deformation energy values of the nanoclusters indicate an endothermic-unfavorable adsorption process, while the negative deformation energies imply an exothermic-favorable adsorption process.

- **Recovery time:** Although the electronic properties of the nanoclusters are sensitive to the X molecule, the high adsorption energy can prevent the desorption process. Exposure to ultraviolet (UV) light[17] can be used to recover a sensor experimentally. The recovery time (τ) can be determined from the following equation of the transition theory:

$$\tau = \upsilon^{-1}\exp(-E_{ad} / kT) \tag{8.5}$$

where T is the temperature, k is the Boltzmann's constant ($\sim 8.31 \times 10^{-3}\ kJ / mol.K$), and υ is the attempt frequency.

- **Electronic properties:** The remarkable contribution in chemical stability and chemical reactivity is related to the frontier orbitals of the highest occupied molecular orbital energies (HOMO) and lowest unoccupied molecular orbital energies (LUMO), which illustrate the electron-donating and -accepting abilities, respectively.[18]
- **Chemical reactivity:** The chemical reactivity of nanoclusters can be characterized by their energy gap, which is an important parameter relying on the HOMO and LUMO energy levels. The large energy gap indicates high stability and consequently low chemical reactivity; a small gap represents low stability and high chemical reactivity.[19–21] The dependency of the E_{gap} on the conduction electron population (N) based on equation (8.6) has been proven many times,[22–25] and its alteration can be used as an appropriate index for the adsorbent sensitivity to a chemical.

$$N = AT^{3/2}\exp\left(-E_{gap} / 2kT\right) \tag{8.7}$$

where k is the Boltzmann's constant, and A (electrons/m^3 K$^{3/2}$) is a constant.

Equation (8.7) estimates an exponential increase for the electrical conductivity of the nanocluster by decreasing the E_{gap}. This alteration can be converted to an electrical signal, assisting in the detection of the molecule.

- **Chemical hardness:** The resistance of a chemical species to alter its electronic configuration is called the chemical hardness, while electronic chemical potential illustrates the desire to escape the electron cloud.[26–28]
- **Density of states (DOS) diagram:** The density of states (DOS) is essentially the number of different states at a particular energy level that electrons are allowed to occupy, i.e., the number of electron states per unit volume per unit energy. Bulk properties such as specific heat, paramagnetic susceptibility, and other transport phenomena of conductive solids depend on this function. DOS calculations allow one to determine the general distribution of states as a function of energy and can also determine the spacing between energy bands in semiconductors. If the examination of the DOS plots reveal that the conduction levels shift toward the lower/higher energies compared to those of their pristine nanoclusters, there will be change in the conductance of the nanocluster based on the adsorption of the X molecule. On the other hand, if the DOS plots do not show significant changes compared to that of the pristine nanocluster, there is no change in the conductance of nanocluster based on the adsorption of X molecule.
- **Thermodynamic parameters:** Thermodynamic parameters such as standard enthalpy (ΔH°) and standard Gibbs free energy (ΔG°) at 298 K and 1 atmosphere are calculated based on equation (8.8).

$$\Delta X^{\circ} = X\left(E_{M \ldots x}\right) - \left(X_M + X_x \right) \qquad X = H, G \tag{8.8}$$

where $X_{M \ldots x}$, X_M, dna X_x are thermodynamic parameters corresponding to the complex (nanocluster/N_2), nanocluster, and N_2 molecule, respectively.

As a summary, according to the different thermodynamics properties, we can say that, during the adsorption process, the negative values of ΔH° indicate an exothermic reaction. The large negative values of $T\Delta S^{\circ}$ denote an entropy reduction during the formation of the complex. The positive values of ΔG° for complexes will predict a thermodynamically disfavored adsorption process. Negative values of ΔG° verify a spontaneous and thermodynamically favorable adsorption process, which leads to an increase in the capability for X storage

8.10 COMPUTATIONAL CHEMISTRY

Computational chemistry is a branch of chemistry that tries to solve chemistry problems with the help of computers. The main goal in this field is to obtain the structure, properties, and reactivity of chemical compounds using theoretical methods and without conducting experiments. In this field, the results of pure chemistry in the form of effective computer programs are used to calculate the structure and properties of molecules and solids. While the computational results usually complement the information obtained from chemical experiments, in some cases it can lead to predicting unobserved chemical phenomena. Therefore, computational chemistry can help laboratory chemistry and can accompany experimental chemistry in finding new chemical topics. Applications of computational chemistry include molecular modeling with computational methods and the design of drugs, catalysts, and new materials.

So far, many computational methods have been invented to solve chemistry problems. These methods use the laws of quantum mechanics, classical mechanics, or a combination of them to solve problems. Choosing the right method for performing calculations depends on the type and complexity of the structure being studied, as well as the type of information that is needed. These methods can be classified as molecular mechanics methods, semiempirical methods, *ab initio* methods, and density functional theory methods. Software tools for computational chemistry are often based on experimental data. To use this tool, we need to understand how the method is used and the basic nature of the information used to determine the method.

Using theoretical calculations and computational chemistry, we next specifically address the role of magnesium nanoclusters in the separation and purification of various air-forming gases.

8.11 MAGNESIUM NANOCLUSTERS FOR PURIFICATION OF AIR-FORMING GASES: A DFT APPROACH

Metal nanoclusters have been considered a new class of chemical sensors due to their unique electronic structures and particular physicochemical properties. As mentioned previously, magnesium nanoclusters consist of 17 atoms (including two square frames [4-MRs] on the top and bottom with an octagon frame [8-MR] at the center [Figure 8.10]), which have been considered an aromatic metal cluster and are very stable so that can be used to store different gases. It should be noted that a good container should be able to establish a strong bond with the absorbed species and should be

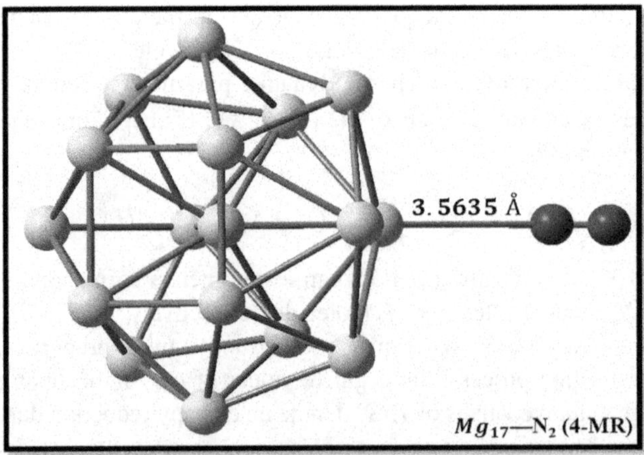

FIGURE 8.10 Optimized structures of the adsorbed nitrogen over the Mg_{17} N_2 nanocluster from the 4-MRs position. The black color represents nitrogen molecule, and the white color represents the nanocluster.

able to easily release the adsorbed species with minimum energy consumption and without changing its electronic structure.

In this section, the ability of magnesium nanoclusters to store different oxygen, nitrogen, carbon monoxide, and hydrogen gases has been investigated using quantum calculations. In continuation, by changing the central magnesium atom in the studied nanocluster with beryllium and calcium atoms, the change in absorption can be checked.

8.12 APPLICATION OF MG_{17} ($MG_{16}M$; M=BE, MG, AND CA) NANOCLUSTER IN PURIFICATION OF N_2 FROM AIR

The neutral Mg_{17} nanocluster with a D_{4d} point group is the most stable among other point groups such as C_s, D_{17h}, and, $D_{\infty h}$.[4, 15] Furthermore, the clusters with 17 atoms and 34 electrons are considered closed-shell clusters according to the jellium superatom model. A filled-cage-like construction, including two square (4-MRs) frames and an octagon (8-MR) frame was applied to stabilize the $Mg_{16}M$ due to the potent interactions between the Mg_4 units and the central M atom (M = Be, Mg, and Ca). More significantly, because of the tendency of local aromaticity, a single N_2 molecule was added to each of these nanoclusters from two different sites of 4-MRs and 8-MR (Figure 8.10). All structures were geometrically optimized using the hybrid exchange-correlation functional CAM-B3LYP.[29]

According to theoretical calculations, accepting or losing an electron directly influences the bond lengths and angles of the nanoclusters. Therefore, the bond lengths and angles of the 4-MRs and 8-RM frames ofMg_{17}^{+1}, and Mg_{17}^{-1} slightly change in comparison to the neutral Mg_{17} nanocluster. On the other hand, the bond lengths and angles in $Mg_{16}M$ (M=Be, Mg, and Ca) nanoclusters are affected by the central atom (M) in both the 4-MRs and 8-MR frames. As such, the bond lengths for 4-MRs frames of $Mg_{16}M$ increased with increasing the size of the central atom, while, for the 8-MR frame, the bond lengths showed an inverse trend. The bond angle for the 8-MR frame, in $Mg_{16}Ca$, is 112.76°, which is smaller than the values of $Mg_{16}Be$ and Mg_{17}, consequently leading to cluster instability. It can be concluded that the 8-MR frame of $Mg_{16}Ca$ with the smallest bond length (2.95 Å) and angle (112.76°) and high strain ring is the most favorable position of the nanocluster for adsorbing the N_2 molecule. The adsorption energy values clearly indicate a strong interaction between the N_2 molecule and $Mg_{16}Ca$ nanocluster, which shows a chemisorption process, whereas the adsorption energy values for the other complexes indicate physical adsorption of the N_2 molecule upon the surface of the investigated nanoclusters. In fact, increasing the central atom radius will decrease the stability of the neutral $Mg_{16}M$ (M = Be, Mg, and Ca) nanoclusters, which leads to more reactivity. It should be noted that the electronegativity of the central atom can be an important factor during the nitrogen adsorptions on the investigated nanoclusters. Hence calcium can be an important factor during nitrogen adsorption on the investigated $Mg_{16}M$ (M=Be, Mg, and Ca) nanoclusters. Hence, calcium with the least electronegativity plays a more effective role in nitrogen adsorption.

Although the electronic properties of the $Mg_{16}M$ (M = Be, Mg, and Ca) nanoclusters are sensitive to the N_2 molecule, the high adsorption energy can prevent the desorption process. The estimated recovery time represents a possible desorption process of the N_2 molecule from the surface of the $Mg_{16}Ca$ nanocluster. Hence the $Mg_{16}Ca$ nanocluster could be an appropriate sensor for the N_2 molecule in comparison to $Mg_{16}Be$ and Mg_{17} nanoclusters.

The remarkable contribution in chemical stability and chemical reactivity is related to the frontier orbitals of the highest occupied molecular orbital energies (HOMO) and the lowest unoccupied molecular orbital energies (LUMO), which illustrate the electron-donating and -accepting ability, respectively. Accordingly, the HOMO and LUMO energy levels for the $Mg_{16}Ca$ nanocluster are significantly affected by the adsorption of the N_2 molecule. The ε_{HOMO} and ε_{LUMO} values for 4-MRs and 8-MR frames of the other complexes were increased during the adsorption process. A large energy gap indicates high stability and consequently low chemical reactivity; a small gap represents low stability and high chemical reactivity. Accordingly, $Mg_{16}Ca$ nanocluster has a smaller energy gap, low stability, and high chemical reactivity in comparison to the Mg_{17} and $Mg_{16}Be$ nanoclusters, which leads to more sensitivity to the presence of the N_2 molecule. The dependency of the E_{gap} on the conduction electron population can be used as an appropriate index for the adsorbent sensitivity to a chemical. Equation (8.7) estimates an exponential increase in the electrical conductivity of $Mg_{16}M$ by decreasing E_{gap}. This alteration can be converted to an electrical signal, assisting in detecting the molecule. Therefore, equation (8.5) predicts a decrease in the electrical conductivity of $Mg_{16}Ca$—N_2 by increasing E_{gap}.

The chemical potential values of the neutral and ionic magnesium nanoclusters $Mg_{17}^{q}(q=0,\pm1)$ show that the neutral nanocluster is more stabilized by losing one electron and that accepting an electron causes instability in the resulting anionic nanocluster. It should be noted that the instability of the anionic nanocluster, compared to the cationic one, does not lead to its high reactivity with the N_2 molecule because the presence of an empty orbital in the cationic nanocluster plays an effective role in the nucleophilic attack by the lone pair of electrons of the N_2 molecule. Accordingly, the N_2 molecule has better adsorption on the cationic cluster.

The negative values of the standard enthalpy for $Mg_{16}Ca$—N_2 (4-MRs and 8-MR frames) and $Mg_{16}Be$—N_2 (8-MR) complexes indicated an exothermic reaction. The large negative values of $T\Delta S^{\circ}$ denote an entropy reduction during the formation of these complexes. The positive values of the standard Gibbs free energy for both 4-MRs and 8-MR frames of the Mg_{17}^{+1}—N_2, Mg_{17}^{-1}—N_2, $Mg_{16}Be$ —N_2, and $Mg_{16}Be$—N_2 complexes predicted a thermodynamically disfavored adsorption process for these complexes. The negative values of the standard Gibbs free energy for $Mg_{16}Ca$—N_2 complex for both 4-MRs and 8-MR frames verify a spontaneous and thermodynamically favorable adsorption process, which leads to an increase in the capability for N_2 storage. These results represent a linear correlation between adsorption energy and standard enthalpy of these complexes in the CAM-B3LYP approach with the corresponding correlation coefficient of 0.99. The $Mg_{16}Ca$—N_2 nanocluster, with the $E_{ad}=-87.617$ (kJ.mol^{-1}) and $\Delta H^{\circ}=-86.5365$ (kJ.mol^{-1}) for 8-MR frame and $E_{ad}=-88.0620$ (kJ.mol^{-1}) and $\Delta H^{\circ}=-84.5727$ (kJ.mol^{-1}) for 4-MR frames, adsorbed the N_2 molecule during a chemisorption reaction.

8.13 APPLICATION OF MG_{17} ($MG_{16}M$; M = BE, MG, AND CA) NANOCLUSTER IN PURIFICATION OF CO FROM AIR

In the study of CO molecule absorption on a magnesium nanocluster, it is necessary to pay attention to the fact that each CO molecule can be absorbed by the nanocluster from both the carbon and oxygen sides. In this section, the results of the calculations performed from both sides are examined. The average bond lengths of the $Mg_{16}Ca$ nanocluster in both 4-MRs and 8-MR frames significantly changed by the adsorption of the CO (via O and C atoms) molecule. The maximum observed bond length for $Mg_{16}Ca$—CO for the 8-MR frame before and after the first CO adsorption is 2.95 and 3.20 Å, respectively. The change of the central atom in the studied nanoclusters has more effect on the bond angles and lengths of the 8-MR frame than those of the 4-MRs frames. Therefore, by decreasing the bond angle (from 115.44° to 112.76°) and bond length (from 3.12 to 2.95 Å) in the 8-MR frame of the $Mg_{16}Ca$ nanocluster, the strain increases. Thus the $Mg_{16}Ca$ nanocluster becomes more unstable.

Considering two possibilities of the carbon monoxide molecule adsorption on the exterior surface of the magnesium nanoclusters causes the generation of four adsorption models for each nanocluster because each CO molecule can adsorb from the 4-MR and 8-MR positions. The negative adsorption energy values for $Mg_{16}Ca$—CO indicate a strong interaction between the CO (via O and C atoms) molecule and the $Mg_{16}Ca$ nanocluster, which represents a chemisorption process. The adsorption energy values for the other complexes ($Mg_{16}Be$—CO, $Mg_{16}Mg$—CO) indicate the physical adsorption for the CO (via O and C atoms) molecule upon the surface of the investigated nanoclusters. A comparison between the adsorption energies of different complexes demonstrates that the adsorption from the carbon atom is more energetically favorable than those that attach from the O atom. It is noteworthy that, negative adsorption energy indicates that the formed complex is stable, and positive adsorption energy belongs to the local minimum in which the adsorption of the gas molecule on the nanocluster is prevented by a barrier. The adsorption energy encompasses both binding (E_{bind}) and deformation (E_{def}) energy contributions, which both occur during the adsorption process. The deformation energy values of the optimized complexes of $Mg_{16}M$—CO and $Mg_{16}M$—CO (M = Be, Mg) are positive and vary with the radius of the central atom from a minimum of 0.0252 kcal/mol for $Mg_{16}Be$—CO (8-MR) to a maximum of +0.4413 kcal/mol for $Mg_{16}Mg$—CO (4-MRs). This indicates an endothermic-unfavorable adsorption process of the CO (via O and C atoms) molecule on the surface of these nanoclusters; negative deformation energies for $Mg_{16}Mg$—CO and $Mg_{16}Mg$—OC imply an exothermic-favorable adsorption process. Hence, for the $Mg_{16}Mg$—CO nanocluster, the adsorption process is thermodynamically spontaneous and leads to a distortion in the nanocluster structure. In all optimized complexes, the Mg-Mg bond lengths in the adsorption regions increased compared to the pristine nanocluster, indicating the weakening of these bonds due to the interaction between the CO (via O and C atoms) molecule and the nanocluster surface. Figure 8.11 represents the optimized structures of the adsorbed CO over the $Mg_{16}M$ (M = Be, Mg, and Ca) nanocluster.

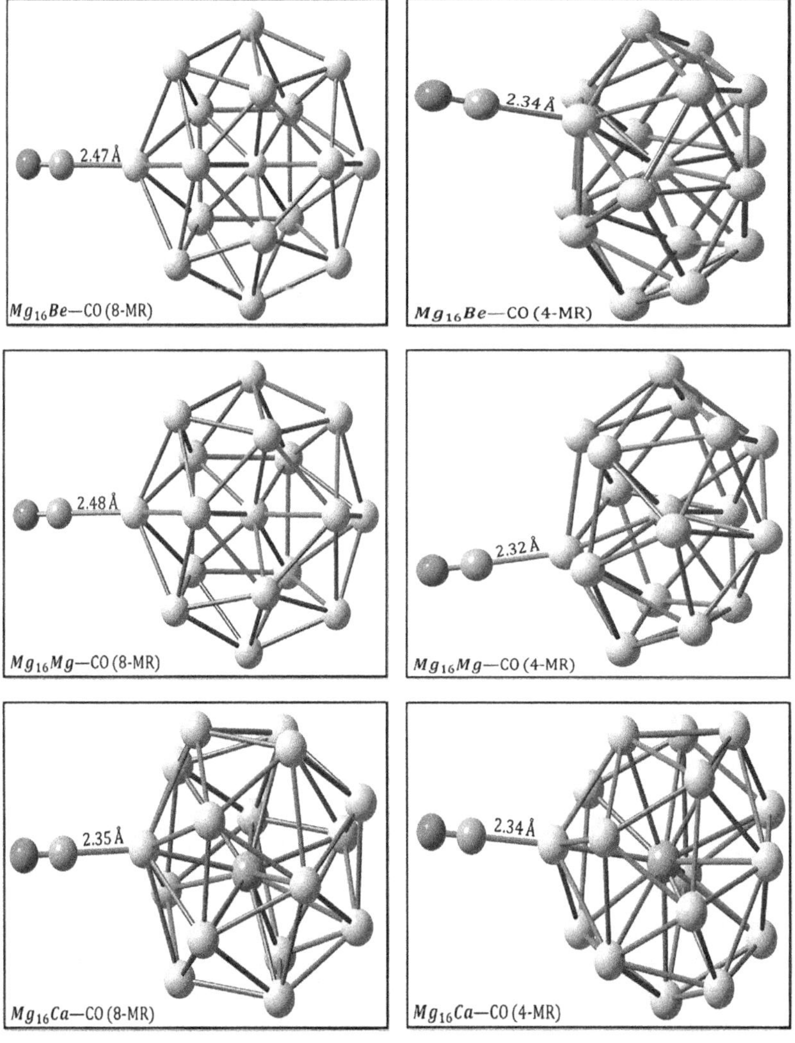

FIGURE 8.11 Optimized structures of the adsorbed CO over the $Mg_{16}M$ (M = Be, Mg, and Ca) nanocluster.

A small charge of 0.15829 e transfers during the adsorption of the CO molecule on the 8-MR position of $Mg_{16}Ca$ nanocluster, which drives the highest interaction energy among the adsorption models. The calculated NBO charges for the nanoclusters of $Mg_{16}M$ (M = Be, Mg, and Ca) represent the greatest accumulation of negative charge on the central atom. The electron density depletion (EDD) map analysis represents a condensed electronic cloud surrounding the Mg-C bond, which is evidence for the chemical adsorption of the CO molecule on the $Mg_{16}Ca$ nanocluster.

8.14 APPLICATION OF MG_{17} ($MG_{16}M$; M = BE, MG, AND CA) NANOCLUSTER IN PURIFICATION OF O_2 FROM AIR

In order to investigate the absorption of O_2 molecules on magnesium nanoclusters, such as N_2 and CO molecules, we consider two positions: approaching from the side of either the 4-MR or 8-MR rings. During the adsorption of the O_2 molecule on the exterior surface of the magnesium nanocluster, the distance between the $Mg_{16}M$ nanocluster and the first adsorbed O_2 molecule decreases by increasing the size of the central atom. The negative values of the adsorption energies indicate a potent interaction among O_2 and different nanoclusters, which revealed a chemisorption mode. The comparison between the adsorption energy values of the 4-MRs and 8-MR frames in each nanocluster demonstrates that the 4-MRs frames are more desirable for O_2 adsorption than the 8-MR frame.

According to equation (8.2), E_{ad} <0 illustrates the stability of the obtained complex, and E_{ad} > 0 is related to the local minimum, which shows the existence of a barrier during the adsorption. E_{ad} includes the share of binding energy (E_{bind}) and deformation (E_{def}), both of which happen within the adsorption process. The E_{def} values of the optimized $Mg_{16}M—O_2$ (M = Be, Mg, and Ca) complexes are positive and vary according to the size of the middle atom from a minority +52.24 to an utmost of +237.42 kcal/mol, for $Mg_{16}Be—O_2$ and $Mg_{16}Ca—O_2$, respectively. This indicates strong adsorption of the first O_2 molecule on the surface of the studied nanoclusters. The chemical adsorption of the first O_2 molecule alters the electronic arrangement of the nanocluster, which in turn may break or weaken some of the bonds in the nanocluster. During the unimolecular O_2 adsorption on the Mg_{17} and $Mg_{16}Be$ nanoclusters, the molecular structure of O_2 is preserved. The (O=O) bond is broken in the $Mg_{16}Ca$ nanocluster, and each oxygen atom adsorbs on different Mg atoms of the nanocluster. The high charge transfer value (3.85914 e) occurred for the adsorption of the oxygen on the surface of the $Mg_{16}Ca$ nanocluster for 4-MR frame, which creates the highest interaction energy among the three nanoclusters. The highest agglomeration of negative charge is on the central M atom for each nanocluster (Figure 8.12).

Investigating the absorption of hydrogen gas on the magnesium nanocluster using theoretical methods with the aim of purifying hydrogen gas from the air, it was observed that hydrogen gas has no attraction on the magnesium nanocluster. As a result, magnesium nanoclusters cannot be used directly for hydrogen purification.

8.15 CONCLUSION

Metal nanoclusters are collections of atoms or molecules that may contain particles from three to tens or hundreds of millions of members. They are the missing link between atoms and metal nanoparticles. The constituent particles of nanoclusters may be the same, which leads to homo-atom (or homo-molecule) nanoclusters, A_n, or they can be of two or more different species, which leads to hetero-atom or (hetero-molecule) nanoclusters, A_aB_b. In terms of bond type, nanoclusters can divided into four types: covalent, van der Waals, ionic, and metallic bonds. Nanoclusters have special properties and potential applications in microelectronics, magnetic storage, optical data storage, sensors, transducers, biological markers, chemical reactors, and catalysts. In recent years, with the expansion of studies in the nano field, because nanomaterials have a high surface-to-volume ratio, the phenomenon of absorption has been investigated and welcomed more than ever. Surface adsorption or just *adsorption*) is the process of atoms, ions, biomolecules, gases, liquid, or dissolved solid

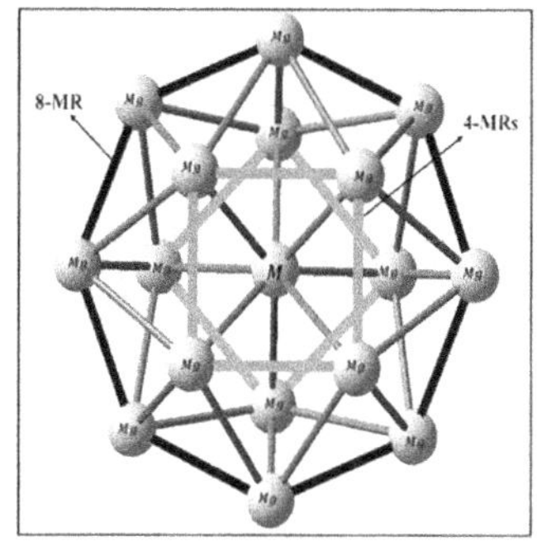

(a) $Mg_{16}M$ (M = Be, Mg and, Ca) nanocluster

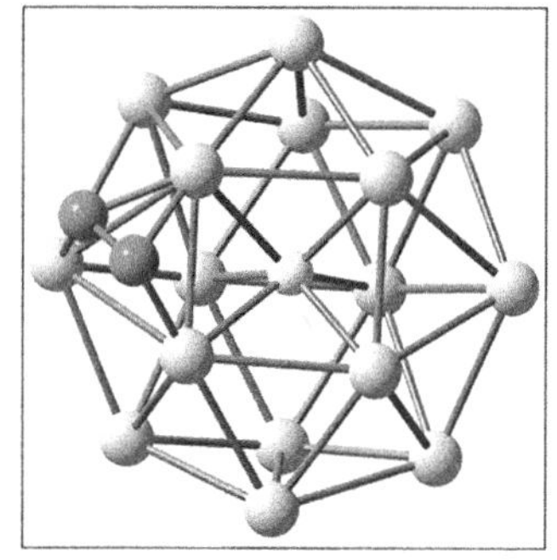

(b) $Mg_{16}Be—O_2$ (4-MRs) complex

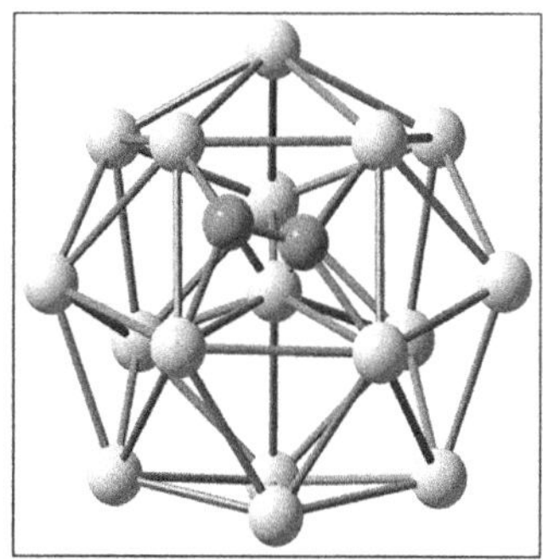

(c) $Mg_{16}Mg—O_2$ (4-MRs) complex

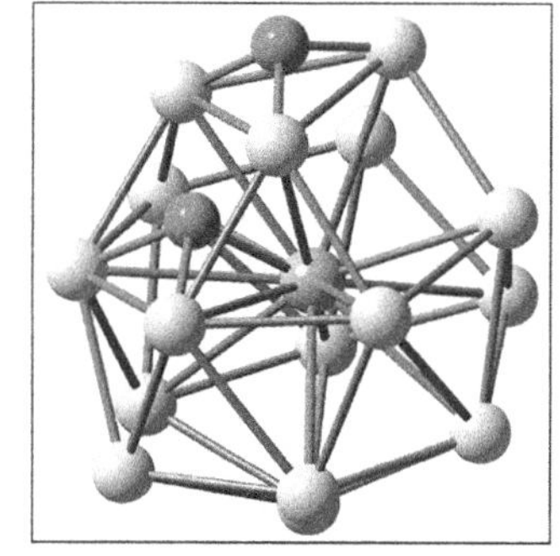

(d) $Mg_{16}Ca—O_2$ (4-MRs) complex

FIGURE 8.12 (a) Schematic view of the $Mg_{16}M$ (M = Be, Mg, and Ca) nanoclusters. The gray bonds represent the two 4-membered rings (4-MRs), and the thick black bonds represent one 8-membered ring (8-MR), M is the central atom (M = Be, Mg, and, Ca). (b), (c), (d) Optimized structures of the unimolecular O_2 adsorption of $Mg_{16}M—O_2$ (M = Be, Mg, and Ca) at the CAM-B3LYP/6–311+G(d) level of theory.

molecules sticking to the surface. In this process, a thin layer of adsorbed material is created on the surface of the adsorbent. This process is present in many physical, biological, and chemical systems. On the other hand, *absorption* is the penetration of the collected materials into the absorbent. Since these two phenomena often occur simultaneously, the word "sorption" is used for these phenomena.

Theoretical calculations are very powerful and intelligent methods that can be used to visualize reactions, and this is especially important for reactions with short half-lives or explosive reactions.

In this chapter, we treated the adsorption of different gases that make up air, by 17-atom magnesium nanocluster with a special structure of filled-cage-like construction, including two square (4-MRs) frames and an octagon (8-MR). We also changed the central atom to Be and Ca to see the effect of the doped atom. Changing the size of the central atom is effective in the rate of absorption of different gases so that the maximum rate of absorption was observed in the nanocluster with the central Ca atom. Resultantly, O_2, N_2, and CO can be separated from the air. There was no attraction between the H_2 and the magnesium nanocluster.

Quantum calculations showed that magnesium Mg_{17} is not a good holder for hydrogen gas and that there is no interaction between hydrogen gas and magnesium nanocluster. As a result, this complex can be used to separate different gases that make up air and finally obtain pure hydrogen gas.

Our studies show that the Mg_{17} nanocluster is more stable than nanoclusters with a smaller or larger size and nanoclusters with different structures. It has good absorption and desorption energy during the adsorption process. The absorption rate of various gases increases by changing the size of the central atom in the nanocluster, and replacing the central magnesium atom with a divalent metal with a larger radius than magnesium can increase the sensitivity of the aromatic metal $Mg_{16}M$ nanoclusters to be used as sensors for various gases.

REFERENCES

1. Mollaamin, F. and M. Monajjemi, *In silico-DFT investigation of nanocluster alloys of Al-(Mg, Ge, Sn) coated by nitrogen heterocyclic carbenes as corrosion inhibitors.* Journal of Cluster Science, 2023: pp. 1–18.
2. Zhu, B.-C., et al., *Computational exploration on the structural and optical properties of gold-doped alkaline-earth magnesium AuMgn (n= 2–12) nanoclusters: DFT study.* Frontiers in Chemistry, 2022: pp. 1–10
3. Shi, D., et al., *A computational study of Mg m H n nanoclusters with n: m≥ 2: 1 for efficient hydrogen storage.* International Journal of Quantum Chemistry, 2023. **123**(6): p. e27058.
4. Lyalin, A., et al., *Evolution of the electronic and ionic structure of Mg clusters with increase in cluster size.* Physical Review A, 2003. **67**(6): p. 063203.
5. Jellinek, J. and P.H. Acioli, *Magnesium clusters: Structural and electronic properties and the size-induced nonmetal-to-metal transition.* The Journal of Physical Chemistry A, 2002. **106**(45): pp. 10919–10925.
6. Yoo, J. and A. Aksimentiev, *Improved parametrization of Li+, Na+, K+, and Mg2+ ions for all-atom molecular dynamics simulations of nucleic acid systems.* The Journal of Physical Chemistry Letters, 2012. **3**(1): pp. 45–50.
7. Acioli, P.H. and J. Jellinek, *Electron binding energies of anionic magnesium clusters and the nonmetal-to-metal transition.* Physical Review Letters, 2002. **89**(21): pp. 213402.
8. Hailstone, R., *Computer simulation studies of silver cluster formation on AgBr microcrystals.* The Journal of Physical Chemistry, 1995. **99**(13): pp. 4414–4428.
9. Fayet, P., et al., *Latent-image generation by deposition of monodisperse silver clusters.* Physical Review Letters, 1985. **55**(27): p. 3002.
10. Chen, L.-Y., et al., *Processing and properties of magnesium containing a dense uniform dispersion of nanoparticles.* Nature, 2015. **528**(7583): pp. 539–543.
11. Jellinek, J. and P.H. Acioli, *Magnesium clusters: Structural and electronic properties and the size-induced nonmetal-to-metal transition.* The Journal of Physical Chemistry A, 2003. **107**(10): pp. 1670–1670.
12. Akola, J., K. Rytkönen, and M. Manninen, *Metallic evolution of small magnesium clusters.* The European Physical Journal D-Atomic, Molecular, Optical and Plasma Physics, 2001. **16**: pp. 21–24.
13. Davidson, E.R. and R.F. Frey, *Density functional calculations for Mg n+ clusters.* The Journal of Chemical Physics, 1997. **106**(6): pp. 2331–2341.
14. Dehghan, M.R. and S. Ahmadi, *CRS.* Journal of Chemical Reactivity and Synthesis, 2021. **11**(2): pp. 62–90.
15. Xia, X., et al., *Deciphering the structural evolution and electronic properties of magnesium clusters: An aromatic homonuclear metal Mg17 cluster.* The Journal of Physical Chemistry A, 2016. **120**(40): pp. 7947–7954.
16. Gutman, I., *Violations of the Hückel (4n+ 2) rule.* Journal of the Chinese Chemical Society, 1993. **40**(1): pp. 7–10.
17. Wang, H., T. Maiyalagan, and X. Wang, *Review on recent progress in nitrogen-doped graphene: Synthesis, characterization, and its potential applications.* ACS Catalysis, 2012. **2**(5): pp. 781–794.
18. Ahmadi, S., et al., *Atomistic simulation studies of the α/β-glucoside and galactoside in anhydrous bilayers: Effect of the anomeric and epimeric configurations.* Journal of Molecular Modeling, 2014. **20**: pp. 1–12.
19. Pearson, R.G., *Absolute electronegativity and hardness: Applications to organic chemistry.* The Journal of Organic Chemistry, 1989. **54**(6): pp. 1423–1430.
20. Zhou, Z. and R.G. Parr, *Activation hardness: New index for describing the orientation of electrophilic aromatic substitution.* Journal of the American Chemical Society, 1990. **112**(15): pp. 5720–5724.
21. Faust, W., *Explosive molecular ionic crystals.* Science, 1989. **245**(4913): pp. 37–42.
22. Eslami, M., V. Vahabi, and A.A. Peyghan, *Sensing properties of BN nanotube toward carcinogenic 4-chloroaniline: A computational study.* Physica E: Low-dimensional Systems and Nanostructures, 2016. **76**: pp. 6–11.

23. Khalafi, H., S. Ahmadi, and Z.M. Kotena, *First-principles study of the adsorption behavior of Octyl-β-D-xyloside surfactant on pristine Al12N12 and B12N12 nanocages.* Journal of Molecular Structure, 2022. **1247**: p. 131360.
24. Dehghan, M.R., et al., *A computational study of N2 adsorption on aromatic metal Mg16M;(M= Be, Mg, and Ca) nanoclusters.* Journal of Molecular Graphics and Modelling, 2021. **105**: p. 107862.
25. Dehghan, M.R., S. Ahmadi, and Z.M. Kotena, *Adsorption behaviors of carbon monoxide (CO) over aromatic magnesium nanoclusters: A DFT study.* Structural Chemistry, 2021. **32**: pp. 1949–1960.
26. Parr, R.G. and R.G. Pearson, *Absolute hardness: Companion parameter to absolute electronegativity.* Journal of the American Chemical Society, 1983. **105**(26): pp. 7512–7516.
27. Parr, R.G. and P.K. Chattaraj, *Principle of maximum hardness.* Journal of the American Chemical Society, 1991. **113**(5): pp. 1854–1855.
28. Pearson, R.G., *Absolute electronegativity and absolute hardness of Lewis acids and bases.* Journal of the American Chemical Society, 1985. **107**(24): pp. 6801–6806.
29. Yanai, T., D.P. Tew, and N.C. Handy, *A new hybrid exchange–correlation functional using the Coulomb-attenuating method (CAM-B3LYP).* Chemical Physics Letters, 2004. **393**(1–3): pp. 51–57.

9 Effect of Confinement in Bonding and Catalysis

Ruchi Jha, Ranita Pal, and Pratim Kumar Chattaraj

9.1 INTRODUCTION

We live in a world where chemical reactions are ubiquitous and are incredibly important in many areas of science as well as in our daily lives. These are fundamental to our understanding of the natural world and have countless practical applications in fields ranging from medicine to industry to environmental science.

A chemical reaction can be spontaneous or nonspontaneous depending on whether or not it needs an external energy input or intervention to be completed. In other words, a spontaneous reaction is one that happens naturally, driven by the thermodynamic properties of the reactants and products. The spontaneity of a chemical reaction is determined by the change in Gibbs free energy (ΔG) that occurs during the reaction. If ΔG is negative, the reaction is spontaneous, and if ΔG is positive, the reaction is nonspontaneous. Although spontaneity of a chemical reaction is a major deciding factor, considering the fate of a reaction it is not the only one. A few reactions are spontaneous in nature but take a considerably long time to be completed, and hence the rate of a reaction should also be considered. For example, the very basic reaction of the breakdown of $C_6H_{12}O_6$ (sugar) into CO_2, H_2O, and energy is spontaneous in nature, but the rate of the reaction is too slow to be observed outside our bodies. The same reaction occurs all the time inside our bodies due to the presence of enzymes such as hexokinase, pyruvate kinase, phosfructo kinase-1, etc. These enzymes lower the activation barrier of the reaction and make them occur at an extremely faster rate. These enzymes act as catalysts for the reaction to occur at a faster rate by lowering the activation barrier and providing an alternative route for the reactants to form products.

Apart from using a completely different chemical substance for reducing the activation barrier and making reactions occur, it has been found recently that there are other ways to make reactions fast and to selectively design the product. One of the ways is to confine the reactants inside a cavity and let the effect of confinement do the work. Such methods have been studied extensively in recent times and can prove to be helpful in many ways. Such a kind of catalysis can be termed confinement-induced catalysis (CIC). Confinement-induced catalysis is a concept that refers to the enhancement of catalytic activity of a material when it is confined within a small space, such as a nanopore or a nanocavity.[1] The idea behind this concept is that, when a catalyst is confined within a small space, it is forced to adopt a particular geometry or structure, which can lead to a change in its electronic structure, resulting in enhanced catalytic activity.[2] A catalyst is a substance that increases the rate of a chemical reaction without being consumed in the process. Catalysts work by lowering the activation energy required for a reaction to occur, which means that reactions can proceed more quickly and with less energy input. In some cases, the presence of a catalyst can even lead to desired products.

Confinement-induced catalysis is a fascinating phenomenon that occurs when chemical reactions are accelerated in confined spaces compared to the same reactions in bulk solutions. This field of study has gained significant attention in recent years due to its potential applications in various fields, including catalysis, materials science, and chemical engineering. The confinement effect can be achieved by using porous materials, such as zeolites,[3–7] metal-organic frameworks (MOFs),[8–12] and carbon nanotubes[13–17] or by confining the reactants in small droplets or nanoscale compartments.[18]

 DOI: 10.1201/9781003441328-9

The idea of CIC is not new, and scientists have been investigating it for decades. However, it has gained renewed interest due to advances in experimental and theoretical techniques that have enabled a more in-depth understanding of the underlying mechanisms of CIC. Theoretical studies have suggested that confinement can alter the reaction pathway, stabilize intermediates, and enhance the reactivity of catalytic sites. Experimental studies have confirmed these predictions and provided evidence for the effectiveness of CIC in various reactions, such as hydrogenation, oxidation, and polymerization.

Theoretical studies on confinement-induced catalysis involve the use of computational modeling and simulation techniques to investigate the electronic and structural properties of catalysts confined within small spaces. These studies can provide insights into the mechanisms behind the enhanced catalytic activity observed in confined catalysts and can help in the design of new, more efficient catalysts. One of the key challenges in the theoretical studies of confinement-induced catalysis is the accurate modeling of the complex interactions between the catalyst and the confined environment. This requires the use of sophisticated computational techniques, such as density functional theory (DFT) and quantum mechanics/molecular mechanics (QM/MM) simulations. Theoretical studies on confinement-induced catalysis have shown that confinement can lead to a range of catalytic enhancements, such as increased selectivity, increased activity,[19] and improved stability. This has led to the development of a range of new catalysts for a variety of applications, including energy production, pollution control, and pharmaceuticals. Overall, theoretical studies on confinement-induced catalysis are a rapidly growing area of research, with significant potential for the development of new, more efficient catalysts.

In this chapter, we will provide a comprehensive overview of confinement-induced catalysis focusing on the latest developments in theoretical models and applications. We will begin by discussing the different types of confinement, followed by a detailed examination of the mechanisms that underlie CIC. Finally, we will highlight the potential applications of CIC in various fields, including energy conversion, environmental remediation, and chemical synthesis.

9.2 DIFFERENT TYPES OF GEOMETRICAL CONFINEMENT

The physical and chemical properties of atoms and molecules can be profoundly affected by geometrical confinement.[20] This effect arises from the fact that the confined environment alters the electronic and molecular structure of the confined species, as well as their interactions with other species and their surroundings. For example, when a molecule is confined to a small space, such as a nanopore or a thin film, the interactions between the molecule and the confining surfaces can modify the electronic structure of the molecule and influence its reactivity and transport properties. Similarly, when atoms are confined within nanoscale pores or clusters, their electronic and magnetic properties can be significantly altered, leading to novel properties and applications.

Research in this area has shown that confinement can affect a wide range of properties, including chemical reactions, adsorption, transport, electronic and magnetic properties, and even the stability of certain compounds. As a result, understanding the effects of confinement is important for a wide range of applications, especially catalysis.

9.2.1 Cucurbit[*n*]uril

One such type of geometrical confinement is imposed by cucurbituril (CB).[21] CB is a type of macrocyclic molecule that is composed of glycoluril units linked by methylene bridges.[22] It has a hollow, barrel-shaped structure that can encapsulate small guest molecules within its cavity. CB is often used as a host molecule for supramolecular chemistry[23] and can be used for applications such as drug delivery and catalysis. Figure 9.1 shows the structures of CB[n] $\{n = 5\text{–}8, 10\}$.

Compared with other organic host molecules, CB[*n*] moieties have been shown to be particularly effective at confining organic guest molecules. This is because CB[*n*] molecules are relatively

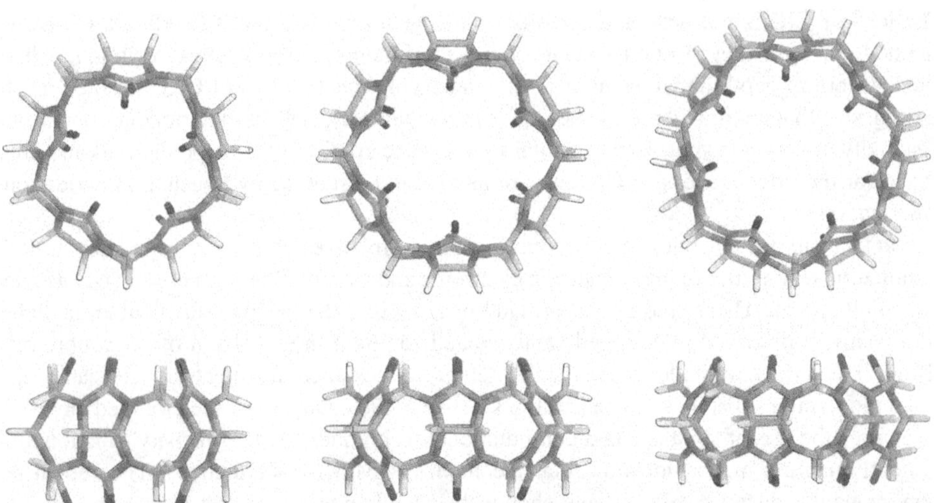

FIGURE 9.1 Crystal structure of CB[5], CB[6], CB[7], CB[8], CB[10] from Cambridge Crystallographic Data Centre.

Source: CCDC deposition numbers 883370, 883372, 1428201, 1859221, 1507270 contain the supplementary crystallographic data for these structures and are provided free of charge by The Cambridge Crystallographic Data Centre.

rigid, and the size and shape of their cavities can be precisely controlled by varying the number of glycoluril units in the molecule. In addition, CB[*n*] molecules have a high binding affinity for a wide range of organic guest molecules, which makes them versatile and useful for a variety of applications.

CB molecules are water soluble and have high stability, making them useful in a variety of biological and chemical applications. They have a unique selectivity for certain types of guest molecules, which makes them valuable tools in separation science and analytical chemistry. CB has also been investigated for its potential in biomedical applications, such as drug delivery and diagnostic imaging. Overall, CB is a promising molecule with a wide range of potential applications in various fields of science and technology.

9.2.2 Fullerene Cages C_N [N = 60, 70, 80, 90, etc.]

Fullerenes are a fascinating class of carbon-based molecules that were first discovered in the mid-1980s. They are composed entirely of carbon atoms arranged in a unique structure, forming a hollow cage-like shape. The most famous fullerene is the buckminsterfullerene, or "buckyball," which has a spherical shape composed of 60 carbon atoms. Fullerene cages come in various shapes and sizes, with the number of carbon atoms ranging from a few to several hundred. These carbon cages have garnered significant interest in various scientific fields due to their unique chemical and physical properties. They are incredibly stable, have high tensile strength, and exhibit excellent electrical conductivity, making them suitable for various applications, including drug delivery, energy storage, and catalysis.

The discovery of fullerene cages has opened up a new realm of research, leading to the development of new materials with improved properties. These cages can be used to encapsulate guest molecules, which can then be used as catalysts for various chemical reactions. The unique properties of fullerenes, such as their high surface area, high stability, and selective permeability, make them ideal for supramolecular catalysis. In particular, the use of fullerene cages allows for the confinement of guest molecules, which can improve their stability and selectivity in catalytic reactions. One example of supramolecular catalysis in fullerene cages is the use of fullerene-based hosts to encapsulate metal ions, which can then be used as catalysts for various reactions. Another example is the use of fullerene-based hosts to encapsulate organic molecules, which can then be used as catalysts for various organic reactions. Overall, supramolecular catalysis in fullerene cages is a promising area of research with potential applications in fields such as organic synthesis, materials science, and drug discovery.

9.3 CATALYSIS USING CUCURBITURIL CAVITIES

CB[*n*]s have been utilized to catalyze and template a variety of chemical reactions because of their capacity to encapsulate a variety of guest molecules. The development of click-chemistry reactions was made possible by the initial publication[24] that described the catalysis of [3 + 2] cycloadditions in the presence of CB[6].

When an alkyene (E1) and an azide (E2) were found to undergo a [3 + 2] cycloaddition inside the cavity of CB[6] molecules, Mock et al. reported the first occurrence of CB catalysis. The formation of substituted triazoles occurs slowly in an aqueous formic acid solution at 40 °C, but the addition of catalytic amounts of CB[6] led to a 6×10^5 increase in reaction rate, with triazole (E3) being the only isomer formed because of the orientation of E1 and E2 within the 1:1:1 complex formed with CB[6]. The schematic of the reaction is shown in Figure 9.2. The ideal orientation was used to overcome entropic restrictions, and strain-induced compression of the reactants within the CB[6] cavity was also credited with the catalytic rate enhancement of the reaction. This groundbreaking breakthrough served as a catalyst for the advancement of azide-alkyene cycloadditions toward "click chemistry" processes, leading to the effective synthesis of rotaxanes, polyrotaxanes, and pseudorotaxanes from [3 + 2] cycloadditions catalyzed by CB[6].

Researchers Reddy and colleagues[25] studied the use of CB[6] complexes as catalysts for halogen-catalyzed reactions. They exposed fine CB[6] powders to either I_2 or Br_2 vapor to form I_2–CB[6]·$4H_2O$ and Br_2–CB[6]·$10H_2O$, respectively, and then performed reactions in the presence of the halogen complex or its complex with CB[6]. They compared the resulting yields to those reported in the literature for uncomplexed halogens. The results showed that the greatest increase in yield was seen for the iodine-catalyzed Prins cyclization, where the yield was 7% higher than that reported in the literature. The remaining reactions tested all showed yields comparable to what had been demonstrated in the literature for the uncomplexed halogens. This represents a novel use of CB[6] complexes in halogen-catalyzed reactions, and the study demonstrates the potential for using supramolecular complexes as catalysts in chemical reactions.

The study by Francis et al.[26] shows how the mild reaction conditions of the CB[6] catalyzed alkyne-azide cycloaddition (AAC) reaction may be used to successfully modify a variety of proteins. This was accomplished by substituting protonated piperidine groups for amino groups and lowering the likelihood of cross-reactions. This study emphasizes the biocompatibility, functional group tolerance, and aqueous solubility of CB[*n*] that make it a promising candidate for bioapplications. A precise and effective approach for altering azidoethylamine derivatives when alkyl azides

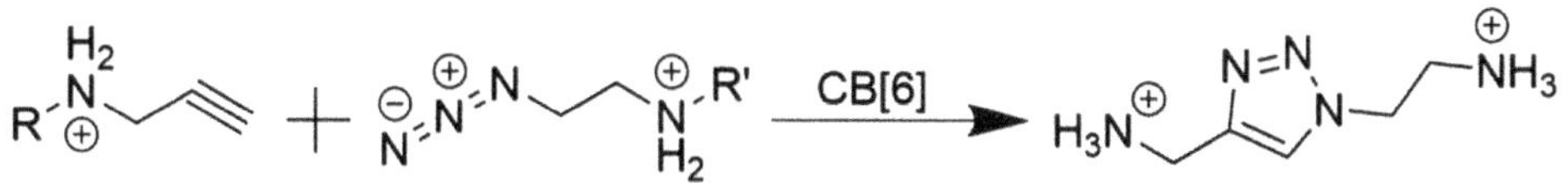

FIGURE 9.2 Schematic of the [3+2] cycloaddition reaction between alkyne (E1) and azide (E2) in the presence of CB[6].

Source: Reproduced from Ref. [27] with permission from American Chemical Society.

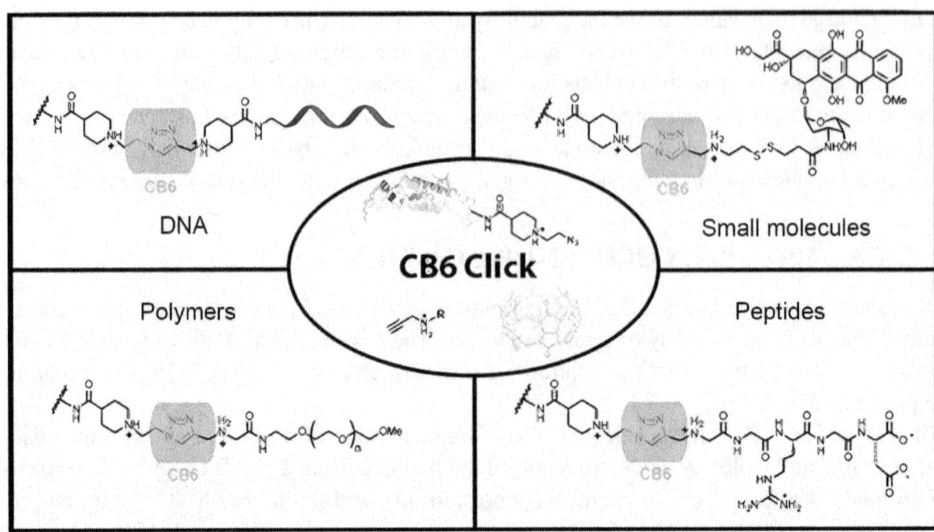

FIGURE 9.3 CB[6] catalyzed AAC reaction for protein modification.

Source: Reproduced from Ref. [26] with permission from American Chemical Society.

are present is the use of the CB[6] click reaction (Figure 9.3). Additionally, the study demonstrated that the CB[6] click reaction exhibited orthogonality toward reactive thiols, which are essential for the widely used bioconjugation method known as thiol-maleimide coupling.

Overall, this discovery offers up possibilities for CB[*n*]-based supramolecular catalysis to be developed further in the realm of biochemistry. The design and synthesis of new materials with specialized biological activities, such as medication delivery systems, biomimetic sensors, and bioactive surfaces, may be significantly impacted by this.

CB[*n*]s are quite successful in facilitating Diels Alder reactions, and some of the first evidences comes from the work of Berta et al.[28] They found that a favorable anisotropic environment is created when the supramolecular host attaches to the substrate, setting up the cycloaddition both conformationally and electrostatically.

Computational analysis has been conducted by Chakraborty et al.[29] to try and find a suitable confining vessel that would facilitate a model Diels–Alder reaction between 1,3-butadiene and ethylene from a kinetic perspective. Results show that, in comparison to their structural equivalent CB[7], neither CB[6] nor CB[8] could speed up the selected reaction under ambient circumstances. The type of the restricting regime itself is a key element that influences how the reaction will be carried out within the hosts. It seems that while CB[6]'s cavity is too tiny to effectively confine the reactants and facilitate the reaction, CB[8] has a considerably larger cavity than what could be appropriate (as in CB[7]). Therefore, one might notice that any restricting state cannot kinetically facilitate any given reaction as a proof of concept. The preorganization of the reactants inside the cavitands appears to be a key factor in determining the process's result. Therefore, the size of the cavity in relation to the reactants should be of utmost significance. Given the significance of confinement-induced kinetic facilitation of chemical reactions from the perspective of diverse applications, it may be beneficial to consider a number of additional reactions inside different organic cavitands in order to further our understanding in this area.

9.4 CATALYSIS USING FULLERENE CAGES

Fullerene cages have been proposed as potential confinement systems for CIC due to their unique properties as hollow nanospheres with well-defined sizes and shapes. In CIC using fullerene cages, the reactants are trapped within the interior of the cage, and the cage serves as a reaction vessel where the reactants can undergo catalytic reactions. The confinement of the reactants within the cage can lead to increased reaction rates and selectivity, as the reactants are brought into close proximity with the catalyst and are shielded from external influences. CIC using fullerene cages has potential applications in various fields, including catalytic synthesis, energy production, and environmental remediation. The use of fullerene cages as confinement systems for CIC could lead to the development of more efficient and selective catalytic processes.

A study conducted by Anafchen[30] describes a method that explored the hydrogenation of CO_2 using a new catalyst called aluminum-ligated NNN pincerfullerene, AlH_2-bis(imino)pyrollidinofullerene. The study used density functional theory calculations to investigate the reaction mechanism and the effect of water on the reaction. The results showed that the presence of a water molecule accelerated the initial step of CO_2 hydrogenation. The study also used NBO and AIM analyses to confirm the presence of hydrogen bonding and charge distributions in the intermediate and transition states. The appropriate charge distributions and inter-residue hydrogen bonding indicated the kinetic and thermodynamic performance of the H_2O-assisted metal-ligand cooperative pathway. In the second hydrogenation catalytic cycle, the rotational isomerization and hydrogen molecule insertion to the NO bond in the metal-ligand H_2O-assisted pathway led to the dissociation of H_2CO_2 and Al-H_2 complex. This suggests that the H_2O-assisted pathway could be a promising approach for catalytic hydrogenation of CO_2. Overall, this study provides insights into the reaction mechanism and the role of water in CO_2 hydrogenation using a new catalyst. The findings could contribute to the development of more efficient and selective catalytic processes for CO_2 conversion.

In acid-base reactions, even when placed in close proximity, a single molecule of hydrogen chloride (HCl) and a single molecule of ammonia (NH_3) do not undergo proton transfer to create the $NH_4^+Cl^-$ ion pair. This is due to the excessively high energy needed to transfer a proton from HCl to NH_3, which is around 120 kcal mol^{-1} in the gas phase. Acid-base reactions typically take place between compounds that may form hydrogen bound complexes in solution, which is consistent with the high energy demand. In this instance, HCl and NH_3 preferentially combine to create this complex, which is held together by hydrogen bonds with energies ranging from 3 to 10 kcal mol^{-1}. These hydrogen bonds have a far lower strength than the energy needed for proton transfer. Overall, the behavior of molecules in solution and in the gas phase is heavily influenced by the laws of acid-base chemistry and the energy needs for proton transfer.

It has been theoretically demonstrated by Varadwaj et al.[31] that this proton transfer reaction can be made facile inside a fullerene cage (C_{70}) and suggests that fullerene cages can act as catalytic vessels for many reactions. They have shown that the reactivity of three significant complex systems, H_2OHX (X = F, Cl, Br), inside the fullerene C_{70} cage using DFT and first-principles research. They have demonstrated that the interior of the C_{70} cage is not hydrophobic and frequently donates charge density to visiting species. Additionally, the C_{70} cage interior has the capacity to filter the electrostatic field of visiting species, reducing the observability of numerous IR-active vibrational bands. Additionally, the interior of the C_{70} cage can operate as a terrace to promote effective proton transfer events between visitor species and to cause charge rearrangement, bond polarization, and ion pair formation.

According to the speculations and the studies of other systems, the interior of the C_{70} cage offers an inventive platform for electrogenerating reactive species between H_2O and HX. Further research into the catalytic profiles of the C_{70} cage interior, such as those involving NH_3 and HX (X = F, Br, Cl), is expected to provide unique physical chemistry and catalytic features of these materials.

9.5 BONDING INSIDE FULLERENE CAGES

Chemical properties of atoms, molecules and clusters change drastically upon confinement in fullerene cages.[32] It has been found by Jha et al.[33] that the electronic properties such as ionization energy and electron affinity of alkali atoms and halogen atoms, respectively, can be improved by choosing proper confinement regime. The size of the confinement is also important. The oxidizing and reducing power of superhalogens and superalkalis is also affected due to confinement.[34] A study conducted by Wang et al.[35] investigated water clusters confined within fullerene cages of various sizes, including C_{60}, C_{140}, C_{180}, and C_{240}, with diameters ranging from 0.73 to 1.41 nm, using *ab initio* computational methods. They examined the relationship between the cavity volume and the maximum number of exothermically confined water molecules and found a linear correlation. It was observed that smaller water clusters, such as monomers and dimers, sit off-center within the C_{180} cage, and a transition from cyclic to cage-like configurations occurs at $(H_2O)_6$. For smaller clusters $(H_2O)_n$ with $n < 11$, the guest–host interaction has minimal impact on cluster geometries. However, for medium-sized clusters ($n > 12$) confined in the C_{180} cage, cage-like configurations become energetically favored over layered structures, which are favorable in the gas phase. The interaction between the enclosed water clusters and the outer fullerene cages was determined to be of van der Waals (vdW) type, with an energy of approximately 1.1 kcal/mol per water molecule. This interaction has negligible effects on the electronic structures (molecular orbitals) of both components. The confined water clusters induce an inverse polarization of the outer cage due to their permanent dipole moment, resulting in a screening effect on the cluster's dipole moment. Consequently, the dipole moments of water clusters inside fullerene cages are significantly reduced compared to those in the gas phase. The findings of this study provide valuable insights into the behavior of water clusters in nonpolar cavities at the nanoscale, contributing to a better understanding of water in biologically relevant environments.

Through a comparative study Srivastava et al.[36] investigated the impact of encapsulating Li or F within C_{60}, namely $Li@C_{60}$ and $F@C_{60}$, on the structure and properties of the fullerene. Charge transfer occurred between the C_{60} cage and the encapsulated species in both cases, with $Li@C_{60}$ exhibiting greater polarity than $F@C_{60}$. This suggests that C_{60} acts better as an electron acceptor than as a donor. The NBO charge, dipole moment, and binding energy values further supported this observation. Khatua et al.[37,38] investigated through density functional theory calculations, the influence of spatial confinement on the strength of the H···F hydrogen bond within $(HF)_2$ dimer encapsulated in fullerene cages of various sizes. It was found that the hydrogen bond distance within the confined cases was consistently shorter than in the free situation, although no clear trend emerged regarding the decrease in bond distance with increased confinement strength. These results suggest that, in a confined environment, a shorter bond does not necessarily indicate a stronger bond. It depends on a delicate balance between attractive electrostatic and orbital terms, as well as the repulsive term. These findings contribute to our understanding of hydrogen bond behavior under spatial confinement and provide valuable insights for designing systems based on these principles.

9.6 CONCLUSION

Theoretical studies have suggested that confinement can alter the reaction pathway, stabilize intermediates, and enhance the reactivity of catalytic sites. Experimental studies have confirmed these predictions and provided evidence for the effectiveness of CIC in various reactions. Studies on confinement-induced catalysis have shown that confinement can lead to catalytic enhancements, such as increased selectivity, increased activity, and improved stability. This has led to the development of new catalysts for various applications, such as energy production, pollution control, and pharmaceuticals.

Confinement-induced catalysis (CIC) is a phenomenon where chemical reactions are accelerated in confined spaces, such as porous materials or nanoscale compartments, compared to bulk solutions. CIC has gained significant attention in recent years due to its potential applications in catalysis, materials science, and chemical engineering. Theoretical and experimental studies have shown that confinement can alter the reaction pathway, stabilize intermediates, and enhance the reactivity of catalytic sites, leading to increased selectivity, activity, and improved stability.

In this chapter, we discussed different types of reaction vessels that can impose geometrical confinement on the reacting species such as cucurbituril[*n*] and fullerene cages of different sizes.

We have also discussed the reactions that have been facilitated inside these reaction vessels. Despite recent advancements in this area, it has also been shown that any confinement regime is not always a necessary requirement for fostering a chemical reaction. While confinement can enhance certain chemical reactions, it is not always necessary for them to occur. Additionally, certain reactions may even be hindered by confinement, as the restricted space can limit the movement of reactants and hinder the formation of intermediates and products. Ultimately, the effect of confinement on a chemical reaction depends on the specific reaction conditions and the properties of the reactants and catalysts involved.

We have also described several bonding scenarios that lead to these behavior within confinement. The confinement of atoms, molecules, and clusters within fullerene cages has a profound impact on their chemical properties. The choice of confinement regime can improve the electronic properties of alkali and halogen atoms, while the size of confinement also plays a significant role. The oxidizing and reducing power of superhalogens and superalkalis is affected by confinement as well. Regarding water clusters confined within fullerene cages, it was observed that the cavity volume correlates linearly with the maximum number of confined water molecules. Smaller clusters exhibit off-center configurations, transitioning to cage-like structures at a certain size. The interaction between the confined clusters and the outer cages is of van der Waals type and has minimal effects on their electronic structures. Furthermore, the strength of hydrogen bonds in confined $(HF)_2$ dimers shows no clear trend with increased confinement strength, highlighting the delicate balance between attractive and repulsive terms. These findings enhance our understanding of confined systems and offer insights for future design strategies.

9.6.1 Acknowledgment

PKC would like to thank DST, New Delhi, for the J. C. Bose National Fellowship, grant number SR/S2/JCB-09/2009. RJ and RP thank IIT Kharagpur and CSIR, respectively, for their fellowships.

9.6.2 Conflict of Interest

The authors declare that they have no conflict of interest, financial and/or otherwise.

REFERENCES

(1) Thomas, J. M.; Raja, R. Nanopore and Nanoparticle Catalysts. *Chem. Rec.* **2001**, *1* (6), 448–466.
(2) Zhuang, T.-T.; Pang, Y.; Liang, Z.-Q.; Wang, Z.; Li, Y.; Tan, C.-S.; Li, J.; Dinh, C. T.; De Luna, P.; Hsieh, P.-L.; Burdyny, T.; Li, H.-H.; Liu, M.; Wang, Y.; Li, F.; Proppe, A.; Johnston, A.; Nam, D.-H.; Wu, Z.-Y.; Zheng, Y.-R.; Ip, A. H.; Tan, H.; Chen, L.-J.; Yu, S.-H.; Kelley, S. O.; Sinton, D.; Sargent, E. H. Copper Nanocavities Confine Intermediates for Efficient Electrosynthesis of C3 Alcohol Fuels from Carbon Monoxide. *Nat. Catal.* **2018**, *1*, 946–951.
(3) Grifoni, E.; Piccini, G.; Lercher, J. A.; Glezakou, V.-A.; Rousseau, R.; Parrinello, M. Confinement Effects and Acid Strength in Zeolites. *Nat. Commun.* **2021**, *12*, 2630.
(4) Zalazar, M. F.; Cabral, N. D.; Romero Ojeda, G. D.; Alegre, C. I. A.; Peruchena, N. M. Confinement Effects in Protonation Reactions Catalyzed by Zeolites with Large Void Structures. *J. Phys. Chem. C* **2018**, *122*, 27350–27359.
(5) Chai, Y.; Dai, W.; Wu, G.; Guan, N.; Li, L. Confinement in a Zeolite and Zeolite Catalysis. *Acc. Chem. Res.* **2021**, *54*, 2894–2904.
(6) Wang, C.; Chu, Y.; Hu, M.; Cai, W.; Wang, Q.; Li, S.; Xu, J.; Deng, F. Influence of Zeolite Confinement Effects on Cation–π Interactions in Methanol-to-Hydrocarbon Conversion. *Chem. Commun.* **2022**, *58*, 9242–9245.
(7) Wu, S.; Yang, X.; Janiak, C. Confinement Effects in Zeolite-Confined Noble Metals. *Angew. Chemie.* **2019**, *131*, 12468–12482.
(8) Yang, D.; Gates, B. C. Catalysis by Metal Organic Frameworks: Perspective and Suggestions for Future Research. *ACS Catal.* **2019**, *9*, 1779–1798.

(9) Alhumaimess, M. S. Metal–Organic Frameworks and Their Catalytic Applications. *J. Saudi Chem. Soc.* **2020**, *24*, 461–473.
(10) Ranocchiari, M.; van Bokhoven, J. A. Catalysis by Metal–Organic Frameworks: Fundamentals and Opportunities. *Phys. Chem. Chem. Phys.* **2011**, *13*, 6388.
(11) Liu, X.; Liu, Z.; Wang, R. Functionalized Metal-Organic Framework Catalysts for Sustainable Biomass Valorization. *Adv. Polym. Technol.* **2020**, *2020*, 1–11.
(12) Shen, Y.; Pan, T.; Wang, L.; Ren, Z.; Zhang, W.; Huo, F. Programmable Logic in Metal–Organic Frameworks for Catalysis. *Adv. Mater.* **2021**, *33*, 2007442.
(13) Melchionna, M.; Marchesan, S.; Prato, M.; Fornasiero, P. Carbon Nanotubes and Catalysis: The Many Facets of a Successful Marriage. *Catal. Sci. & Technol.* **2015**, *5*, 3859–3875.
(14) Harmon, N. J.; Rooney, C. L.; Tao, Z.; Shang, B.; Raychaudhuri, N.; Choi, C.; Li, H.; Wang, H. Intrinsic Catalytic Activity of Carbon Nanotubes for Electrochemical Nitrate Reduction. *ACS Catal.* **2022**, *12*, 9135–9142.
(15) Yan, Y.; Miao, J.; Yang, Z.; Xiao, F.-X.; Yang, H. Bin; Liu, B.; Yang, Y. Carbon Nanotube Catalysts: Recent Advances in Synthesis, Characterization and Applications. *Chem. Soc. Rev.* **2015**, *44*, 3295–3346.
(16) Pan, X.; Bao, X. The Effects of Confinement inside Carbon Nanotubes on Catalysis. *Acc. Chem. Res.* **2011**, *44*, 553–562.
(17) Serp, P. Carbon Nanotubes and Nanofibers in Catalysis. *Appl. Catal. A Gen.* **2003**, *253* (2), 337–358. https://doi.org/10.1016/S0926-860X(03)00549-0.
(18) Wordsworth, J.; Benedetti, T. M.; Alinezhad, A.; Tilley, R. D.; Edwards, M. A.; Schuhmann, W.; Gooding, J. J. The Importance of Nanoscale Confinement to Electrocatalytic Performance. *Chem. Sci.* **2020**, *11*, 1233–1240.
(19) Grochala, W.; Hoffmann, R.; Feng, J.; Ashcroft, N. W. The Chemical Imagination at Work in Very Tight Places. *Angew. Chemie Int. Ed.* **2007**, *46*, 3620–3642.
(20) Sabin J. R.; Brändas E. J. *Advances in Quantum Chemistry: Theory of Confined Quantum Systems.* Academic, Waltham, 2009.
(21) Behrend, R.; Meyer, E.; Rusche, F. I. Ueber Condensationsproducte Aus Glycoluril Und Formaldehyd. *Justus Liebig's Ann. der Chemie* **1905**, *339*, 1–37.
(22) Jon, S. Y.; Ko, Y. H.; Park, S. H.; Kim, H.-J.; Kim, K. A Facile, Stereoselective [2 + 2] Photoreaction Mediated by Cucurbit[8]Uril. *Chem. Commun.* **2001**, 19, 1938–1939.
(23) Tang, B.; Zhao, J.; Xu, J.; Zhang, X. Cucurbit[n]Urils for Supramolecular Catalysis. *Chem. –A Eur. J.* **2020**, *26*, 15446–15460.
(24) Mock, W. L.; Irra, T. A.; Wepsiec, J. P.; Adhya, M. Catalysis by Cucurbituril. The Significance of Bound-Substrate Destabilization for Induced Triazole Formation. *J. Org. Chem.* **1989**, *54*, 5302–5308.
(25) Reddy, K. R. K. K.; Cavallini, T. S.; Demets, G. J. F.; Silva, L. F. Bromine and Iodine–Cucurbit[6]Uril Complexes: Preparation and Applications in Synthetic Organic Chemistry. *New J. Chem.* **2014**, *38*, 2262.
(26) Finbloom, J. A.; Han, K.; Slack, C. C.; Furst, A. L.; Francis, M. B. Cucurbit[6]Uril-Promoted Click Chemistry for Protein Modification. *J. Am. Chem. Soc.* **2017**, *139*, 9691–9697.
(27) Barrow, S. J.; Kasera, S.; Rowland, M. J.; Del Barrio, J.; Scherman, O. A. Cucurbituril-Based Molecular Recognition. *Chem. Rev.* **2015**, *115*, 12320–12406.
(28) Berta, D.; Szabó, I.; Scherman, O. A.; Rosta, E. Toward Understanding CB[7]-Based Supramolecular Diels-Alder Catalysis. *Front. Chem.* **2020**, *8*, 587084.
(29) Chakraborty, D.; Chattaraj, P. K. Confinement Induced Thermodynamic and Kinetic Facilitation of Some Diels-Alder Reactions inside a CB[7] Cavitand. *J. Comput. Chem.* **2018**, *39*, 151–160.
(30) Anafcheh, M.; Zahedi, M. Hydrogenation of Carbon Dioxide into Formic Acid by Aluminum Ligated NNN Pincer Fullerene Through Metal–Ligand H2O-Assisted Pathway: A Computational Study. *Catal. Letters* **2022**, *152*, 1224–1232.
(31) Varadwaj, P. R.; Varadwaj, A.; Marques, H. M. C70 Fullerene Cage as a Novel Catalyst for Efficient Proton Transfer Reactions between Small Molecules: A Theoretical Study. *Sci. Rep.* **2019**, *9*, 10650.
(32) Ramachandran, C.N.; Sathyamurthy, N. Atoms and molecules confined inside C_{60} *Chem. Phys. Lett.* **2005**, 410, 348.
(33) Jha, R.; Giri, S.; Chattaraj, P. K. Does confinement alter the ionisation energy & electron affinity of atoms? *Eur. Phys. J. D* **2021**, 75, 1.
(34) Jha, R.; Giri, S.; Chattaraj, P. K. Effect of confinement on the behaviour of superhalogen & superalkali. *Comput. Theor. Chem.* **2021**, 1206, 113491.
(35) Wang, L.; Zhao, J.; Fang, F. Water Clusters Confined in Nonpolar Cavities by Ab Initio Calculations. *J. Phys. Chem. C.* **2008,** *112*, 11779.

(36) Srivastava, A. K.; Pandey, S. K.; Misra, N. C_{60} as Electron Acceptor and Donor: A Comparative DFT Study of Li@C_{60} and F@C_{60} *Chem. Phys. Lett.* **2016,** *662*, 240.
(37) Khatua, M.; Pan, S.; Chattaraj, P.K. Confinement of $(HF)_2$ in C_n(n = 60, 70, 80, 90) cages. *Chem. Phys. Lett.* **2014,** *49*, 616.
(38) Poddar, A.; Ramachandran, C. N.; Chattaraj, P.K. Behavior of HF and (HF)2 inside a fullerene cage: An in silico study using different density functionals. *Int. J. Quantum Chem.* **2022,** *123*, 7.

10 Computational Studies on the NLO Properties of Molecules and Clusters Containing Excess Electrons

Wei-Ming Sun

10.1 INTRODUCTION

Nonlinear optics has been the focus of scientists and technologists since the first functional laser was invented by Maiman in 1960.[1] The following several decades have witnessed great developments in the field of nonlinear optics, including new nonlinear optical phenomena and extensive applications for modern optical data storage, optical communication, dynamic image processing, optical switches, optical computing, frequency mixing, sensing, photodynamic therapy, and so forth.[2–7] Hence a lot of new strategies have been proposed to design and synthesize various types of nonlinear optical (NLO) materials, such as designing donor-π-conjugated bridge-acceptor (D-π-A) molecules[7] and extended[8] or twisted[9] π-electron systems, applying bond length alternation (BLA) theory,[10] doping metal atoms,[11] synthesizing octupolar molecules,[12] enhancing push-pull effects,[13] utilizing a multidecker sandwich cluster as a building block,[14] considering specific molecules with diradical character,[15,16] and so on.

In 2004, Li and coworkers[17,18] have proposed that the introduction of loosely bound excess electrons into molecular systems can endow them with large NLO responses. To be specific, they found that the static first hyperpolarizabilities (β_0) of hydrogen fluoride trimer cluster anion $(FH)_2\{e\}(HF)$[17] and the noncentrosymmetrical water trimer cluster anion $(H_2O)_3\{e\}$[18] are greatly (about 10^6 times) larger than those of corresponding neutral clusters. This indicates that the molecules or clusters containing diffuse excess electrons could be considered new candidates for potential NLO materials. As a result, a large number of molecular systems with excess electrons have been reported in the past two decades.[2–4] This chapter reviews the history and current progress (before June 1, 2023) of the computational studies on the NLO properties of clusters and molecules containing excess electrons to present a broad overview of the development of excess electron strategy for NLO material design.

10.2 BACKGROUND OF THE EXCESS ELECTRON

Excess electrons in various media can be loosely bound to a cluster, molecule, or aggregate of metal atoms, and so forth.[4] The history of the studies on the compounds containing excess electrons can be traced to the work of Humphry Davy in 1808.[19–21] He firstly synthesized potassium-ammonia compounds with beautiful colors by reacting potassium with dry and gaseous ammonia. More than 50 years after Davy, a very similar observation on the solubility of alkali metals in liquid ammonia was documented by Weyl.[22] In 1879 and 1880, Hannay and Hogarth also reported that gaseous ammonia could dissolve metallic sodium.[23,24] The first crystallization of alkali metals (Na and K) dissolved in ammonia was carried out by Joannis in 1889.[25] This work was extended to the case of Li and Ca by Moissan,[26] who first reported the existence of the "lithium ammonium" solid.[27] Until 1916, Gibson and Argo were aware that the free excess electron from alkali metals in liquid

 DOI: 10.1201/9781003441328-10

ammonia is the crucial factor that leads to the blue color and other fascinating properties of alkali ammonia solution.[28] This is the first time that the term "solvated electrons" was invoked.[27]

Inspired by the solvated electrons, Dye's group synthesized the first compound with excess electrons, $Cs^+(18\text{-crown-}6)_2{\cdot}e^-$ in 1983, which was named electride.[29,30] Electrides are a representative type of excess electron compounds, in which the cations are usually complexed by organic complexants and anionic sites are occupied by trapped excess electrons.[31–33] In 2003, Matsuishi et al.[34] reported the successful synthesis of the first inorganic electride, $12CaO{\cdot}7Al_2O$, which is thermally stable at room temperature and can be exposed to air. The unique physical state of the excess electron endows such compounds with many important applications in catalysis, nanodevices, and conductive materials.[35]

The crucial role of excess electrons in enhancing the NLO response of molecular systems was first revealed in 2004 by Li and his coworkers.[17,18] Li et al.[17] performed a theoretical study to prove that the hydrogen fluoride trimer anion $(FH)_2\{e\}(HF)$ possesses exceptionally large β_0 of 8.1×10^6 au, which is significantly larger than that of its neutral core. In the same year, Chen et al.[18] predicted that the β_0 of the molecular cluster dipole-bound excess electron system $(H_2O)_3\{e\}$ is also as large as 1.7×10^7 au, which is about 10^6 times that of the corresponding molecular cluster $(H_2O)_3$. The dramatic effect of excess electron on the β_0 values of the cluster anions paves a new way to the theoretical design of novel NLO molecular materials and sets up a novel NLO material field for neutral molecules or clusters with an excess electron. In the past two decades, this strategy has been successfully used to design a great amount of excess electron compounds with high NLO response.

10.3 COMPUTATIONAL METHODOLOGY

10.3.1 Theoretical Background

Nonlinear optics illustrates the light–matter nonlinear relationship in a medium under an intense light beam (laser) and electromagnetic fields.[36] When a beam of light propagates through a macroscopic material, the interaction of the electromagnetic field with the valence electrons of molecules in the material will induce a medium polarization. The relationship between macroscopic polarization intensity (P) of medium and electric field strength (F) of the incident light can be expressed as:

$$P = \chi^{(1)}F + \chi^{(2)}F \cdot F + \chi^{(3)}F \cdot F \cdot F + \ldots \tag{10.1}$$

where P is the polarization strength, F is the applied electric field strength, and the coefficient $\chi^{(n)}$ is the nth-order susceptibility of the macroscopic material. Usually, a higher-order susceptibility $\chi^{(n)}$ is much smaller than a lower one $\chi^{(n-1)}$. When the electric field strength is weak, the polarization of the material can be expressed as the following linear form:

$$P = \chi^{(1)}F \tag{10.2}$$

However, when materials interact with a strong electric field, the nonlinear coefficients $\chi^{(2)}$ and $\chi^{(3)}$ are large enough to be considered. At the microscopic level, the relationship between the molecular electric polarization (p) and the electric field intensity F can be represented as:

$$p = \alpha_{ij}F_j + \beta_{ijk}F_j \cdot F_k + \gamma_{ijkl}F_j \cdot F_k \cdot F_l + \ldots . \tag{10.3}$$

where α_{ij} is the ij component of the molecular polarizability tensor (α_0), β_{ijk} is the ijk component of the first hyperpolarizability tensor (β_0), and γ_{ijkl} is the $ijkl$ component of the second hyperpolarizability tensor (γ_0). Commonly, the macroscopic polarization (P) of the material can be obtained by making the statistical average of the microscopic polarization (p).

For computational studies, the microscopic polarization nonlinear coefficients, namely the first and second hyperpolarizabilities (i.e., β_0 and γ_0) of molecules are usually computed to measure the quality of new NLO materials in theoretical design because they are the origins of the macroscopic second- and third-order nonlinear optical (NLO) susceptibilities $\chi^{(2)}$ and $\chi^{(3)}$, respectively. Therein, β_0 is the mostly popular one that is applied for the molecular systems containing excess electrons.

There are four second-order NLO effects in experiment: the second-harmonic generation [SHG, $\beta(-2\omega;\ \omega,\omega)$], linear electro-optic effect [LEOE, $\beta(-\omega;\ \omega,0)$], sum-frequency generation [SFG, $\beta(2\omega_1-\omega_2;\ \omega_1,\ \omega_2)$], and optical rectification [OR; $\beta(0;\ \omega,-\omega)$], where ω is the frequency of the applied electric field or the wavelength of device operation. Among these, $\beta(-2\omega;\ \omega,\omega)$ has been most frequently studied in experiments, which can be described accurately by the "sum-over-states" expression in theory.[7] For convenience, Oudar and Chemla[37,38] have established a simple relationship between the β value and a low-lying charge-transfer transition through the following two-level model:

$$\beta \propto \frac{\left(\hbar\omega_{gn}\right) f_{gn} \Delta\mu_{gn}}{\left[\left(\hbar\omega_{gn}\right)^2 - \left(2\hbar\omega\right)^2\right]\left[\left(\hbar\omega_{gn}\right)^2 - \left(\hbar\omega\right)^2\right]} \tag{10.4}$$

where ω is the frequency of the applied electric field. The subscripts g and n represent the ground state and the crucial low-lying excited state with the largest oscillator strength, respectively. The f_{gn}, $\hbar\omega_{gn}$, and $\Delta\mu_{gn}$ are the oscillator strength, the excited energy corresponding to the transition from g to n states, and the difference of the dipole moment between the g and n states, respectively.

Generally, computing the hyperpolarizabilities of molecules have been regarded as an important approach to predict their application in electro-optic switching/modulation or frequency doubling. For a special device, researchers can directly assess the NLO response of material at some specific ω values.[4] However, the comparison of frequency-dependent hyperpolarizability is quite complicated in molecular-engineering studies. The static β_0 quality is known as the total intrinsic hyperpolarizability, serving as a frequency-independent scale. Thus it has been widely adopted in computational studies on the NLO properties of molecular species with excess electrons.

For such a static β_0 ($\omega = 0.0$) value, the two-level model[37,38] can be written as:

$$\beta_0 \propto \frac{\Delta\mu \cdot f_0}{\Delta E^3} \tag{10.5}$$

where ΔE, f_0, and $\Delta\mu$ are the transition energy, oscillator strength, and the difference in the dipole moment between the ground state and crucial excited state, respectively. From the two-level expression, β_0 is proportional to f_0 and $\Delta\mu$ and is inversely proportional to the third power of ΔE. This indicates that small transition energies, large transition moments, and large dipole moment differences are required for large second-order NLO properties of molecular systems. In particular, the low transition energy has been proved to be the decisive factor for large first hyperpolarizabilities of many previously reported species containing excess electrons.[3,4]

10.3.2 Computational Methods

Ab initio techniques coupled with the finite field approach are commonly employed to calculate the β_0 values in most computational studies. In this frame, when a molecule is subjected to the static electric field (F), its energy can be expressed:[39,40]

$$E = E^0 - \mu_i F_i - \frac{1}{2}\alpha_{ij} F_i F_j - \frac{1}{6}\beta_{ijk} F_i F_j F_k - \ldots \tag{10.6}$$

where E^0 is the molecular energy without the applied field, F_i is a component of the strength on the i direction of the applied electrostatic field, while μ_i, α_{ij}, and β_{ijk} are the components of μ_0, α_0, and β_0 tensors, respectively.

For a molecular system, μ_0, α_0, and β_0 are defined as follows:

$$\mu_0 = (\mu_x^2 + \mu_y^2 + \mu_z^2)^{1/2} \tag{10.7}$$

$$\alpha_0 = \frac{1}{3}\left(\alpha_{xx} + \alpha_{yy} + \alpha_{zz}\right) \tag{10.8}$$

$$\beta_0 = (\beta_x^2 + \beta_y^2 + \beta_z^2)^{1/2} \tag{10.9}$$

where $\beta_i = \frac{3}{5}(\beta_{iii} + \beta_{ijj} + \beta_{ikk})$, $i, j, k = x, y, z$

As for the calculation of the first hyperpolarizability, choosing a proper computational method is essential to estimate the NLO responses of studied systems. The sophisticated quadratic configuration interaction, including the single and double substitutions (QCISD) method,[41] is suitable for small clusters with dipole-bound electrons, but it is too costly for relatively large systems. Therefore, the second-order Møller-Plesset perturbation (MP2) method[42] has been extensively used for several molecules with an excess electron[43–45] because it can provide β_0 results that are close to those obtained from the more sophisticated correlation methods, such as QCISD[43] and coupled-cluster singles and doubles (CCSD)[46] with acceptable economic cost. However, the cost of MP2 computations for large molecules or clusters is still too expensive.

Besides the expensive computational methods, some density functional theory (DFT) approaches have been adopted as effective alternatives to study the NLO properties of large systems in view of the compromise of computational efficiency and accuracy. For the reported excess electron compounds, a few density functional theory (DFT) methods, such as M06-2X,[47] CAM-B3LYP,[48] and BHandHLYP[15,49] were recommended for reproducing the results of post-SCF methods (such as MP2). For example, the BHandHLYP functional with more percentages of Hartree–Fock exchange than B3LYP can almost reproduce the (hyper)polarizabilities from the sophisticated CCSD(T) method.[15] However, there is no universal DFT functional for all the reported molecular systems with excess electrons up to now. Therefore, a computational test is usually performed on different DFT functionals and one of the sophisticated correlation functionals, such as MP2, CCSD, and QCISD, before carrying out a further investigation, so that the selected methods can be reliable in evaluating the NLO response of the studied NLO molecules and clusters. Besides, it should be mentioned that the relatively large basis set including the diffuse functions can be advantageous to accurately calculate the hyperpolarizabilities in view of the highly diffuse characteristic of excess electrons.[50]

Even though such DFT methods have been widely used to compute the β_0 values of large systems, developing new methods to accurately simulate the special electronic structures of excess electron compounds remains a big challenge. Among others, Wang et al.[51] proposed an effective intelligent computing method, named extreme learning machine-neural network (ELM-NN) to accurately predict the β_0 values of alkalides from low-accuracy β_0 values. Their results proved that ELM-NN possesses the ability to obtain calculated values of high accuracy level with less computing cost, suggesting its potential as a powerful tool to predict the β_0 values of large-scale molecules that are difficult to obtain by means of high-accuracy theoretical methods due to dramatically increasing computational costs.

10.3.3 Characterization of the Excess Electron

To prove the existence of excess electron in the studied systems, their highest occupied molecular orbitals (HOMOs) were usually plotted and analyzed. Herein, the solvated electron cluster

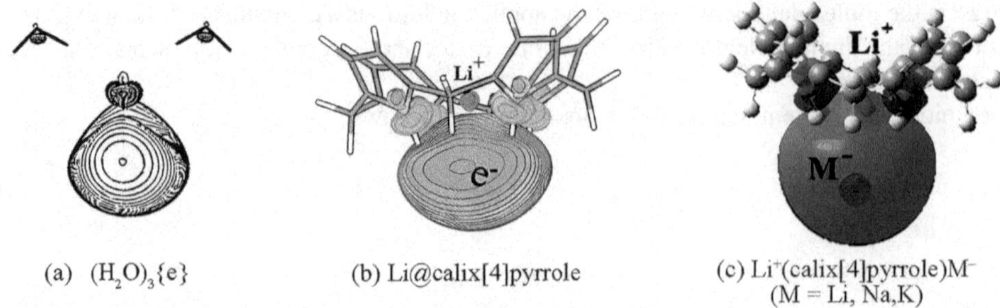

FIGURE 10.1 HOMO orbitals of three typical systems: (a) solvated electron cluster (H2O)3{e}, (b) electride Li@calix[4]pyrrole, and (c) alkalideLi$^+$(calix[4]pyrrole)M$^-$ (M = Li, Na, and K).

Sources: Plates (a), (b), and (c) are reproduced from Ref. [18] with permission (Copyright 2004 American Institute of Physics), Ref. [52] with permission (Copyright 2005 American Chemical Society), and Ref. [53] with permission (Copyright 2006 American Chemical Society), respectively.

$(H_2O)_3\{e\}$,[18] the electride Li@calix[4]pyrrole,[52] and the alkalide Li$^+$(calix[4]pyrrole)M$^-$ (M = Li, Na, and K)[53] were taken as three typical representatives to show the existence of delocalized diffuse excess electrons in their HOMOs (see Figure 10.1). In the $(H_2O)_3\{e\}$ anion, a solvated excess electron is bound by the dipole of a water trimer $(H_2O)_3$. The *s* valence electron of the Li atom is pushed by calix[4]pyrrole complexant to form an isolated excess electron in Li@calix[4]pyrrole and to wrap the M atom, forming an M$^-$ in Li$^+$(calix[4]pyrrole)M$^-$ (M = Li, Na, and K).

In addition, the molecules and clusters involving excess electrons usually show quite large polarizability (α_0) as the loosely bound excess electrons are more easily polarized. Usually, the atomic charges are also calculated for the excess electron compounds, such as electrides and alkalides, to characterize the metal cation or anion in them. In particular, Luis and Matito have proposed an unambiguous computational means to distinguish electrides from similar species.[54–56] They pointed out that the presence of a non-nuclear attractor (NNA) of the electron density, an electron localization function (ELF) basin, and negative values of the Laplacian of the electron density ($\nabla^2\rho$) are necessary conditions to assert the existence of an isolated electron.

10.4 STRATEGIES FOR DESIGNING MOLECULES AND CLUSTERS WITH EXCESS ELECTRONS

Due to the various applications of nonlinear optics, great attention has been focused on improving the NLO responses of molecules and clusters by using various strategies.[2] According to the two-level model (see equation 10.5), it can be seen clearly that the β_0 value is inversely proportional to the third power of the transition energy (ΔE), suggesting that transition energy is a crucial factor to determine the NLO response of a molecular system. Introducing excess electrons into a molecular system significantly decreases the ΔE value of its crucial excited state because such diffuse electrons can be easily excited to the unoccupied orbital. As a result, the introduction of an excess electron has been proposed, by Li and coworkers, as an important method to effectively enhance the first hyperpolarizabilities of a hydrogen fluoride trimer cluster anion, $(FH)_2\{e\}(HF)$,[17] and a noncentrosymmetric water trimer cluster anion, $(H_2O)_3\{e\}$, with a surface state excess electron.[18]

In the past two decades, much attention has been focused on how to introduce an excess electron into molecular materials by doping metal atoms to enhance their first hyperpolarizabilities. Initially, alkali metal atoms are usually adopted as the source or acceptor of excess electrons for different molecules, inspired by the successful synthesis of electrides[29–34] and alkalides[57–59] in experiments. Besides the extensively used alkali metals, the alkaline-earth metals, transition metals, and even clusters have been also employed as new source or acceptor of excess electrons in past years. In this chapter, the previously reported molecules and clusters with excess electrons have been classified into five groups according to the involved source or acceptors of excess electrons. Brief discussions on some crucial works have been performed to promote further theoretical design and experimental synthesis of NLO materials involving excess electrons.

10.4.1 Alkali-Metal-Based Excess Electron Compounds

10.4.1.1 Electrides

The first alkali-metal-based electride, Cs@18-crown-6, was successfully synthesized in 1983.[29] However, the studies on designing such electrides with considerable NLO responses began after revealing the crucial role of the excess electron on the NLO response of the molecular system in 2004.[17,18] In 2005, Chen et al.[52] theoretically designed a cup-like Li@calix[4]pyrrole (see Figure 10.2a) by intercalating one Li atom into a calix[4]pyrrole molecule. As shown in the HOMO of Li@calix[4]pyrrole, the *s* valence electron of the Li atom overflows from the bottom of the

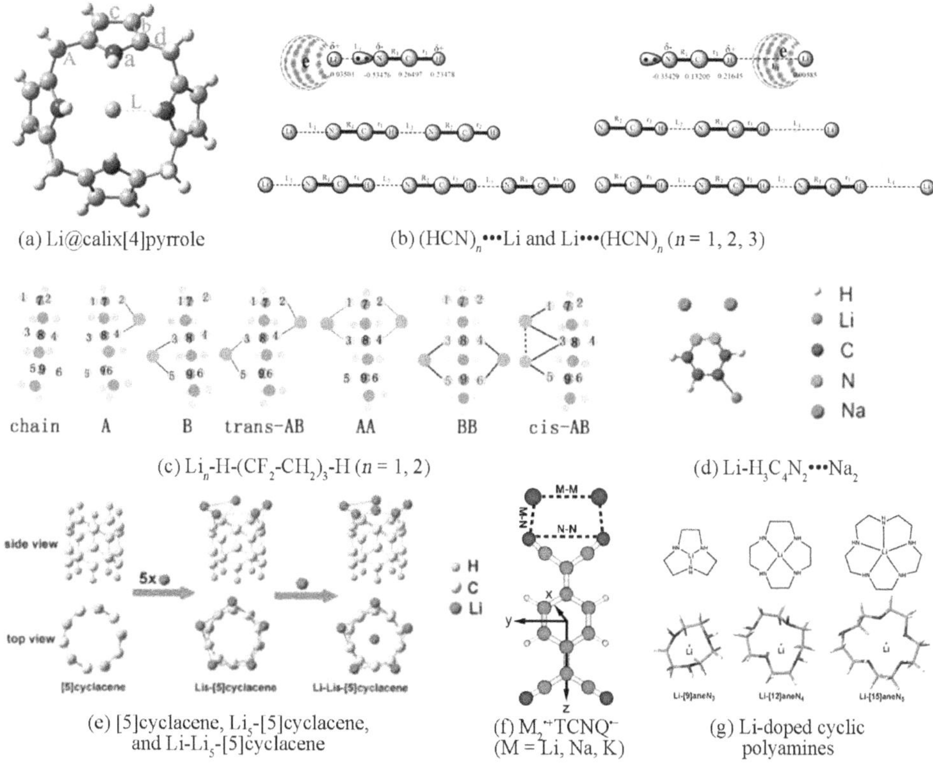

FIGURE 10.2 Optimized structures of typical electrides: (a) Li@calix[4]pyrrole, (b) (HCN)$_n$···Li and Li···(HCN)$_n$ (n = 1, 2, 3), (c) Lin-H-(CF$_2$-CH$_2$)$_3$-H (n = 1, 2), (d) Li-H$_3$C$_4$N$_2$···Na$_2$, (e) [5]cyclacene, Li$_5$-[5]cyclacene, Li-Li$_5$-[5]cyclacene, (f) $M_2^{\cdot+}$TCNQ$^{\cdot-}$ (M = Li, Na, K), and (g) Li-doped cyclic polyamines.

Sources: Plates (a), (b), (c), (d), (e), (f), and (g) are reproduced from Ref. [52] with permission (Copyright 2005 American Chemical Society), Ref. [43] with permission (Copyright 2005 American Chemical Society), Ref. [61] with permission (Copyright 2007 American Chemical Society), Ref. [45] with permission (Copyright 2008 American Chemical Society), Ref. [62] with permission (Copyright 2009 American Chemical Society), Ref. [63] with permission (Copyright 2009 Royal Society of Chemistry), and Ref. [44] with permission (Copyright 2009 American Chemical Society), respectively.

cup-like complexant to become an excess electron under the action of four N atoms. So Li@calix[4]pyrrole exhibits the electride characteristics and can be expressed as Li^+(calix[4]pyrrole)e^-. Due to the formation of excess electron, this electride shows a large β_0 of 7326 au, while the β_0 value of the related calix[4]pyrrole system is only 390 au.

In the same year, Chen et al.[43] also designed two classes of $(HCN)_n\cdots Li$ and $Li\cdots(HCN)_n$ ($n = 1, 2, 3$) clusters with electride characteristics by attaching a Li atom to the $(HCN)_n$ ($n = 1, 2, 3$) clusters (see Figure 10.2b). Under the interaction between the Li atom and the $(HCN)_n$ part, the 2*s* valence electron of Li atom becomes a loosely bound excess electron. As a result, these new clusters show greatly larger β_0 values than the corresponding $(HCN)_n$ ($n = 1, 2, 3$) clusters. Obviously, the excess electron from the Li atom plays a crucial role in the large first hyperpolarizabilities of these clusters. These seminal studies[43,52] paved a new way for designing electrides with significant NLO responses by doping one alkali metal atom into various complexants. Also, it must be mentioned that, in the same year, the first thermally stable organic electride at room temperature was successfully synthesized by inserting a Na atom into a per-aza analogue of cryptand[2.2.2] by Dye's group.[60] This achievement further stimulates interest in designing a variety of electrides on the basis of organic or inorganic complexants.

Intrigued by these findings, Xu et al.[61] studied the alkali-metal-doped effect on the nonlinear optical (NLO) property by doping Li atoms into a fluorocarbon chain $H\text{-}(CF_2\text{-}CH_2)_3\text{-}H$. They obtained six stable structures of $Li_n\text{-}H\text{-}(CF_2\text{-}CH_2)_3\text{-}H$ ($n = 1, 2$) at the MP2/6–31+G (d) level (see Figure 10.2c). It was found that the β_0 of 112 au for $H\text{-}(CF_2\text{-}CH_2)_3\text{-}H$ is sharply increased to 76,978 au for the trans-AB isomer of $Li_2\text{-}H\text{-}(CF_2\text{-}CH_2)_3\text{-}H$, which indicates a dramatic Li-atom-doped effect on the first hyperpolarizability of organic molecules. Moreover, three other interesting relationships between the geometric structure and β_0 value have been observed, which may be helpful for the design and synthesis of electrides with large NLO response by using the alkali-metal-doped effect.

In 2008, a new lithium salt electride with an excess electron pair was designed by doping two sodium atoms into the lithium salt of pyridazine (see Figure 10.2d).[45] First, one H atom of a pyridazine molecule $H_4C_4N_2$ is substituted by Li to generate the lithium salt, i.e., $Li\text{-}H_3C_4N_2$. The lithium salt effect increased the β_0 value of $H_4C_4N_2$ from 5 to 859 au for $Li\text{-}H_3C_4N_2$ by about 170 times. For the electride effect, an electride $H_3C_4N_2\cdots Na_2$ is formed by doping two Na atoms into pyridazine, whose β_0 value is increased by about 3000 times from 5 to 1.5×10^4 au. Finally, a new lithium salt electride $Li\text{-}H_3C_4N_2\cdots Na_2$ has been designed by doping two Na atoms into the lithium salt $Li\text{-}H_3C_4N_2$. They found that this proposed lithium salt electride shows an extremely large β_0 value of 1.412×10^6 au, which is 2.8×10^5 times larger than that of $H_4C_4N_2$, 1644 times larger than that of $Li\text{-}H_3C_4N_2$, and still 93 times larger than that of the electride $H_4C_4N_2\cdots Na_2$. According to the two-level model,[37,38] this considerable value of $Li\text{-}H_3C_4N_2\cdots Na_2$ is attributed to its small transition energy and large difference in the dipole moments between the ground state and the excited state. In addition, they also calculated the frequency-dependent hyperpolarizability, which shows almost the same trends as $H_4C_4N_2 \ll Li\text{-}H_3C_4N_2 \ll H_4C_4N_2\cdots Na_2 \ll Li\text{-}H_3C_4N_2\cdots Na_2$. This study proposes a new idea to design potential candidates for high-performance NLO molecules by simultaneously combining the lithium salt effect and electride effect.

Similar strategy has been used to design a new electride $Li\text{-}Li_5$-[5]cyclacene by Xu et al.[62] As shown in Figure 10.2e, five edge H atoms of the tubiform[5]cyclacene were substituted by five Li atoms to form the noncentrosymmetric Li_5-[5]cyclacene, a new multilithium salt. Then, one Li atom was further doped into the center of Li_5 ring in Li_5-[5]cyclacene, yielding a new tubiform electride $Li\text{-}Li_5$-[5]cyclacene with a very large β_0 value of 35,384 au, which is about 5 times larger than that of 7938 au for Li_5-[5]cyclacene. This work demonstrates that both of the lithiation and Li-doped effect can greatly increase the first hyperpolarizability of other molecules. In view of the significant alkali-metal doped effect on β_0, they also designed a series of three-propeller-blade-shaped electrides[64] by doping alkali metal atoms (Li, Na, and K) into effective multinitrogen complexant tris[(2-imidazolyl)methyl]amine (TIMA). These A_m-TIMA (A_m = Li, Na, and K) complexes have more diffuse excess electrons than the reported electrides due to the strong interaction between the complexant TIMA and

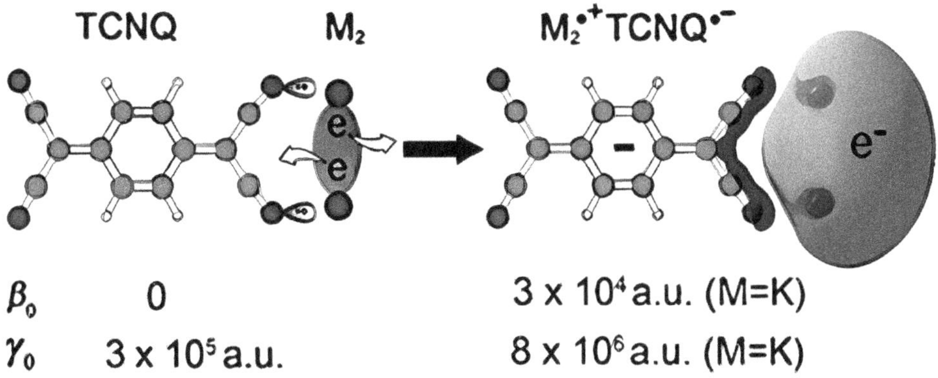

FIGURE 10.3 Excess electron generated in $M2^{\bullet+}TCNQ^{\bullet-}$ (M = Li, Na, K).

Source: Reproduced from Ref. [63] with permission (Copyright 2009 Royal Society of Chemistry).

alkali metal atoms, resulting in the greatly larger β_0 values (3464–29,705 times) than that (338 au) of TIMA. In particular, the obtained K-TIMA has a β_0 value of 1.00×10^7 au. Considering the possibility of assembling molecules into material, the β_0 values of optimized Li-TIMA-dimer and Li-TIMA-tetramer have been also calculated, and the results revealed that the order of β_0 value is Li-TIMA-monomer < Li-TIMA-dimer < Li-TIMA-tetramer.

Furthermore, two alkali metal atoms have been co-doped into a tetracyanoquinodimethane (TCNQ) molecule to design a series of radical ion pair salts $M_2^{\bullet+}TCNQ^{\bullet-}$ (M = Li, Na, K) (see Figure 10.2f), as a particular kind of charge transfer complexes with excess electrons.[63] In these complexes, as shown in Figure 10.3, a TCNQ molecule accepts one electron from M_2 and becomes an anion $TCNQ^{\bullet-}$. The anion $TCNQ^{\bullet-}$ powerfully polarizes the cation M_2^+ to form an excess electron. Due to the existence of excess electrons, these complexes exhibit large nonlinear optical (NLO) responses (β_0 = 19,203 ~ 29,065 au). Besides, the authors also studied a complexant effect on β_0 by doping Li atoms into higher flexible cyclic polyamines with many amine unit petals (ethyleneimine) to form a class of n-petal-shaped Li-doped cyclic polyamines (n = 3 ~ 5) (see Figure 10.2g).[44] Because the chemical doping with Li and the deformation of complexant produce more diffuse excess electron, the n-petal-shaped Li-doped cyclic polyamines (n = 3 ~ 5) exhibit extremely large β_0 values. In particular, the β_0 values increases with the increasing petal number (n), namely 52,282 (n = 3) < 65,505 (n = 4) < 127,617 au (n = 5). This study shows a dependence of β_0 on the petal number (n) because the flexibility of the complexants increases with the petal number.

The bound state of excess electron significantly depends on the features of the complexant of electrides, which serves as an important factor in affecting the β_0 values. As shown in Table 10.1, all of these reported electrides contain pushed excess electrons. In 2009, Su and coworkers[65] first designed a new inorganic electride $Li@B_{10}H_{14}$ by doping Li atom into a pulled electron complexant, namely the basket-like decaborane ($B_{10}H_{14}$). The most intriguing feature of this electride is its excess electron bound state, which is diffused and pulled into the cavity of the $B_{10}H_{14}$ basket with four electron-deficient terminal hydrogen atoms and two electron-deficient terminal boron atoms, as shown in Figure 10.4a. Strikingly, the β_0 of $Li@B_{10}H_{14}$ is computed to be 23,075 au, which is about 340 times larger than that of 68 au for $B_{10}H_{14}$. They further proposed a synergistic effect of conical push, and the inward pull in the fluoro derivative $Li@F_4B_{10}H_{10}$ induces a larger β_0(181624 au) and higher VIE(6.45 eV) than $Li@B_{10}H_{14}$, which makes the fluorinated decaborane a better choice for designing electrides.[66] It is noted that these electride molecules with pulled excess electrons have high stability in view of their high VIE values, which may open a new thriving area, i.e., alkali metal-boranes for future NLO application.

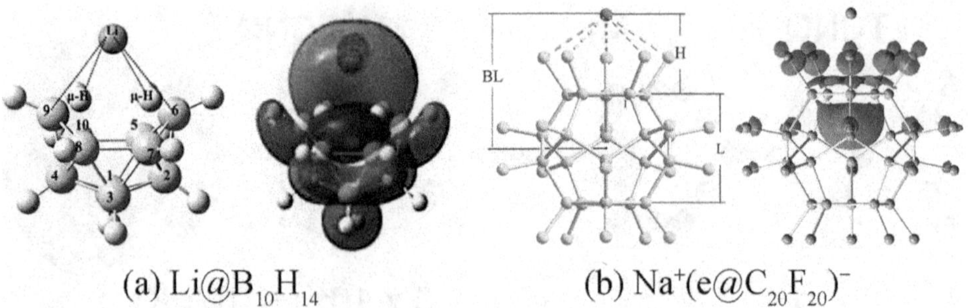

FIGURE 10.4 Structures and HOMOs of (a) $Li@B_{10}H_{14}$ and (b) $Na^+(e@C_{20}F_{20})^-$.

Sources: Plates (a) and (b) are reproduced from Ref.[65] with permission (Copyright 2009 American Chemical Society), reproduced from Ref. [68] with permission (Copyright 2012 Royal Society of Chemistry).

TABLE 10.1
Static First Hyperpolarizabilities (β_0, au), Bound State of Excess Electron, Transition Energies (ΔE, eV) of Crucial Excited States, Vertical Ionization Energies (VIEs, eV) of Selected Alkali-Metal-Based Electrides

Species	β_0	Bound State	ΔE	VIE	Ref.
HCN···Li	1.57×10^4	Pushed	1.33	4.15	43
Li-[15]aneN_5	1.28×10^5	Pushed	0.56	2.32	44
Li-$H_3C_4N_2$···Na_2	1.41×10^6	Pushed	1.32	—	45
Li@calix[4]pyrrole	7.33×10^3	Pushed	1.98	4.12	52
Li_2-H-(CF_2-CH_2)$_3$-H (trans-AB isomer)	7.70×10^4	Pushed	1.30	—	61
Li-Li_5-[5]cyclacene	3.54×10^4	Pushed	1.10	—	62
Li-TIMA	1.17×10^6	Pushed	0.52	—	63
$Li_2^{\bullet+}TCNQ^{\bullet-}$	1.92×10^4	Pushed	2.80	6.19	64
$Li@B_{10}H_{14}$	2.31×10^4	Pulled	1.55	6.12	65
$Li@F_4B_{10}H_{10}$	1.82×10^5	Pulled	1.20	6.45	66
$e^-@C_{20}F_{19}$-$(CH_2)_4$-NH_2···Na^+	9.50×10^6	Pulled	2.38	4.76	69
$K^+[3C_8(O)]^-$	7.1×10^5	Pulled	0.68	—	72
$^2L_3$ with 3Na@9C3	2.2×10^5	Pushed	0.49	3.12	73
Li@**N**-BNNT	3.40×10^4	Pulled	0.99	—	74
K@pyrrole	1.14×10^4	Pushed	1.32	3.64	80
$K@C_6O_6Li_6$	2.9×10^5	Pushed	1.02	—	81
Li@r_6-$Al_{12}N_{12}$	8.89×10^5	Pushed	1.21	—	82
Li-$F_6C_6H_6$	7.00×10^5	Pushed	1.04	3.42	93
K@N-AlNNT	1.37×10^6	Pulled	0.66	—	95
Na@(TriPip222)	1.13×10^6	Pushed	0.23	—	98

More recently, the structure and electronic properties of $Li@B_{10}H_{14}$ complex in chloroform and water in normal thermodynamic conditions have been investigated using sequential QM/MM calculations by means of the average solvent electrostatic configuration (ASEC) and the free energy gradient (FEG) methods.[67] Results demonstrated that this $Li@B_{10}H_{14}$ cluster is stable in chloroform but unstable in water.

The effect of electronic polarization due to solvents on the β_0 of $Li@B_{10}H_{14}$ was especially explored. In water, the contribution of the excess electron for NLO responses is significantly affected by the electrostatic polarization effects, leading to the disappearance of NLO responses. The electrostatic polarization effect is also obvious in chloroform. Therefore, the influence of solvent environment must be considered in the design of new NLO materials for use in solutions.

As mentioned previously, confining excess electrons into a specific space to reduce its dispersivity seems to be a useful strategy to achieve stable NLO molecules with excess electrons. Thereby, Li and coworkers[68,69] have tried to design stable molecular NLO materials by introducing nanostructure-confined excess electrons. Among their strategies, the perfluorinated fullerene $C_{20}F_{20}$ was used as the electron container to protect the excess electron from alkali metal atoms or superalkali clusters because they have found that perfluorinated fullerene cages with sufficient attractive potentials at their inner space can effectively confine the excess electrons.[70,71] By doping alkali metal atoms (M = Na, K) and superalkalis (M_3O, M = Na, K) onto $C_{20}F_{20}$, a series of single-caged electrides $M^+(e@C_{20}F_{20})^-$ and $(M_3O)^+(e@C_{20}F_{20})^-$ resulted due to the long-range charge transfer from the (super)alkali to the inner cavity of the cage (see $Na^+[e@C_{20}F_{20}]^-$ for an example in Figure 10.4b).[68] These newly designed complexes have large VIE values of 6.53 ~ 7.38 eV, reflecting their high electronic stability. Moreover, the β_0 value of $(K_3O)^+(e@C_{20}F_{20})^-$ is calculated to be 1.30×10^5 au, which is greatly larger than that (6.00×10^2 au) of $K^+(e@C_{20}F_{20})^-$, indicating that the interaction between superalkalis and perfluorinated fullerenes is an effective strategy to construct stable NLO molecules with confined excess electrons.

To further increase the β_0 value of cage-confined excess electron systems, the specific σ bridge $(CH_2)_4$ was used to construct a new kind of electrides, $e^-@C_{20}F_{19}\text{-}(CH_2)_4\text{-}NH_2\cdots M^+/M_3O^+$ (M = Li, Na, and K) with an excess electron anion inside the $C_{20}F_{19}$ cage.[69] Due to the long-range charge transfer in these designed electrides, they exhibit large NLO responses. For instance, the β_0 of $e^-@C_{20}F_{19}\text{-}(CH_2)_4\text{-}NH_2\cdots Na^+$ is as large as 9.50×10^6 au. In addition, multicage electride molecules with an excess electron were also designed by Li and coworkers[72] to reveal the effects of age unit size, number of cage units, and bridge unit on their β_0 values. Based on their calculations, they concluded that the small C_8 cage with an O-bridge exhibits a larger β_0 and that the β_0 value can be enhanced by increasing the number of involved cage units. As a result, the designed multicage electride with three small C_8 cage units connected by the O-bridge, namely $K^+[3C_8(O)]^-$, shows a large β_0 of 7.1×10^5 au. Furthermore, by multidoping alkali atoms on a multiunit complexant, the unusual manipulative effects of spin multiplicity and excess electron number on the β_0 of electride molecules with multi-excess electrons were investigated.[73] For the studied models, the β_0 values increased rapidly with increasing excess electron number, and the β_0 value of the low-spin structure is significantly larger than that of the high-spin structure due to one excess electron spin reversal.

Also, the excess electron can be also pulled to a boron nitride nanotube (BNNT). In 2012, Zhong et al.[74] found that an excess electron could be effectively bound by the boron atoms of BNNT, which is inverted pyramidally and distributed from the boron-rich edge to the nitrogen-rich edge. By using Li atom as excess electron source, the Li@**B**-BNNT and Li@**N**-BNNT were designed by doping Li to the two edges of BNNT, respectively. Under the interaction between Li and BNNT, the 2*s* valence electron of Li became a loosely bound excess electron. They found that the distribution of excess electrons in Li@**N**-BNNT is more diffuse and pyramidal from B-rich edge to N-rich edge, whereas the excess electron in Li@**B**-BNNT is inverted pyramidally distributed and less diffuse. Hence, the transition energy of 0.99 eV for Li@**N**-BNNT is significantly smaller than that of 2.65 eV for Li@**B**-BNNT. So the β_0 value (3.40×10^4 au) of Li@**N**-BNNT is dramatically larger (25 times) than that (1.35×10^3 au) of Li@**B**-BNNT. Moreover, they proposed that the pyramidal distribution of excess electrons is the key factor to determine β_0, providing some useful information for developing new BNNT-based electro-optic materials.

Besides the BNNT, the interaction between boron-carbon-nitride (BCN) heteronanotubes and lithium atoms has been also explored by Zhong et al.[75] by means of systematically studying a class of Li@BCN models. Their results reveal that the crucial electron population in the BCNs of Li@B-BCN and Li@N-BCN series is dramatically different. Consequently, the β_0 value of Li@B-BCN series is enhanced as the carbon proportion increases, whereas the β_0 of Li@N-BCN significantly decreases with the increasing carbon proportion. This study suggests that the physical properties of Li@BCN models significantly depend on the different chemical environment of the tube termination. Hou et al.[76] carried out a comparative theoretical study on the NLO properties of Li-adsorbed AlN and BN single-walled nanotubes instead of the tube termination. Their results demonstrate that the formed AlNNT–Li and BNNT–Li systems containing diffuse excess electrons can be regarded as inorganic electrides with considerable NLO responses. In fact, different BN nanomaterials have been applied to design NLO materials by doping alkali metal atoms, such as boron nitride nanocone,[77] boron-nitrogen (BN) edge-doped grapheme,[78] and Al/P-substituted BN nanosheets.[79]

When interacting the alkali metal atom with the π-conjugated aromatic rings, such as pyrrole, benzene, and indole, several new electrides have been achieved as the *s* valence electron of the alkali metal atom can be pushed out to produce the diffuse excess electron under the action of the π electron cloud of these aromatic rings, resulting in the considerable β_0 values.[80] This provides a method to improve the NLO response of materials or molecules containing π-conjugated aromatic ring, such as carbon nanomaterials and biomolecules, by introducing excess electrons. In 2021, Wajid et al.[81] investigated the alkali metal-doped $C_6O_6Li_6$, and found that the generated e^-@M@$C_6O_6Li_6$ (M = Li, Na, and K) showed the electride characteristics and large NLO responses.

By doping the alkali metal atoms onto the fullerene-like $Al_{12}N_{12}$ nanocage, several inorganic electride compounds, namely M@*x*-$Al_{12}N_{12}$ (M = Li, Na, and K; $x = b_{66}$, b_{64}, and r_6), with high stability were designed by Niu et al.,[82] where the alkali atom is attached to the Al-N bond (b_{66}/b_{64} site) or six-membered ring (r_6 site). They found that doping the alkali atom can significantly reduce the wide HOMO-LUMO gap (6.12 eV) of the $Al_{12}N_{12}$ nanocage to 0.49–0.71 eV because a molecular orbital containing the excess electron with high energy is introduced as the new HOMO orbital of M@*x*-$Al_{12}N_{12}$. The diffuse excess electron endows these M@*x*-$Al_{12}N_{12}$ nanostructures considerable β_0 values of $1.09 \times 10^4 \sim 8.89 \times 10^5$ au. These interesting findings are of great significance in promoting the applications of AlN-based nanosystems in the new type of electronic nanodevices and high-performance NLO materials. After this work, some other fullerene-like nanocages were also doped by one or two alkali metal atoms to design inorganic electrides, including $B_{12}N_{12}$,[83] B_{40},[84] $Si_{12}C_{12}$,[85] $Ca_{12}O_{12}$,[86] and $B_{12}P_{12}$.[87] Bi-alkali doping on $Al_{12}N_{12}$ nanocage has been also studied to explore NLO materials with high stability.[88] Besides the doping effect, the substituent effect of alkali metal atoms has been also employed to enhance the NLO responses of $B_{12}N_{12}$[89] and $Al_{12}N_{12}$.[90]

Other efforts have been devoted to exploring new complexants for the design of electrides. For example, the synthesized all-cis 1,2,3,4,5,6-hexafluorocyclohexane ($F_6C_6H_6$),[91] a facially polarized cyclohexane greatly attracted our attention and, in 2017, was used to design an intriguing kind of alkalides with considerable NLO responses and high stability.[92] One year later, Hou et al.[93] used this Janus molecule to design a series of alkali-metal-atom-doped M-$F_6C_6H_6$ (M = Li, Na, and K) compounds with electride characteristics. Among these electrides, the largest β_0 of the Li-$F_6C_6H_6$ is 7.00×10^5 au, which is about 3030 times larger than pure $F_6C_6H_6$. Zheng et al.[94] extended such a $F_6C_6H_6$ molecule to a three-ring $C_{13}H_{10}F_{12}$ molecule with a larger Janus face, where multi-alkali metal atoms can be doped. By tuning the number of alkali metal atoms doped on both faces of $C_{13}H_{10}F_{12}$, three types of excess electron compound have been generated: electrides, alkalides, and alkaline-earthide-like molecules. They proposed that the number of Li atoms doped on the hydrocarbon face determined the type of the excess electron compound, whereas the number of alkali metal atoms on the fluorocarbon face could adjust the distribution of excess electrons. Aluminum nitride nanotubes were doped by alkali metal atoms to obtain a series of M@N-AlNNT (M = Li, Na, K) with large β_0 up to 1.37×10^6 au.[95] Teng et al.[96] theoretically designed a novel organic molecular electride model, $Li^+(C_{20}H_{15}Li_5)e^-$, by modifying the lithiation and Li-doping based on dodecahedrane ($C_{20}H_{20}$).

Of course, the complexants used for synthesizing electrides or alkalides are also a good choice for the design of electrides by doping alkali metal atoms. Based on the synthesized H@3^6Adz (3^6Adz = 3^6 adamanzane), Zhou et al.[97] designed a novel kind of three-dimensional (3D) octupolar molecule Li@3^6Adz with electride characteristics, in which the *s* electron of the encapsulated Li atom is extruded from the cage of 3^6Adz to form a diffused excess electron. The β_0 value of room-temperature stable organic electride Na@(TriPip222) synthesized by the Dye group[60] was evaluated to be 1.13×10^6 au,[98] revealing its potential as a high-performance NLO material. Moreover, the substituent effects of different functional groups on the structure, electride character, and NLO response of this electride were systemically studied by Sun et al.[98] It was revealed that the β_0 of this electride could be further enhanced to 8.30×10^6 au by introducing a fluoro substituent, whereas its NLO response completely disappeared when one nitryl group was introduced to deprive its electride identity.

In addition, some attention has been focused on improving the performance of previously reported electrides. For instance, Chen et al.[99] found that the β_0 (7326 au) of Li@calix[4]pyrrole[52] could be enhanced to those of 7500 and 17,304 au for M@calix[4]pyrrole (M = Na and K), respectively. Thus introducing heavier alkali atom into organic electride compounds is an efficient approach to achieve larger NLO response. Besides, the substituent effects on the electride characteristics and NLO properties of Li@calix[4]pyrrole were studied by Zhou and his coworkers.[100] They found that electron-donating and electron-withdrawing groups can effectively increase and decrease β_0, respectively, without affecting the electride characteristics. Ma et al.[101] combined the electron deficient molecule BX_3 (as a Lewis acid) with Li@calix[4]pyrrole (as a Lewis base) to design a new type of Lewis acid–base complex BX_3···Li@Calix[4]pyrrole (X = H and F) with an interesting push-excess-electron-pull (P-e-P) framework, which enhances the stability and nonlinear optical (NLO) response. To obtain better thermostability and NLO response molecules, inorganic complexants $B_{20}H_{26}$ containing two cavities were proposed by Ma et al.,[102] which is derived from the $B_{10}H_{14}$ used in Li@$B_{10}H_{14}$.[65] Doped by one or two Li atoms, a series of electrides, namely Li@$B_{20}H_{26}$ and Li_2@$B_{20}H_{26}$, are generated. In particular, Li_2@$B_{20}H_{26}$ exhibits the largest β_0 value of 108,846 au, which is 850 times larger than that of the corresponding $B_{20}H_{26}$.

Wang et al.[103,104] designed several molecular NLO switches by manipulating the excess electrons in the reported electrides by external electric fields. For Li@AR (AR = benzene and naphthalene), the hopping of the Li atom under the external electric field between electride and lithium salt configurations brings entirely different charge transfer directions to the direction of dipole moment orientation, and thus considerably different β_0 is exhibited.[103] From electride to lithium salt configuration, the β_0 value is increased approximately 34-fold from 1425 (electride) to 49,746 au (lithium salt) for Li@benzene. This large difference in β_0 values between them reinforce the potential of these simple systems for application in the nonlinear optics switch. For the electride K-$F_6C_6H_6$,[93] the oriented external electric field can drive the excess electron transfer from the region outside the K atom to that outside $F_6C_6H_6$.[104] Subsequently, the electride-like molecule K-$F_6C_6H_6$ switches to the molecular electride K-$F_6C_6H_6$···e^-, and the significantly increased β_0 value constitutes an order of magnitude improvement. The different β_0 values between them proves K-$F_6C_6H_6$ to be a good candidate for multiple-response nonlinear optics switch.

10.4.1.2 Alkalides

Another typical kind of alkali-metal-based excess electron compound is alkalide, in which one more alkali metal atom is introduced to accept the excess electron pushed from the alkali metal atom by the complexants. Such unique complexes have been successfully synthesized at room temperature by Dye's group in 1999.[54] However, the potential of alkalides as novel NLO materials was first proposed by Chen et al. in 2006.[53] They designed a new type of alkalide compound by using the cup-like complexant, namely Li^+(calix[4]pyrrole)M^- (M = Li, Na, and K), which may also be stable at room temperature. In this kind of compound, as shown in Figure 10.1c, the lone pairs of four N atoms push out the 2*s* electron of the Li atom inside calix[4]pyrrole to form the excess electron around the M atom, forming an M^- anion. Due to the existence of their diffused excess electrons, these alkalides show considerably large NLO responses. Moreover, it is found that the alkali metal anion plays a crucial role in the large β_0; that is, the β_0 increases as the atomic number of the

TABLE 10.2
Static First Hyperpolarizabilities (β_0, au), Transition Energies (ΔE, eV) of Crucial Excited States, Vertical Ionization Energies (VIEs, eV) of Selected Alkali-Metal-Based Alkalides

Species	β_0	ΔE	VIE	Ref.
Li^+(calix[4]pyrrole)K^-	3.59×10^4	1.757	—	53
K^+·**1**·K^-	1.45×10^6	1.747	2.96	92
$Li^+(NH_3)_4Na^-$	7.79×10^4	1.534	—	107
$Li^+(NH_3)_4K^-$	1.86×10^5	1.259	—	108
AdzH$^+$Na$^-$	6.21×10^5	2.143	—	111
(K^+@2^6Adz)K^-	3.18×10^5	1.657	2.51	112
Li^+aza222K^-	1.00×10^6	1.228	2.60	113
$Li^+(N_3H_3)K^{\cdot-}$	1.48×10^5	1.896	3.70	114
(Li-HMHC)$^+$-K^-	1.34×10^5	1.642	—	116

alkali anion increases. Therefore, introducing a larger alkali anion into such alkalides is an effective method to further enhance their NLO responses. They also explored the nonlinear optical properties of the Li^+(calix[4]pyrrole)M^- dimer, trimer, and its polymer and found that the number of excess electrons plays an important role in increasing the β_0.[105]

The effects of substituents on the geometric structure, alkalide character, and NLO responses of Li^+(calix[4]pyrrole)Li^- have been investigated by Sun et al.[106] Their results reveal that electron-donating substituents strengthen the alkalide character of Li^+(calix[4]pyrrole)Li^- and thus are beneficial for a larger β_0 value. However, electron-withdrawing substituents show the opposite effect. The dependence of NLO properties on the number of substituents and their relative position was detected by systemically studying the multisubstituted Li^+(calix[4]pyrrole)Li^- compounds. The β_0 values of multisubstituted derivatives increase as more pyrrole β-H atoms are substituted by amino and methyl groups. Also, distribution of the substituents so that they are as far away from one another as possible resulted in an increase in the β_0 value. These new knowledge may provide an effective strategy to enhance the NLO responses of alkalides using pyrrole derivatives as complexants.

To explore the coordination number (around the cation) dependence of the nonlinear optical (NLO) properties in alkalides, Jing et al. investigated the NLO properties of model alkalides, Li$(NH_3)_n$Na (n = 1–4).[107] A prominent dependence of β_0 value on the coordination number, i.e., 13,669 (n = 1) < 26,840 (n = 2) < 39,764 (n = 3) < 77,921 au (n = 4), is found for these Li$(NH_3)_n$Na (n = 1–4). In particular, with the same coordination number (four N atoms) of Li^+ cations, the β_0 value (77,921 au) of the "small" inorganic Li$(NH_3)_n$Na is more than five times larger than that of the "big" organic alkalide Li@Calix[4]pyrrole-Na (14,772 au), which may be because the flexibility of $(NH_3)_4$ is significantly greater than that of the cup-like Calix[4]pyrrole. This study indicates that two important factors, i.e., the coordination number around the cation and the flexibility of the complexant, should be considered to enhance the NLO responses of the alkalide. They also studied the effect of complexant shape effect on the β_0 value of the alkalides $Li^+(NH_3)_4M^-$ (M = Li, Na, K).[108] The β_0 values of $Li^+(NH_3)_4M^-$ with the "smaller," inorganic, T_d-symmetric $(NH_3)_4$ complexant are more than four times larger than those of Li^+(calix[4]pyrrole)M^- with the "larger," organic C_{4v}-symmetric calix[4]pyrrole complexant, again verifying the strong complexant shape effects. This demonstrates that using highly symmetric complexant is an important factor that should be taken into account when one tries to enhance the β_0 of alkalides by chemical doping. Besides the NH_3 molecule, the HF and HCN molecules have been also employed to design field-induced inorganic NLO molecular switches M$(HF)_3$M (M = Na or Li)[109] and Na$(HCN)_3$Na,[110] respectively.

Focusing on the synthesized alkalides, the special "inverse sodium hydride" (AdzH$^+$Na$^-$) taking of 3^6 adamanzane (Adz) as the complexant attracts the attention of theoretical chemists. Chen et al.[1111] reported that this hydrogen-based alkalide exhibits a considerably large NLO response β_0 of 5.77×10^4 au. By substituting the centered hydrogen atom with a Li atom, the β_0 value is increased to 6.21×10^5 au for the Li-based AdzLi$^+$Na$^-$. Obviously, such adamanzane cages are good complexants for the design of alkalides. Hence new organic alkalides (M$^+$@n^6Adz)M′$^-$ (M, M′ = Li, Na, K; n = 2, 3), with the alkali-metal cation M$^+$ lying near the center of the Adz cage and the alkali-metal anion M′$^-$ located outside, were designed by Wang et al.[112] These alkalides show very large β_0 values up to 3.2×10^5 au. Additionally, the effects of complexant cage size and alkali-metal anion on β_0 have been also revealed in this study. Also, the larger azacryptand[2.2.2] (aza222) complexant used in synthesizing the room-temperature stable alkalides has been also adapted to constructing a series of organic alkalides M$^+$aza222M′$^-$ (M, M′ = Li, Na, K) with considerable β_0 up to 1.0×10^6 au.[113]

By doping the model complexant N_3H_3 with one or two lithium atoms, an in-depth understanding on the origin of electride and alkalide and the large nonlinear optical properties of the resulting Li(N_3H_3), (N_3H_3)Li′, and Li(N_3H_3)Li′ complexes was revealed by Li and coworkers.[114] In this work, the dependance of electric properties of alkalide Li(N_3H_3)Li′ on the involved alkali atoms and the complexant layer number were detected by studying the related M(N_3H_3)Li′ and Li(N_3H_3)M′ (M = Na and K), and Li$(N_3H_3)_n$Li′ (n = 2, 3) systems. It was found that the β_0 values of M(N_3H_3)M′ can be enhanced by increasing the atomic number of both the M′$^-$ anion and the M$^+$ cation, which is different from the previously reported cases. Moreover, the β_0 of the Li(N_3H_3)Li′ alkalide is increased by increasing the complexant layers of Li$(N_3H_3)_n$Li′ (n = 2, 3).

To obtain novel alkalides with higher performance, Sun et al.[92] focused on the facially polarized all-*cis*-1,2,3,4,5,6-hexafluorocyclohexane ($F_6C_6H_6$, also written as **1**), which is an excellent complexant to simultaneously stabilize the alkali-metal cation and anion. As shown in Figure 10.5, this molecule has one hydrocarbon face and one fluorocarbon face, and thus it shows an unusually large facial polarization and the largest molecular dipole moment of 6.2 Debye. In view of the separate positive and negative facial polarities, this Janus molecule was adopted as a new complexant to design a new kind of alkalide by the simultaneous binding of alkali metal cations and anions with its negative and positive faces, respectively.[92] Under the action of the lone pairs of the axial F atoms, the *s* electron of the lower alkali-metal atom is ejected to the upper alkali-metal atom (see Figure 10.5), yielding a novel series of alkalides, namely M$^+$·**1**·M′$^-$ (M, M′ = Li, Na, K). The results

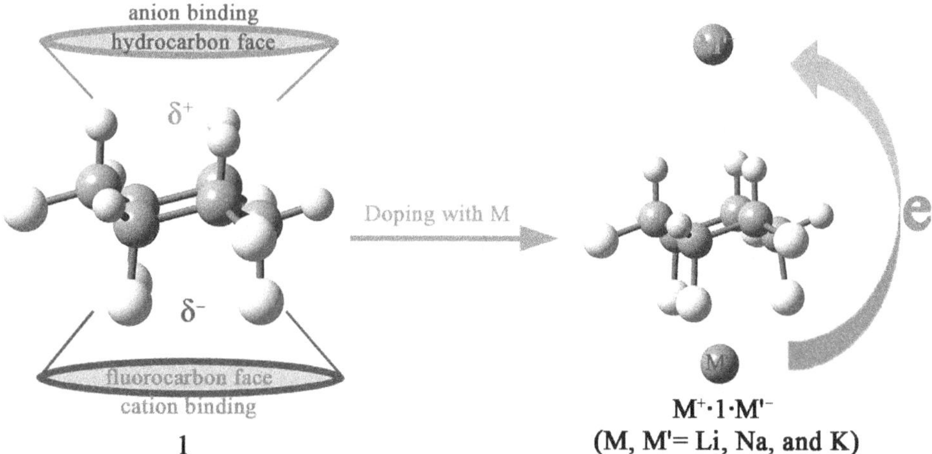

FIGURE 10.5 Structure and two-faced Janus polarity of 1 as well as the resulting geometric structure of M+·1·M′− (M, M′ = Li, Na, K).

Source: Reproduced from Ref. [92] with permission (Copyright 2017 American Chemical Society).

showed that all of the designed alkalides not only possess strong electrostatic interaction between **1** and alkali-metal ions but also exhibit remarkably β_0 values up to 1.45 × 10^6 au. Furthermore, the possibility to use **1** to design superalkali-based alkalides and superalkalides was also explored in this study. As the first attempt to design molecular NLO materials based on such an interesting Janus molecule, this work has attracted much attention and stimulated a lot of theoretical studies to design other excess compounds.

In addition, the crown-ether and peralkylated polyamine are two typical representatives of ligands for the development of more stable alkalides. For the latter, a recent work has demonstrated that the hexa-aza-crown 1,4,7,10,13,16-hexamethyl-1,4,7,10,13,16-hexaazacyclooctadecane (hexamethylhexacyclen, HMHC) can significantly outperform the reported crown-ether, i.e., 15-crown-5 in its resistance to reductive decomposition.[115] Riedel et al.[115] proved that the nature of anionic alkali metals in the alkalide solution is not "gas-like" and unperturbed, which it is traditionally thought to be, but superalkali-alkalide ion pairs in the $(M\text{-HMHC})^{\delta+}\text{-}M'^{\delta-}$ (M′ = Na or K) solutions. Thereafter, a DFT study was performed by Wang et al.[116] to investigate the NLO responses of $(M\text{-HMHC})^{\delta+}\text{-}M'^{\delta-}$ (M, M′ = Li, Na, and K) to promote their application in the field of molecular electronics. They found that the NLO responses of $(M\text{-HMHC})^{\delta+}\text{-}M'^{\delta-}$ display obvious M′ atomic number dependence.

10.4.2 Alkaline-Earth-Based Excess Electron Compounds

Like alkali metals, the alkaline-earth atoms also possess *s* electrons in their outermost shells. Hence, these *s* block elements can also serve as the excess electron source of electrides, alkalides, and some other compounds.[117–128] For example, Dye and coworkers have synthesized the first alkalide with an alkaline-earth cation, namely $Ba^{2+}(H_5Aza222)^-Na^-\cdot 2MeNH_2$ in 2003.[59] More recently, the alkaline-earth metal atoms have been also used as excess electron acceptors[129–136] to stabilize the excess electrons from other metal atoms. In this section, recent progress on these compounds containing alkaline-earth metal atoms is reviewed.

10.4.2.1 Alkaline-Earth Metals as the Source of Excess Electrons

Inspired by the first synthesis of alkaline-earth-based alkalide $Ba^{2+}(H_5Aza222)^-Na^-\cdot 2MeNH_2$,[59] Fan et al.[117] theoretically studied the structural and NLO properties of $M^{2+}(H_5Aza222)^-M'^-$ (M = Be, Mg, Ca; M′ = Li, Na, K) alkalides using alkaline-earth atoms as the source of excess electrons. Such alkaline-earth-based alkalides show extraordinarily large β_0 values up to 1.98 × 10^5 au. The comparison results demonstrate that the Aza222 cage is preferable to the previously investigated calix[4]pyrrole and n^6adamanzane (n = 2, 3) complexants in enhancing the NLO response of alkalides. Additionally, the relationships between the β_0 values of these $M^{2+}(H_5Aza222)^-M'^-$ compounds and the atomic number of M'^- and M^{2+} cation, and the M–M′ distance have been also revealed by them.[117]

To gain more alkaline-earth-based excess electron compounds, Wang et al. proposed several alkaline-earth-doped compounds, including Be-IAD, Be-PAD, and Be-IPD, where the diffuse excess electrons are near the Be atom. These alkaline-earth-doped compounds show very large β_0 values up to 2.76 × 10^4 au, indicating that alkaline-earth atom doping is also a useful approach to achieve NLO molecules.[118] They also replaced the two Na atoms in $Li\text{-}H_3C_4N_2\cdots Na_2$[45] with Ca atoms to obtain a kind of alkaline-earth-based alkaline salt electride $Li\text{-}H_3C_4N_2\cdots Ca$ with a large β_0 value of 2.69 × 10^5 au.[119] One year later, they studied the effect of an alkaline-earth metal atom on the NLO responses of alkaline-earth-based alkalides $Be(NH_3)_nM$ (M = Be and Ca)[120] in comparison with traditional alkalides $Li(NH_3)_nNa$ (n = 1–3).[107] Similar to $Li(NH_3)_nNa$, the newly proposed Be $(NH_3)_nM$ not only exhibited unusual alkaline-earth-based alkalide's features but also had considerable β_0 values up to 1.26 × 10^5 au.[120] Except for $Be(NH_3)Be$, the other $Be(NH_3)_nM$ always shows larger β_0 than the corresponding $Li(NH_3)_nNa$, which again confirms that, besides the alkali metal doping, the alkaline-earth doping is also an effective method to enhance the NLO responses of the resulting complexes with excess electrons.

On the basis of experimentally synthesized alkaline-earth metal ammines $Mg(NH_3)_6Cl_2$, a series of excess electron compounds, namely $M(NH_3)_6NaCl$ and $M(NH_3)_6Na_2$ (M = Mg and Ca) were theoretically constructed and studied by Sun et al.[121] The electride or alkalide characteristics of these compounds are verified by their electronic structures, HOMOs, and small VIE values, which resulted in large β_0 values up to 1.23×10^5 au. In particular, the proposed inorganic alkalides $M(NH_3)_6Na_2$ were proved to show novel electronic structures that contain double alkali metal anions. Based on these alkalides, the external electric field (EEF) inducing second-order NLO switch molecules has been reported by Gu and coworkers.[122] In the presence of an EEF, the centrosymmetric structure of $M(NH_3)_6Na_2$ is destroyed and then results in a long-range charge transfer from one Na atom to the other Na atom through the inorganic $M(NH_3)_6$ cluster. Finally, the excess electrons are completely located on one of the Na atoms, leading to the On form of such a molecular switch with large β_0 values of 5.95×10^6 au (EEF = 58×10^{-4} au) for $Mg(NH_3)_6Na_2$ and 1.83×10^7 au for $Ca(NH_3)_6Na_2$ (EEF = 53×10^{-4} au). This study demonstrates that these inorganic alkalides $M(NH_3)_6Na_2$ can also serve as potential candidates for NLO switches.

Based on the facially polarized *all-cis*1,2,3,4,5,6-hexafluorocyclohexan (**1**), a class of excess electron compounds, namely, Ca^+-**1**-M'^- (M'^- = Li, Na, and K) has been achieved and investigated by Sun and coworkers.[123] Unlike Be and Mg, one 4*s* electron of the Ca atom with lower ionization energy can be transferred to the upper alkali metal atoms by the instinctive facial polarization of **1** or with the assistance of oriented external electric fields (OEEFs), resulting in the unique excess electron compounds Ca^+-**1**-M'^- (M'^- = Li, Na, and K) simultaneously containing typical alkalide features and electride-like characteristics. As a result, these novel compounds exhibit both high stability and remarkably large β_0 up to 1.81×10^6 au, indicating that these Ca^+-**1**-M'^- compounds have great potential in serving as high performance NLO molecules. Ayub and coworkers[124] designed a series of AE-**1**-AE (AE = Be, Mg, and Ca) complexes by simultaneously doping two alkaline-earth metal atoms to the two different faces of one **1** molecule. These bi-alkaline-metal-doped compounds show β_0 up to 2.91×10^4 au, which paves a new path for the design and synthesis of novel NLO materials.

Besides, two alkalides M-LCaL-M (M = Li or Na, L = $C_6F_6H_6$), were designed by doping alkali (Li and Na) and alkaline-earth (Ca) atoms to two **1** molecules (here named L molecules by the authors).[125] The two 4*s* electrons of Ca atom are pushed out by the negative fluorocarbon faces of two L molecules to combine with the *s* electron of the two separated alkali metal atoms to form excess electron pairs, respectively, and forming the novel alkalides M^--$LCa^{2+}L$-M^-. Under continuously increasing *EEF*, the centrosymmetric M^--$LCa^{2+}L$-M^- with two excess electron pairs is obviously broken, and a long-range charge transfer from the alkali atom at one end through the middle LCaL to the other end one is observed. Meanwhile, the introduction of *EEF* also brings a large change in β_0 from 0 (*EEF* = 0, Off form) to 59,826 (M = Li, *EEF* = 19×10^{-4} au, On form) or 64,231 au (M = Na, *EEF* = 12×10^{-4} au, On form), indicating these alkalides can serve as *EEF*-induced NLO switches.

The alkaline-earth metal atoms have been also used as an excess electron source to design a series of endohedral all-metal electride cages $2e^-Mg^{2+}(M@E_{12})^{2-}Ca^{2+}$ (E = Ge, Sn, and Pb; M = Ni, Pd, and Pt) by He et al.[126] in 2019. In these electrides, the metal cage $M@E_{12}$ first pulls valence electrons from the Ca atom to form a polyanion $(M@E_{12})^{2-}$, and then the formed polyanion pushes valence electrons of the Mg atom out of its valence shell to yield the isolated excess electrons characterizing this species as all-metal electrides. Even so, the endohedral all-metal electride cage $2e^-Mg^{2+}(Ni@Sn_{12})^{2-}Ca^{2+}$ exhibits the largest β_0 value of 1.69×10^4 au (see Table 10.3). They also studied the substituent effect of alkaline-earth metal on the geometric structure and electronic properties of an aluminum nitride nanocage $Al_{12}N_{12}$ by designing two series of compounds: $M@Al_{12}N_{11}$ and $M@Al_{11}N_{12}$ (M = Be, Mg, and Ca).[127] These cages exhibit high stability though they contain diffuse excess electrons. Also, these alkaline-earth-metal-substituted compounds exhibit larger β_0 value than pure $Al_{12}N_{12}$ nanocage. More recently, alkaline-earth metal atoms have been also doped into benzocryptand to serve as the excess electron source for the alkali metal anion in a series of compounds with remarkable NLO properties.[128]

TABLE 10.3
Static First Hyperpolarizabilities (β_0, au), Transition Energies (ΔE, eV) of Crucial Excited States, Vertical Ionization Energies (VIEs, eV) of Typical Alkaline-Earth-Based Excess Electron Compounds

Species	β_0	ΔE	VIE	Ref.
$Be^{2+}(H_5Aza222)^-K^-$	1.98×10^5	1.02	2.60	117
$Li\text{-}H_3C_4N_2\cdots Ca$	2.69×10^5	1.92	3.70	119
$Be(NH_3)_3Be$	1.26×10^5	0.91	3.41	120
Ca^+-**1**-K^-	1.81×10^6	0.37	4.70	123
Ca-**1**-Ca	2.91×10^4	3.07	4.70	124
$2e^-Mg^{2+}(Ni@Sn_{12})^{2-}Ca^{2+}$	1.69×10^4	2.29	5.93	126
Li-**1**-Ca	3.51×10^6	0.52	3.50	129
K-**1**-Ca	5.82×10^6	1.22	3.01	130
Li-**3**-Be	1.46×10^7	1.10	2.71	131
$Na^+(2^6Adz)Mg^-$	1.00×10^6	0.65	2.58	132
$Li^+(3^6Adz)Mg^-$	5.10×10^8	0.30	2.24	133
$Li^+(NH_3)_6Mg^-$	6.40×10^5	3.16	2.77	134
K^+(Calix[4]pyrrole)Ca^-	1.98×10^4	2.95	3.50	135
Li^+(15-crown-5)Be^-	2.50×10^4	2.54	3.42	136

10.4.2.2 Alkaline-Earthides

In addition to serving as an excess electron source, alkaline-earth metals can also act as excess electron acceptors for the design of excess electron compounds with large NLO responses. In 2018, a new class of excess electron compounds, alkaline-earthides, Li-**1**-M (M = Be, Mg, and Ca) with excellent NLO responses have been designed by Hou et al.[129] based on the previously mentioned all-cis-1,2,3,4,5,6-hexafluor-ocyclohexane (**1**). As shown in Figure 10.6, the α highest occupied molecular orbital (αHOMO) of these Li-**1**-M (M = Be, Mg and Ca) compounds manifest that partial *s* electron of the Li atom is pushed into the *np* orbital of the alkaline-earth atom by the lone pairs of the axial F atoms and becomes the partial polarized *p*-electron, which is apparently different from the polarized *s* electron in electrides and alkalides (see Figure 10.1). Such an excess electron on the *np* orbital of the alkaline-earth atom creates an alkaline-earth anion in each Li-**1**-M compound. By analogy with alkalide, these compounds were named alkaline-earthides. As the polarized *p*-electron is more easily excited, these alkaline-earthides have pretty low transition energies (0.520–0.621 eV) of the crucial excited state, resulting in their large β_0 values up to 3.51×10^6 au. These results suggest that designing alkaline-earthides may be a new strategy to obtain NLO materials of high performance.

Sajjad et al. also used the sodium and potassium as a source of excess electrons to design the alkaline-earthides A-**1**-AE (A = Na, K, and AE = Be, Mg, Ca) based on the same complexant **1**.[130] These alkaline-earthides also exhibit small excitation energies (0.54 ~1.64 eV) and extremely large β_0 values of up to 5.82×10^6 au. In 2022, Hou and coworkers designed a series of alkaline-earthides MA-**2**-MAE and MA-**3**-MAE (MA = Li, Na, and K; MAE = Be, Mg, and Ca) by using the dimer (**2**) and trimer (**3**) of molecule **1** in their parallel-stacked conformations as two ligands to research the aggregate effect on β_0.[131] The largest β_0 of the Li-**3**-Be is calculated to be as large as 1.46×10^7 au.

To obtain more alkaline-earthides, the complexants in previously reported alkalides have been also used to combine with alkali metal and alkaline-earth metal atoms by Ayub and coworkers.[132–136] For example, the 2^6 adamanzane (2^6Adz) and 3^6 adamanzane (3^6Adz) cages have been utilized to design the alkaline-earthides, $M^+(2^6Adz)M^-$[132] and $M^+(3^6Adz)M^-$ (M^+ = Li and Na; M^- = Be, Mg,

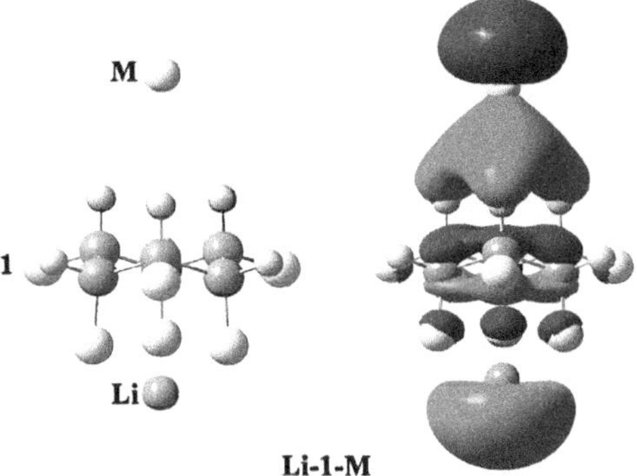

FIGURE 10.6 Geometric structures and αHOMOs of Li-1-M (M = Be, Mg, and Ca).

Source: Reproduced from Ref. [129] with permission (Copyright 2018 Elsevier B.V.).

and Ca),[133] respectively, where alkali metals act as the sources of electrons for alkaline-earth metals. By using hexa-ammine as the complexant, nine alkaline-earthides M^+(Hexaammine)M^- (M^+ = Li, Na, and K; M^- = Be, Mg, and Ca) have been also studied.[134] The calix[4]pyrrole and 15-crown-5 ether complexants were also employed to constructed the alkaline-earthides M^+(calix[4]pyrrole)M^- (M^+ = Li, Na, and K; M^- = Be, Mg, and Ca)[135] and M1(15-crown-5)M2 (M1 = Li and Na; M2 = Be and Mg),[136] respectively. The alkaline-earthide identities of these compounds were determined by the highest occupied molecular orbitals (HOMOs), partial density of states (PDOS), and natural bond orbital (NBO) analyses. It is noted from Table 10.3 that all these alkaline-earthides show much larger NLO responses but very small ΔE and VIE values.

10.4.3 Transition Metal-Based Excess Electron Compounds

In view of the fact that transition metals usually contain unpaired valence electrons, theoretical chemists also considered the possibility to use them as the source or acceptor of excess electrons. In particular, as the coinage metal atoms (Cu, Ag, and Au) have the similar ns^1 valence electron configuration with alkali metal atoms, they were first tested in 2017 to serve as a novel excess electron source for the design of alkalides[137] and in 2020 as the excess electron acceptor of coinage metalides[138] by my group. These pioneer works have inspired a large amount of endeavor to apply transition metals in the design of excess compounds with high NLO responses in the ensuing several years.[139–146]

10.4.3.1 Transition Metal Atoms as the Excess Electron Source

The first attempt to employ transition metal atoms as the excess electron source was the one utilizing the 3^6 adamanzane (3^6Adz) as complexant to encapsulate the coinage metal atoms to design a series of alkalides, i.e., (M@3^6Adz)M′ (M = Cu, Ag, and Au; M′ = Li, Na, and K).[137] As shown in Figure 10.7, the valence electrons on the outmost *s* atomic orbitals of Cu, Ag, and Au atoms can be pushed out of the 3^6Adz cage to form diffuse excess electrons around the outer alkali metal atoms, which yields a novel series of (M^+@3^6Adz)M'^- (M = Cu, Ag, and Au; M′ = Li, Na, and K) alkalides. The alkalide characteristics of these compounds have been further confirmed by their low VIE values and NBO analysis. Moreover, all the proposed alkalides have considerable β_0 up to 6.16×10^4 au, suggesting that they can be regarded as novel NLO molecules of high performance. This study paves a new way to design various excess electron compounds by using transition metal atoms as an excess electron source. Just by embedding the 3D transition metal atoms into the 3^6Adz complexant, a series of organometallic M@3^6Adz (M = Sc ~ Zn) complexes with magnetic properties, superalkali characteristics, and diffuse excess electrons were obtained.[139] Due to the existence of

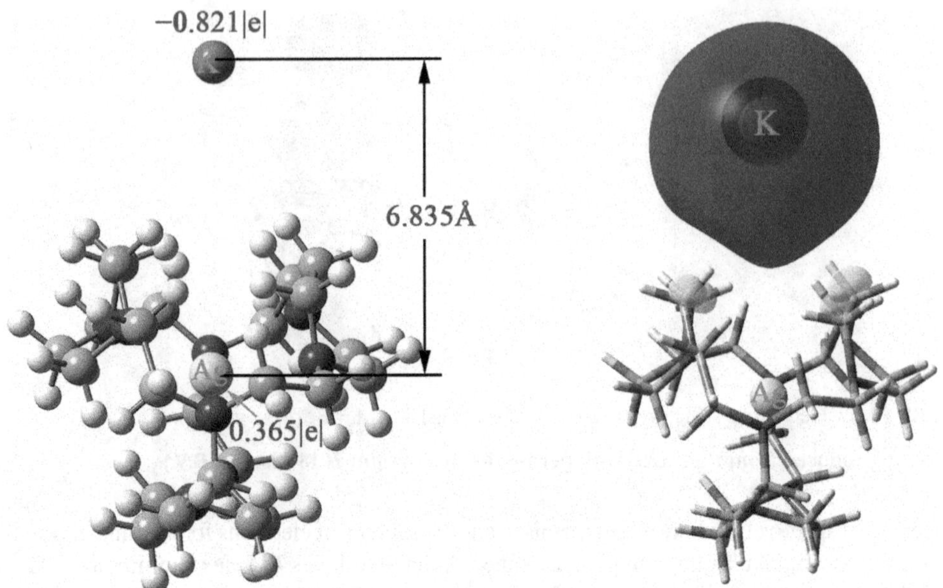

FIGURE 10.7 Geometric structure (a) and HOMO (b) of (M^+@3^6Adz)M'^- (M = Cu, Ag, and Au; M′ = Li, Na, and K).[137]

Source: Reproduced with permission (Copyright 2017 American Chemical Society).

diffuse electrons in their αSOMO, most of these M@3^6Adz complexes also exhibit remarkable NLO responses with a large β_0 of $1.56 \times 10^6 \sim 6.33 \times 10^7$ au. These computational results provide intriguing inspiration for experimental chemists to synthesize such potential NLO materials with high stability in the laboratory

In 2019, Liang et al.[140] designed a class of excess electron electrides, M[Cu(Ag)@$(NH_3)_n$] (M = Be, Mg and Ca; n = 1–3) with transition metal (Cu and Ag) doped in a NH_3 cluster. The natural population and the highest occupied molecular orbital analyses demonstrate that these M[Cu(Ag)@$(NH_3)_n$] compounds contain diffuse excess electrons and thus have large β_0 values up to 2.17×10^5 au. In another study, an Ag metal atom is encapsulated by 15-crown-5 and is exohedrally doped with an alkali metal (Li, Na, and K) to obtain a class of new complexes Ag(CR)M (M = Li, Na, and K) with excess electrons.[141] The Ag(CR)K possessed the maximum β_0 value of 3.69×10^4 au and therefore showed good NLO response. In addition, different transition metals have also been introduced to nanocages, such as the C_{20}[142] and $Al_{12}P_{12}$[143] to bring forth the excess electron compounds with considerable NLO properties. It is highly hoped that more transition metal-based excess electron compounds could be obtained to be used as NLO materials in the future.

10.4.3.2 Transition Metal Atoms as the Excess Electron Acceptor

To improve the stability of excess electron compounds, attention has been focused exploring novel species to replace unstable *s* block metal anions in the alkalides or alkaline-earthides.[138,144–146] As coinage metal atoms (Cu, Ag, and Au) not only possess a similar ns^1 valence electron configuration to alkali metal atoms but also have higher electron affinities (EAs) than those of alkali metal atoms, they have been also used as excess electron acceptors to obtain a new kind of M^+-**1**-M′ (M = Li, Na, and K; M′ = Cu, Ag, and Au) compounds on the basis of the Janus-type **1** molecule.[138] As shown in Figure 10.8, under the large facial polarization of **1**, the outermost ns^1 electrons of alkali metal atoms can be transferred to coinage metal atoms, forming diffuse excess electrons around them. By analogy with alkalide, these designed complexes were termed "coinage metalides" by Sun and

TABLE 10.4
Static First Hyperpolarizabilities (β_0, au), Transition Energies (ΔE, eV) of Crucial Excited States, Vertical Ionization Energies (VIEs, eV) or Adiabatic Ionization Energies (AIE, eV) of Typical Transition Metal-Based Excess Electron Compounds

Species	β_0	ΔE	VIE/AIE	Ref.
$(Ag^+@3^6Adz)K^-$	6.16×10^4	1.45	2.48	137
$Mn@3^6Adz$	6.33×10^7	0.24	1.78	139
$Be[Ag@(NH_3)_2]$	2.17×10^5	0.11	—	140
Ag(CR)K	3.69×10^4	2.12	4.10	141
K^+-**1**-Ag^-	9.02×10^4	2.91	4.57	138
K-**1**-Zn	9.89×10^5	—	3.10	144
Li-**1**-V	4.95×10^8	—	4.08	145
Ag^+-**1**-Au^-	8.97×10^4	0.63	7.90	146

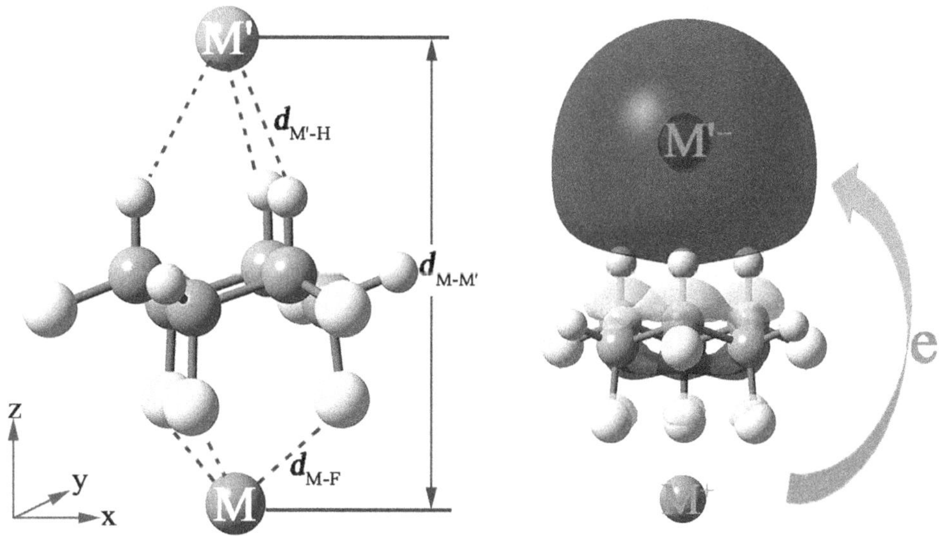

FIGURE 10.8 Geometric structure and the HOMO of M-1-M′ (M = Li, Na, and K; M′ = Cu, Ag, and Au).[138]
Source: Reproduced with permission (Copyright 2020 the Owner Societies).

coworkers.[138] It is exciting to find that these novel M^+-**1**-M′ compounds not only exhibit considerable NLO responses but also show much higher stability (larger VIEs and E_c values) than that of the corresponding M^+-**1**-M′ (M, M′ = Li, Na, and K) alkalides.[92] So this work opens up new possibilities for designing NLO material by using transition metal atoms as excess electron acceptors.

To replace the alkaline-earth metals in alkaline-earthides, a group of 12 elements, Zn, Cd, and Hg with paired *ns* electrons, were chosen by Zhang et al.[144] to design a class of excess electron compounds MF-**1**-MH (MF = Li, Na, and K, MH = Zn, Cd, and Hg) with remarkably large NLO response. In 2023, Ahsan et al.[145] further designed a group of AM-**1**-TM (where AM = Li, Na, and K, and TM = Sc, Ti, V, Cr, Mn, Fe, Co, Ni, Cu, and Zn) containing alkali metals as a source of excess electrons for the first row transition metals. In a similar vein, these compounds were named transition metalides. The metalide nature of these AM-**1**-TM complexes was validated through HOMO and NBO analyses, and their stability was guaranteed by the values of interaction energies, vertical ionization energies (VIE), and vertical electron affinity (VEA). The extremely large β_0 of 4.95×10^8 au is observed for the vanadium-based Li-**1**-V complex, which indicates the excellent NLO response of the designed transition metalides.

Though the introduction of transition metal atoms indeed increases the stability of the transition metalides, one alkali metal atom is still involved in them, which makes it a great challenge in synthesizing such compounds. Thus the possibility of designing alkali-metal-free coinage metalides has been explored by constructing a series of M-**1**-M′ (M; M′ = Cu, Ag, and Au) by Sun and coworkers.[146] Under the synergetic action of inherent polarization of **1** and the electronegativity difference between M′ and M, the *s* electron of Cu/Ag atom can be ejected to the Au atom, forming the diffuse excess electron around the Au^- anions in the Cu^+-O-Au^- and Ag^+-O-Au^- compounds. Hence, these two complexes can be considered the first examples of alkali-metal-free coinage metalides. It is exciting to find that they not only have higher stability (larger VIE and E_c values) than previously reported alkalides[92] but also exhibit remarkably larger β_0 values than the reported coinage metalides M^+-O-Au^- (M = Li, Na, and K),[138] suggesting their intriguing potential as stable molecular NLO materials with high performance. Furthermore, they also found that the rest of the M-O-M′ compounds can also be converted into coinage metalides under oriented external electric fields (OEEFs) by taking Cu^+-O-Cu^- and Cu^+-O-Ag^- as examples, which offers a new strategy to achieve novel NLO molecular switches. This work paves a novel way to design stable NLO materials just by using coinage metal atoms instead of alkali or alkaline-earth metals, which are more likely to be synthesized in the laboratory in the future.

10.4.4 Superalkali-Based Excess Electron Compounds

Superalkalis are clusters or molecules that have lower ionization energies (IEs) than that of cesium atoms[147] and that thus show very strong reducing ability. These special species exhibit great potential to be applied in the synthesis of various nanomaterials with tailored properties and thus have attracted a lot of the attention of chemists.[148] The superalkali clusters have been successfully used as building blocks to design charge-transfer compounds with high NLO responses, such as $(BLi_6)^+(X)^-$ (X = F, LiF_2, BeF_3, BF_4),[149,150] $(M)^+(BF_4)^-$ (M = FLi_2, OLi_3, NLi_4),[151] and $(M_3O)^+(Al_{13})^-$ (M = Na and K).[152] With a higher tendency to lose a valence electron than alkali metal atoms, superalkalis can replace the alkali metal atoms to serve as an excellent source of excess electrons for the design of excess electron compounds, such as electrides[68] and alkalides[153] with high NLO responses. As alkali-metal-like superatoms, superalkalis can also be used as the excess electron acceptors of superalkalides.[154] Next is a brief review on the progress of using superalkalis to design NLO materials in the past 10 years.

10.4.4.1 Superalkalis as the Source of Excess Electrons

In 2012, Li and coworkers[68] first designed two superalkali-based electrides $(M_3O)^+(e@C_{20}F_{20})^-$ (M = Na, K), where the excess electron of superalkali M_3O is encapsulated in a $C_{20}F_{20}$ cage with a sufficient interior electron attractive potential (see Figure 10.9a). In the same year, the dinuclear superalkli Li_2CN was wrapped into a boron nitride nanotube (BNNT), which serves as a protective shield molecule to obtain more stable NLO materials by Zhong et al.[155] After being encapsulated into the tube, the stability of Li_2CN was enhanced, and the resulting Li_2CN@BNNT complex exhibits a considerable β_0 value of 10,645 au. One year later, Li and coworkers[69] reported a specific acceptor-bridge-donor strategy to enhance the NLO response of superalkali-based electrides, where an interesting push-pull mechanism of excess electron generation and its long-range transfer are exhibited. Afterward, the $Li_3O@P_4$ and $Li_3O@C_3H_6$,[156] as well as a series of superalkali-boron-heterofullerene dyads M_3O-BC_{59} (M = Li, Na, K)[157] with considerable β_0 values, were reported by Chen's group.

By doping superalkali Li_3O onto the $Al_{12}N_{12}$ nanocage, Sun et al.[158] proposed a series of stable inorganic $Li_3O@Al_{12}N_{12}$ electrides. The calculated results demonstrate that the most stable isomer

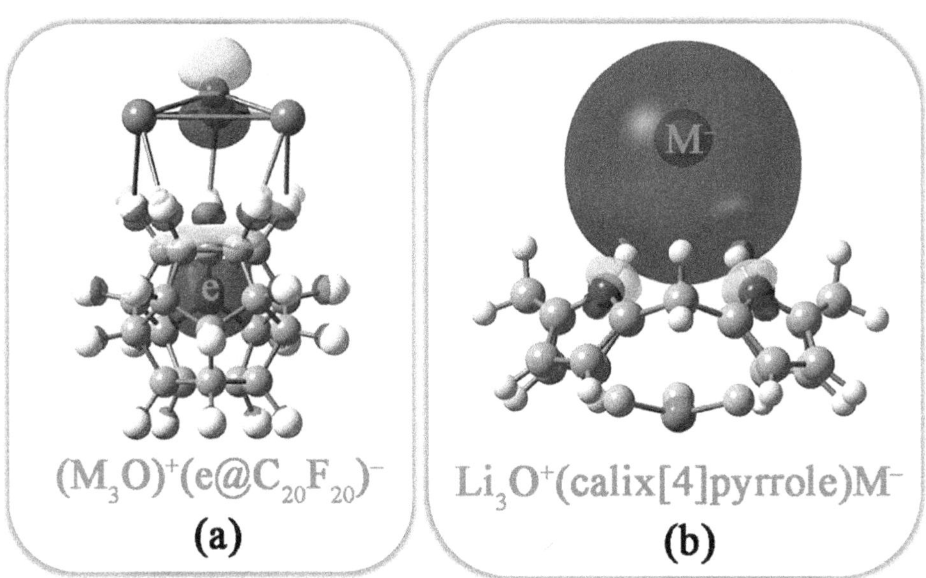

FIGURE 10.9 HOMOs of (a) electride $(M_3O)^+(e@C_{20}F_{20})$–(M = Na, K) with the excess electron mainly confined in the $C_{20}F_{20}$ cage and (b) alkalide $Li_{3O}{}^+$(calix[4]pyrrole)M^- (M = Li, Na, K)[8h] with an alkali metal anion M^-.

Source: Reproduced from Ref. [148]. Copyright 2019 Wiley-VCH Verlag GmbH & Co. KGaA, Weinheim.

of $Li_3O@Al_{12}N_{12}$ shows a large β_0 of 8.73 × 10^5 au. They also considered the effects of superalkali and nanocage subunits on the NLO responses of $M_3O@Al_{12}N_{12}$ (M = Li, Na, K) and $Li_3O@X_{12}Y_{12}$ (X = B, Al; Y = N, P) and found that the $Na_3O@Al_{12}N_{12}$ and $Li_3O@B_{12}P_{12}$ have a larger β_0 of 9.31 × 10^5 and 2.85 × 10^6 au, respectively. This indicates that selecting a suitable combination of nanocage and superalkali can bring forth more excellent NLO molecules. Afterward, continuous efforts have been devoted to designing stable NLO molecules by doping typical superalkalis (such as Li_2F, Li_3O, and Li_4N) onto different nanocages, including B_{40},[159] $Si_{12}C_{12}$,[160,161] $C_{24}N_{24}$,[162] $Zn_{12}O_{12}$,[163] $B_{12}P_{12}$,[164] $Al_{12}P_{12}$,[165] C_{20},[166] C_{24},[167] $Si_nAl_{12-n}N_{12}$ (n = 1, 2),[168] $B_{12}N_{12}$,[169] and $Al_{12}N_{12}$.[170] As shown in Table 10.5, the β_0 values of these superalkali@nanocage complexes are in the range of 1.18 × 10^4 ~ 7.10 × 10^5 au,[159–170] further confirming that doping the superalkali cluster onto nanocages is a promising strategy to design new NLO materials. In 2022, Gilani and coworkers[171] explored the geometric, electronic, and NLO properties of $B_{12}N_{12}$ nanocage doped by mixed superalkalis (Li_2NaO, Li_2KO, Na_2LiO, Na_2KO, K_2LiO, K_2NaO, and LiNaKO). They found that all these complexes possess higher stability and β_0 values than the $Li_3O@B_{12}N_{12}$[158] and thus proposed that mixed superalkalis are better choice than pure superalkalis for a $B_{12}N_{12}$ nanocage to design NLO materials.

In addition to various nanocages, a lot of nanomaterials with different structures have also been doped by superalkalis to design various NLO materials. For instance, the planar molecules, such as hexalithiobenzene (C_6Li_6),[172] Si_6Li_6,[173] $C_6S_6Li_6$,[174] triphenylene derivatives,[175] phenalenyl,[176,177] and borazine,[178] have been used to design several compounds with NLO properties. Moreover, a lot of NLO materials can be obtained by doping superalkalis onto the surface of two dimensional materials, including graphdiyne,[179–183] g-C_3N_4,[184,185] pyridinic vacancy graphene,[186] phosphorene,[187,188] graphene quantum dot,[189] macrocyclic [hexa-]thiophene,[190] biphenylene-based sheets,[191] carbon nitride (C_2N),[192] and boron nitride,[193] as well as some special molecules, such as the previously mentioned Janus molecule $C_6H_6F_6$,[194,195] hexamethylenetetramine (HMT),[196]

TABLE 10.5
Static First Hyperpolarizabilities (β_0, au), Transition Energies (ΔE, eV) of Crucial Excited States, Vertical Ionization Energies (VIEs, eV) of Typical Superalkali-Based Excess Electron Compounds

Species	β_0	ΔE	VIE	Ref.
$(Na_3O)^+(e@C_{20}F_{20})^-$	1.30×10^4	2.26	6.91	68
$Li_3O^+ \cdot \mathbf{1} \cdot K^-$	7.60×10^4	—	3.28	92
K_3O^+(calix[4]pyrrole)K^-	3.47×10^4	1.77	3.34	153
$K^+(en)_3K_3O^-$	9.16×10^4	0.82	2.83	154
$Li_2CN@BNNT$	1.06×10^4	2.14	4.14	155
$Li_3O@Al_{12}N_{12}$	8.73×10^5	1.24	3.55	158
$K_3O@B_{40}$	5.40×10^4	1.53	—	159
$K_3O@Si_{12}C_{12}$	2.17×10^4	1.37	—	160
$Li_4N@Si_{12}C_{12}$	1.99×10^4	1.12	4.30	161
$Li_3O@C_{24}N_{24}$	1.02×10^4	—	—	162
$Li_3O@Zn_{12}O_{12}$	1.67×10^4	—	3.59	163
$Li_3O@B_{12}P_{12}$	1.17×10^5	0.71	3.31	164
$Li_4N@Al_{12}P_{12}$	1.18×10^4	1.28	—	165
$Na_3O@C_{20}$	2.34×10^4	2.51	—	166
$K_3O@C_{24}$	3.66×10^4	1.07	—	167
$Na_3O@Si_2Al_{10}N_{12}$	7.10×10^5	0.77	4.24	168
$Li_3O@B_{12}N_{12}$	1.26×10^5	0.35	2.51	169
$Li_2F@Al_{12}N_{12}$	1.20×10^4	2.28	5.35	170
$Na_2LiO@B_{12}N_{12}$	1.70×10^7	1.16	4.02	171
$Li_3^+(NH_3)_4Na^-$	3.82×10^4	1.36	3.46	200
Li_3^+(aza222)K^-	6.00×10^4	—	2.42	201
Li_3O(12-crown-4)K^-	5.26×10^6	1.53	2.62	202
L-Li-L-Li_3O	1.31×10^6	0.29	—	203

teetotum boron clusters,[197] and macrocyclic oligofuran rings.[198] In all these materials, the doped superalkalis serve as the source of excess electrons, which bring with them the large NLO responses.

The superalkali K_3O has been also used as the excess electron source to obtain a special kind of cup-saucer-cage-shaped sandwich electride molecules calix[4]pyrrole…K_3O^+…$e@C_nF_n^-$ ($n = 8$, 10, 14, 20, and 36).[199] In these sandwich structures, as shown in Figure 10.10, the below calix[4]pyrrole cup pushes the valence electron of the sandwiched saucer-like superalkali K_3O to form an excess electron, which is further pulled and protected inside the above C_nF_n cage, thus an electron-transfer relay occurs. Moreover, it is found that increasing the C_nF_n cage size enhances not only the nonlinear optical response (β_0), but also the electron stability (VIE), which provides a new strategy to achieve novel cup-saucer-cage-shaped sandwich electride molecules with high performance.

Superalkalis can also serve as excess electron source of alkalides.[153] In 2014, Sun et al. first revealed this application prospect of superalkali by studying a series of superalkali-based alkalides, i.e., Li_3^+(calix[4]pyrrole)M^-, Li_3O^+(calix[4]pyrrole)M^-, and M_3O^+(calix[4]pyrrole)K^- (M = Li, Na, and K).[153] Taking the Li_3O^+(calix[4]pyrrole)M^- as an example, the excess electron of Li_3O is pushed to form a diffuse electron cloud wrapping the M atom and creating an M^- anion (see Figure 10.9b). As compared with traditional alkali-metal-based alkalides, these novel alkalides exhibit structural diversity as well as higher stability and NLO responses. Thereafter, several superalkali-based alkalides have been reported by them by using different complexants, such as $C_6H_6F_6$,[192] NH_3,[200] aza222.[201] These studies open a new door to obtaining novel alkalides by using superalkalis as building blocks. More recently, this strategy has been used by Ahsin et al.[202] to design a series

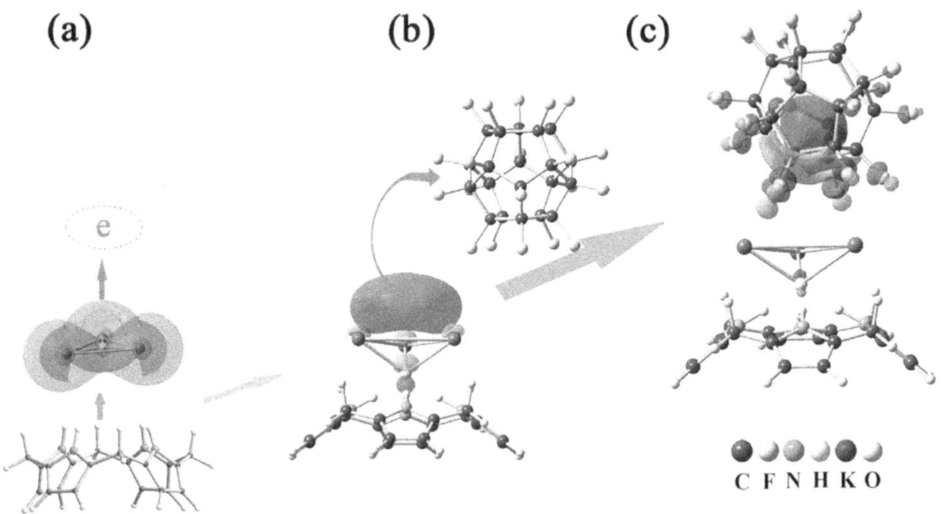

FIGURE 10.10 Example of long-range excess electron transfer relay in the formation of the calix[4]pyrrole...K_3O^+...e@$C_{20}F_{20}$: (a) calix[4]pyrrole complexant pushes the valence electron of superalkali metal K_3O and forms an excess electron; (b) the excess electron flies toward the $C_{20}F_{20}$ cage; (c) the excess electron has arrived at the $C_{20}F_{20}$ cage.

Source: Reproduced from Ref. [199]. Copyright 2015 The Royal Society of Chemistry.

of superalkali-based alkalides Li_3O@[12-crown-4]M (M = Li, Na, and K) with remarkable static and dynamic NLO properties.

10.4.4.2 Superalkalis as the Acceptors of Excess Electrons

As a kind of superatoms with similar properties to alkali metal atoms, superalkalis can be also utilized as excess electron acceptors to design a new kind of excess electron compounds, in which superalkalis occupy the anionic sites.[154,203,204] This new kind of excess electron compound was named superalkalides by Mai et al.[154] In 2015, they first proposed a series of $M^+(en)_3M'_3O^-$ (M, M′ = Li, Na, and K; en = ethylenediamine) superalkalides with the β_0 values of $7.80 \times 10^3 \sim 9.16 \times 10^4$ au. They found that the superalkalides $M^+(en)_3M'_3O^-$ show higher stability than the corresponding $M^+(en)_3M'^-$ alkalides due to the existence of the hydrogen bonds in them. The designed superalkalides, with their excellent NLO responses and high stabilities, are promising candidates for NLO materials. In another work, the superalkalis M'_3O (M′ = Li and Na) were combined with two $C_6H_6F_6$ (L) ligands to design novel superalkalides L-M-L-M'_3O with high thermodynamic stability and NLO responses.[203] This work found that the number of ligands L considerably impacts β_0 value; namely, the β_0 value (7.76×10^4 au) of Li-L-Li_3O can be improved two orders of magnitude by adding additional ligands for L-Li-L-Li_3O (1.31×10^6 au).

10.4.5 Clusters with Excess Electrons

Besides these excess compounds using different metal atoms as excess electron sources or acceptors, some precisely designed clusters can also contain diffuse excess electrons, which gives them large NLO responses. Among various clusters, the superalkalis usually contain diffuse excess electrons, and therefore they themselves also exhibit high NLO responses[139,204–211] Srivastava et al.[204] theoretically explored the NLO properties of Li_nF (n = 2–5) superalkali clusters and found that these clusters may also possess alkalide and/or electride characteristics in view of the existence of excess electrons. Ahsin et al. performed a series of studies on the NLO properties of bimetallic superalkalis,[205] NM_3M' (M, M′ = Li, Na, K),[206] M_2OCN and M_2NCO (M = Li, Na, K),[207] $C_3X_3Y_3$ (X = O, S; Y = Li, Na, K),[208] germanium-based superalkalis,[209] Li_n (n = 3, 5, and 7),[210] and P_7M_2 (M = Be, Mg, Ca).[211]

In 2022, Sun and coworkers proposed two kinds of superalkalis, namely the magnetic superalkalis[139] and nonmetallic superalkalis[212] of M@3^6Adz type, which are found to show significant NLO responses owing to the diffuse excess electrons involved. It should be mentioned that the B@3^6Adz exhibits a remarkably large β_0 of 1.35×10^6 au,[212] suggesting its potential as a suitable NLO molecule. In particular, as far as I know, this is the first metal-free inorganic molecule with diffuse excess electrons. As shown in Table 10.6, all these superalkalis possess very small ΔE and VIE values but have considerably large β_0 values of $1.23 \times 10^4 \sim 3.90 \times 10^7$ au. In addition, two superalkali units can be simultaneously combined with a central atom or group to design NLO molecules, such as OLi_3-M-Li_3O (M = Li, Na, and K) complexes[213] and sandwich compounds M_3-NO_3-M'_3 (M, M′ = Li, Na, K).[214]

Moreover, the alkali metal clusters have been doped onto the fullerene (C_{60}) to enhance the electric β_0.[215] In the design complexes, C_{60} unconventionally functions as an electron donor at the excited states, and the easily polarizable excess electrons are maintained in the framework of the adsorbed clusters. From a certain point of view, such systems can be considered hybrids that combine the basic characteristics of a classical donor/acceptor superstructure and systems with easily polarizable excess electrons. Interestingly, Luo and coworkers[216] evaluated the potential of using the multivalent Al_7 superatom as an excess electron acceptor to construct NLO materials with cluster-trapped excess electrons. They found that, when alkali metal approaches the Al_7 superatom, its outermost *s* electron can be trapped by Al_7 to give the superatom compound MAl_7 (M = Li, Na, K) with an excess electron. The self-assemblies $(MAl_7)_n$ (n = 2, 3) have been optimized to explore the stability of MAl_7 in the solid state, which indicates that only $NaAl_7$ can keep its structural integrity as a building block upon self-assembling, while serious aggregations between Al_7 clusters occur in the dimers and trimers of $LiAl_7$ and KAl_7. Born–Oppenheimer molecular dynamics (BOMD) simulations for $(NaAl_7)_n$ (n = 2, 3) indicate that these species are stable toward fragmentation at 300 K.

TABLE 10.6
Static First Hyperpolarizabilities (β_0, au), Transition Energies (ΔE, eV) of Crucial Excited States, Vertical Ionization Energies (VIEs, eV) of Clusters with Excess Electrons

Clusters	β_0	ΔE	VIE	Ref.
Li_2F	1.23×10^4	1.26	3.85	204
Li_7Mg	4.30×10^4	1.15	3.61	205
NK_3Li	3.90×10^7	0.96	1.71	206
K_2NCO	2.20×10^4	1.51	4.12	207
$C_3S_3K_3$	1.60×10^6	0.39	2.06	208
Ge_9K_5	1.14×10^5	0.75	2.15	209
Li_5	7.70×10^4	2.39	4.07	210
P_7Ca_2	3.27×10^4	0.11	5.12	211
B@3^6Adz	1.35×10^6	0.30	2.18	212
$(Na_{12})C_{60}$	1.08×10^5	—	—	215
$(NaAl_7)_3$	7.27×10^4	1.41	—	216
CuAg@Ca_7Mg	1.43×10^4	—	4.37	217
$[(Ni@Ge_9)Ca_3]_4$	5.42×10^4	1.16	4.46	219
CaN_3Ca	2.35×10^4	2.24	3.90	220
Li_3N_3Mg	3.50×10^4	0.96	5.30	221

The calculated β_0 values of $(NaAl_7)_n$ (n = 1, 2, and 3) are in the range of 1.6×10^4 to 7.5×10^4 au. This work demonstrates that this $NaAl_7$ is a promising building block for high-performance NLO materials.

All-metal CuAg@Ca_7M (M = Be, Mg, and Ca) clusters with electride characteristics have been designed and investigated by He et al.[217] in 2016. In these clusters, a pull-push electron relay occurs. Unusually, the all-metal polyanions of fourfold negatively charged $[Cu–Ag–Be/Mg]^{4-}$ and $[Cu–Ag]^{4-}$ with four extra electrons gained from Ca atoms push the remaining valence electrons of the Ca atoms, forming multiexcess electrons (Ne = 10/12). Therefore, these molecules can be described as salt-like $[(Ca^{2+})_7(CuAgM)^{4-}] + 10e^-$ (M = Be and Mg) and $[(Ca^{2+})_8(CuAg)^{4-}] + 12e^-$. Their results show that these all-metal electrides have large β_0, which can be further enhanced by manipulating the atomic number and position of M atoms. Resultantly, a CuAg@Ca_7Mg isomer exhibits a large β_0 of 1.43×10^4 au. Moreover, they found that these all-metal electrides have the infrared (IR) transparent region of 1.3–6 μm and hence can be regarded as novel IR NLO molecules.

Focusing on the interesting concept of an all-metal electride, He et al.[218] further designed a group of centrosymmetric molecules $e^-+M^{2+}(Ni@Pb_{12})^{2-}M^{2+}+e^-$ (M = Be, Mg, and Ca) with two anionic excess electrons located at the opposite ends of each cluster. They proved that these novel molecular all-metal electrides can act as IR NLO switches. For example, a small *EEF* of 8×10^{-4} au (0.04 V/Å) can drive a long-range excess electron transfer from one end of the $e^-+Ca^{2+}(Ni@Pb_{12})^{2-}Ca^{2+}+e^-$ cluster through the middle all-metal anion cage $(Ni@Pb_{12})^{2-}$ to the other end. This long-range electron transfer is confirmed by a prominent change of excess electron orbitals from double lobes to a single lobe, forming an excess electron lone pair. As a result, a small *EEF* induces a dramatic β_0 contrast from 0 (off form) to 2.2×10^6 au (on form) for Ca($Ni@Pb_{12}$)Ca without altering the valence and chemical bond nature. Therefore, this switching mechanism is a distinct nonbonding evolution, called electronic structure isomerization, which means that such switching has the advantages of being fast and reversible. Besides, these all-metal electride molecules also have an IR transparent characteristic (1.5–10 μm) in NLO electride molecules and hence are commendable molecular IR NLO switches.

He et al.[219] also proposed a class of multicage chain structures $[(Ni@Ge_9)Ca_3]_n$ (n = 1–4), in which an interesting pull-push electron relay occurs. In these nanostructures, the chain skeleton pulls valence electrons from nonbridge Ca atoms to form a skeleton polyanion, and then the resulting polyanion further pushes the remaining valence electrons of the Ca atoms to generate excess electrons. The excess electron numbers are 2, 0, 4, 4 for n = 1, 2, 3, 4, respectively. Results show that the electride chains have large β_0, which has been strongly increased from 9321 (n = 1, Ne = 2) to 54,232 au (n = Ne = 4), which exhibits that significant cage number and excess electron number effects on NLO response. This provides a new design strategy to enhance NLO response by increasing the metal cage number for such cluster chains. Hence, these molecular all-metal electride multicage chains, as novel nanorods, are promising new NLO nanomaterials.

To obtain innovative single-pole double-throw nonlinear optical (NLO) molecular switches, three isomeric configurations of bipyramidal CaN_3Ca have been obtained by Wang et al.[220] in 2020. These CaN_3Ca clusters can be regarded as both alkaline-earth-based aromatic superalkalis and electrides. Under the small external electric field of 0.011 au (0.565 V·Å^{-1}), the small-distance hopping of Ca atoms in CaN_3Ca brings half an electron transfer long range from one end of the Ca atom to another, resulting in a large β_{zzz} contrast between two configurations. This reinforces the intriguing potential of these inorganic Robin–Day-type superalkali electrides as high-sensitive multistate nonlinear optics switch. In the same year, they also proposed another intramolecular self-redox switch, Li_3N_3Mg.[221] They found that a suitably oriented external electric field (OEEF) can make a long-range excess electron transfer from a Mg atom to a Li_3 ring, leading to an interesting intramolecular self-redox from $Li_3^{2+}N_3^{3-}Mg^+$ to $Li_3^+N_3^{3-}Mg^{2+}$. This also results in an obvious rise of the β_0 value from $Li_3^{2+}N_3^{3-}Mg^+$ (3.50×10^4 au) to $Li_3^+N_3^{3-}Mg^{2+}$ (1.01×10^5 au), suggesting that this Li_3N_3Mg cluster is a good candidate for a self-redox NLO molecular switch.

10.5 CONCLUDING REMARKS

Various NLO materials have been widely investigated in view of the applications in optoelectronic devices and the like. The molecules and clusters with excess electrons usually show very large NLO responses. Therefore, such species have attracted more and more attention and have achieved significant advances. In this chapter, we have briefly introduced the background of the excess electron and the computational methodology for studying the molecules and clusters containing excess electrons. In particular, previous investigations on the theoretical design of species with excess electrons in the past two decades have been systemically summarized by dividing them into five groups according to the source and acceptor of the excess electrons. This chapter highlighted the crucial role of excess electron in enhancing the NLO properties of systems and presented some valuable rules to design the NLO species by introducing different excess electron sources or acceptors to various complexants.

Despite the remarkable progress in the design of excess electron compounds, great challenges still remain to move these reported species with fascinating NLO properties from academic studies to practical applications. Although several electrides and alkalides have been successfully synthesized at room temperature, they have not been truly applied in the nonlinear optical fields up to now. Hence more and more attention from experimentalists should be paid to synthesize the theoretically designed molecules and clusters with excess electrons and used them as NLO materials in the future. To achieve this goal, the stability of superalkalis, rather than just the NLO response, should be addressed in future designs. How to balance the relationship between stability and nonlinear optical response of the molecular system is very crucial. Confining excess electrons in an environment restricting its dispersity looks to be a promising way to obtain excess electron compounds with both high stability and NLO responses. Inorganic materials with excess electrons, such as the all-metal electrides seem to be the most promising NLO species that can be synthesized by experimentalists at ambient temperature and environment. Besides, except for the previously used alkali-metal, alkaline-earth, transition metal, and superalkalis, more and more source or acceptors of excess electrons, such as nonmetallic atoms and various superatoms should be explored. Through multidisciplinary endeavors, fruitful theoretical results and truly applicable NLO materials based on molecules and clusters with excess electrons will appear in the near future. It is highly hoped that this chapter will be helpful for the future theoretical design and experimental synthesis of new NLO materials based on the concept of the excess electron.

10.5.1 ACKNOWLEDGMENT

The author thanks Professors Li Zhi-Ru, Wu Di, and Li Ying from Jilin University for their guidance and help during his study on the theoretical design of nonlinear optical materials.

REFERENCES

(1) Maiman, T. H. Stimulated Optical Radiation in Ruby Masers. *Nature.* **1960**, *187*, 493.
(2) Bano, R.; Asghar, M.; Ayub, K.; Mahmood, T.; Iqbal, J.; Tabassum, S.; Zakaria, R.; Gilani, M. A. Theoretical Perspective on Strategies for Modeling High Performance Nonlinear Optical Materials. *Front. Mater.* **2021**, *8*, 783239.
(3) Zhong, R.-L.; Xu, H.-L.; Li, Z.-R.; Su, Z.-M. Role of Excess Electrons in Nonlinear Optical Response. *J. Phys. Chem. Lett.* **2015**, *6*, 612.
(4) Yu, G.; Huang, X.; Li, S.; Chen, W. Theoretical Insights and Design of Intriguing Nonlinear Optical Species Involving the Excess Electron. *Int. J. Quantum Chem.* **2015**, *115*, 671.
(5) Muhammad, S.; Xu, H.-L.; Zhong, R.-L.; Su, Z.-M.; Al-Sehemi, A. G.; Irfan, A. Quantum Chemical Design of Nonlinear Optical Materials by sp^2-Hybridized Carbon Nanomaterials: Issues and Opportunities. *J. Mater. Chem. C.* **2013**, *1*, 5439.
(6) Ray, P. C. Size and Shape Dependent Second Order Nonlinear Optical Properties of Nanomaterials and Their Application in Biological and Chemical Sensing. *Chem. Rev.* **2010**, *110*, 5332.

(7) Kanis, D. R.; Ratner, M. A.; Marks, T. J. Design and Construction of Molecular Assemblies with Large Second-Order Optical Nonlinearities. Quantum Chemical Aspects. *Chem. Rev.* **1994**, *94*, 195.
(8) Coe, B. J.; Jones, L. A.; Harris, J. A.; Brunschwig, B. S.; Asselberghs, I.; Clays, K.; Persoons, A. Highly Unusual Effects of π-Conjugation Extension on the Molecular Linear and Quadratic Nonlinear Optical Properties of Ruthenium(II) Ammine Complexes. *J. Am. Chem. Soc.* **2003**, *125*, 862.
(9) Yang, J.-S.; Liau, K.-L.; Li, C.-Y.; Chen, M.-Y. Meta Conjugation Effect on the Torsional Motion of Aminostilbenes in the Photoinduced Intramolecular Charge-Transfer State. *J. Am. Chem. Soc.* **2007**, *129*, 13183.
(10) Meyers, F.; Marder, S. R.; Pierce, B. M.; Brédas, J. L. Electric Field Modulated Nonlinear Optical Properties of Donor-Acceptor Polyenes: Sum-Over-States Investigation of the Relationship between Molecular Polarizabilities (α, β, and γ) and Bond Length Alternation. *J. Am. Chem. Soc.* **1994**, *116*, 10703.
(11) Champagne, B. T.; Spassova, M.; Jadin, J.-B.; Kirtman, B. Ab Initio Investigation of Doping-Enhanced Electronic and Vibrational Second Hyperpolarizability of Polyacetylene Chains. *J. Chem. Phys.* **2002**, *116*, 3935.
(12) Ja Lee, M.; Piao, M.; Jeong, M.-Y.; Hae Lee, S.; Min Kang, K.; Jeon, S.-J.; Gun Lim, T.; Rae Cho, B. Novel Azo Octupoles with Large First Hyperpolarizabilities. *J. Mater. Chem.* **2003**, *13*, 1030.
(13) Liu, Z.-B.; Zhou, Z.-J.; Li, Y.; Li, Z.-R.; Wang, R.; Li, Q.-Z.; Li, Y.; Jia, F.-Y.; Wang, Y.-F.; Li, Z.-J.; Cheng, J.-B.; Sun, C.-C. Push-Pull Electron Effects of the Complexant in a Li Atom Doped Molecule with Electride Character: A New Strategy to Enhance the First Hyperpolarizability. *Phys. Chem. Chem. Phys.* **2010**, *12*, 10562.
(14) Wang, S.-J.; Wang, Y.-F.; Cai, C. Multidecker Sandwich Cluster V_nBen_{n+1}(n = 1, 2, 3, 4) as a Polarizable Bridge for Designing 1D Second-Order NLO Chromophore: Metal-π Sandwich Multilayer Structure as a Particular Charge Transfer Axis for Constructing Multi-Dimensional NLO Moleculars. *J. Phys. Chem. C.* **2015**, *119*, 16256.
(15) Nakano, M.; Kishi, R.; Nitta, T.; Kubo, T.; Nakasuji, K.; Kamada, K.; Ohta, K.; Champagne, B.; Botek, E.; Yamaguchi, K. Second Hyperpolarizability (γ) of Singlet Diradical System: Dependence of γ on the Diradical Character. *J. Phys. Chem. A.* **2005**, *109*, 885.
(16) Nakano, M.; Kubo, T.; Kamada, K.; Ohta, K.; Kishi, R.; Ohta, S.; Nakagawa, N.; Takahashi, H.; Furukawa, S.-I.; Morita, Y.; Nakasuji, K.; Yamaguchi, K. Second Hyperpolarizabilities of Polycyclic Aromatic Hydrocarbons Involving Phenalenyl Radical Units. *Chem. Phys. Lett.* **2006**, *418*, 142.
(17) Li, Y.; Li, Z. R.; Wu, D.; Li, R. Y.; Hao, X. Y.; Sun, C. C. An Ab Initio Prediction of the Extraordinary Static First Hyperpolarizability for the Electron-Solvated Cluster (FH)(2){e}(HF). *J. Phys. Chem. B.* **2004**, *108*, 3145.
(18) Chen, W.; Li, Z. R.; Wu, D.; Gu, F. L.; Hao, X. Y.; Wang, B. Q.; Li, R. J.; Sun, C. C. The Static Polarizability and First Hyperpolarizability of the Water Trimer Anion: Ab Initio Study. *J. Chem. Phys.* **2004**, *121*, 10489.
(19) Edwards, P. P. The Electronic Properties of Metal Solutions in Liquid Ammonia and Related Solvents. *Adv. Inorg. Chem. Radiochem.* **1982**, *25*, 135.
(20) Holton, D.; Edwards, P. Metals in Nonaqueous Solvents. *Chem. Br.* **1985**, *21*, 1007.
(21) Thomas, J. M.; Edwards, P. P.; Kuznetsov, V. L. Boundless Chemist, Physicist, Poet and Man of Action. *ChemPhysChem.* **2008**, *9*, 59.
(22) Weyl, W. *Ann. Phys.* **1864**, *121*, 606.
(23) Hannay, J. B.; Hogarth, J. On the Solubility of Solids in Gases. *Proc. R. Soc. London* **1879**, *29*, 324
(24) Hannay, J. B.; Hogarth, J. On the Solubility of Solids in Gases. *Proc. R. Soc. London.* **1880**, *30*, 178
(25) Joannis, M. A. C. R. *Acad. Sci.* **1889**, *109*, 900.
(26) Moissan, H. C. R. *Acad. Sci.* **1898**, *127*, 685.
(27) Zurek, E.; Edwards, P. P.; Hoffmann, R. A Molecular Perspective on Lithium-Ammonia Solutions. *Angew. Chem. Int. Ed.* **2009**, *48*, 8198.
(28) Gibson, G. E.; Argo, W. L. The Absorption Spectra of the Blue Solutions of Sodium and Magnesium in Liquid Ammonia. *Phys. Rev.* **1916**, *7*, 33.
(29) Ellaboudy, A.; Dye, J. L.; Smith, P. B. Cesium 18-Crown-6 Compounds. A Crystalline Ceside and a Crystalline Electride. *J. Am. Chem. Soc.* **1983**, *105*, 6490.
(30) Dawes, S. B.; Ward, D. L.; Huang, R. H.; Dye, J. L. First Electride Crystal Structure. *J. Am. Chem. Soc.* **1986**, *108*, 3535.
(31) Dye, J. L. Anionic Electrons in Electrides. *Nature* **1993**, *365*, 10
(32) Dye, J. L. Chemistry: Electrons as Anions. *Science* **2003**, *301*, 607.

(33) Dye, J. L. Electrides: Early Examples of Quantum Confinement. *Acc. Chem. Res.* **2009**, *42*, 1564.

(34) Matsuishi, S.; Toda, Y.; Miyakawa, M.; Hayashi, K.; Kamiya, T.; Hirano, M.; Tanaka, I.; Hosono, H. High-Density Electron Anions in a Nanoporous Single Crystal:$[Ca_{24}Al_{28}O_{64}]_4(4e^-)$. *Science.* **2003**, *301*, 626.

(35) Hosono, H.; Kitano, M. Advances in Materials and Applications of Inorganic Electrides. *Chem. Rev.* **2021**, *121*, 3121.

(36) Becker, P. Borate Materials in Nonlinear Optics. *Adv. Mater.* **1998**, *10*, 979.

(37) Oudar, J. L.; Chemla, D. S. Hyperpolarizabilities of the Nitroanilines and Their Relations to the Excited State Dipole Moment. *J. Chem. Phys.* **1977**, *66*, 2664.

(38) Oudar, J. L. Optical Nonlinearities of Conjugated Molecules. Stilbene Derivatives and Highly Polar Aromatic Compounds. *J. Chem. Phys.* **1977**, *67*, 446.

(39) McLean, A. D.; Yoshimine, M. Theory of Molecular Polarizabilities. *J. Chem. Phys.* **1967**, *47*, 1927.

(40) Buckingham, A. D. Permanent and Induced Molecular Moments and Long-Range Intermolecular Forces, John Wiley and Sons Ltd, 1967, pp. 107–142.

(41) Pople, J. A.; Head-Gordon, M.; Raghavachari, K. Quadratic Configuration Interaction. A General Technique for Determining Electron Correlation Energies. *J. Chem. Phys.* **1987**, *87*, 5968.

(42) Head-Gordon, M.; Pople, J. A.; Frisch, M. J. MP2 Energy Evaluation by Direct Methods. *Chem. Phys. Lett.* **1988**, *153*, 503.

(43) Chen, W.; Li, Z. R.; Wu, D.; Li, R. Y.; Sun, C. C. Theoretical Investigation of the Large Nonlinear Optical Properties of $(HCN)_n$ Clusters with Li Atom. *J. Phys. Chem. B.* **2005**, *109*, 601.

(44) Li, Z. J.; Li, Z. R.; Wang, F. F.; Luo, C.; Ma, F.; Wu, D.; Wang, Q.; Huang, X. R. Dependence on the Petal Number of the Static and Dynamic First Hyperpolarizability for Electride Molecules: Many-Petal-Shaped Li-Doped Cyclic Polyamines. *J. Phys. Chem. A.* **2009**, *113*, 2961.

(45) Ma, F.; Li, Z. R.; Xu, H. L.; Li, Z. J.; Li, Z. S.; Aoki, Y.; Gu, F. L. Lithium Salt Electride with an Excess Electron Pair-A Class of Nonlinear Optical Molecules for Extraordinary First Hyperpolarizability. *J. Phys. Chem. A.* **2008**, *112*, 11462.

(46) Sekino, H.; Maeda, Y.; Kamiya, M.; Hirao, K. Polarizability and Second Hyperpolarizability Evaluation of Long Molecules by the Density Functional Theory with Long-Range Correction. *J. Chem. Phys.* **2007**, *126*, 014107.

(47) Zhao, Y.; Truhlar, D. The M06 Suite of Density Functionals for Main Group Thermochemistry, Thermochemical Kinetics, Non-Covalent Interactions, Excited States, and Transition Elements: Two New Functionals and Systematic Testing of Four M06-Class Functionals and 12 Other Functionals. *Theor. Chem. Acc.* **2008**, *120*, 215.

(48) Iikura, H.; Tsuneda, T.; Yanai, T.; Hirao, K. A Long-Range Correction Scheme for Generalized-Gradient-Approximation Exchange Functionals. *J. Chem. Phys.* **2001**, *115*, 3540.

(49) Champagne, B.; Botek, E.; Nakano, M.; Nitta, T.; Yamaguchi, K. Basis Set and Electron Correlation Effects on the Polarizability and Second Hyperpolarizability of Model Open-Shell π-Conjugated Systems. *J. Chem. Phys.* **2005**, *122*, 114315.

(50) Maroulis, G.; Xenides, D.; Hohm, U.; Loose, A. Dipole, Dipole–Quadrupole, and Dipole–Octopole Polarizability of Adamantane, $C_{10}H_{16}$, from Refractive Index Measurements, Depolarized Collision-Induced Light Scattering, Conventional Ab Initio and Density Functional Theory Calculations. *J. Chem. Phys.* **2001**, *115*, 7957.

(51) Wang, J.-N.; Xu, H.-L.; Sun, S.-L.; Gao, T.; Li, H.-Z.; Li, H.; Su, Z.-M. An Effective Method for Accurate Prediction of the First Hyperpolarizability of Alkalides. *J. Comput. Chem.* **2012**, *33*, 231.

(52) Chen, W.; Li, Z. R.; Wu, D.; Li, Y.; Sun, C. C.; Gu, F. L. The Structure and the Large Nonlinear Optical Properties of Li@Calix[4]Pyrrole. *J. Am. Chem. Soc.* **2005**, *127*, 10977.

(53) Chen, W.; Li, Z. R.; Wu, D.; Li, Y.; Sun, C. C.; Gu, F. L.; Aoki, Y. Nonlinear Optical Properties of Alkalides Li^+(Calix[4]Pyrrole)M^- (M = Li, Na, and K): Alkali Anion Atomic Number Dependence. *J. Am. Chem. Soc.* **2006**, *128*, 1072.

(54) Postils, V.; Garcia-Borras, M.; Sola, M.; Luis, J. M.; Matito, E. On the Existence and Characterization of Molecular Electrides. *Chem. Commun.* **2015**, *51*, 4865.

(55) El Bakouri, O.; Postils, V.; Garcia-Borràs, M.; Duran, M.; Luis, J. M.; Calvello, S.; Soncini, A.; Matito, E.; Feixas, F.; Solà, M. Metal Cluster Electrides: A New Type of Molecular Electride with Delocalised Polyattractor Character. *Chem. Eur. J.* **2018**, *24*, 9853.

(56) Sitkiewicz, S. P.; Ramos-Cordoba, E.; Luis, J. M.; Matito, E. How Many Electrons Does a Molecular Electride Hold? *J. Phys. Chem. A* **2021**, *125*, 4819.

(57) Kim, J.; Ichimura, A. S.; Huang, R. H.; Redko, M.; Phillips, R. C.; Jackson, J. E.; Dye, J. L. Crystalline Salts of Na and K (Alkalides) That Are Stable at Room Temperature. *J. Am. Chem. Soc.* **1999**, *121*, 10666.

(58) Redko, M. Y.; Vlassa, M.; Jackson, J. E.; Misiolek, A. W.; Huang, R. H.; Dye, J. L. "Inverse Sodium Hydride": A Crystalline Salt That Contains H^+ and Na^-. *J. Am. Chem. Soc.* **2002**, *124*, 5928.

(59) Redko, M. Y.; Huang, R. H.; Jackson, J. E.; Harrison, J. F.; Dye, J. L. Barium Azacryptand Sodide, the First Alkalide with an Alkaline Earth Cation, Also Contains a Novel Dimer, $(Na_2)^{2-}$. *J. Am. Chem. Soc.* **2003**, *125*, 2259.

(60) Redko, M. Y.; Jackson, J. E.; Huang, R. H.; Dye, J. L. Design and Synthesis of a Thermally Stable Organic Electride. *J. Am. Chem. Soc.* **2005**, *127*, 12416.

(61) Xu, H. L.; Li, Z. R.; Wu, D.; Wang, B. Q.; Li, Y.; Gu, F. L.; Aoki, Y. Structures and Large NLO Responses of New Electrides: Li-Doped Fluorocarbon Chain. *J. Am. Chem. Soc.* **2007**, *129*, 2967.

(62) Xu, H. L.; Li, Z. R.; Wu, D.; Ma, F.; Li, Z. J. Lithiation and Li-Doped Effects of 5 Cyclacene on the Static First Hyperpolarizability. *J. Phys. Chem. C.* **2009**, *113*, 4984.

(63) Li, Z. J.; Wang, F. F.; Li, Z. R.; Xu, H. L.; Huang, X. R.; Wu, D.; Chen, W.; Yu, G. T.; Gu, F. L.; Aoki, Y. Large Static First and Second Hyperpolarizabilities Dominated by Excess Electron Transition for Radical Ion Pair Salts $M_2^{\bullet+}TCNQ^{\bullet-}$ (M = Li, Na, K). *Phys. Chem. Chem. Phys.* **2009**, *11*, 402.

(64) Xu, H.-L.; Sun, S.-L.; Muhammad, S.; Su, Z.-M. Three-Propeller-Blade-Shaped Electride: Remarkable Alkali-Metal-Doped Effect on the First Hyperpolarizability. *Theor. Chem. Acc.* **2010**, *128*, 241.

(65) Muhammad, S.; Xu, H.; Liao, Y.; Kan, Y.; Su, Z. Quantum Mechanical Design and Structure of the Li@ $B_{10}H_{14}$ Basket with a Remarkably Enhanced Electro-Optical Response. *J. Am. Chem. Soc.* **2009**, *131*, 11833.

(66) Muhammad, S.; Xu, H.; Su, Z. Capturing a Synergistic Effect of a Conical Push and an Inward Pull in Fluoro Derivatives of Li@$B_{10}H_{14}$ Basket: Toward a Higher Vertical Ionization Potential and Nonlinear Optical Response. *J. Phys. Chem. A.* **2011**, *115*, 923.

(67) Brandão, I.; Fonseca, T. L.; Georg, H. C.; Castro, M. A.; Pontes, R. B. Assessing the Structure and First Hyperpolarizability of Li@$B_{10}H_{14}$ in Solution: A Sequential QM/MM Study Using the ASEC–FEG Method. *Phys. Chem. Chem. Phys.* **2020**, *22*, 17314.

(68) Wang, J. J.; Zhou, Z. J.; Bai, Y.; Liu, Z. B.; Li, Y.; Wu, D.; Chen, W.; Li, Z. R.; Sun, C. C. The Interaction between Superalkalis (M_3O, M = Na, K) and a $C_{20}F_{20}$ Cage Forming Superalkali Electride Salt Molecules with Excess Electrons Inside the $C_{20}F_{20}$ Cage: Dramatic Superalkali Effect on the Nonlinear Optical Property. *J. Mater. Chem.* **2012**, *22*, 9652.

(69) Bai, Y.; Zhou, Z.-J.; Wang, J.-J.; Li, Y.; Wu, D.; Chen, W.; Li, Z.-R.; Sun, C.-C. New Acceptor-Bridge-Donor Strategy for Enhancing NLO Response with Long-Range Excess Electron Transfer from the NH_2...M/M_3O Donor (M = Li, Na, K) to Inside the Electron Hole Cage $C_{20}F_{19}$ Acceptor through the Unusual Sigma Chain Bridge $(CH_2)_4$. *J. Phys. Chem. A.* **2013**, *117*, 2835.

(70) Wang, Y.-F.; Li, Z.-R.; Wu, D.; Sun, C.-C.; Gu, F.-L. Excess Electron Is Trapped in a Large Single Molecular Cage $C_{60}F_{60}$. *J. Comput. Chem.* **2010**, *31*, 195.

(71) Wang, Y.-F.; Chen, W.; Yu, G.-T.; Li, Z.-R.; Wu, D.; Sun, C.-C. Evolution of Lone Pair of Excess Electrons Inside Molecular Cages with the Deformation of the Cage in e_2@$C_{60}F_{60}$ Systems. *J. Comput. Chem.* **2011**, *32*, 2012.

(72) Liu, Z. B.; Li, Y. C.; Wang, J. J.; Bai, Y.; Wu, D.; Li, Z. R. Effects of the Cage Unit Size and Number of Cage Units as Well as Bridge Unit on the Second Order Nonlinear Optical Response in Multicage Electride Molecules. *J. Phys. Chem. A.* **2013**, *117*, 6678.

(73) He, H.-M.; Li, Z.-R.; Li, Y.; Sun, W.-M.; Wang, J.-J.; Liu, J.-Y.; Wu, D. Unusual Manipulative Effects of Spin Multiplicity and Excess Electron Number on the Structure and Nonlinear Optical Response in New Linear and Cyclic Electride Molecules with Multiexcess Electrons. *J. Phys. Chem. C.* **2014**, *118*, 23937.

(74) Zhong, R.-L.; Xu, H.-L.; Sun, S.-L.; Qiu, Y.-Q.; Su, Z.-M. The Excess Electron in a Boron Nitride Nanotube: Pyramidal NBO Charge Distribution and Remarkable First Hyperpolarizability. *Chem. Eur. J.* **2012**, *18*, 11350.

(75) Zhong, R.-L.; Xu, H.-L.; Su, Z.-M. The Interaction between Boron-Carbon-Nitride Heteronanotubes and Lithium Atoms: Role of Composition Proportion. *Chem. Phys. Lett.* **2016**, *658*, 230.

(76) Hou, N.; Wu, Y.-Y.; Wei, Q.; Liu, X.-L.; Ma, X.-J.; Zhang, M.; Zhang, J.-J.; Wang, B.-Q.; Wu, H.-S. A Comparative Theoretical Study on the Electrical and Nonlinear Optical Properties of Li Atom Adsorbed on AlN and BN Single-Walled Nanotubes. *J. Mol. Model.* **2017**, *23*, 286.

(77) Wang, W. Y.; Ma, N. N.; Wang, C. H.; Zhang, M. Y.; Sun, S. L.; Qiu, Y. Q. Enhancement of Second-Order Nonlinear Optical Response in Boron Nitride Nanocone: Li-Doped Effect. *J. Mol. Graph. Model.* **2014**, *48*, 28.

(78) Song, Y. D.; Wu, L. M.; Chen, Q. L.; Liu, F. K.; Tang, X. W. How Lithium Atoms Affect the First Hyperpolarizability of BN Edge-Doped Graphene. *J. Mol. Model.* **2016**, *22*, 27.

(79) Hou, N.; Wu, Y.; Wu, H. The Influence of Alkali Metals Interaction with Al/P-Substituted BN Nanosheets on Their Electronic and Nonlinear Optical Properties: A DFT Theoretical Study. *Chemistry Select.* **2019**, *4*, 1441.
(80) Yu, G.; Huang, X.-R.; Chen, W.; Sun, C.-C. Alkali Metal Atom-Aromatic Ring: A Novel Interaction Mode Realizes Large First Hyperpolarizabilities of M@AR (M = Li, Na, and K, AR = Pyrrole, Indole, Thiophene, and Benzene). *J. Comput. Chem.* **2011**, *32*, 2005.
(81) Wajid, S.; Kosar, N.; Ullah, F.; Gilani, M. A.; Ayub, K.; Muhammad, S.; Mahmood, T. Demonstrating the Potential of Alkali Metal-Doped Cyclic $C_6O_6Li_6$ Organometallics as Electrides and High-Performance NLO Materials. *ACS Omega.* **2021**, *6*, 29852.
(82) Niu, M.; Yu, G.; Yang, G.; Chen, W.; Zhao, X.; Huang, X. Doping the Alkali Atom: An Effective Strategy to Improve the Electronic and Nonlinear Optical Properties of the Inorganic $Al_{12}N_{12}$ Nanocage. *Inorganic Chemistry* **2014**, *53*, 349.
(83) Hou, N.; Wu, Y.-Y.; Liu, J.-Y. Theoretical Studies on Structures and Nonlinear Optical Properties of Alkali Doped Electrides $B_{12}N_{12}$-M (M=Li, Na, K). *Int. J. Quantum Chem.* **2016**, *116*, 1296.
(84) Shakerzadeh, E.; Biglari, Z.; Tahmasebi, E. M@B_{40} (M=Li, Na, K) Serving as a Potential Promising Novel NLO Nanomaterial. *Chem. Phys. Lett.* **2016**, *654*, 76.
(85) Solimannejad, M.; Rahimi, R.; Kamalinahad, S. Nonlinear Optical (NLO) Response of $Si_{12}C_{12}$ Nanocage Decorated with Alkali Metals (M=Li, Na and K): A Theoretical Study. *J. Inorg. Organomet. P.* **2017**, *27*, 1234.
(86) Ahsan, A.; Khan, S.; Gilani, M. A.; Ayub, K. Endohedral Metallofullerene Electrides of $Ca_{12}O_{12}$ with Remarkable Nonlinear Optical Response. *RSC Adv.* **2021**, *11*, 1569.
(87) Baloach, R.; Ayub, K.; Mahmood, T.; Asif, A.; Tabassum, S.; Gilani, M. A. New Strategy of Bi-Alkali Metal Doping to Design Boron Phosphide Nanocages of High Nonlinear Optical Response with Better Thermodynamic Stability. *J. Inorg. Organomet. P.* **2021**, *31*, 3062.
(88) Sohail, M.; Khaliq, F.; Mahmood, T.; Ayub, K.; Tabassum, S.; Gilani, M. A. Influence of Bi-Alkali Metals Doping over $Al_{12}N_{12}$ Nanocage on Stability and Optoelectronic Properties: A DFT Investigation. *Radiat. Phys. Chem.* **2021**, *184*, 109457.
(89) Maria; Iqbal, J.; Ayub, K. Enhanced Electronic and Non-Linear Optical Properties of Alkali Metal (Li, Na, K) Doped Boron Nitride Nano-Cages. *J. Alloy. Compd.* **2016**, *687*, 976.
(90) Maria; Iqbal, J.; Ayub, K. Theoretical Study of the Non Linear Optical Properties of Alkali Metal (Li, Na, K) Doped Aluminum Nitride Nanocages. *RSC Adv.* **2016**, *6*, 94228.
(91) Keddie, N. S.; Slawin, A. M.; Lebl, T.; Philp, D.; O'Hagan, D. All-Cis 1,2,3,4,5,6-Hexafluorocyclohexane Is a Facially Polarized Cyclohexane. *Nat. Chem.* **2015**, *7*, 483.
(92) Sun, W.-M.; Ni, B.-L.; Wu, D.; Lan, J.-M.; Li, C.-Y.; Li, Y.; Li, Z.-R. Designing Alkalides with Considerable Nonlinear Optical Responses and High Stability Based on the Facially Polarized Janus All-Cis-1,2,3,4,5,6-Hexafluorocyclohexane. *Organometallics* **2017**, *36*, 3352.
(93) Hou, N.; Wu, Y.-Y.; Wang, B.-Q.; Wu, H.-S. Investigation on the Electronic Structures and Nonlinear Optical Properties of Alkali Metal Atom Doped All-Cis 1,2,3,4,5,6-Hexafluorocyclohexane. *Int. J. Quantum Chem.* **2018**, *118*, e25619.
(94) Zheng, R.; Zhang, B.; Wang, C.; Hou, J. DFT Studies of Electronic and Nonlinear Optical Properties of a Novel Class of Excess Electron Compounds Based on Multi-Alkali Metal Atoms-Doped Janus Face $C_{13}H_{10}F_{12}$. *New J. Chem.* **2022**, *46*, 15334.
(95) Yuan, T.-M.; Liu, S.-L.; Liu, Z.-B.; Wang, X.; Li, W.-Z.; Cheng, J.-B.; Li, Q.-Z. Nonlinear Optical Properties of Aluminum Nitride Nanotubes Doped by Excess Electron: A First Principle Study. *J. Mol. Model.* **2018**, *24*, 205.
(96) Teng, Y.; Sheng, Q.; Weng, H.; Zhou, Z.; Huang, X.; Li, Z.; Zhang, T. Theoretical Study of a Novel Organic Electride with Large Nonlinear Optical Responses. *Int. J. Quantum Chem.* **2020**, *120*, e26235.
(97) Zhou, Z.-J.; Li, H.; Huang, X.-R.; Wu, Z.-J.; Ma, F.; Li, Z.-R. The Structure and Large Nonlinear Optical Properties of a Novel Octupolar Electride Li@3^6Adz. *Comput. Theor. Chem.* **2013**, *1023*, 99.
(98) Sun, W.-M.; Li, X.-H.; Li, Y.; Ni, B.-L.; Chen, J.-H.; Li, C.-Y.; Wu, D.; Li, Z.-R. Theoretical Study of the Substituent Effect on the Nonlinear Optical Properties of a Room-Temperature-Stable Organic Electride. *Chem. Phys. Chem.* **2016**, *17*, 3907.
(99) Chen, W.; Yu, G.; Jin, P.; Li, Z.-R.; Huang, X.-R. The Heavier, the Better—Increased First Hyperpolarizabilities in M@Calix[4]Pyrrole (M = Na and K). *J. Comput. Theor. Nanos.* **2011**, *8*, 2482.
(100) Weng, H.; Teng, Y.; Sheng, Q.; Zhou, Z.; Huang, X.; Li, Z.; Zhang, T. Theoretical Study of Substituent Effects on Electride Characteristics and the Nonlinear Optical Properties of Li@Calix[4]Pyrrole. *RSC Adv.* **2019**, *9*, 37919.

(101) Ma, F.; Miao, T.; Zhou, Z.; Sun, D. Design of Lewis Acid-Base Complex: Enhancing the Stability and First Hyperpolarizability of Large Excess Electron Compound. *J. Mol. Model.* **2013**, *19*, 4805.

(102) Ma, N.; Gong, J.; Li, S.; Zhang, J.; Qiu, Y.; Zhang, G. Second-Order NLO Responses of Two-Cavity Inorganic Electrides $Li_n@B_{20}H_{26}$ (n = 1, 2): Evolutions with Increasing Excess Electron Number and Various B-B Connection Sites of $B_{20}H_{26}$. *Phys. Chem. Chem. Phys.* **2017**, *19*, 2557.

(103) Wang, Y.-F.; Qin, T.; Huang, J.; Yang, K.; Li, Z.-R. Hopping of Li@AR (AR = Benzene and Naphthalene) between Electride and Lithium Salt Configurations Brings Considerably Different First Hyperpolarizabilities: Candidate for High-Performance NLO Molecular Switches. *J. Phys. Chem. C.* **2019**, *123*, 24248.

(104) Wang, Y.-F.; Qin, T.; Wang, J.-J.; Liu, X.-X.; Wang, Z.-J.; Huang, J.; Li, J.; Li, Z.-R. Switching from Electride-like Molecule to Molecular Electride $K\text{-}F_6C_6H_6$ Driven by Oriented External Electric Field. *Phys. Chem. Chem. Phys.* **2020**.

(105) Yu, G. T.; Chen, W.; Gu, F. L.; Aoki, Y. Theoretical Study on Nonlinear Optical Properties of the Li^+[Calix[4]Pyrrole]Li^- Dimer, Trimer and Its Polymer with Diffuse Excess Electrons. *J. Comput. Chem.* **2009**, 31, 863.

(106) Sun, W.-M.; Wu, D.; Li, Y.; Li, Z.-R. Substituent Effects on the Structural Features and Nonlinear Optical Properties of the Organic Alkalide Li^+(Calix[4]Pyrrole)Li^-. *Chem. Phys. Chem.* **2013**, *14*, 408.

(107) Jing, Y. Q.; Li, Z. R.; Wu, D.; Li, Y.; Wang, B. Q.; Gu, F. L. What Is the Role of the Complexant in the Large First Hyperpolarizability of Sodide Systems $Li(NH_3)_nNa$ (n = 1-4)? *J. Phys. Chem. B.* **2006**, *110*, 11725.

(108) Jing, Y. Q.; Li, Z. R.; Wu, D.; Li, Y.; Wang, B. Q.; Gu, F. L.; Aoki, Y. Effect of the Complexant Shape on the Large First Hyperpolarizability of Alkalides $Li^+(NH_3)_4M^-$. *Chem. Phys. Chem.* **2006**, *7*, 1759.

(109) Wang, Y.-F.; Wang, J.-J.; Li, J.; Liu, X.-X.; Wang, Z.-J.; Huang, J.; Li, Z.-R. From an Electride-Like Super Alkali Earth Atom to a Superalkalide or Superalkali Electride: $M(HF)_3M$ (M = Na or Li) as Field-Induced Excellent Inorganic NLO Molecular Switches. *J. Mater. Chem. C.* **2021**.

(110) Wang, Z.-J.; Wang, Y.-F.; Zeng, J.-M.; Zhang, W.; Liu, X.-X.; Li, J.; Li, Z.-R. Theoretical Insights into Single-Pole Quadruple-Throw (SP4T) Inorganic Nonlinear Optics Molecular Switch of $Na(HCN)_3Na$: From Superalkali to Superalkalides. *J. Mater. Chem. C.* **2022**, *10*, 16789.

(111) Chen, W.; Li, Z. R.; Wu, D.; Li, Y.; Li, R. Y.; Sun, C. C. Inverse Sodium Hydride: Density Functional Theory Study of the Large Nonlinear Optical Properties. *J. Phys. Chem. A.* **2005**, *109*, 2920.

(112) Wang, F. F.; Li, Z. R.; Wu, D.; Wang, B. Q.; Li, Y.; Li, Z. J.; Chen, W.; Yu, G. T.; Gu, F. L.; Aoki, Y. Structures and Considerable Static First Hyperpolarizabilities: New Organic Alkalides ($M^+@n^6$adz)M'^- (M, M' = Li, Na, K; n = 2, 3) with Cation Inside and Anion Outside of the Cage Complexants. *J. Phys. Chem. B.* **2008**, *112*, 1090.

(113) Fan, L.-T.; Li, Y.; Wu, D.; Li, Z.-R.; Sun, C.-C. Structures and Nonlinear Optical Properties of the Alkalides M^+aza222M'^- (M, M' = Li, Na, K). *Acta Phys. Chim. Sin.* **2012**, *28*, 555.

(114) Liang, W. M.; Zhao, Z. X.; Wu, D.; Sun, W. M.; Li, Y.; Li, Z. R. Theoretical Study on Alkali-Metal Doped N_3H_3 Complexes: An In-Depth Understanding of the Origin of Electride and Alkalide and Their Large Nonlinear Optical Properties. *J. Mol. Model.* **2015**, *21*, 311.

(115) Riedel, R.; Seel, A. G.; Malko, D.; Miller, D. P.; Sperling, B. T.; Choi, H.; Headen, T. F.; Zurek, E.; Porch, A.; Kucernak, A.; Pyper, N. C.; Edwards, P. P.; Barrett, A. G. M. Superalkali-Alkalide Interactions and Ion Pairing in Low-Polarity Solvents. *J. Am. Chem. Soc.* **2021**, *143*, 3934.

(116) Wang, Y.-F.; Huang, J.; Wang, Z.-J.; Liu, X.-X.; Li, J.; Li, Z.-R. Superalkali-Alkalide Ion Pairs $^{\delta+}$(M-HMHC)-$M'^{\delta-}$ (M, M' = Li, Na and K) Serving as High-Performance NLO Molecular Materials. *J. Mol. Liq.* **2022**, *349*, 118101.

(117) Fan, L. T.; Li, Y.; Wu, D.; Li, Z. R.; Sun, C. C. Structural Characteristics and Large Non-Linear Optical Responses of New Alkaline Earth-Based Alkalides. *Aust. J. Chem.* **2012**, *65*, 138.

(118) Wang, Y.-F.; Huang, J.; Zhou, G. Coordination Number Effect of Nitrogen Atom on the Structures and NLO Responses: Alkaline Earth-Based Alkalides. *Struct. Chem.* **2013**, *24*, 1545.

(119) Wang, Y. F.; Huang, J.; Jia, L.; Zhou, G. Theoretical Investigation of the Structures, Stabilities, and NLO Responses of Calcium-Doped Pyridazine: Alkaline-Earth-Based Alkaline Salt Electrides. *J. Mol. Graph. Model.* **2014**, *47*, 77.

(120) Wang, Y.-F.; Li, B.; Huang, J.; Li, J.; Li, Z.-R. Effect of Alkaline Earth Metal Atom on the Large Static First Hyperpolarizabilities of Alkaline Earth-Based Alkalides $Be(NH_3)_nM$ (M = Be and Ca) in Comparison with Alkalides $Li(NH_3)_nNa$ (n = 1–3). *Comput. Theor. Chem.* **2015**, *1051*, 10.

(121) Sun, W.-M.; Wu, D.; Li, Y.; Liu, J.-Y.; He, H.-M.; Li, Z.-R. A Theoretical Study on Novel Alkaline Earth-Based Excess Electron Compounds: Unique Alkalides with Considerable Nonlinear Optical Responses. *Phys. Chem. Chem. Phys.* **2015**, *17*, 4524.

(122) Li, B.; Peng, D.; Gu, F. L.; Zhu, C. A Nonlinear Optical Switch Induced by an External Electric Field: Inorganic Alkaline–Earth Alkalide. *RSC Adv.* **2019**, *9*, 16718.
(123) Li, X.-H.; Zhang, L.; Zhang, X.-L.; Ni, B.-L.; Li, C.-Y.; Sun, W.-M. Designing a New Class of Excess Electron Compounds with Unique Electronic Structures and Extremely Large Non-Linear Optical Responses. *New J. Chem.* **2020**, *44*, 6411.
(124) Ahsin, A.; Ali, A.; Ayub, K. Alkaline Earth Metals Serving as Source of Excess Electron for Alkaline Earth Metals to Impart Large Second and Third Order Nonlinear Optical Response; a DFT Study. *J. Mol. Graph. Model.* **2020**, *101*, 107759.
(125) Wang, Y.-F.; Li, J.; Huang, J.; Qin, T.; Liu, Y.-M.; Zhong, F.; Zhang, W.; Li, Z.-R. Long-Range Charge Transfer Driven by External Electric Field in Alkalides M-LCaL-M (M = Li or Na, L = All-Cis 1,2,3,4,5,6-Exafluorocyclohexane): Facially Polarized Janustype Second Order Nonlinear Molecular Optical Switches. *J. Phys. Chem. C.* **2019**, *123*, 23610.
(126) He, H.-M.; Luis, J. M.; Chen, W.-H.; Yu, D.; Li, Y.; Wu, D.; Sun, W.-M.; Li, Z.-R. Nonlinear Optical Response of Endohedral All-Metal Electride Cages $2e^{-}Mg^{2+}(M@E_{12})^{2-}Ca^{2+}$ (M = Ni, Pd, and Pt; E = Ge, Sn, and Pb). *J. Mater. Chem. C.* **2019**, *7*, 645.
(127) He, H.-M.; Yang, H.; Li, Y.; Li, Z.-R. Theoretical Study of Alkaline-Earth Metal (Be, Mg, and Ca)-Substituted Aluminum Nitride Nanocages with High Stability and Large Nonlinear Optical Responses. *Front. Chem.* **2022**, *10*, 918704.
(128) Maqsood, N.; Asif, A.; Ayub, K.; Iqbal, J.; Elnaggar, A. Y.; Mersal, G. A. M.; Ibrahim, M. M.; El-Bahy, S. M. DFT Study of Alkali and Alkaline Earth Metal-Doped Benzocryptand with Remarkable NLO Properties. *RSC Adv.* **2022**, *12*, 16029.
(129) Hou, J.; Linsheng; Zhu; Jiang, D.; Qin, J.; Duan, Q. Alkaline-Earthide: A New Class of Excess Electron Compounds Li-$C_6H_6F_6$-M (M = Be, Mg and Ca) with Extremely Large Nonlinear Optical Responses. *Chem. Phys. Lett.* **2018**, *711*, 55.
(130) Sajjad, S.; Ali, A.; Mahmood, T.; Ayub, K. Janus Alkaline Earthides with Excellent NLO Response from Sodium and Potassium as Source of Excess Electrons; a First Principles Study. *J. Mol. Graph. Model.* **2020**, *100*, 107668.
(131) Zhang, B.; Zheng, R.; Wang, C.; Hou, J. The Alkaline-Earthides Based Parallel-Stacked Dimer and Trimer of Janus Face $C_6H_6F_6$ Showing Extremely Large Nonlinear Optical Responses. *Polyhedron.* **2022**, *227*, 116119.
(132) Ahsan, A.; Ayub, K. Adamanzane Based Alkaline Earthides with Excellent Nonlinear Optical Response and Ultraviolet Transparency. *Opt. Laser Technol.* **2020**, *129*, 106298.
(133) Ahsan, A.; Sarfaraz, S.; Fayyaz, F.; Asghar, M.; Ayub, K. Theoretical Study of 36Adz Based Alkaline Earthides $M^{+}(3^{6}Adz)M^{-}$ (M^{+} = Li & Na; M^{-} = Be, Mg & Ca) with Remarkable Nonlinear Optical Response. *Mat. Sci. Semicon. Proc.* **2023**, *153*, 107119.
(134) Ahsan, A.; Ayub, K. Extremely Large Nonlinear Optical Response and Excellent Electronic Stability of True Alkaline Earthides Based on Hexaammine Complexant. *J. Mol. Liq.* **2020**, *297*, 111899.
(135) Ahsan, A.; Sarfaraz, S.; Fayyaz, F.; Asghar, M.; Ayub, K. Enhanced Non-Linear Optical Response of Calix[4]Pyrrole Complexant Based Earthides in the Presence of Oriented External Electric Field. *J. Mol. Liq.* **2022**, *350*, 118504.
(136) Ahsan, A.; Sarfaraz, S.; Gilani, M. A.; Mahmood, T.; Ahmad, Z.; Ayub, K. Alkaline Earthides Based on 15-Crown-5 Ether with Remarkable NLO Response. *Eur. Phys. J. Plus* **2022**, *137.*
(137) Sun, W.-M.; Li, X.-H.; Wu, J.; Lan, J.-M.; Li, C.-Y.; Wu, D.; Li, Y.; Li, Z.-R. Can Coinage Metal Atoms Be Capable of Serving as an Excess Electron Source of Alkalides with Considerable Nonlinear Optical Responses? *Inorg. Chem.* **2017**, *56*, 4594.
(138) Li, X.-H.; Zhang, X.-L.; Chen, Q.-H.; Zhang, L.; Chen, J.-H.; Wu, D.; Sun, W.-M.; Li, Z.-R. Coinage Metalides: A New Class of Excess Electron Compounds with High Stability and Large Nonlinear Optical Responses. *Phys. Chem. Chem. Phys.* **2020**, *22*, 8476.
(139) Sun, W.-M.; Cheng, X.; Wang, W.-L.; Li, X.-H. Designing Magnetic Superalkalis with Extremely Large Nonlinear Optical Responses. *Organomet.* **2022**, *41*, 2406.
(140) Liang, Y. Y.; Li, B.; Xu, X.; Long Gu, F.; Zhu, C. A Density Functional Theory Study on Nonlinear Optical Properties of Double Cage Excess Electron Compounds: Theoretically Design M[Cu(Ag)@$(NH_3)_n$](M = Be, Mg and Ca; n = 1-3). *J. Comput. Chem.* **2019**, *40*, 971.
(141) Maqbool, H.; Rafique, A.; Bhatti, I. A.; Shawky, A. M.; Ayub, K.; Iqbal, J. Novel Endohedrally and Exohedrally Metals (Li, Na, and K, Ag) Doped (15-Crown-5) with Remarkable Electronic, Static and Dynamic NLO Response. *Optik.* **2022**, *271*, 170169.
(142) Rad, A. S.; Ayub, K. Change in the Electronic and Nonlinear Optical Properties of Fullerene through Its Incorporation with Sc-, Fe-, Cu-, and Zn Transition Metals. *Appl. Phys. A.* **2019**, *125.*

(143) Ullah, F.; Irshad, S.; Khan, S.; Hashmi, M. A.; Ludwig, R.; Mahmood, T.; Ayub, K. Nonlinear Optical Response of First-Row Transition Metal Doped $Al_{12}P_{12}$ Nanoclusters; a First-Principles Study. *J. Phys. Chem. Solids.* **2021**, *151*, 109914.
(144) Zhang, B.; Wen, J.; Zhang, Y.; Xiong, Y.; Huang, X.; Hou, J.; Wang, X.; Guan, J.; Zhi, Q. Design a Novel Type of Excess Electron Compounds with Large Nonlinear Optical Responses Using Group 12 Elements (Zn, Cd and Hg). *J. Mol. Graph. Model.* **2021**, *109*, 108003.
(145) Ahsan, F.; Ayub, K. Transition Metalides Based on Facially Polarized All-Cis-1,2,3,4,5,6-Hexafluorocyclohexane—A New Class of High Performance Second Order Nonlinear Optical Materials. *Phys. Chem. Chem. Phys.* **2023**.
(146) Li, X.-H.; Cheng, X.; Wang, W.-L.; Yu, D.; Ni, B.-L.; Sun, W.-M. Alkali-Metal-Free Coinage Metalides: Specific Pairing and Location of Doping Atoms Bring Forth High Stability and Considerable Nonlinear Optical Response. *Organomet.* **2022**, *41*, 3697.
(147) Gutsev, G. L.; Boldyrev, A. I. DVM-Xα Calculations on the Electronic Structure of "Superalkali" Cations. *Chem. Phys. Lett.* **1982**, *92*, 262.
(148) Sun, W. M.; Wu, D. Recent Progress on the Design, Characterization, and Application of Superalkalis. *Chem. Eur. J.* **2019**, *25*, 9568.
(149) Li, Y.; Wu, D.; Li, Z.-R. Compounds of Superatom Clusters: Preferred Structures and Significant Nonlinear Optical Properties of the BLi_6-X (X= F, LiF_2, BeF_3, BF_4) Motifs. *Inorg. Chem.* **2008**, *47*, 9773.
(150) Sun, W.-M.; Li, C.-Y.; Kang, J.; Wu, D.; Li, Y.; Ni, B.-L.; Li, X.-H.; Li, Z.-R. Superatom Compounds under Oriented External Electric Fields: Simultaneously Enhanced Bond Energies and Nonlinear Optical Responses. *J. Phys. Chem. C.* **2018**, *122*, 7867.
(151) Yang, H.; Li, Y.; Wu, D.; Li, Z. R. Structural Properties and Nonlinear Optical Responses of Superatom Compounds BF_4-M (M = Li, FLi_2, OLi_3, NLi_4). *Int. J. Quantum Chem.* **2012**, *112*, 770.
(152) Omidvar, A. Design of a Novel Series of Donor–Acceptor Frameworks via Superalkali–Superhalogen Assemblage to Improve the Nonlinear Optical Responses. *Inorg. Chem.* **2018**, *57*, 9335.
(153) Sun, W. M.; Fan, L. T.; Li, Y.; Liu, J. Y.; Wu, D.; Li, Z. R. On the Potential Application of Superalkali Clusters in Designing Novel Alkalides with Large Nonlinear Optical Properties. *Inorg. Chem.* **2014**, *53*, 6170.
(154) Mai, J.; Gong, S.; Li, N.; Luo, Q.; Li, Z. A Novel Class of Compounds-Superalkalides: $M^+(en)_3M_3'O^-$ (M, M' = Li, Na, and K; en = Ethylenediamine)-with Excellent Nonlinear Optical Properties and High Stabilities. *Phys. Chem. Chem. Phys.* **2015**, *17*, 28754.
(155) Zhong, R.-L.; Xu, H.-L.; Muhammad, S.; Zhang, J.; Su, Z.-M. The Stability and Nonlinear Optical Properties: Encapsulation of an Excess Electron Compound LiCN...Li Within Boron Nitride Nanotubes. *J. Mater. Chem.* **2012**, *22*, 2196.
(156) Zhao, X.; Yu, G.; Huang, X.; Chen, W.; Niu, M. (Super)alkali Atoms Interacting with the Sigma Electron Cloud: A Novel Interaction Mode Triggers Large Nonlinear Optical Response of M@P_4 and M@C_3H_6 (M=Li, Na, K and Li_3O). *J. Mol. Model.* **2013**.
(157) Tu, C.; Yu, G.; Yang, G.; Zhao, X.; Chen, W.; Li, S.; Huang, X. Constructing (Super)alkali-Boron-Heterofullerene Dyads: An Effective Approach to Achieve Large First Hyperpolarizabilities and High Stabilities in M_3O-BC_{59} (M = Li, Na and K) and K@n-BC_{59} (n = 5 and 6). *Phys. Chem. Chem. Phys.* **2014**, *16*, 1597.
(158) Sun, W.-M.; Li, X.-H.; Wu, D.; Li, Y.; He, H.-M.; Li, Z.-R.; Chen, J.; Li, C. A Theoretical Study on Superalkali-Doped Nanocages: Unique Inorganic Electrides with High Stability, Deep-Ultraviolet Transparency, and Considerable Nonlinear Optical Response. *Dalton Trans.* **2016**, *45*, 7500.
(159) Li, Z.; Yu, G.; Zhang, X.; Huang, X.; Chen, W. Bonding the Superalkali M_3O (M = Li and K): An Effective Strategy to Improve the Electronic and Nonlinear Optical Properties of the Inorganic B_{40} Nanocage. *Physica E.* **2017**, *94*, 204.
(160) Lin, Z.; Lu, T.; Ding, X.-L. A Theoretical Investigation on Doping Superalkali for Triggering Considerable Nonlinear Optical Properties of $Si_{12}C_{12}$ Nanostructure. *J. Comput. Chem.* **2017**, *38*, 1574.
(161) Ullah, F.; Kosar, N.; Arshad, M. N.; Gilani, M. A.; Ayub, K.; Mahmood, T. Design of Novel Superalkali Doped Silicon Carbide Nanocages with Giant Nonlinear Optical Response. *Opt. Laser Technol.* **2020**, *122*, 105855.
(162) Shakerzadeh, E.; Tahmasebi, E.; Solimannejad, M.; Chigo Anota, E. Tuning the Electronic-Optical Properties of Porphyrin-Like Porous $C_{24}N_{24}$ Fullerene with $(Li_3O)_{n\,=\,(1\text{-}5)}$ Decoration. A Computational Study. *Appl. Organomet. Chem.* **2019**, *33*, e4654.
(163) Kosar, N.; Mahmood, T.; Ayub, K.; Tabassum, S.; Arshad, M.; Gilani, M. A. Doping Superalkali on $Zn_{12}O_{12}$ Nanocage Constitutes a Superior Approach to Fabricate Stable and High-Performance Nonlinear Optical Materials. *Opt. Laser Technol.* **2019**, *120*, 105753.

(164) Ullah, F.; Kosar, N.; Ayub, K.; Gilani, M. A.; Mahmood, T. Theoretical Study on a Boron Phosphide Nanocage Doped with Superalkalis: Novel Electrides Having Significant Nonlinear Optical Response. *New J. Chem.* **2019**, *43*, 5727.
(165) Ullah, F.; Kosar, N.; Ayub, K.; Mahmood, T. Superalkalis as a Source of Diffuse Excess Electrons in Newly Designed Inorganic Electrides with Remarkable Nonlinear Response and Deep Ultraviolet Transparency: A DFT Study. *Appl. Surf. Sci.* **2019**, *483*, 1118.
(166) Noormohammadbeigi, M.; Shamlouei, H. R. The Effect of Superalkali M_3O (M=Li, Na and K) on Structure, Electrical and Nonlinear Optical Properties of C_{20} Fullerene Nanocluster. *J. Inorg. Organomet. P.* **2017**, *28*, 110.
(167) Omidi, M.; Sabzehzari, M.; Shamlouei, H. R. Influence of Superalkali Oxides on Structural, Electrical and Optical Properties of C_{24} Fullerene Nanocluster: A Theoretical Study. *Chin. J. Phys.* **2020**, *65*, 567.
(168) Li, Y.; Ruan, M.; Chen, H. Decorating Electron Redundant $Si_nAl_{12-n}N_{12}$ (n=1, 2) Nanocages with Superalkalis M_3O (M=Li, Na, K): Excess Electron D-A Frameworks and Nonlinear Optical Properties. *Mol. Phys.* **2021**, *119*, e1909161.
(169) Raza Ayub, A.; Aqil Shehzad, R.; Alarfaji, S. S.; Iqbal, J. Super Alkali (OLi_3) Doped Boron Nitride with Enhanced Nonlinear Optical Behavior. *J. Nonlinear Opt. Phys.* **2021**, *29*, 2050004.
(170) Bano, R.; Arshad, M.; Mahmood, T.; Ayub, K.; Sharif, A.; Tabassum, S.; Gilani, M. A. Superalkali (Li_2F, Li_3F) doped $Al_{12}N_{12}$ Electrides with Enhanced Static, Dynamic Nonlinear Optical Responses and Refractive Indices. *Mat. Sci. Semicon. Proc.* **2022**, *143*, 106518.
(171) Bano, R.; Ayub, K.; Mahmood, T.; Arshad, M.; Sharif, A.; Tabassum, S.; Gilani, M. Mixed Superalkalis Are Better Choice Than Pure Superalkalis for $B_{12}N_{12}$ Nanocage to Design High Performance Nonlinear Optical Materials. *Dalton Trans.* **2022**, *51*, 8437.
(172) Ullah, F.; Ayub, K.; Mahmood, T. Remarkable Second and Third Order Nonlinear Optical Properties of Organometallic C_6Li_6–M_3O Electrides. *New J. Chem.* **2020**, *44*, 9822.
(173) Khaliq, F.; Afzaal, A.; Tabassum, S.; Mahmood, T.; Ayub, K.; Khan, A. L.; Yasin, M.; Gilani, M. A. Surface Functionalization of Si_6Li_6 Cluster with Superalkalis to Achieve High Nonlinear Optical Response: A DFT Study. *Colloid. Surface. A.* **2022**, *653*, 129985.
(174) Kosar, N.; Zari, L.; Ayub, K.; Gilani, M. A.; Mahmood, T. Static, Dynamic Nonlinear Optical (NLO) Response and Electride Characteristics of Superalkalis Doped Star Like $C_6S_6Li_6$. *Surf. Interfaces.* **2022**, *31*, 102044.
(175) Sajid, H.; Mahmood, T. A DFT Study on M_3O (M = Li & Na) Doped Triphenylene and Its Amino-, Hydroxy- and Thiol-Functionalized Quantum Dots for Triggering Remarkable Nonlinear Optical Properties and Ultra-Deep Transparency in Ultraviolet Region. *Physica E.* **2021**, *134*, 114905.
(176) Chen, S.; Xu, H. L.; Sun, S. L.; Zhao, L.; Su, Z. M. Superalkali Atoms Bonding to the Phenalenyl Radical: Structures, Intermolecular Interaction and Nonlinear Optical Properties. *J. Mol. Model.* **2015**, *21*, 209.
(177) Yi, X. G.; Wang, Y. F.; Zhang, H. R.; Cai, J. H.; Liu, X. X.; Li, J.; Wang, Z. J.; Bai, F. Q.; Li, Z. R. Can a Molecular Switch Exist in Both Superalkali Electride and Superalkalide Forms? *Phys. Chem. Chem. Phys.* **2022**, *24*, 5690.
(178) Roy, R. S.; Ghosh, S.; Hatua, K.; Nandi, P. K. Superalkali-Doped Borazine and Lithiated Borazine Complexes: Diffuse Excess Electron and Large First-Hyperpolarizability. *J. Mol. Model.* **2021**, *27*, 74.
(179) Li, X.; Zhang, Y.; Lu, J. Remarkably Enhanced First Hyperpolarizability and Nonlinear Refractive Index of Novel Graphdiyne-based Materials for Promising Optoelectronic Applications: A First-Principles Study. *Appl. Surf. Sci.* **2020**, *512*, 145544.
(180) Hou, N.; Du, F. Y.; Feng, R.; Wu, H. S.; Li, Z. R. Effects of the Atomic Number of Alkali Atom and Pore Size of Graphyne on the Second-Order Nonlinear Optical Response of Superalkali Salts of Graphynes $OM_3^+@GYs^-$ (M = Li, Na, and K). *Int. J. Quantum Chem.* **2020**, *121*.
(181) Kosar, N.; Shehzadi, K.; Ayub, K.; Mahmood, T. Nonlinear Optical Response of Sodium Based Superalkalis Decorated Graphdiyne Surface: A DFT Study. *Optik.* **2020**, *218*, 165033.
(182) Song, Y.-D.; Wang, Q.-T. The Structures and Nonlinear Optical Responses of Superalkali-Doped Graphyne and Boron-Doped Graphyne: A Density Functional Study. *Optik.* **2020**, *220*, 164947.
(183) Kosar, N.; Shehzadi, K.; Ayub, K.; Mahmood, T. Theoretical Study on Novel Superalkali Doped Graphdiyne Complexes: Unique Approach for the Enhancement of Electronic and Nonlinear Optical Response. *J. Mol. Graph. Model.* **2020**, *97*, 107573.
(184) Hou, N.; Wei-Ming, S.; Du, F.-Y.; Wu, H.-S. Effect of (Super)alkali Doping on the Electronic and Second-Order Nonlinear Optical Properties of Graphitic C_3N_4. *Optik.* **2019**, *183*, 455.
(185) Khan, A. U.; Khera, R. A.; Anjum, N.; Shehzad, R. A.; Iqbal, S.; Ayub, K.; Iqbal, J. DFT Study of Superhalogen and Superalkali Doped Graphitic Carbon Nitride and Its Non-Linear Optical Properties. *RSC Adv.* **2021**, *11*, 7779.

(186) Song, Y.-D.; Wang, L.; Wang, Q.-T. Structures and Nonlinear Optical Properties of Alkali Atom/Superalkali Doped Pyridinic Vacancy Graphene. *Optik.* **2018**, *154*, 411.
(187) Hanif, A.; Kiran, R.; Khera, R. A.; Ayoub, A.; Ayub, K.; Iqbal, J. Tuning the Optoelectronic Properties of Superalkali Doped Phosphorene. *J. Mol. Graph. Model.* **2021**, *107*, 107973.
(188) Kiran, R.; Khera, R. A.; Khan, A. U.; Ayoub, A.; Iqbal, N.; Ayub, K.; Iqbal, J. Study of Nonlinear Optical Properties of Superhalogen and Superalkali Doped Phosphorene. *J. Mol. Struct.* **2021**, *1236*, 130348.
(189) Umar, A.; Yaqoob, J.; Khan, M. U.; Hussain, R.; Alhadhrami, A.; Almalki, A. S. A.; Janjua, M. R. S. A. Doping of Superalkali and Superhalogen on Graphene Quantum Dot Surfaces to Enhance Nonlinear Optical Response: An Efficient Strategy for Fabricating Novel Electro-Optical Materials. *J. Phys. Chem. Solids.* **2022**, *169*, 110859.
(190) Sajid, H.; Ullah, F.; Khan, S.; Ayub, K.; Arshad, M.; Mahmood, T. Remarkable Static and Dynamic NLO Response of Alkali and Superalkali Doped Macrocyclic [Hexa-]thiophene Complexes; a DFT Approach. *RSC Adv.* **2021**, *11*, 4118.
(191) Song, Y.-D.; Wang, Q.-T. Investigations of Electronic and Nonlinear Optical Properties of Superalkali Adsorbed Biphenylene Based Sheets by First-Principles Calculations. *Optik.* **2021**, *242*, 166830.
(192) Ishfaq, T.; Ahmad Khera, R.; Zahid, S.; Yaqoob, U.; Aqil Shehzad, R.; Ayub, K.; Iqbal, J. Enhancement in Non-Linear Optical Properties of Carbon Nitride (C_2N) by Doping Superalkali (Li_3O): A DFT Study. *Comput. Theor. Chem.* **2022**, *1211*, 113654.
(193) Shafiq, S.; Shehzad, R. A.; Yaseen, M.; Ayub, K.; Ayub, A. R.; Iqbal, J.; Mahmoud, K. H.; El-Bahy, Z. M. DFT Study of OLi_3 and MgF_3 Doped Boron Nitride with Enhanced Nonlinear Optical Behavior. *J. Mol. Struct.* **2022**, *1251*, 131934.
(194) Bano, R.; Arshad, M.; Mahmood, T.; Ayub, K.; Sharif, A.; Perveen, S.; Tabassum, S.; Yang, J.; Gilani, M. A. Face Specific Doping of Janus All-Cis-1,2,3,4,5,6-Hexafluorocyclohexane with Superalkalis and Alkaline Earth Metals Leads to Enhanced Static and Dynamic NLO Responses. *J. Phys. Chem. Solids.* **2022**, *160*, 110361.
(195) Kosar, N.; Zari, L.; Ayub, K.; Gilani, M. A.; Arshad, M.; Rauf, A.; Ans, M.; Mahmood, T. NLO Properties and Electride Characteristics of Suparalkalis Doped All-Cis-1,2,3,4,5,6-Hexafluorocyclohexane Complexes. *Optik.* **2022**, 170139.
(196) Hou, N.; Wu, Y.-Y.; Wu, H.-S.; He, H.-M. The Important Role of Superalkalis on the Static First Hyperpolarizabilities of New Electrides: Theoretical Investigation on Superalkali-Doped Hexamethylenetetramine (HMT). *Synthetic Met.* **2017**, *232*, 39.
(197) Omidvar, A. Nonlinear Optical Response of Teetotum Boron Clusters. *Comput. Theor. Chem.* **2021**, *1198*, 113178.
(198) Sajid, H.; Ayub, K.; Mahmood, T. Exceptionally High NLO Response and Deep Ultraviolet Transparency of Superalkali Doped Macrocyclic Oligofuran Rings. *New J. Chem.* **2020**, *44*, 2609.
(199) Wang, J. J.; Zhou, Z. J.; Bai, Y.; He, H. M.; Wu, D.; Li, Y.; Li, Z. R.; Zhang, H. X. A New Strategy for Simultaneously Enhancing Nonlinear Optical Response and Electron Stability in Novel Cup-Saucer(+)-Cage(-)-Shaped Sandwich Electride Molecules with an Excess Electron Protected Inside the Cage. *Dalton Trans.* **2015**, *44*, 4207.
(200) Sun, W. M.; Wu, D.; Li, Y.; Li, Z. R. Theoretical Study on Superalkali (Li_3) in Ammonia: Novel Alkalides with Considerably Large First Hyperpolarizabilities. *Dalton Trans.* **2014**, *43*, 486.
(201) Sun, W. M.; Li, Y.; Li, X. H.; Wu, D.; He, H. M.; Li, C. Y.; Chen, J. H.; Li, Z. R. Stability and Nonlinear Optical Response of Alkalides That Contain a Completely Encapsulated Superalkali Cluster. *ChemPhysChem.* **2016**, *17*, 2672.
(202) Ahsin, A.; Ayub, K. Superalkali-Based Alkalides Li_3O@[12-Crown-4]M (Where M= Li, Na, and K) with Remarkable Static and Dynamic NLO Properties; a DFT Study. *Mat. Sci. Semicon. Proc.* **2022**, *138*, 106254.
(203) Li, B.; Peng, D.; Gu, F. L.; Zhu, C. Facially Polarized Molecule for Alkalides and Superalkalides with Considerable Nonlinear Optical Response. *Chemistry Selec.* **2018**, *3*, 12782.
(204) Srivastava, A. K.; Misra, N. Nonlinear Optical Behavior of Li_nF(n = 2-5) Superalkali Clusters. *J. Mol. Model.* **2015**, *21*, 305.
(205) Ahsin, A.; Ayub, K. Remarkable Electronic and NLO Properties of Bimetallic Superalkali Clusters: A DFT Study. *J. Nanostruct. Chem.* **2021**, *12*, 529.
(206) Ahsin, A.; Ayub, K. Extremely Large Static and Dynamic Nonlinear Optical Response of Small Superalkali Clusters NM_3M' (M, M' = Li, Na, K). *J. Mol. Graph. Model.* **2021**, *109*, 108031.
(207) Ahsin, A.; Ayub, K. Theoretical Investigation of Superalkali Clusters M_2OCN and M_2NCO (Where M = Li, Na, K) as Excess Electron System with Significant Static and Dynamic Nonlinear Optical Response. *Optik.* **2021**, *227*, 166037.

(208) Ahsin, A.; Ayub, K. Oxacarbon Superalkali $C_3X_3Y_3$ (X = O, S and Y = Li, Na, K) Clusters as Excess Electron Compounds for Remarkable Static and Dynamic NLO Response. *J. Mol. Graph. Model.* **2021**, *106*, 107922.
(209) Ahsin, A.; Shah, A. B.; Ayub, K. Germanium-Based Superatom Clusters as Excess Electron Compounds with Significant Static and Dynamic NLO Response; a DFT Study. *RSC Adv.* **2022**, *12*, 365.
(210) Ahsin, A.; Ayub, K. Theoretical Investigation of Lithium-Based Clusters Lin (Where n=3, 5, 7) with Remarkable Electronic and Frequency-Dependent NLO Properties. *Eur. Phys. J. Plus.* **2022**, *137*, 803.
(211) Ahsin, A.; Ayub, K. Zintl Based Superatom P_7M_2 (M=Li, Na, K & Be, Mg, Ca) Clusters with Excellent Second and Third-Order Nonlinear Optical Response. *Mat. Sci. Semicon. Proc.* **2021**, *134*, 105986.
(212) Ye, Y.-L.; Pan, K.-Y.; Ni, B.-L.; Sun, W.-M. Designing Special Nonmetallic Superalkalis Based on a Cage-Like Adamanzane Complexant. *Front. Chem.* **2022**, *10*, 853160.
(213) Srivastava, A. K.; Misra, N. Competition between Alkalide Characteristics and Nonlinear Optical Properties in OLi_3-M-Li_3O (M=Li, Na, and K) Complexes. *Int. J. Quantum Chem.* **2017**, *117*, 208.
(214) Yu, J.; Xiao, W.; Xin, J.; Jin, R. Theoretical Research on Nonlinear Optical Properties of Sandwich Compounds M_3-NO_3-M'_3 (M, M' = Li, Na, K). *Comput. Theor. Chem.* **2020**, *1185*, 112853.
(215) Karamanis, P.; Pouchan, C. Fullerene–C_{60} in Contact with Alkali Metal Clusters: Prototype Nano-Objects of Enhanced First Hyperpolarizabilities. *J. Phys. Chem. C.* **2012**, *116*, 11808.
(216) Huang, S.; Liao, K.; Peng, B.; Luo, Q. On the Potential of Using the Al Superatom as an Excess Electron Acceptor to Construct Materials with Excellent Nonlinear Optical Properties. *Inorg. Chem.* **2016**, *55*, 4421.
(217) He, H.-M.; Li, Y.; Sun, W.-M.; Wang, J.-J.; Wu, D.; Zhong, R.-L.; Zhou, Z.-J.; Li, Z.-R. All-Metal Electride Molecules CuAg@Ca_7M (M= Be, Mg, and Ca) with Multi-Excess Electrons and All-Metal Polyanions: Molecular Structures and Bonding Modes as Well as Large Infrared Nonlinear Optical Responses. *Dalton Trans.* **2016**, *45*, 2656.
(218) He, H.-M.; Li, Y.; Yang, H.; Yu, D.; Li, S.-Y.; Wu, D.; Hou, J.-H.; Zhong, R.-L.; Zhou, Z.-J.; Gu, F.-L.; Luis, J. M.; Li, Z.-R. Efficient External Electric Field Manipulated Nonlinear Optical Switches of All-Metal Electride Molecules with Infrared Transparency: Nonbonding Electron Transfer Forms an Excess Electron Lone Pair. *J. Phys. Chem. C.* **2017**, *121*, 958.
(219) He, H.-M.; Li, Y.; Yang, H.; Yu, D.; Wu, D.; Zhong, R.-L.; Zhou, Z.-J.; Li, Z.-R. Effects of the Cage Number and Excess Electron Number on the Second Order Nonlinear Optical Response in Molecular All-Metal Electride Multicage Chains. *J. Phys. Chem. C.* **2017**, *121*, 25531.
(220) Wang, Y.-F.; Qin, T.; Tang, J.-M.; Liu, Y.-J.; Xie, M.; Li, J.; Huang, J.; Li, Z.-R. Novel Inorganic Aromatic Mixed-Valent Superalkali Electrides CaN_3Ca: Alkaline-Earth-Based High-Sensitive Multi-State Nonlinear Optical Molecular Switch. *Phys. Chem. Chem. Phys.* **2020**, *22*, 5985.
(221) Yi, X.-G.; Wang, Y.-F.; Qin, T.; Liu, X.-X.; Jiang, S.-L.; Huang, J.; Yang, K.; Li, J.; Li, Z.-R. Electric Field Induced Intra-Molecular Self-Redox: Superalkali Li_3N_3Mg as Candidate for NLO Molecular Switches. *Phys. Chem. Chem. Phys.* **2020**, *22*, 21928.

11 Organic Semiconducting Materials in Electronic Devices

Shamoon Ahmad Siddiqui, Ankit Kargeti, and Tabish Rasheed

11.1 INTRODUCTION

Organic semiconductors are widely utilized now as alternative silicon-based inorganic semiconducting materials in electronic devices. The 2000's Nobel Prize in chemistry was given to Alan Heeger, Alan MacDiarmid, and Hideki Shirakawa for the discovery and development of conductive polymer. Their experiments conducted in 1970–1980 with trans-polyacetylene showed that it was possible to determine the conductivity of covalent organic material by exposing it to vapors of chlorine, bromine, or iodine.[1] Also in the 1960s, Weiss and coworkers already demonstrated electrical conductivity in iodine-doped oxidized polypyrrole, and in 1965 electroluminescence was observed in anthracene crystals by Kallmann et al.,[2] Today, interest in organic electronics is increased due to the demand for alternating substitutes of Si-based electronics or inorganic materials-based electronics that could offer easier synthesis and processability.[3] One of the key differences between inorganic and organic semiconductors lies in their processing techniques. Inorganic semiconductors are processed through an expensive technique that requires highly pure-quality crystalline substrates. Organic semiconductors don't need such processing techniques, and they can be deposited over inexpensive substrates such as glass, plastic, etc.[4] Synthesis techniques for organics are readily available like spin-coating, drop-casting, and direct printing via stamps or inkjets. Their compatibility with lightweight, mechanically flexible plastic substrates and innovative fabrication routes makes them possible candidates for future electronic devices. Moreover, due to different molecular tailoring possibilities via chemical synthesis, organic materials present a wide variety in functionality.[5] The design of the molecular structures can be engineered to enhance properties like band gap modification of the materials. Modification of the chemical end groups also allows the fabrication of large transistor arrays for organic sensors.[6] Different methods are developed to increase the sensitivity and selectivity of organic sensing transistors to different chemical or biological species. There has been a rapid progress in the industrial development of organic electronic devices. Organic devices are suitable for large-scale application areas where their operation can offer high performance, reliability, stability, long lifetime, good control, and reproducibility.[7] As a constituent of plastic electronic components, organic semiconductors have arrived at a good place.

Organic materials can be divided broadly into two categories: small molecules and polymers.[8] *Small molecules* are the simplest type of organic solids, and, despite their small size, they can be relatively massive, with typical masses of several hundred atomic mass units (amu). Regardless of their size, all small molecules are distinct units with identical structures. *Polymers* are the long chains of repeating units based on a backbone of carbon-containing bonds. Polymer masses are very large, and their masses vary from tens to thousands of repeating units up to a million amu. The complexity of semiconducting polymers also varies greatly, from relatively simple polythiophenes to intricate donor–acceptor complexes.[9]

Charge transport within organic materials is a combination of two processes: intramolecular carrier movement and intermolecular charge transfer. In the organic molecular complexes, π-bond conjugation helps in charge carriers to move freely but, due to the weak van der Waal forces, the charge transport is limited, and it lowers charge carrier mobility to typical values of 10^{-5} to 10^{-2} $cm^2 V^{-1} s^{-1}$ in organic photovoltaics.[10]

Charge carriers are strongly localized on individual molecules, and the intermolecular transport occurs through a hopping process as a charge carrier overcomes an energy barrier to move from

DOI: 10.1201/9781003441328-11

one molecule to the next. The charge transport is well described by the Marcus theory[11] in organic semiconductors, and device performance is affected by the stability of holes and electrons and the energy barrier for the charge injection process. The charge transfer rate (k_{ct}) between two adjacent monomers can be described by this formulation. The relevant expression for k_{ct} when there is no barrier can be given as:[12]

$$k_{ct} = \sqrt{\frac{\pi}{\lambda k_B T}} \frac{V^2}{\hbar} e^{\frac{-\lambda}{4k_B T}}$$

where T is the temperature at absolute scale, V is the electronic coupling matrix element between two molecules, k_B is the Boltzmann constant, λ is the reorganization energy associated with the geometry relaxation process required for charge transfer, and $\hbar$ is the reduced Planck's constant. k_{ct} mainly depends on the charge transfer integral (V) and reorganization energy (λ).

The reorganization energy for hole transfer, λ_h, is calculated as the sum of two contributions $\lambda_1 + \lambda_2$, and the reorganization energy for electron transfer, λ_e, is calculated as the sum of two contributions $\lambda_3 + \lambda_4$, which are defined as:[13]

$$\lambda_h = \lambda_1 + \lambda_2 = \left[E^+M^0 - E^+M^+\right] + \left[E^0M^+ - E^0M^0\right]$$

$$\lambda_e = \lambda_3 + \lambda_4 = \left[E^0M^- - E^0M^0\right] + \left[E^-M^0 - E^-M^-\right]$$

where $E^0, E^+, E^-, M^0, M^+, M^-$ represents ground state, charged state, whose representation is done by the energy term, E, and molecule is represented by M.

11.2 APPLICATION OF ORGANIC SEMICONDUCTING MATERIALS IN DESIGNING ELECTRONIC DEVICES AND THEIR PROPERTIES

We have discussed the modeling of organic semiconducting materials for application in organic molecular diodes as single molecule diodes, as organic field effect transistors, and as organic solar cells, specifically dye-sensitized solar cells where the novel molecules have been reviewed and their behavior is observed on applying the external electric field. Further, the characteristics of the organic molecular diode, organic field effect transistor, and dye-sensitized solar cell are discussed next.

11.2.1 Characteristic of the Single Molecular Diode

- A molecular diode contains two terminals and functions like a semiconductor *p-n* junction diode or a rectifier, that conducts current only in one direction as forward-biased condition or reverse-biased condition.
- A diode is the fundamental block of any three terminal semiconductor electronic devices such as a bipolar transistor or a field effect transistor.
- Diode-based logic circuits using AND/OR gates are widely utilized for building logic families by using the rectifying diodes.
- Aviram and Ratner[14] suggested that electron-donating constituents cause conjugated molecular groups to have a large electron density (also called *N*-type), and electron withdrawing constituents make conjugated molecular groups poor in electron density (also called as P-type). According to their paper, an asymmetric molecule, having appropriate donor and acceptor moieties linked with a sigma-bridge (also known as saturated alkyl chain) and connected with suitable electrodes, will conduct current only in one direction—acting as a rectifier.

11.2.2 Characteristic of the Organic Field Effect Transistor

- Field effect transistors are the main logic units in electronic circuits, where they usually function as either a switch or an amplifier. A field effect transistor can be described as a three-terminal device in which the current through the semiconductor connected to two terminals (namely source and drain) is controlled at the third terminal (the gate) by a voltage that creates an electric field through the dielectric on which the semiconductor is deposited.[15]
- The nature and quality of the organic semiconductor are crucial for achieving high OFET performance, which is mainly determined by the charge carrier mobility (*m*) that is a measure of the charge carrier drift velocity per unit of electric field.
- The other important parameter of OFET is described through its on/off ratio, which is the ratio of current in the accumulation mode over the current in the depletion mode, and the threshold voltage (V^T), that is, the gate voltage from which the conduction channel starts to form.
- At room temperature, the charge mobility of semiconducting organic materials is determined by a hopping transport process. Hopping transport can be depicted as an electron or hole transfer reaction in which an electron or hole is transferred from one molecule to the neighboring one. The charges localize on a molecule for a long enough time for the nuclei to relax to their optimum geometry. Two major parameters determine self-exchange rates and thus the charge mobility: (1) the electronic coupling between adjacent molecules, which needs to be maximized, and (2) the reorganization energy (reorg), which needs to be small for efficient charge transport as described in the preceding section.

11.2.3 Characteristic of the Organic Solar Cells (Dye-Sensitized Solar Cells)

- The role of a sensitizer in DSSC is very crucial. It helps in absorbing the visible light, pumping an electron into the conduction band of the semiconducting layer (e.g., TiO_2), accepting an electron from the redox couple, and then repeating the cycle. A dye sensitizer should cover a large part of the visible spectrum and extend up to the NIR (near infrared) region.[16]
- It should have high stability in the oxidized, ground, and excited states.
- The excited state level of the dye sensitizer should be higher in energy than the conduction band edge of the *n*-type semiconductor so that an efficient electron injection process from the excited dye into the conduction band of the semiconductor can occur.
- Dye should have good efficiency in the charge injection and regeneration process. The HOMO of the photosensitizer should lie below the energy level of the redox mediator to promote dye regeneration.

11.2.4 Computational Methodology

We utilized density functional theory[17, 18] for the modeling and simulation of all the organic semiconducting materials. The time-dependent density functional theory is utilized for obtaining the absorption spectra of the dye-based sensitizers. For the molecular diodes, we utilized the B3LYP/6–31G(d,p)[19] basis set for the calculation. The electronic structure calculations are performed utilizing Gaussian 09 software.[20] To investigate the charge transfer characteristics of these molecular species, the electric field was applied along the Cartesian axis (±X) using the Gaussian keyword for electric field calculations "field=X±EF (au)." Further the molecules for the field effect transistor application are optimized at the same level of theory as for molecular diodes. All the dyes from 1 to 8 based on the D-π-A and D-D-π-A models, i.e., dyes with mono- and di-substituted donors,[21] were modeled using Gaussview[22] at the same level of theory as for molecular diodes. Further, to compute the optoelectronic properties at the excited state, a long-range corrected B3LYP known as CAM-B3LYP functional,[23] along with the 6–311++G(d,p) basis set, was employed using the TD-DFT approach. For dyes 9 and10[49] which are based on D-π-A and D-π-A-A model have been optimized at the B3LYP/cc-pVDZ basis set, and their excited state is simulated using the TD-DFT at CAM-B3LYP/cc-pVDZ basis set.

11.3 ORGANIC MOLECULAR DIODES

Organic molecular diodes work on the same principle as the inorganic *p-n* junction diode. The use of organic conjugated materials as rectifier started the field of molecular electronics.[9] The primary objective of moletronics is to exploit the intrinsic properties of molecules/molecular ensembles for the creation of ultraminiaturized electronic circuitry like silicon-based microelectronics. Single-molecule diodes (SMDs) are one of the active components of moletronics, which can mimic the current–voltage characteristics of conventional inorganic *p-n* junction diodes.[24] Electron transport in SMDs is primarily governed by basic principles of chemistry and physics. On the application of an external electric field, SMDs display asymmetrical electron transfer along its length in two directions. The movement of electrons is enabled in a particular direction but inhibited in the opposite direction. The basic mechanism of electron tunneling relevant to moletronics was first discovered by Mann and Kuhn between metal-to-metal monolayers separated by organic compounds. Later, Aviram and Ratner used this concept to revolutionize the field by predicting the concept of SMD for current rectification. Since their pioneering work, the Aviram–Ratner model (AR model) has inspired us to work on single-molecule diodes.[14]

In our published papers on molecular diodes,[19, 25] we have discussed five molecules based on the donor-sigma-acceptor model (i.e., D-σ-A) for better understanding of the mechanism of organic semiconductors to be utilized as single-molecule diodes (see Figures 11.1 and 11.2):

S1 [p- Sexiphenyl-σ- TCNQ]
S2 [p-Sexiphenyl-σ-NTCDA]
S3 [TCNQ-σ-Tetrathiafulvalene (TTF)]
S4 [TCNQ-σ-2,7-Diphenylbenzothieno(3,2-b)benzothiophene (DPh-BTBT)]
S5 [TCNQ-σ-Bis(ethylenedithio) tetrathiafulvalene (BEDT-TTF)]

11.3.1 Analysis of Molecular System Taken from Figure 11.1: (a) S1 [p-Sexiphenyl-σ-TCNQ], (b) S2 [p-Sexiphenyl-σ-NTCDA]

The optimized structures of **S1** and **S2** are shown in Figures 11.1a and b. The values of HOMO energies (E_{HOMO}), LUMO energies (E_{LUMO}) and HOMO-LUMO gaps (HLG = $E_{LUMO} - E_{HOMO}$) as the electric field is increased for both molecules are discussed here. To further confirm their diode behavior, both molecules were tested for HLG reduction and induced dipole moment effect on applying the electric field calculation. This is discussed next.

Figure 11.3a shows that the electric field applied on **S1** in the +*X* direction (D to A) is slowly increased from 0 to 0.08 V/Å in the interval of 0.02 V/Å. Further, the electric field was also reversed (i.e., applied along the –*X* direction) from 0 to – 0.20 V/Å with the interval of – 0.05 V/Å. Here, a graphical depiction of the variation in the HLG values with the applied external electric field is provided. A strong correlation between HLG values and applied electric field strengths is observed here. HLG of S1 without a biasing condition or at zero electric field of 1.04 eV, which is even smaller than inorganic semiconducting material silicon's band gap of (1.12 eV) at 300 K. The molecular system S1 displays a favorable electron transfer channel between donor and acceptor moieties when connected via σ "bridge," but in the acceptor-to-donor direction, the condition is reversed, i.e., it is not a favorable path for electron transfer. But at higher electric fields, this channel again becomes possible for the transfer of charges. These calculations confirm that, upon applying the electric field along +*X* direction

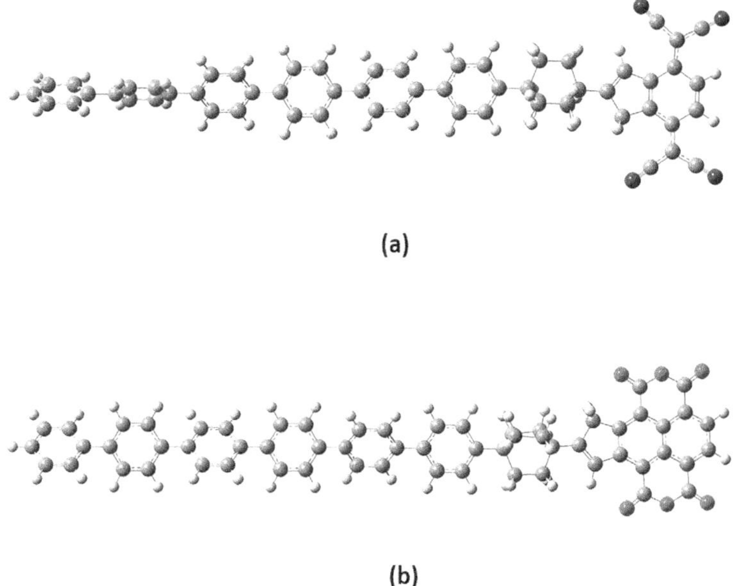

FIGURE 11.1 (a) S1 [p-Sexiphenyl-σ-TCNQ] and (b) S2 [p-Sexiphenyl-σ-NTCDA].

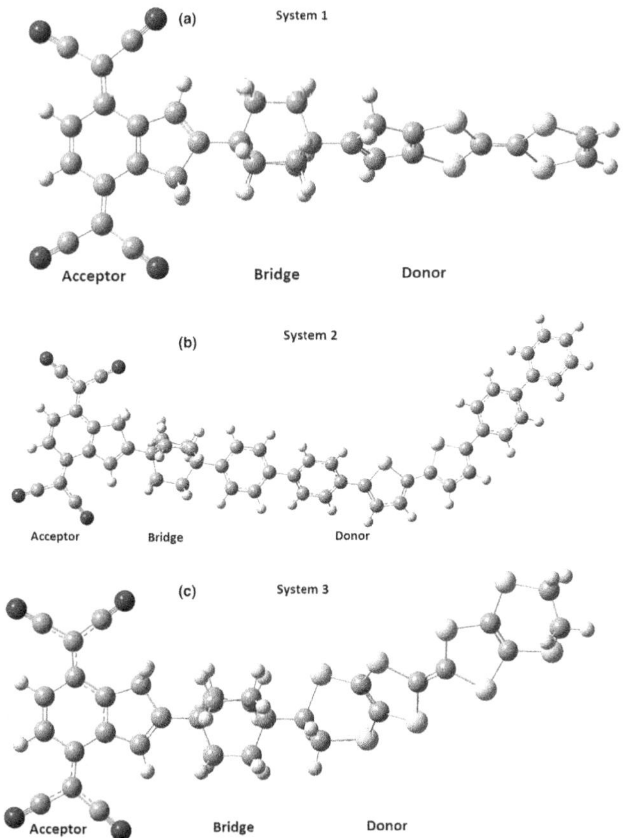

FIGURE 11.2 (a) S3 [TCNQ-σ-(TTF)], (b) S4 [TCNQ-σ-(DPh-BTBT)], and (c) S5 [TCNQ-σ-(BEDT-TTF)].

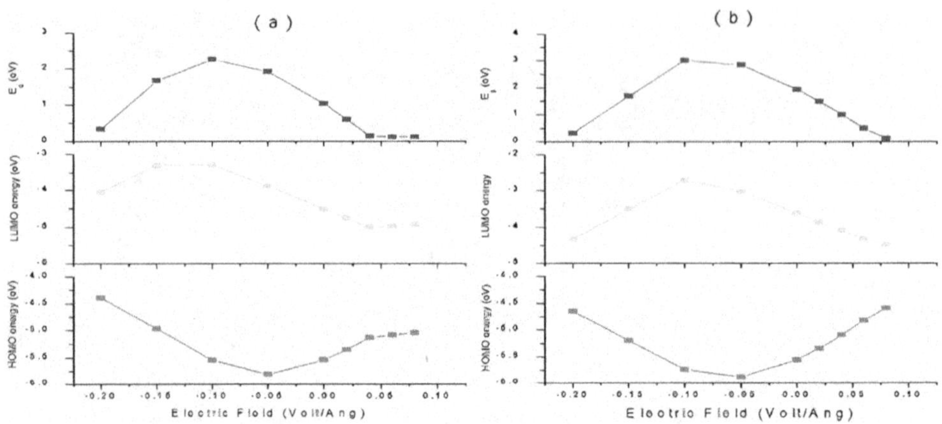

FIGURE 11.3 Electric field dependence of HOMO, LUMO and HOMO-LUMO gap (HLG) energies for (a) S1 and (b) S2.

(D to A), the condition is equivalent to a forward bias potential. As the electric field is increased, HLG values show a significant reduction. The minimum HLG energy for this system in the forward-biased condition was 0.11 eV, obtained at 0.08 V/Å. Therefore, it is safe to say that this system has a conducting channel for the transport of electrons between D and A moieties at high electric fields. And as the electric field is applied along the –*X* direction (A to D), a reverse-biased condition is obtained. HLG values first increase to 2.26 eV and then decrease to 0.33 eV. This type of behavior is like the conventional *p-n* junction diode. And from Figure 11.4a, the dipole moment for S1 at zero electric field was 3.62 Debye. As the electric field was increased toward the +*X* direction (D to A) from 0 to 0.08 V/Å, the dipole moment also increased and reached a maximum value of 34.19 Debye at 0.08 V/Å. Similarly, when the electric field direction was reversed, changes in the induced dipole moment were observed. On reversing the electric field direction up to –0.05 V/Å, the induced dipole moment decreased to a minimum value of 0.72 Debye. On increasing the reverse electric field further, an increase in the dipole moment was observed, reaching a maximum value of 10.82 Debye at –0.20 V/Å. This result indicates preferential electron flow in the D to A direction in the forward-biased condition. Therefore, the behavior of the designed molecule is suitable for its application as a *p-n* junction diode.

From Figure 11.3b of molecular system S2, when there is zero electric field or no biasing condition, the HLG was 1.94 eV, confirming its semiconducting nature. As the electric field was increased along the +*X* direction, HLG values are reduced. At an electric field of 0.08 V/Å, the HLG value was reduced to 0.11 eV. This shows that, on increasing the electric field toward the +*X* direction, S2 begins to offer a favorable channel for electron transport from D to A moieties. Hence, in this process, S2 also resembles a forward-biased *p-n* junction diode. As the electric field direction is reversed, the HLG values start to increase, reaching a maximum value of 3.01 eV (at electric field of – 0.10 V/Å). After this point, the HLG values exhibit reduction. When the magnitude of electric field in –X direction was further increased, HLG value tend to decrease. At an electric field of –0.20 V/Å, the HLG value reduces to 0.32 eV. When the electric field is changed from –0.15 V/Å to –0.20 V/Å, a sharp drop in HLG value is observed resembling the I–V characteristics of the breakdown region of the diode. At this point, a high probability of electron transport along the reverse direction exists for sufficiently high values of applied electric field mimicking *p-n* junction diode behavior. And from Figure 11.4b, the changes in induced dipole moment of S2 for different values of electric field ranging from –0.20 to 0.08 V/Å are given. At zero electric field, the magnitude of dipole moment was 4.54 Debye. As the electric field was increased toward +*X* direction, dipole moment also increased, reaching a maximum value of 14.82 Debye at an electric field of 0.08 V/Å. When the electric field was reversed and applied along the –*X* direction, from 0.0 V/Å to – 0.05 V/Å, the

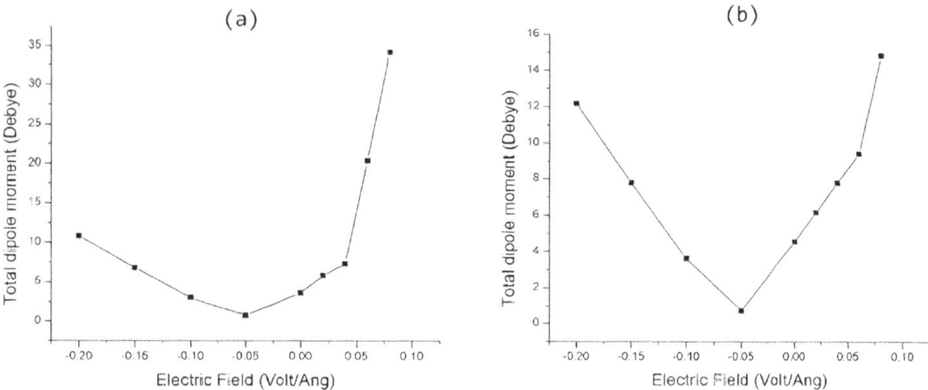

FIGURE 11.4 Electric field dependence of total dipole moment for (a) S1 and (b) S2.

dipole moment abruptly reduced from 4.54 to 0.72 Debye. On further decreasing the electric field, the induced dipole moment increased again. The reason for this change is the reversal of the dipole moment vector of S2 under the influence of electric field in the A to D direction. At electric field of −0.20 V/Å, the dipole moment attained a high value of 12.17 Debye. These results indicate clearly that the preferential electron flow path is available along the D to A direction for system S2.

11.3.2 Analysis of Molecular System Taken from Figure 11.2: (a) S3 [TCNQ-σ-(TTF)], (b) S4 [TCNQ-σ-(DPh-BTBT)], (c) S5 [TCNQ-σ-(BEDT-TTF)]

The optimized structures of S3, S4, and S5 are shown in Figures 11.2a–c. The frontier molecular orbital energy levels, which are HOMO energies (E_{HOMO}), LUMO energies (E_{LUMO}) and HOMO-LUMO gaps (HLG = $E_{LUMO} - E_{HOMO}$) for all three molecules, are discussed here. To further confirm their diode behavior, both molecules were investigated for HLG reduction and induced dipole moment effect on applying the electric field strength calculation, as discussed here.

In the S3 [TCNQ-σ-TTF] shown in Figure 11.5a, variations of HOMO, LUMO, and HLG energy are shown with respect to the applied electric field. At zero electric field, the HLG energy is 0.21 eV. Such a small HOMO-LUMO energy gap strongly suggests that a good semiconducting channel is available between donor and acceptor moieties. The electric field applied toward the −*X*-axis (from D to A moieties) resulted in the lowering of HOMO-LUMO gap energy to 0.20 eV at an electric field of −0.30 V/Å. When the electric field was increased from 0.05 to 0.30 V/Å, toward the +*X*-axis (from A to D moieties), HLG energy was reduced sharply beyond 0.25 V/Å, which resembles the forward bias connections of a *p–n* junction diode, while the +*X* direction of the electric field is analogous to the reverse bias condition of the *p–n* junction diode. Charge transfer between donor and acceptor moieties via a σ insulating bridge can be further understood by studying the effect of an applied electric field on the induced electric dipole moment of the molecule. Therefore, from Figure 11.6a, the variation of dipole moment w.r.t. to the electric field is shown. At zero electric field, the dipole moment was 7.05 Debye. As the electric field was applied in the forward direction, the dipole moment reached a peak value of 65.81 D at an electric field of −0.30 V/Å. As the electric field was reversed, the dipole moment drops significantly. The effect of the electric field on the induced dipole moment indicates that, when the HOMO-LUMO gap is low or decreasing, the electric dipole moment is high, mimicking the diode behavior.

From Figure 11.5b's S4 of [TCNQ-σ-DPh-BTBT] system, at zero electric field, the HOMO-LUMO gap shown is 0.62 eV. As the electric field is applied along the −*X* direction, that results in a reduction of the HOMO-LUMO gap. At an electric field of −0.10 V/Å in −*X* direction, the HOMO-LUMO gap is reduced to 0.13 eV. This clearly shows that the electric field in the −*X* direction

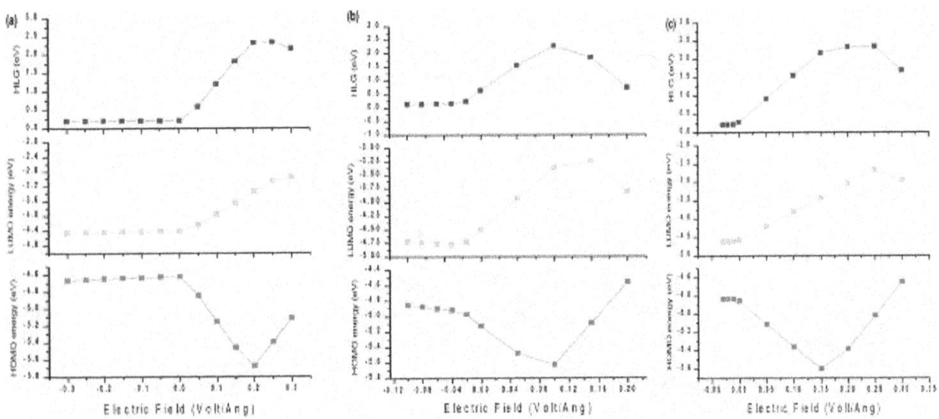

FIGURE 11.5 Variation of HOMO energy, LUMO energy, and HOMO-LUMO gap with respect to the change in electric field for (a) TCNQ-σ-TTF system, (b) TCNQ-σ-DPh-BTBT system, and (c) TCNQ-σ-BEDT-TTF system.

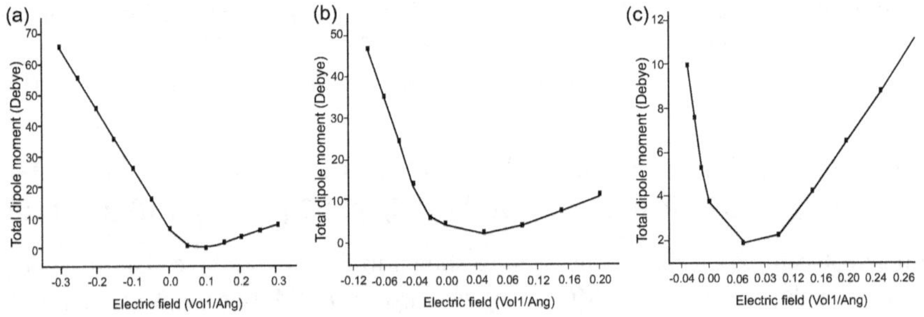

FIGURE 11.6 Variation of total dipole moment with respect to change in electric field: (a) TCNQ-σ-TTF system, (b) TCNQ-σ-DPh-BTBT system, and (c) TCNQ-σ-BEDT-TTF system.

provides a good semiconducting route for the charge transport, which resembles the forward-biased condition of the *p–n* junction diode. When the electric field was applied in the opposite direction (+*X*-axis), at 0.10 V/Å, the HOMO–LUMO gap was increased to 2.26 eV. This condition is like the reverse bias condition of a *p–n* junction diode. When the electric field goes over 0.10 V/Å, a rapid reduction was observed in the HOMO-LUMO gap. At an electric field of 0.20 V/Å, the HOMO-LUMO gap was found to be 0.73 eV. This suggests that, with the extremely high electric field in the +*X* direction, a semiconducting channel is formed, which confirms the behavior of a designed molecule like a *p-n* junction diode. Further, from Figure 11.6b, the variation of the dipole moment w.r.t. the electric field is observed. The total electric dipole moment at zero electric field was 4.80 Debye. As the electric field is applied along the –*X*-axis, the dipole moment significantly increases and reaches a peak value of 46.91 D at –0.10 V/Å electric field. Along the +*X*-axis, the electric dipole moment reduces initially. With the applied electric field of 0.05 V/Å and 0.10 V/Å, the dipole moment was 2.38 D and 3.97 D, respectively. When the electric field reaches 0.15 V/Å and 0.20 V/Å, the dipole moment increases to values of 7.52 D and 11.71 D. This indicates that, when the HOMO-LUMO gap drops or the possibility of charge transport increases, the induced electric dipole moment also increases. This type of behavior confirms its diode nature.

From Figure 11.5c's S5 [TCNQ-σ-BEDT-TTF] system, the HLG energy without electric field is reported to be 0.28 eV, confirming its semiconducting nature. As the electric field is applied along the –*X*-axis (from donor to acceptor site), there seems to be a small reduction in the HLG energy as reported to be 0.21 eV at –0.03 V/Å. This reveals that there is a good semiconducting channel between donor and acceptor moieties, which again resembles the forward bias connection of a *p–n* junction diode. As the electric field is applied along the +*X*-axis (from acceptor to donor site), notable changes were observed in the HOMOLUMO gap energy as it was increased with respect to the electric field with a peak value of 2.35 eV at 0.25 V/Å. When this electric field was further increased, a sharp reduction was observed in the HOMO-LUMO gap at 0.30 V/Å, and the HOMO-LUMO gap was calculated as 1.69 eV. This explains that, with the excessive electric field in the reverse direction, energy levels are realigned, come closer to each other, and result in the reduction of HLG energy. From Figure 11.6c, the electric dipole moment is calculated as 3.80 Debye at zero electric field. In the forward direction of the electric field (from donor to acceptor), the induced electric dipole moment reached a peak value of 9.99 Debye. In the reverse direction of the electric field (from acceptor to donor), the induced electric dipole moment decreased initially and then increased continuously with the applied electric field. At an electric field of 0.30 V/Å in the reverse direction, the electric dipole moment was simulated as 11.18 Debye with a HOMO-LUMO gap of 1.69 eV. This confirms that this kind of behavior shown by the designed system mimics the diode nature of the conventional *p-n* junction diode.

11.4 ORGANIC FIELD EFFECT TRANSISTORS

Today organic field effect transistors are one of the more highly researched domains in the field of organic electronics. The first reported OFET was made of polythiophene in 1986.[15] Since then, good progress has happened in the field of OFET. With the transistor performance that is readily available today, a wide range of applications can already be addressed: e-paper displays, simple circuits, and chemical and biological sensors, electronic papers, identification tags, etc.[26] Three long decades of progress on conjugated polymer and small-molecule organic semiconductors have greatly contributed to understanding the fundamentals of the charge transport nature of organic materials.[27] The improved physical understanding of transport and of structure–property relationships has supported the development of better organic semiconductor materials for organic field effect transistor (OFET) applications as well as other devices that rely on charge transport, such as organic solar cells and light-emitting devices. Nowadays, charge carrier mobilities of the same order as amorphous silicon (0.1–1 $cm^2\ V^{-1}\ s^{-1}$)[10] are achieved in the best OFETs. Thiophene and especially acene derivatives are the benchmark in OFETs, and most of the best mobilities have been found in these two families of compounds.[28] The fundamental material characteristics of organic semiconductors are most clearly measured in single crystals, where the highest mobilities have been observed due to their higher molecular order and to the fact that they do not present grain boundaries (a mobility of up to 18 $cm^2\ V^{-1}\ s^{-1}$ has been reported for a rubrene single crystal).[29] To promote, therefore, the development and utility of organic semiconductors, there is a clear need to find materials that can be solution-processed and simultaneously can achieve a high OFET mobility. Spin-coated films of di-alkyl end substituted derivatives of benzothienobenzothiophene Cn-BTBT (n = 11–13) with a highly ordered, crystalline microstructure and alternating layers of aliphatic side chains and conjugated layers parallel to the substrate exhibit high mobilities of (1–3 $cm^2\ V^{-1}\ s^{-1}$).[10] Cn-BTBT with longer sides chains has slightly higher mobility than Cn -BTBT with shorter side chains because the stronger hydrophobic interaction induced by the longer side chains tends to enhance the conjugated molecular overlap. By drop-casting onto an inclined substrate, the film morphology for C8 -BTBT can be optimized, and bottom-gate FETs with mobilities of 5 $cm^2\ V^{-1}\ s^{-1}$ were reported.

We will next review our published papers on the relevant parameters required to define organic field effect transistors.[30, 27]

We reviewed three molecules:

M1 (2,2-bis(4-trifluoromethylphenyl)-5,5–bithiazole)
M2 [coumarin_102] (2,3,6,7-tetrahydro-9-methyl-1H,5H-quinolizino[9,1-gh] coumarin)
M3 [coumarin_153] (2,3,6,7-tetrahydro-9-(trifluoromethyl)-1H,5H,11H-[1]-benzopyrano (6,7,8-ij)quinolizin-11-one)

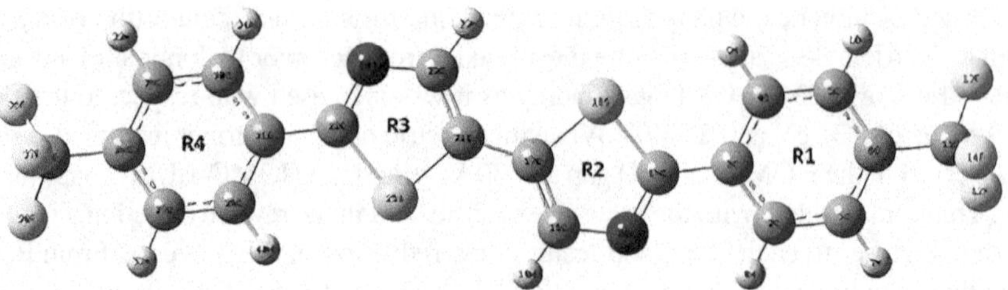

FIGURE 11.7 M1: (2,2-bis(4-trifluoromethylphenyl)-5,5–bithiazole).

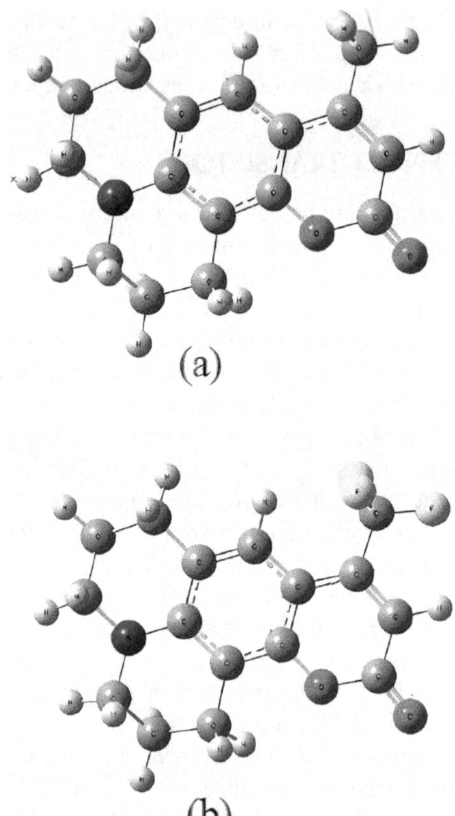

FIGURE 11.8 (a) M2: coumarin (102) and (b); M3: coumarin (153).

Here, the relationship between the important OFET parameters and molecular structure is discussed. More specifically, we focused on the geometrical and electronic structure of the molecule, including HOMO, LUMO, and HOMO-LUMO gap (HLG), reorganization energy of holes and electrons to define the charge carrier nature of the modeled molecules.

11.4.1 Analysis of Molecular System Taken from Figure 11.7: M1(2,2-bis(4-trifluoromethylphenyl)-5,5–bithiazole)

The optimized structure of M1 is discussed here with various parameters like the frontier molecular orbital energy levels, which are HOMO energies (E_{HOMO}), LUMO energies (E_{LUMO}),

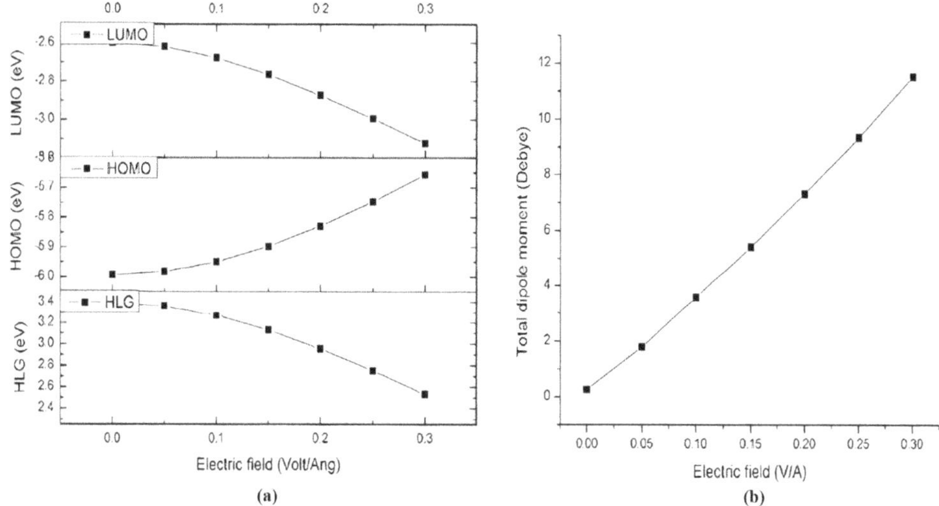

FIGURE 11.9 (a) Electric field dependence of HOMO, LUMO, and HLG energy and (b) EF dependence of dipole moment for system M1.

and HOMO-LUMO gaps (HLG = $E_{LUMO} - E_{HOMO}$), dipole moment. To further confirm their usage as an organic field effect transistor, the reorganization energy of holes and electrons is calculated.

From Figure 11.9a, it is shown that HLG energy is affected by the increasing electric field, discussed here. The HLG energy at zero electric field is reported to be 3.39 eV; as the electric field is increased from 0.05 to 0.30 V/Å, a noticeable change is observed in the HLG energy, which reaches 2.53 eV at 0.30 V/Å electric field. This shows that the increase in electric field leads to a decrease in HLG by nearly 0.86 eV. Further, from Figure 11.9b, the dipole moment is also plotted for the system as we increase the electric field strength. At zero electric field strength, the dipole moment observed is 0.25 Debye, while as the electric field is increased from 0.00 to 0.30 V/A° in steps of 0.05 V/A°, the induced dipole moment reaches a peak value of 11.51 Debye at 0.30 V/A° of electric field strength.

Further to know the type of organic semiconductor, the reorganization energy of holes and electrons is also calculated, which is shown in Table 11.1. These values are calculated utilizing the Marcus theory whose formulation is already discussed in section 11.1.1. The molecules with higher λ_h and lower λ_e are suitable for electron transport, whereas the molecules with higher λ_e and lower λ_h are suitable for hole transport in organic semiconductors. From Table 11.1, the system M1 [2,2 -bis (4-trifluoromethylphenyl)-5,5-bithiazole] has a smaller value of λ_e than that of λ_h, i.e., 0.32 and 0.35 eV, respectively, and thus it will be a good electron transport material.

11.4.2 Analysis of Molecular System Taken from Figures 11.8a and 11.8b

The molecular systems are denoted as M2 [coumarin 102] (2,3,6,7-tetrahydro-9-methyl-1H,5H-quinolizino[9,1-gh] coumarin) and M3 [coumarin 153] (2,3,6,7-tetrahydro-9-(trifluoromethyl) -1H,5H,11H-[1]-benzopyrano(6,7,8-ij)quinolizin-11-one)(2,3,6,7-tetrahydro-9-methyl-1H,5H-quinolizino[9,1-gh] coumarin).

The optimized structure of M2 and M3 have been discussed here with various parameters like the frontier molecular orbital energy levels, which are HOMO energies (E_{HOMO}), LUMO energies (E_{LUMO}) and HOMO-LUMO gaps (HLG = $E_{LUMO} - E_{HOMO}$). To further confirm their usage as organic field effect transistors, the reorganization energy of holes and electrons is calculated.

TABLE 11.1
System M1 Ionization Potential (adiabatic) and (vertical), Electron Affinity (adiabatic) and (vertical), Reorganization Energy for Holes λ_h, and Reorganization Energy for Electrons λ_e (all values in eV)

IP(a)	IP(v)	EA(a)	EA(v)	λ_h	λ_e
7.13	7.31	1.50	1.34	0.35	0.32

TABLE 11.2
HOMO, LUMO, HLG, and the Total Dipole Moment of M2 [Coumarin 102] under Various Electric Fields

EF (in V/Å)	HOMO (in eV)	LUMO in (eV)	HLG energy in (eV)	Dipole Moment (Debye)
0.00	−5.14	−1.22	3.92	6.72
0.05	−5.16	−1.17	3.99	6.04
0.10	−5.18	−1.13	4.05	5.42
0.15	−5.20	−1.09	4.11	4.85
0.20	−5.22	−1.04	4.18	4.35
0.25	−5.24	−1.00	4.24	3.97
0.30	−5.25	−0.95	4.29	3.71

From Table 11.2, the parameters for molecular system M2[Coumarin 102] is reported. The electron transport channel was described by applying the EF along the +*X* direction. At zero electric field, the high value of total dipole moment for coumarin 102 clearly indicates that this molecule has an asymmetric charge distribution. A notable difference in the frontier molecular orbitals has been observed due to the change in the applied EF along the +*X* direction in Coumarin 102. Hence, it is of great significance to understand how the HOMO, LUMO, and HLG are sensitive to the variation in the EF. The HLG energy at zero electric field was reported to be 3.92 eV, as the electric field is increased to +0.30 V/Å; then the HLG energy rises to 4.29 eV. It has been observed that in Coumarin 102 the variation of the EF from 0.0 to +0.30 V/Å leads to the increase in HLG by 0.37 eV.

Further, from the Table 11.4, the reorganization energy calculated for Coumarin 102 shows that the λ_h value is 0.201 eV, and the λ_e value is 0.489 eV. This molecule has lower reorganization energy for holes as compared to the electrons. Therefore, this molecule can be utilized as a *p*-type semiconducting channel in the organic field effect transistor.

From Table 11.3, the molecular system M3 [Coumarin 153] has reported values of HOMO, LUMO, HLG energy, and dipole moment. Here, in Coumarin 153, the HLG value at zero electric field is 3.59 eV, which drops to 3.18 eV as the electric field is increased from 0.0 to +0.30 V/Å. The variation of the EF from 0.0 to +0.30 V/Å leads to the decrease in HLG by 0.41 eV. And the dipole moment reported to be 7.02 Debye at zero electric field increases to 12.03 Debye at the electric field strength of 0.30 V/Å. This increase in the dipole moment also suggests that the asymmetric distribution of the charges helps in reducing the HLG energy, and that results in a higher value of dipole moment in the molecule.

Further, from the Table 11.4, the reorganization energy calculated for Coumarin 153 shows that the λ_h value is 0.171 eV, and λ_e value is 0.639 eV. This molecule has lower reorganization energy for holes as compared to the electrons. Therefore, this molecule can be utilized as *p*-type semiconducting channel in the organic field effect transistor.

TABLE 11.3
HOMO, LUMO, HLG, and the Total Dipole Moment of M3 [Coumarin 153] under Various Electric Fields

EF (V/Å)	HOMO (eV)	LUMO (eV)	HLG energy (eV)	Dipole Moment (Debye)
0.00	−5.40	−1.81	3.59	7.02
0.05	−5.35	−1.83	3.52	7.82
0.10	−5.30	−1.85	3.45	8.63
0.15	−5.25	−1.87	3.38	9.46
0.20	−5.21	−1.89	3.32	10.31
0.25	−5.16	−1.92	3.24	11.16
0.30	−5.12	−1.94	3.18	12.03

Table 11.4
Systems M2 and M3: Ionization Potential (adiabatic) and (vertical), Electron Affinity (adiabatic) and (vertical), Reorganization Energy for holes λ_h, and Reorganization Energy for Electrons λ_e (all values in eV)

Molecule	IP(a)	IP(v)	EA(a)	EA(v)	λ_h	λ_e
M2 [Coumarin 102]	4.96	5.07	1.67	1.43	0.201	0.489
M3 [Coumarin 153]	5.14	5.23	2.24	1.93	0.171	0.639

11.5 DYE-SENSITIZED SOLAR CELLS

Due to the environmentally friendly nature and cheaper fabrication cost, dye-sensitized solar cells are now a trending research area in organic photovoltaics. DSSC was first introduced in 1991 by Gratzel's group using ruthenium (Ru) complex as a photosensitizer.[31] These days, DSSC has been the front-runner in the renewable energy source for low-powered devices or for those that can perform in ambient light conditions.[32] Since the work of Gratzel and O'Regan in 1991, efficiency has improved a lot, and the progress on metal-free organic dyes is significant. DSSC has seen improvement from 10 to 15% in the last decade.[33]–[40] Some of the important reported results showed improved efficiency were developed by Nazeeruddin et al., with efficiency of 11.18%,[38] Chen et al., with efficiency of 11.50%,[39] and Wang et al. with efficiency of the cell at 11.16%.[40] Mathew et al., reported efficiency of 13%;[41] further, we have seen that the group of Kakiage et al. reported efficiency of the cell at more than 14%.[42] The state-of-the-art donor-bridge-acceptor (D-π-A) model for designing photosensitizers has been an efficient strategy that enhances the charge transfer from donor to acceptor moiety through a bridging compound.[43] An increase in the maximum absorption region of the dye or red shift is generally achieved by improving the electron donating strength of donor molecules or by expanding the bridging unit, or it could be done by enhancing the electron withdrawing ability of the acceptor. D-π-A molecular design has led to the development of unique structures such as D-D-π-A, D-A-π-A, $(D\text{-}\pi\text{-})_2D\text{-}\pi\text{-}A$, and various others.[44, 45]

The mechanism part of the DSSC includes three main steps: the first step is to absorb the photon energy to excite the dye electron, the second is to transfer this electron to the semiconducting material (generally TiO_2 as the reference material for a photoanode), and the third step is to regenerate the oxidized dye.[32] For better performance of the device, the designed sensitizer should have certain characteristics: it's a wide absorption region from visible to near infrared and chemical stability of the dye molecule, and the frontier molecular energy orbitals should be aligned with the semiconducting material's energy levels and with the electrolyte utilized in the cell.[47] The best known DSSCs are based on mainly ruthenium-based dye complex and porphyrins-based metal complexes.[34, 35] The disadvantage with metal organic dyes is in their high cost of synthesis route, and their toxicity is also not good for the environment. Therefore, metal-free organic dyes are of interest to researchers in the quest for an efficient DSSC. Some of the photosensitizers that are widely utilized for preparing metal-free organic dyes are coumarin, diphenylamine, triphenylamine, indoline, phenothiazine, naphthalene, among many other organic molecules.[50] The advantage with these materials is that they can be tuned according to the intended use of the devices.

To improve the performance of the DSSC, we may choose to alter donor moiety or to play with acceptor moiety so that the new molecular energy orbitals formed due to the donor–acceptor interaction are aligned with the semiconducting material's energy levels. We have seen from past published papers that, to improve the efficiency of the cell, more than one donor is appropriate in designing the dye complex.[44] And some papers have reported that increasing the acceptor unit in the donor–acceptor template also helps in improving the open circuit voltage and short circuit current density, which ultimately increases the efficiency of the designed cell.[51] Lowering the LUMO energy levels improves the electron injection rate and stops the recombination of the charges. Therefore, acceptor moieties become important in designing the photosensitizers because LUMO mostly gets a contribution from the acceptor moiety.

In the present work, we have chosen ten dyes from our published papers for review, Dyes 1–8 are modeled using coumarin (C) and triphenylamine (TPA) as electron donor moieties, which are related to four different electron acceptor units—barbituric acid (Ba), cyanoacetic acid (Cy), hydantoin (Hy), and rhodamine-n-acetic acid (Rh)—via thiophene bridge. Dyes 1–4 were modeled using coumarin as electron donor and Ba, Cy, Hy, and Rh as an electron acceptor anchoring group connected through a thiophene bridge. In dyes 5–8, TPA is utilized as second electron donor, along with coumarin, both of which are connected with the same electron acceptor moieties via thiophene bridge.[48] Therefore, the first four D-π-A dyes (dyes 1–4) are C-π-Ba, C-π-Cy, C-π-Hy, and C-π-Rh. The next four (dyes 5–8) are based on the D-D-π-A model: T-C-π-Ba, T-C-π-Cy, T-C-π-Hy, and T-C-π-Rh. Further, two more dyes (9 and 10) have been reviewed that exhibit absorption in the near infrared region. The dyes have been designed on the D-π-A and D-π-A-A architectures using thiophene as the π-bridge. Here, the donor moiety is carbazole, and the acceptors are naphthalimide and benzothiadiazole moieties. Naphthalimide and benzothiadiazole are strong electron acceptors; therefore, these are chosen for designing the efficient photosensitizers. Dye 9 (C-π-NI) is based on the D-π-A model [carbazole-thiophene-naphthalimide], and dye 10 (C-π-NI-BT) is based on the D-π-A-A model [carbazole-thiophene-naphthalimide-benzothiadiazole].[49] Optimized structures of all the dyes 1–10 are shown in Figure 11.10.

11.5.1 Photovoltaic Performance Analysis

The opto-electronic performance parameters for dyes are calculated utilizing the following formulae.[53]

The efficiency of the DSSC can be calculated from the integral of the J_{SC}, V_{oc}, The fill factor (FF), and the intensity of the light (I_s):

$$\eta = \frac{V_{oc} J_{sc} FF}{Is} * 100\%$$

The J_{SC} in DSSCs is determined by the following equation:

$$J_{sc} = \int LHE(\lambda) \phi_{inject} * \eta_{collect} \, d\lambda$$

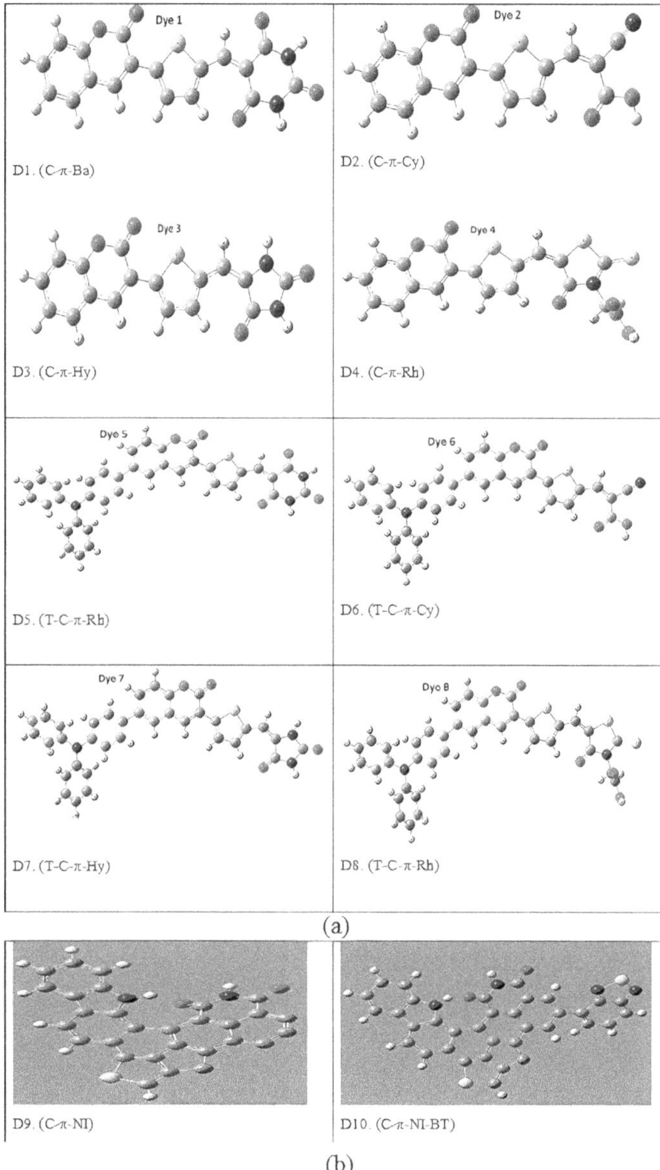

FIGURE 11.10 (a) Optimized geometries of D1–D8 obtained at B3LYP/6–31G(d,p), (b) dyes 9 and10 optimized at B3LYP/cc-pVDZ level of theory.

where LHE (λ) is the light harvesting efficiency at a given wavelength, ϕ_{inject} is the electron injection efficiency, and $\eta_{collect}$ is the charge collection efficiency. Absorption is also one of the important factors in determining the short circuit photo current density; therefore the red-shifted dyes would show better performance as compared to the blue-shifted dyes. The charge collection is dependent on the electrode material and charge recombination process, which in the ideal case is supposed to be 1.

Thus the LHE and ϕ_{inject} play the key role in the control of the *Jsc*. The other way to improve the *Jsc* is by improvement in the ΔG^{inject} values, which are the driving force for the electron injection from the photo-induced excited states of the dyes to the semiconducting conduction band surface.

The larger the values of the ΔG^{inject}, the better the charge injection efficiency ϕ_{inject} will be. The electron injection is significant for the better photovoltaic properties, and the free energy change in eV for the electron injection can be expressed by the following formula, where it is assumed that the electron injection occurs from the unrelaxed excited states.[12]

$$\Delta \mathbf{G}^{inject} = \mathbf{E}_{ox}^{\ dye*} - \mathbf{E}_{CB}^{\ TiO_2}$$

where E_{ox}^{dye*} is the oxidation potential of the dye in the excited state, $|E_{CB}^{TiO2}|$ is the reduction potential of the conduction band of the $TiO_2 = -4.0$ eV.[54] To calculate the E_{ox}^{dye*}, we have the following formula.

$$\mathbf{E}_{\mathbf{ox}}^{\mathbf{dye}\,*} = \mathbf{E}_{\mathbf{ox}}^{\mathbf{dye}} - \lambda_{\mathbf{max}}{}^{\mathrm{ICT}} = -\mathbf{E}_{\mathrm{HOMO}} - \lambda_{\mathbf{max}}{}^{\mathrm{ICT}} \uparrow$$

where E_{ox}^{dye} is the redox potential of the ground state. E_{ox}^{dye} from the Koopman's theorem is equal to the negative of the HOMO energy. And λ_{max}^{ICT} is the energy of the intramolecular charge transfer, which is equal to the excited state transition energy of the dye.

11.5.1.1 Dye Regeneration Driving Force (ΔG_{reg})

The excited state dye regeneration driving force (ΔG_{reg}) can be approximated:

$$\left(\Delta G_{reg}\right) = \left(E_{0x}^{\ \mathrm{dye}} - \mathbf{4.8eV}\right) \ \mathbf{or} \ \left(-\mathbf{E}_{\mathbf{HOMO}}^{\mathbf{DYE}} - \mathbf{4.8}\right)\mathbf{eV} \cdot$$

where −4.8 eV stands for the redox potential of Iodide electrolyte[52] taken as a reference electrolyte here.

11.5.1.2 Open Circuit Photo Voltage (V_{OC})

The open circuit voltage, V_{OC}, is an experimentally determined parameter, although the theoretical relationship between V_{OC} and electronic structure of the dye is not defined properly. Due to the various previous studies on the DSSCs, which have shown that some analytical relationship exists between V_{OC} and LUMO energy levels of the dye:

$$\mathbf{Voc} = -\mathbf{E}_{LUMO}^{dye} - \mathbf{C.B.}\left(\mathbf{E}_{TiO2}\right)$$

where C.B. is the energy of the semiconductor's conduction band edge. It is known that a higher E_{LUMO} value leads to a larger V_{OC}.

11.5.2 Effect of Double Donor Moieties on the Performance of DSSC for Dyes 1–8

From Figure 11.11, for the favorable charge conduction from the dye molecule to the semiconductor site, it is important that the HOMO and LUMO of the dye molecule are aligned with the semiconductor material. It is shown that all the dyes from 1 to 8 have their HOMO energy level of the dye sensitizer below the redox potential of the iodine/iodide electrolyte and that the LUMO energy level lies above the conduction band edge (E_{CB}) of a TiO_2 semiconductor. This is the favored condition for charge transfer through the dye molecule. Further, the performance of eight modeled dye sensitizers was also analyzed based on three important parameters: (1) open-circuit voltage (V_{OC}), (2) light-harvesting efficiency (LHE), and (3) injection driving force (ΔG_{inject}). For the sake of simplicity, the discussion is divided into four parts, and the performance of dyes with the same electron acceptor anchoring groups is discussed as a single subject.

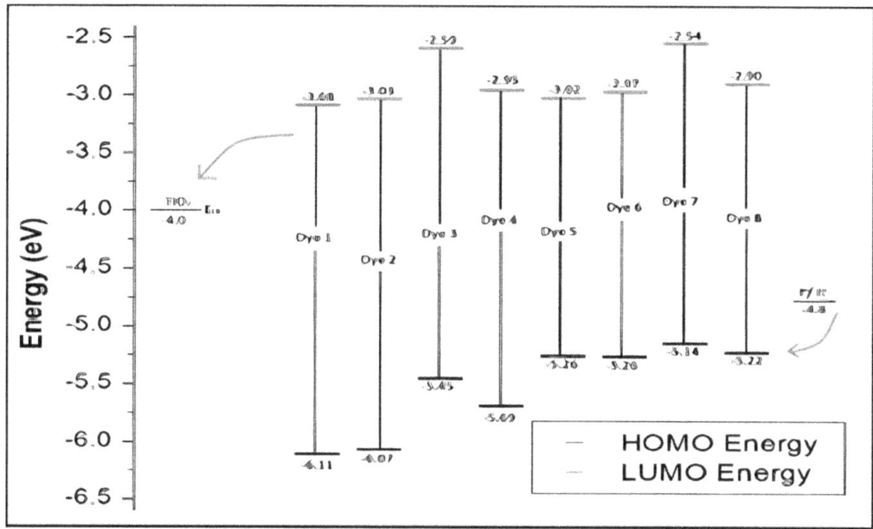

FIGURE 11.11 Energy level diagram of dyes–8. Electrolyte is iodide solution, TiO_2 is photoanode.

11.5.2.1 Dyes with a Barbituric Acid (Ba) Anchoring Group

The absorption wavelength of a dye is of prime importance for the performance of DSSCs. The short-circuit photocurrent density (J_{SC}) is affected by the absorption behavior of the dye photosensitizers. The absorption wavelength, oscillator strength, and LHE of all dye sensitizers are presented in Table 11.5. The absorption wavelength of dye 1 (C-π-Ba) was calculated as 437 nm. When another electron donor moiety TPA was added to the other end of coumarin, a small redshift of 8 nm was seen in dye 5 (T-C-π-Ba) as compared to dye 1. The addition of another donor moiety had a significant effect on the absorption wavelength. Therefore, we suggest that double donor approach is better in comparison to the single donor approach for designing a photosensitizer.

11.5.2.2 Dyes with a Cyanoacetic Acid (Cy) Anchoring Group

With cyanoacetic acid (Cy) as an electron acceptor group, the absorption wavelength of dye 2 (C-π-Cy) was seen as 431 nm (Table 11.5). When triphenylamine (TPA) was added to the other end of coumarin, the absorption wavelength of dye 6 (T-C-π-Cy) was increased to 438 nm, indicating a redshift of 7 nm. The oscillator strength of dye 2 was simulated as 1.1748, which was increased to 1.5235 with the addition of TPA (dye 6). The LHE of dye 2 was calculated as 0.9331, and the LHE of dye 6 was calculated as 0.9700. That means the dye with a di-substituted donor moiety shows better LHE due to the increase in oscillator strength. As provided in Table 11.6, the driving force of electron injection (ΔG_{inject}) was calculated as –0.80 eV for dye 2 and –1.57 eV for dye 6. The open-circuit voltage (V_{OC}) was calculated as 0.969 eV for dye 2 and 1.034 eV for dye 6. The enhancement of ΔG_{inject} and V_{OC} of dye 6 in comparison to dye 2 reveals that di-substitution of the donor moiety is a good approach to enhance the short-circuit photocurrent density (J_{SC}) since J_{SC} mainly depends on LHE and ΔG_{inject}.

11.5.2.3 Dyes with a Hydantoin (Hy) Anchoring Group

With another electron acceptor unit, hydantoin (Hy), we focused on the effect of a di-substituted donor moiety on the efficiency of the DSSC, which seems to be identical to that of the barbituric acid and cyanoacetic acid anchoring group, as shown in Tables 11.5 and 11.6. The absorption wavelengths of dye 3 (C-π-Hy) and dye 7 (T-C-π-Hy) were calculated as 460 nm and

TABLE 11.5
For Dyes 1–8, Maximum Absorption Wavelength (nm), Oscillator Strength, and Light Harvesting Efficiency (eV) Calculated at CAM-B3LYP/6–311++G(d,p) Level

Photosensitizers	λ_{max} (in nm)	F	LHE (in eV)
Dye 1 (C-π-Ba)	437	1.306	0.951
Dye 2 (C-π-Cy)	431	1.174	0.933
Dye 3 (C-π-Hy)	460	1.081	0.917
Dye 4 (C-π-Rh)	487	1.314	0.951
Dye 5 (T-C-π-Ba)	445	1.516	0.969
Dye 6 (T-C-π-Cy)	438	1.523	0.970
Dye 7 (T-C-π-Hy)	463	1.382	0.958
Dye 8 (T-C-π-Rh)	490	1.611	0.975

TABLE 11.6
For Dyes 1–8, Computed ΔG^{inject}, $E_{OX}{}^{dye*}$, $E_{OX}{}^{dye}$, and $\lambda_{max}{}^{ICT}$ in (eV)

Photosensitizers	ΔG^{inject} (eV)	$E_{OX}{}^{dye*}$ (eV)	$E_{OX}{}^{dye}$ (eV)	$\lambda_{max}{}^{ICT}$ (eV)	V_{OC} (eV)
Dye 1 (C-π-Ba)	−0.73	3.27	6.11	2.84	0.916
Dye 2 (C-π-Cy)	−0.80	3.20	6.07	2.87	0.969
Dye 3 (C-π-Hy)	−1.24	2.76	5.45	2.69	1.410
Dye 4 (C-π-Rh)	−0.85	3.15	5.69	2.54	1.047
Dye 5 (T-C-π-Ba)	−1.53	2.47	5.26	2.79	0.979
Dye 6 (T-C-π-Cy)	−1.57	2.43	5.26	2.83	1.034
Dye 7 (T-C-π-Hy)	−1.54	2.46	5.14	2.68	1.456
Dye 8 (T-C-π-Rh)	−1.31	2.69	5.22	2.53	1.102

463 nm, respectively, indicating a small increase in the absorption wavelength due to the addition of another electron donor moiety TPA. The oscillator strength and LHE of dye 3 were simulated as 1.0809 and 0.9170, respectively, which shows improvement to 1.3829 and 0.9586 for dye 7. The two other important parameters, ΔG_{inject} and V_{OC}, were calculated as −1.24 eV and 1.410 eV for dye 3, respectively. Again, with the di-substituted donor moiety, the ΔG_{inject} and V_{OC} values were significantly enhanced to −1.54 eV and 1.456 eV, respectively, for dye 7. These results confirm that di-substitution of an electron donor moiety will probably have a positive effect on the efficiency of DSSCs.

11.5.2.4 Dyes with a Rhodanine-N-Acetic Acid (Rh) Anchoring Group

The absorption wavelength of dye 4 (C-π-Rh) was computed as 487 nm, and that of dye 8 (T-C-π-Rh) was computed as 490 nm (Table 11.5), which means again a small increase in absorption wavelength with a di-substituted donor moiety. The oscillator strength and LHE also increased with the di-substitution of an electron donor moiety: for dye 4, the oscillator strength and LHE were simulated as 1.3144 and 0.9515, respectively, and for dye 8, both were significantly increased to 1.6111 and 0.9755. The electron injection driving force and open-circuit voltage for dye 4 were calculated as −0.85 eV and 1.047 eV, respectively. The same parameters were simulated as −1.31 eV and 1.102 eV for dye 8. These results reveal that the parameters responsible for the regulation of PCE can be increased with a di-substituted donor moiety.

11.5.3 Effect of Double Acceptor Moieties on the Performance of DSSC for Dyes 9 and 10

To design the higher wavelength absorption dyes, the LUMO levels play an important role. Apart from a simple D-π-A model, the introduction of additional donors, acceptors, and extension of *p*-conjugation reduces the HOMO-LUMO energy gap and redshifts the absorption maximum. Here, we used the double acceptor model strategy to design an efficient photosensitizer that would have a higher absorption wavelength or near infrared absorbers. The two acceptors have been attached with the donor site through π-bridge. We reviewed the two sets of molecules where the absorption is found in the near infrared region, which are based on the double acceptor approach. These are denoted as dye 9 and dye 10, as shown in Figure 11.10. Dye 9 is (carbazole-thiophene-naphthalimide), and dye 10 is (carbazole-thiophene-naphthalimide-benzothiadiazole). In the designed molecules, the chosen acceptor site of naphthalimide shows a low LUMO level value, which makes it suitable for designing infrared absorber dyes. In Dye 10, there are two acceptors that exhibit lower LUMO levels as compared to the conduction band of TiO_2. Therefore, to make an efficient solar cell, a different semiconducting material needs to be chosen. As a substituent of TiO_2 photoanode, In_2O_3 has been taken as the reference anode due to its better energy alignment with respect to the LUMO levels of dye molecules. In_2O_3 has been utilized earlier as a probable substituent for photoanode in DSSC. The bandgap of In_2O_3 is 3.5 eV,[49] which is larger compared to the bandgap of TiO_2 (3.2 eV). Various other properties of dyes 9 and 10 are discussed and analyzed in the ensuing subsections.

From Table 11.7, maximum absorption wavelength is shown. In designing DSSC, the absorption spectrum plays a critical role. Efficient light harvesting in DSSC usually takes place in the visible and near infrared region. In dye 9, the absorption maximum is observed at 1432 nm, and in dye 10, the absorption maximum is shown at 1351 nm. The probable reason for high absorption wavelengths in the two dyes may be attributed to the presence of strong electron acceptors. In both dyes, the energy levels of LUMO lie close to the conduction band edge of the semiconducting material In_2O_3. The value (energy of LUMO = −5.4 eV) represents the CB edge of the In_2O_3 semiconductor. It may

TABLE 11.7
For Dyes 9 and 10, HOMO, LUMO, HLG Energy Calculated at B3LYP/cc-pVDZ Level; Maximum Absorption Wavelength (in nm), Oscillator Strength, and Light Harvesting Efficiency (eV) Calculated at CAM-B3LYP/cc-pVDZ Basis Set Level

Photosensitizer	HOMO (in eV)	LUMO (in eV)	HLG (in eV)	λ_{max} (in nm)	f	LHE (in eV)
Dye 9 (C-π-NI)	−5.76	−4.04	1.72	1432	0.0008	0.002
Dye 10 (C-π-NI-BT)	−5.66	−3.91	1.75	1351	0.0009	0.003

TABLE 11.8
For Dyes 9 and 10, Computed ΔG_{inject}, E_{OX}^{dye*}, E_{OX}^{dye}, and λ_{max}^{ICT} in (eV), ΔG_{reg} (eV), eV_{OC} (eV)

Photosensitizer	ΔG_{inject} (in eV)	E_{OX}^{dye*} (in eV)	E_{OX}^{dye} (in eV)	λ_{max}^{ICT} in (eV).	ΔG_{reg} (eV)	eV_{OC} (in eV)
Dye 9 (C-π-NI)	−0.5	4.9	5.766	0.866	0.96	1.36
Dye 10 (C-π-NI-BT)	−0.651	4.749	5.667	0.918	0.86	1.49

also be observed that both the designed dyes have HOMO energy levels (for dye 9, −5.76 eV; for dye 10, −5.66 eV) lower than the CB edge of redox couple of iodide solution (−4.8 eV). The HLGs of both dyes are found to be in the semiconducting region i.e., 1.72 eV and 1.75 eV for dye 9 and dye 10, respectively. The low values of HLGs suggest that there is a better chance of photon absorption in the near the infrared region of the electromagnetic spectrum.

Further, the ΔG_{inject} values for dyes 9 and 10 are −0.5 eV and −0.651 eV, respectively, where dye 10 exhibits more negative value, indicating that it has higher electron injection efficiency as compared to dye 9. The same can be attributed to the presence of double acceptor moieties present in dye 10. From the LHE values, which are shown in Table 11.7, the values are reported as 0.002 eV and 0.003 eV for dyes 9 and 10, respectively. It may be concluded from the results that both the designed dyes have similar LHE values or that the dye with double acceptor moiety shows a higher LHE value as compared to the single acceptor dye molecule. From Table 11.8, we have reported the dye regeneration (ΔG_{reg}) energy, open circuit voltage, and other parameters. The dye regeneration energy (ΔG_{reg}) calculated for dye 9 is 0.96 eV, and for dye 10 it is 0.86 eV. This tells us that dye 10 will take less time to regenerate due to its lower value of ΔG_{reg} energy as compared to dye 9; therefore we can infer from this result that the double acceptor approach is better for designing the photosensitizer, which has absorption in the near infrared region. Further, the open circuit voltage, eV_{OC}, is calculated for both the dyes, with dye 9 and dye 10 showing values of 1.36 eV and 1.49 eV, respectively. This suggests that the double acceptor approach proves to be better as compared to the single acceptor approach in designing the photosensitizer due to the higher value of the open circuit voltage of dye 10 as compared to the dye 9. Moreover, from our results, it is seen that the designed molecules show good enough values to indicate their feasibility for use as effective dye photosensitizers.

11.6 LIMITATION OF ORGANIC SEMICONDUCTING MATERIALS AND FUTURE SCOPE

The limitation of OSM is about their low computing speed and their lower mobility, which make their application limited. The mobility of the organic materials like anthracene, thiophene, and spiro-OMeTAD is found to be in the range of 0.005–10 $cm^2 V^{-1} s^{-1}$, which is quite low as compared to inorganic materials like silicon and germanium, showing electron mobility of 1000–4000 $cm^2 V^{-1} s^{-1}$, respectively. Further, in the organic solar cells, the efficiency is quite low as compared to that of inorganic materials like silicon-based solar cells. Therefore, there is a huge potential in the field of organic electronics to improve the performance of organic semiconducting materials so that an organic material alternative to the conventional inorganic materials (like silicon and germanium, which are mostly used in all types of semiconducting devices) could be found.

11.7 CONCLUSION

The diode-reviewed molecules, based on the Aviram and Ratner model of [D-σ-A] molecular systems, show excellent features for their possible use in organic electronics as organic *p–n* junction diodes. These molecules were exposed to the external electric field and the direction of the field was kept along both sides (+*X* and −*X*). At zero electric field, a small HOMO-LUMO gap indicates their semiconducting nature. When these moieties were reoptimized in the presence of an electric field from the D to A and A to D directions, it was clearly seen that the HOMO-LUMO gap further decreased when the direction of the electric field was from D to A. These results strongly suggests that electric field direction from D to A resembles the forward bias condition of the organic *p–n* junction diode, whereas the electric field direction from A to D resembles the reverse bias condition. Therefore, modeled [D-σ-A] molecules may be good contenders for the design of efficient organic *p–n* junction diodes as charge transport is expected in one direction only.

Further, for the organic field effect transistor, we reviewed three molecules that were denoted by M1, M2, and M3. For M1, the reorganization energy λ_e was found to be smaller as compared to the reorganization energy of the holes λ_h. And for the molecular systems M2 and M3 (coumarin derivatives M2 and M3), the reorganization energy of holes is smaller as compared to the reorganization energy of the electrons. Thus it is expected that M2 and M3 molecular systems are appropriate for the *p*-channel OFETs.

For dye-sensitized solar cells, we reviewed ten dyes where dyes from 1 to 8 strongly suggest that the optoelectronic parameters of dye sensitizers can be significantly improved by increasing electron donor strength. The optoelectronic parameters included LHE, electron injection driving force, short-circuit photocurrent density, and open-circuit voltage. And for dyes 9 and 10, the results suggested that the double acceptor approach has proved to be a significant contributor in enhancing the performance of the photosensitizers for DSSC also. Therefore, we infer that the double acceptor approach is also better in designing efficient photosensitizers.

11.7.1 Acknowledgment

Dr. Shamoon Ahmad Siddiqui and coauthors would like to thank the editor of the book for giving an opportunity to make a contribution as a chapter writing on organic electronics.

REFERENCES

[1] N. Hall, "Twenty-five years of conducting polymers," *Chem. Commun.*, vol. 2, no. 1, pp. 1–4, 2003, doi: 10.1039/b210718j.

[2] M. Sano, M. Pope, and H. Kallmann, "Electroluminescence and band gap in anthracene," *J. Chem. Phys.*, vol. 43, no. 8, pp. 2920–2921, 1965, doi: 10.1063/1.1697243.

[3] J. Yang, Z. Zhao, S. Wang, Y. Guo, and Y. Liu, "Insight into high-performance conjugated polymers for organic field-effect transistors," *Chempr*, vol. 4, no. 12, pp. 2748–2785, 2018, doi: 10.1016/j.chempr.2018.08.005.

[4] A. Facchetti, "Polymer donor-polymer acceptor (all-polymer) solar cells," *Mater. Today*, vol. 16, no. 4, pp. 123–132, 2013, doi: 10.1016/j.mattod.2013.04.005.

[5] M. Forsberg, E. A. Serban, C. L. Hsiao, M. Junaid, J. Birch, and G. Pozina, "Near band gap luminescence in hybrid organic-inorganic structures based on sputtered GaN nanorods," *Sci. Rep.*, vol. 7, 2017, doi: 10.1038/s41598-017-01052-4.

[6] D. Singh, A. Kumar, and D. Kumar, "Adsorption of small gas molecules on pure and Al-doped graphene sheet: A quantum mechanical study," *Bull. Mater. Sci.*, vol. 40, no. 6, pp. 1263–1271, 2017, doi: 10.1007/s12034-017-1478-x.

[7] W. Y. Lo, N. Zhang, Z. Cai, L. Li, and L. Yu, "Beyond molecular wires: Design molecular electronic functions based on dipolar effect," *Acc. Chem. Res.*, vol. 7, no. 9, pp. 1852–1863, 2016, doi: 10.1021/acs.accounts.6b00305.

[8] A. Facchetti, "Semiconductors for organic transistors," *Mater. Today*, vol. 10, no. 3, pp. 28–37, 2007, doi: 10.1016/S1369-7021(07)70017-2.

[9] M. J. Kumar, "Molecular Diodes and Applications," *Recent Pat. Nanotechnol.*, vol. 1, no. 1, pp. 51–57, 2008, doi: 10.2174/187221007779814790.

[10] H. Sirringhaus, "25th Anniversary article : Organic field-effect transistors : The path beyond amorphous silicon," *Adv Mater.*, pp. 1319–1335, 2014, doi: 10.1002/adma.201304346.

[11] R. A. Marcus, "On the theory of electrontransfer reactions. VI. Unified treatment for homogeneous and electrode reactions," *J. Chem. Phys.*, vol. 43, pp. 679–701, 1965, doi: 10.1063/1.1696792.

[12] D. J. Kalita, and S. Ahmed, "Charge transport in isoindigo- dithiophenepyrrole based D-A type oligomers : A DFT/TD-DFT study for the fabrication of fullerene-free organic solar cells," *J. Chem. Phys.*, vol. 5648, no. 1993, 2018, doi: 10.1063/1.5055306.

[13] A. Kargeti, T. Rasheed, and S. A. Siddiqui, "Quantum chemical calculation for the efficient charge transfer molecules in the organic electronic devices : DFT insight," *Mater. Today Proc.*, 2022.

[14] R. Aviram, "Molecular rectifiers," *Chem. Phys. Lett.*, vol. 29, no. 2, pp. 277–283, 1974.

[15] A. Tsumura, H. Koezuka, and T. Ando, "Macromolecular electronic device : Field effect transistor with a polythiophene thin film," *Appl. Phys. Lett.*, vol. 49, pp. 1210–1212, 1986, doi: 10.1063/1.97417.

[16] T. Rasheed, A. Kargeti, and S. Ahmad, "Design of two novel dyes having maximum absorption in infrared region : A DFT investigation," *Mater. Today Proc.*, vol. 68, no. 68, pp. 2715–2719. 2022, doi: 10.1016/j.matpr.2022.06.267.

[17] P. Hohenberg and W. Kohn, "Inhomogeneous electron gas," *Phys. Rev.*, vol. 136, p. B864, 1964.

[18] W. Kohn and L. J. Sham, "Self-consistent equations including exchange and correlation effects," *Phys. Rev.*, vol. 140, 1965.

[19] A. Kargeti, T. Rasheed, and S. Ahmad, "Utilization of asymmetrical electron transport as strategy for modelling and design of efficient single molecule diodes : A DFT investigation," *Comput. Theor. Chem.*, vol. 1205, p. 113441, 2021, doi: 10.1016/j.comptc.2021.113441.

[20] M. J. Frisch et al., "Gaussian 09, Revision D.01," Wallingford: Gaussian Inc., 2013, doi: 10.1017/CBO9781107415324.004.

[21] S. A. Siddiqui, "Molecular modeling and simulation for the design of dye sensitizers with mono- and di-substituted donor moieties," *J. Comput. Electron.*, vol. 21, no. 1, pp. 52–60, 2022, doi: 10.1007/s10825-021-01839-9.

[22] R. Dennington, T. A. Keith, and J. M. Millam, "GaussView 6," Shawnee Mission, KS: Semichem Inc., 2016.

[23] T. Yanai, D. P. Tew, and N. C. Handy, "A new hybrid exchange-correlation functional using the Coulomb-attenuating method (CAM-B3LYP)," *Chem. Phys. Lett.*, vol. 393, no. 1–3, pp. 51–57, 2004, doi: 10.1016/j.cplett.2004.06.011.

[24] A. Salleo, "Charge transport in polymeric transistors," *Mater. Today*, vol. 10, no. 3, pp. 38–45, 2007, doi: 10.1016/S1369-7021(07)70018-4.

[25] S. A. Siddiqui, "In silico design of organic p–n junction diodes using quantum chemical calculations," *J. Comput. Electron.*, vol. 19, no. 1, pp. 80–90, 2020, doi: 10.1007/s10825-020-01447-z.

[26] H. Sirringhaus, P. J. Brown, R. H. Friend, M. M. Nielsen, K. Bechgaard, and A. J. H. Spiering, "Two-dimensional charge transport in conjugated polymers,"Nature, vol. 401, pp. 685–688, 1999.

[27] S. A. Siddiqui, "In silico investigation of the coumarin-based organic semiconductors for the possible use in organic electronic devices," *J. Phys. Org. Chem.*, vol. 32, no. 3, pp. 1–10, 2019, doi: 10.1002/poc.3905.

[28] B. S. Ong, Y. Wu, P. Liu, and S. Gardner, "High-performance semiconducting polythiophenes for organic thin-film transistors," *J. Am. Chem. Soc.*, vol. 126, no. 11, pp. 3378–3379, 2004, doi: 10.1021/ja039772w.

[29] J. Huang et al., "P-type sensitized organic solar cells with cascade energy alignment," *J. Phys. D. Appl. Phys.*, vol. 45, p. 195101, 2012, doi: 10.1088/0022-3727/45/19/195101.

[30] S. A. Siddiqui, A. Al-Hajry, and M. S. Al-Assiri, "Ab initio investigation of 2,2′-bis (4-trifluoromethylphenyl)-5,5′-bithiazole for the design of efficient organic field-effect transistors," *Int. J. Quantum Chem.*, vol. 116, no. 5, pp. 339–345, 2016, doi: 10.1002/qua.25034.

[31] M. G. Brian O' Regan, "A low-cost, high efficiency solar cell based on dye-sensitized colloidal TiO2 films," *Nature*, vol. 354, pp. 56–58, 1991.

[32] S. Shalini, R. Balasundaraprabhu, T. S. Kumar, N. Prabavathy, S. Senthilarasu, and S. Prasanna, "Status and outlook of sensitizers/dyes used in dye sensitized solar cells (DSSC): a review," *Int. J. energy Res.*, 2016, doi: 10.1002/er.

[33] J. Gong, K. Sumathy, Q. Qiao, and Z. Zhou, "Review on dye-sensitized solar cells (DSSCs): Advanced techniques and research trends," *Renew. Sustain. Energy Rev.*, vol. 68, pp. 234–246, 2017, doi: 10.1016/j.rser.2016.09.097.

[34] A. Abate et al., "Energy & environmental science," *Energy Environ. Sci.*, 2015, doi: 10.1039/C5EE02014J.

[35] J. Chen et al., "Backbone conformation tuning of carboxylate-functionalized wide band gap polymers for efficient non-fullerene organic solar cells," *Macromolecules*, 2019, doi: 10.1021/acs.macromol.8b02360.

[36] H. Wang, B. Bao, X. Hu, and J. K. Fang, "Efficient small molecule organic dyes containing different bridges in donor moieties for dye-sensitized solar cells," *Electrochim. Acta*, vol. 250, pp. 278–284, 2017, doi: 10.1016/j.electacta.2017.08.089.

[37] S. Mathew et al., "Dye-sensitized solar cells with 13% efficiency achieved through the molecular engineering of porphyrin sensitizers," *Nat. Chem.*, vol. 6, no. 3, pp. 242–247, 2014, doi: 10.1038/nchem.1861.

[38] K. Kakiage, Y. Aoyama, T. Yano, K. Oya, J. I. Fujisawa, and M. Hanaya, "Highly-efficient dye-sensitized solar cells with collaborative sensitization by silyl-anchor and carboxy-anchor dyes," *Chem. Commun.*, vol. 51, no. 88, pp. 15894–15897, 2015, doi: 10.1039/c5cc06759f.

[39] M. Bourass *et al.*, "DFT and TD-DFT calculation of new thienopyrazine-based small molecules for organic solar cells," *Chem. Cent. J.*, 2016, doi: 10.1186/s13065-016-0216-6.

[40] S. Jungsuttiwong, R. Tarsang, T. Sudyoadsuk, V. Promarak, P. Khongpracha, and S. Namuangruk, "Theoretical study on novel double donor-based dyes used in high efficient dye-sensitized solar cells: The application of TDDFT study to the electron injection process," *Org. Electron.*, 2013, doi: 10.1016/j.orgel.2012.12.018.
[41] S. ElKhattabi *et al.*, "Theoretical investigation of electronic, optical and photovoltaic properties of alkylamine-based organic dyes as sensitizers for application in DSSCs," *J. Mater. Environ. Sci.*, vol. 9, pp. 841–853, 2018, doi: 10.26872/jmes.2018.9.3.93.
[42] Z. M. E. Fahim, Y. A. Aicha, S. M. Bouzzine, M. Bouachrine, and M. Hamidi, "Modulation on dye/TiO2 bending energy and charge transfer to high performance triphenylamine based sensitizers in solar cells: A DFT study," Jul. 2018, doi: 10.1109/IRSEC.2018.8702924.[43] G. D. Sharma et al., "Stepwise co-sensitization as a useful tool for enhancement of power conversion efficiency of dye-sensitized solar cells: The case of an unsymmetrical porphyrin dyad and a metal-free organic dye," *Org. Electron.*, 2014, doi: 10.1016/j.orgel.2014.03.033.
[44] T. D. Nguyen and C. G. Wu, "Non-classical design of high-efficiency sensitizers for dye-sensitized solar cells," *J. Chinese Chem. Soc.*, 2018, doi: 10.1002/jccs.201700399.
[45] M. G. Aswani Yella, H.-W. Lee, H. N. Tsao, C. Yi, A. K. Chandiran, Md.Khaja Nazeeruddin, E. W.-G. Diau, C.-Yu Yeh, S. M. Zakeeruddin, "Porphyrin-sensitized solar cells with yield shorter electron lifetimes when used in the DSC. This faster interfacial charge recombina- cobalt (II/III)–based redox electrolyte tion, when compared with iodide/triiodide–based redox electrolytes (12–18), lowe," *Science,* vol. 334, no. 6056, pp. 629–633, 2011.
[46] L. Zheng *et al.*, "Novel D-A-\$π\$-a-type organic dyes containing a ladderlike dithienocyclopentacarbazole donor for effective dye-sensitized solar cells," *ACS Omega*, vol. 2, no. 10, pp. 7048–7056, 2017, doi: 10.1021/acsomega.7b01387.
[47] M. J. Frisch et al., "Gaussian 16," Gaussian, Inc., Wallingford CT, 2016.
[48] V. V. Divya and C. H. Suresh, "Density functional theory study on the donating strength of donor systems in dye-sensitized solar cells," *New J. Chem.*, vol. 44, no. 17, pp. 7200–7209, 2020, doi: 10.1039/d0nj00723d.
[49] E. Zahedi, "Transport properties of a single-molecular diode with one backbone, and two backbones in parallel: Frontier orbital analysis and NEGF-DFT study," *Eur. Phys. J. Plus*, vol. 130, no. 5, 2015, doi: 10.1140/epjp/i2015-15096-2.
[50] A. D. Becke, "A new mixing of hartree-fock and local density-functional theories," *J. Chem. Phys.*, vol. 98, no. 2, pp. 1372–1377, 1993, doi: 10.1063/1.464304.
[51] A. Kargeti et al., "Exploration of superacidic properties of HMnF using density functional theory," *Macr. Symposia*, vol. 388, no. 1, pp. 1900040, 2019.
[52] M. Mas-Torrent, "Novel small molecules for organic field-effect transistors: Towards processability and high performance," *Chem Soc Rev*, vol. 37, no. 4, 2008, doi: 10.1039/b614393h.
[53] Ankit Kargeti et al.,"DFT investigation on D-pi-A structures for n-type semiconducting materials for the application in organic field effect transistors," *Mat. Today: Proc.,* vol. 57, pp. 569–576, 2022.
[54] G. Koyyada et al., "A new series of EDOT based co-sensitizers for enhanced efficiency of cocktail DSSC: A comparative study of two different anchoring groups," *Molecules*, 2019, doi: 10.3390/molecules24193554.
[55] A. Irfan, R. Jin, A. G. Al-Sehemi, and A. M. Asiri, "Quantum chemical study of the donor-bridge-acceptor triphenylamine based sensitizers," *Spectrochim. Acta—Part A Mol. Biomol. Spectrosc.*, vol. 110, pp. 60–66, 2013, doi: 10.1016/j.saa.2013.02.045.
[56] M. Wagner, J. Hofinger, M. Setvín, L. A. Boatner, M. Schmid, and U. Diebold, "Prototypical organic-oxide interface: Intramolecular resolution of sexiphenyl on In2O3(111)," *ACS Appl. Mater. Interfaces*, vol. 10, no. 16, pp. 14175–14182, 2018, doi: 10.1021/acsami.8b02177.

12 Hydrogen Storage Efficiency of Isomeric Cu(I)-Triazine Complexes

In Quest of New Hydrogen Storage Material

Abhishek Bag, Mrinal Kanti Dash, Santanab Giri, Gobinda Chandra De, and Gourisankar Roymahapatra

12.1 INTRODUCTION

Green energy is the current emergent issue in our modern civilization, and it is gradually increasing. The increasing human population and simultaneous needs signal the increasing demand for energy. Different countries use renewable energy in addition to fossil fuel to meet their needs. Fossil fuel, which is main source of greenhouse gases generation, leads to environmental pollution and global warming.[1,2] So developing countries are trying hard to decrease dependency on fossil fuels and reach the goal of net-zero countries within the oncoming decades. But it is not so easy a task to stop the use of conventional fossil fuel immediately. We need to establish alternative green energy sources to meet the global need. In recent years, hydrogen energy has gained popularity as a green fuel, as an alternative to conventional fuels, with its high energy density, with its optimal working pressure and temperature, and without carbon combustion, all of which produce a net-zero environmental impact.[3] The hydrogen atom contains one electron and one proton. It is a colorless and odorless compound. The gravimetric density of hydrogen energy is seven times higher than that of fossil fuel. In 2040, it is predicted that about 35% of the vehicles in Europe will be powered by hydrogen energy. The Government of India declared 2020 to be the year to introduce hydrogen as the fuel as their mission. India wants to become energy independent by 2047 and achieve "Net Zero" by 2070. To achieve this target, we need to increase the use of renewable energy. Green hydrogen is considered to be a promising alternative for enabling this transition. Hydrogen can be utilized for long-duration storage of renewable energy. It replaces the use of fossil fuels in industry, clean transportation, and potential element for decentralized power generation, aviation, and marine transport. The mission has been approved by the Union Cabinet on January 4, 2022. But due to some problems with storage and transportation of gaseous H_2, more technology has to be developed. So the main challenge is to develop hydrogen storage materials with high gravimetric density and cost-effective transportation at ambient conditioned.[4,5] According to the United States Department of Energy (US-DOE), a good hydrogen storage system should have high hydrogen holding capacity with a gravimetric density greater than 5.5 wt%.[6] Until now, the conventional hydrogen storage methods have been used to hold hydrogen as a gas in storage cylinders at extremely high pressure (700–800 bar) or as a liquid at the cryogenic temperature of 20 K.[7] So these two processes are therefore associated with energy loss, maintenance issues, insufficient storage capacity, and safety concerns. This has motivated researchers to investigate material-based hydrogen storage techniques in which hydrogen molecules are held by adsorption phenomena. There are mainly two types of adsorption process: the physisorption and chemisorption processes. In chemisorption, hydrogen can adsorb on

 DOI: 10.1201/9781003441328-12

the storage material in atomic form with high adsorption energy (2–4 eV), which yields potentially high gravimetric density. But in this process, the desorption temperature is very high. On the other hand, in the case of the physisorption process, hydrogen adsorption energy is comparatively low (~meV), which causes easy desorption at a much lower temperature than in the former case.[8] The detailed studies show that a hydrogen storage material should be considered ideal for practical use at near ambient conditions only if the storage material holds hydrogen in a quasi-molecular fashion via weak noncovalent interaction with an intermediate binding energy, which falls between the chemisorbed and physisorbed states.[9,10]

The development of new hydrogen storage materials has become a challenging area of research. Many researchers have introduced different types of H_2 storage systems like metal-organic frame (MOF) work, clathrates, metal hydrides, and metal clusters.[11–23] Both theoretical and experimental observation shows that hydrogen hydrates (clathrates hydrates) are a new way to store hydrogen gas in the future.[24] Due to low energy consumption for the charge and discharge, safety, cost-effectiveness, and favorable environmental features, hydrogen hydrates are looking like the most important materials for hydrogen storage.[25,26] Due to decreases of fossil fuel supply and increased demand for natural gas, methane hydrates take a vital role to act as a future energy resource.[27–31] But most of the materials are not up to the mark for fulfilling the target set by the U.S. DOE.[32] Different types of H_2 storage materials have different problems. It is shown that the surface of C_{60} and $Li_{12}C_{60}$ strongly bonded with Li metals and that results were excellent for trapping hydrogen molecules with high gravimetric wt%. Light-metal- (Li, Be, Mg, and Al) doped MOFs show high stability during hydrogen adsorption with high binding energy.[33] On the other hand, transition metal- (Sc, Ti, V) or boron-decorated silsesquioxane frameworks are found to be good hydrogen-capturing systems with very low gravimetric wt%.[34] It is reported in the literature that alkali-, alkaline-earth-, and transition-metal-doped with B_6H_6 complexes show higher hydrogen adsorption. Magnesium-doped boron clusters (Mg_2B_n, $n = 4–14$) are very good hydrogen trapping materials with an average adsorption energy of 0.13–0.22 eV/H_2.[35–37]

Our small group is continually studying the importance of M-AH (metal-aromatic heterocyclic) systems, and we have reported many Li-AH systems,[38–44] which meet the target set by DOE in all aspects. But very little work has been done on transition-metal-doped aromatic heterocyclic systems because cluster formation may occur, which lowers the H_2 storage capacity.[45] Based on a review of the different hydrogen storage systems, adsorbing material must have certain properties; (a) H_2 can adsorb in the molecular state, not in the atomic state; (b) it will not form a cluster; (c) it has to be light weight; (d) it should meet the target set by DOE; (e) the reversible adsorption should be preferred in the quasi-adsorption process.

In our present study, we have investigated the storage capacity of isomeric triazine doped with coinage metal, and we have selected Cu(I) due to its easy availability and low cost compared to the other two counterparts. The whole chapter is organized into four major sections. In Section 12.1, we have discussed in detail why hydrogen storage research is required, what are the challenges in this field, their probable solutions, and various studies done in this field. In Section 12.2, we discuss the theoretical studies and computations related to the calculation and analysis of the results. Section 12.3 mainly involves the discussion and analysis of the bonding nature of the mono- and di-Cu(I)- decorated isomeric triazine using the electron localization function (ELF) study, electrostatic potential (ESP) mapping, noncovalent interaction (NCI) study, partial density of state (PDOS) studies, EDA (energy decomposition analysis) calculation study, the effect of temperature on H_2 adsorption, and their gravimetric wt% analysis. Finally, we summarize the study in Section 12.4.

12.2 THEORY AND COMPUTATIONAL DETAILS

The designed systems are optimized to their local minimum on the potential energy surface with zero imaginary frequency with the help of the Gaussian 16w quantum chemistry program package[46] within the density functional theory (DFT)[47–49] framework. We have taken different functionals like

CAM-B3LYP, M05, and different basis sets like 6–311+g (d,p), 6–31+G*, and LanL2DZ. We have optimized the bare ring at CAM-B3LYP/6–311+g (d,p) level of theory and the H_2 trapped metal decorated system at M05 functional with mixed basis sets 6–31+G* and LanL2DZ. The stability and chemical reactivity of the Cu(I)-triazine system and their H_2 trapped complexes have been analyzed by different conceptual DFT- (CDFT)[50,51] based descriptors like hardness (η) and electrophilicity (ω), calculated with standard techniques.[52–54] The average binding energy (E_b) value of each Cu(I)-triazine systems is determined using equation (11.1):

$$E_b = \left[E_{(triazine\text{-}Cu)^+ @nH_2} - \left(E_{(triazine\text{-}Cu)^+} + nEH_{H_2} \right) \right] / n \qquad (11.1)$$

where $E_{(triazine\text{-}Cu)^+ @nH2}$, $E_{(triazine\text{-}Cu)^+}$, and E_{H2} represent the energy of the gradual H_2 trapped Cu(I)-doped triazine system, only Cu(I) doped triazine system, and the energy of the H_2 molecule. Here, n symbolizes the number of H_2 molecules trapped.

The calculation of average adsorption energy (E_{ads}) of H_2 molecules is done by using equation (11.2):

$$E_{ads} = \left[\left(E_{(triazine\text{-}Cu)^+} + nE_{H_2} \right) - E_{(triazine\text{-}Cu)^+ @nH_2]} \right] / n \qquad (11.2)$$

To calculate the hydrogen storage efficiency of our studied systems, we have performed gravimetric wt% analysis with the help of equation (11.3):

$$\begin{aligned} &\text{Gravimetric wt\%} = [\text{Molecular weight of trapped } H_2 / \\ &\text{Molecular weight of} (\text{heterocyclic ring} - \text{Cu})^+ @ nH_2] \times 100 \end{aligned} \qquad (11.3)$$

Gibbs free energy change with variation of temperature has been calculated to know about the spontaneous nature of the hydrogen adsorption process by using the following equation:

$$\Delta G_{(triazine\text{-}Cu\,)^+ @nH_2} = \left[G_{(triazine\text{-}Cu @nH_2\,)^+} - G_{(triazine\text{-}Cu\,)^+} - nG_{H_2} \right] \qquad (11.4)$$

We have also calculated the nucleus independent chemical shift, which is also called NICS. It is a very essential computational tool for the assessment of aromaticity. This method was invented by the late Paul V. R. Schleyer in 1996.[55] Currently, it is the most widely used method for determining and justifying aromaticity and antiaromaticity. At the time of the calculation of NICS, a ghost atom, Bq, is placed at the center of the ring plane (denoted as NICS(0)), which generally accounted for σ aromaticity and took on another ghost atom Bq1 to 1 Å above the center of the molecular plane (denoted as NICS(1)), which can denote π aromaticity. If the NICS(0) and NICS(1) values of a ring become negative, then the systems are both σ and π aromatic in nature, but if the value is positive, then the systems become antiaromatic.

We have calculated the NICS (0,1) (nucleus-independent chemical shift) values to check the pattern of aromaticity and stability of the Cu(I)-doped isomeric triazine systems as well as for Cu(I) isomeric triazine@nH_2 systems.[56–59]

To know the bonding nature inside the different systems, we have performed the topological analysis with the help of the Multiwfn package,[60] followed by shaded surface map and electron localization function (ELF) analysis, which indicates the information about the localized electron. NCI (noncovalent interaction) analysis provides a link for the identification of noncovalent interaction in H_2-trapped Cu(I) doped different isomeric triazine systems.[61] The NCI results are also generated by using the Multiwfn software package.[60] Energy decomposition analysis (EDA) has been performed to explain the stability of model systems by employing S. Liu's method.[60] In hydrogen trapped complexes, we took metal-decorated triazine as one fragment and molecular H_2 as another fragment. The total interaction energy (ΔE_{total}) of the hydrogen-trapped system decomposed into the steric repulsion term (ΔE_{steric}), electrostatic term ($\Delta E_{electrostatic}$), and quantum term ($\Delta E_{Quantum}$). This will help in understanding the stability of the model systems.[43, 62, 63]

12.3 RESULT AND DISCUSSION

12.3.1 Mono- and Di-Cu(I)-Decorated Isomeric Triazine Systems

We have optimized the isomeric triazine (123,124, 135) decorated with Cu(I). At first, we optimized mono-metal-decorated systems and then introduced another Cu(I) to get di-metal-decorated isomeric triazine systems. where all the metal has connected to the more electronegative ring N atom, as in Figure 12.1. The third Cu(I) was not introduced due to the steric hindrance and low gravimetric wt% of the systems.

In Table 12.1 shows the hardness (η), electrophilicity (ω), and NICS value of mono- and di-Cu(I)-decorated systems. From Table 12.1, it has been observed that 135-triazine (bare) is the most stable isomer as it has a high hardness value. Among mono-metal-decorated isomers, 123 gave the higher stability as it contains high hardness value compared to other mono-metal-decorated systems. In the case of di-metal-decorated systems, 135 gives the higher stability as it contains high hardness value among

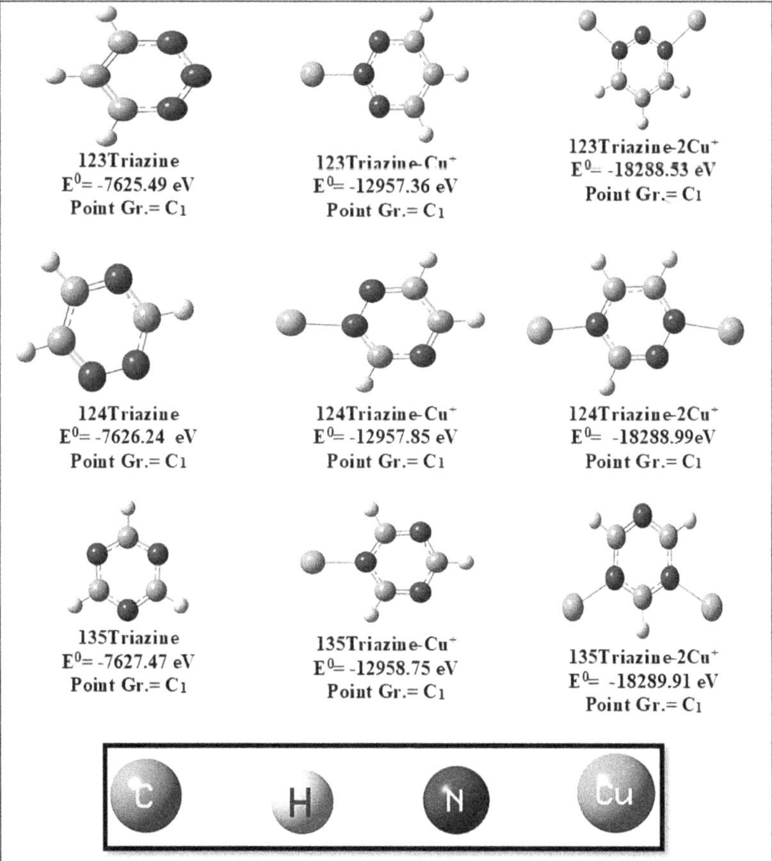

FIGURE 12.1 Different isomers of triazine and their mono- and di-Cu(I)-decorated optimized geometries at a different function like CAM-B3LYP, M05 and a different basis set like 6–311+g (d,p), 6–31+G*, LanL2DZ.

TABLE 12.1
Hardness (η); (eV), Electrophilicity (ω); (eV) and NICS(0,1) of Different Isomers of Triazine

Systems	Electrophilicity	Hardness	NICS(0)	NICS(1)
123 Triazine	1.55	8.01	−4.37	−11.00
123 Triazine-Cu^+	8.36	5.43	−3.60	−11.52
123 Triazine-2Cu^+	17.41	5.18	−3.63	−10.92
124 Triazine	1.62	7.47	−3.72	−10.49
124 Triazine-Cu^+	9.03	4.93	−3.15	−10.49
124 Triazine-2Cu^+	20.64	4.40	−3.75	−10.52
135 Triazine	1.45	9.02	−4.02	−9.74
135 Triazine-Cu^+	9.63	5.00	−4.43	−9.96
135 Triazine-2Cu^+	15.75	5.71	−4.38	−9.54

other di-metal-decorated isomer. These observations also more or less are justified by the electrophilicity value of the different systems. In our studied systems, i.e., bared ring and mono- and di-metal-decorated systems, we have negative NICS (0) and NICS (1) values, indicating that they are both σ and π aromatic in nature and that the aromaticity of the triazine ring does not change very much after metal insertion.

12.3.2 H_2 Adsorption on Mono- and Di-Cu(I)-Decorated Isomeric Triazine Systems

For the study of adsorption, H_2 molecules are gradually added to the Cu center, and resultant complexes are optimized without any symmetry constraint. This process continues until no more H_2 molecules can be added by the mono- and di-Cu(I)-decorated isomeric triazine systems. It has been found that a maximum of five H_2 molecules can be adsorbed on each Cu center without distorting the metal-decorated systems. The geometric parameters of the isomeric ring, metal-decorated isomeric rings, and their hydrogenated form are shown in Table 12.2, where it is shown that H_2 molecules are adsorbed in molecular form, and the bond distance is same or very close to the H-H distance in an isolated H_2 molecule (0.74 Å).[64] From Table 12.2, it is also shown that the adsorption of H_2 molecules at the Cu center causes almost minimal changes in the bond distance like C-C, C-N, the distance between ring to the Cu center, and the distance between Cu to molecular H_2, indicating that the adsorption process is a quasi-molecular binding type.[65,66]

The stability of H_2 adsorbed mono- and di-metal-decorated isomeric triazine systems have been analyzed using global chemical reactivity descriptors such as hardness (η, eV) and electrophilicity (ω, eV). The variation of η and ω for metal-decorated isomeric triazine and its hydrogenated systems are shown in Table 12.3. It can be observed that the electrophilicity/hardness slightly decreases/increases due to the gradual adsorption of H_2 molecule for mono-Cu(I)-decorated isomeric triazine systems as compared to the nonhydrogenated metal-decorated systems. The same pattern of electrophilicity/hardness has been observed in case of di-Cu(I)-decorated isomeric triazine systems.[67,68]

To study the adsorption sequence, H_2 molecules are sequentially added to the Cu center and the successive adsorption of H_2 molecules over mono- and di-metal-decorated isomeric triazine (123-isomer), as presented in Figure 12.2. It can be observed that a maximum five H_2 molecules are adsorbed on each Cu center.

The average adsorption energy (E_{ads}) of mono- and di-metal-decorated isomeric triazine systems is found between 0.1 and 0.5 eV/H_2, as shown in Figure 12.3. This implies that the adsorptions are of a quasi-molecular type,[8] and this fact is also established during their geometrical analysis (see Figure 12.3). It is observed that the average adsorption energy gradually decreases with the gradual addition of H_2 molecules in both mono- and di-metal-decorated systems due to enhanced steric hindrance.

TABLE 12.2
Average Bond Lengths of C-C, C-N, H-H, the Distance of Cu(I) from the Ring on Which It Is Decorated (Cu(I)-ring), and Distance of Cu(I) from Adsorbed H_2 Molecules (Cu(I)-H_2) in Å

System	C-C Å	C-N Å	Ring-Cu Å	Cu-H Å
123 Triazine	1.37	1.33	—	—
123 Triazine-Cu^+	1.38	1.33	1.99	—
123 Triazine-Cu^+@5H_2	1.38	1.33	2.02	1.95
123 Triazine-2Cu^+@10H_2	1.38	1.34	2.13	2.08
124 Triazine	1.39	1.31	—	—
124 Triazine-Cu^+	1.41	1.31	1.98	—
124 Triazine-Cu^+@5H_2	1.40	1.32	2.01	2.10
124 Triazine-2Cu^+@10H_2	1.40	1.32	2.10	2.05
135 Triazine	—	1.32		
135 Triazine-Cu^+	—	1.31	1.98	—
135 Triazine-Cu^+@5H_2	—	1.31	2.02	2.03
135 Triazine-2Cu^+@10H_2	—	1.32	2.04	1.97

TABLE 12.3
Hardness (η) and Electrophilicity (ω) for Mono- and Di-Metal-Decorated Isomeric Triazines and Their Hydrogen-Adsorbed Systems

	Mono-Metal-Decorated Systems					
	123-Triazine		124-Triazine		135-Triazine	
No. of H_2 Trapped	Hardness (η)	Electrophilicity (ω)	Hardness (η)	Electrophilicity (ω)	Hardness (η)	Electrophilicity (ω)
0	5.42	8.35	4.93	9.03	5.00	9.63
1	5.45	8.28	5.06	8.64	5.85	7.45
2	5.46	8.00	5.04	8.46	5.92	7.09
3	5.40	7.88	5.02	8.24	5.94	6.95
4	5.34	7.85	5.02	8.23	5.94	6.93
5	5.18	8.00	5.02	8.19	5.94	6.87

	Di-Metal-Decorated Systems					
	123-Triazine		124-Triazine		135-Triazine	
No. of H_2 Trapped	Hardness (η)	Electrophilicity (ω)	Hardness (η)	Electrophilicity (ω)	Hardness (η)	Electrophilicity (ω)
0	5.18	17.41	4.40	20.64	5.71	15.75
2	5.10	17.38	4.37	20.52	5.66	15.66
4	5.27	16.24	4.46	19.26	5.76	14.82
6	5.15	16.04	4.41	18.83	5.70	14.42
8	5.00	16.13	4.40	18.62	5.51	14.51
10	4.66	16.64	4.24	18.83	5.21	14.71

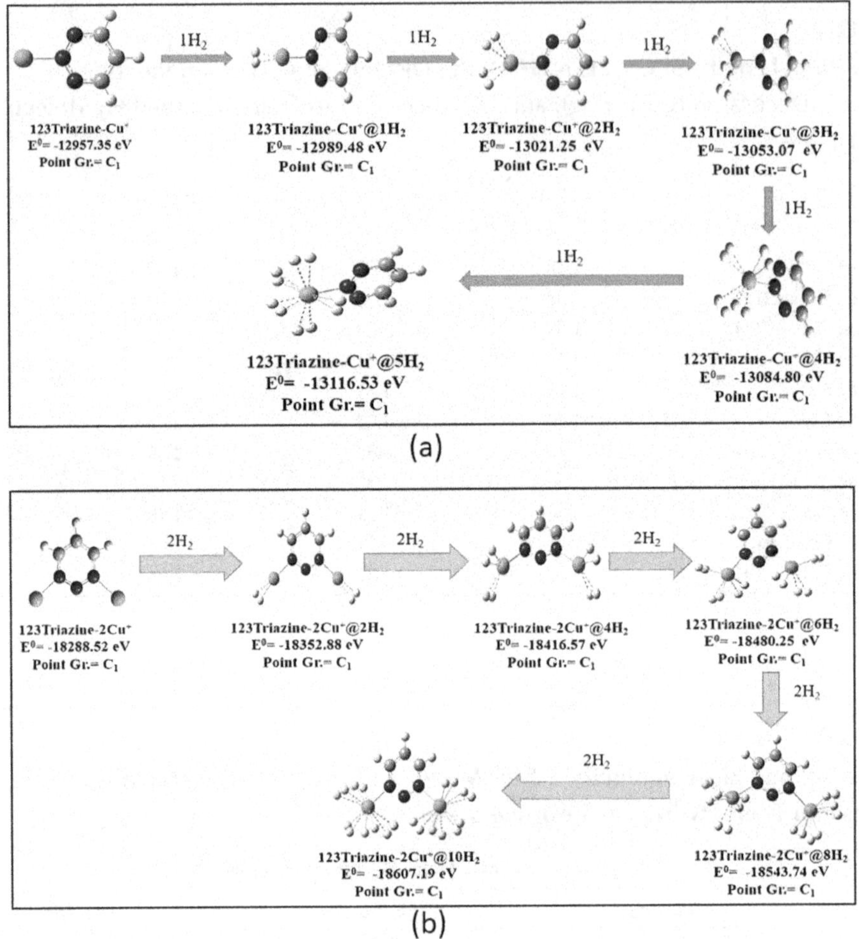

FIGURE 12.2 Sequential adsorption of H_2 molecules over (a) mono- and (b) di-metal-decorated isomeric triazine (123-isomer).

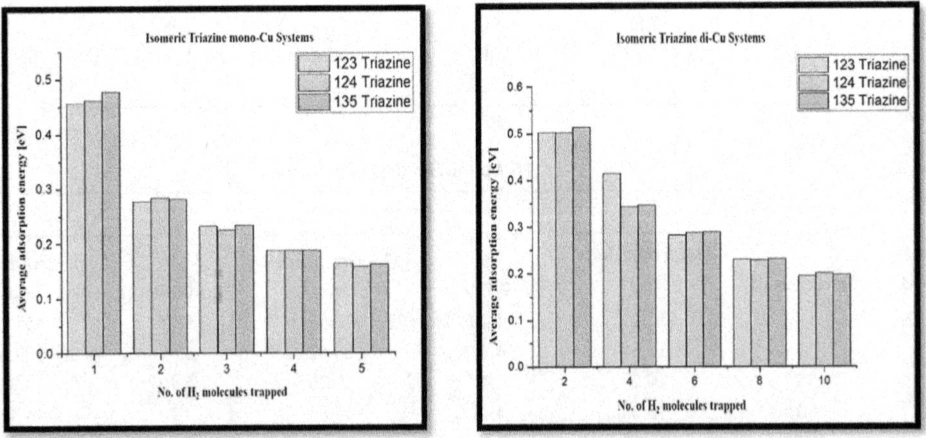

FIGURE 12.3 Variation of E_{ads} with the number of adsorbed H_2 molecules for gradual H_2 trapped mono- and di-metal-decorated isomeric triazine.

12.3.3 ESP and NBO Analysis

Adsorption of H_2 molecules over mono- and di-metal-decorated isomeric triazine causes the redistribution of the total electron density over the whole system and can be well understood by the ESP map of the hydrogenated systems, shown in Figure 12.4. The ESP map gives a three-dimensional pictorial visualization of total electron density over the molecular complexes. The variation of electron density of the complex systems can be understood by ESP mapping. In Figure 12.4, the bare isomer, the border black color appears near the ring N- atom. After the addition of metal, faded black color appears, and the intensity of the black color becomes increased after the trapping of molecular H_2, which indicates the adsorption of molecular H_2. The NBO charges on the Cu center decrease during the gradual adsorption of H_2 molecules (Figure 12.5.) in the triazine-Cu(I) complex, and this sequence also indicates the successive addition of H_2 molecules.

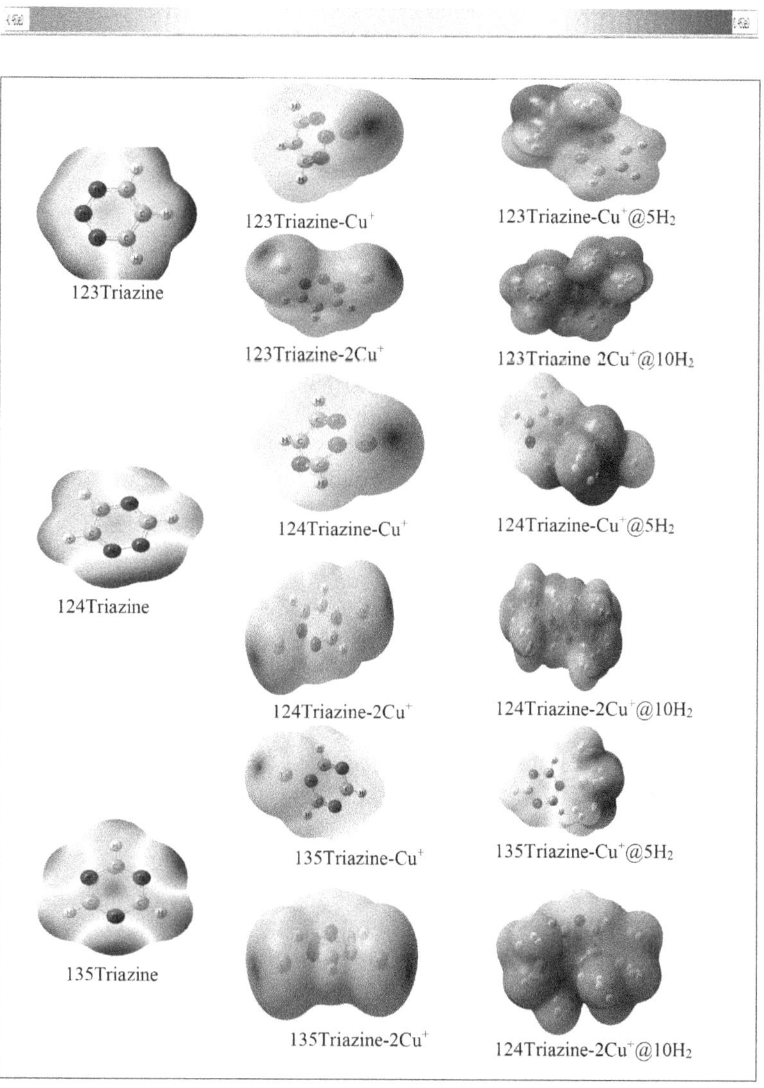

FIGURE 12.4 Electrostatic potential maps of the bare ring, host complexes, and their hydrogen-adsorbed derivatives.

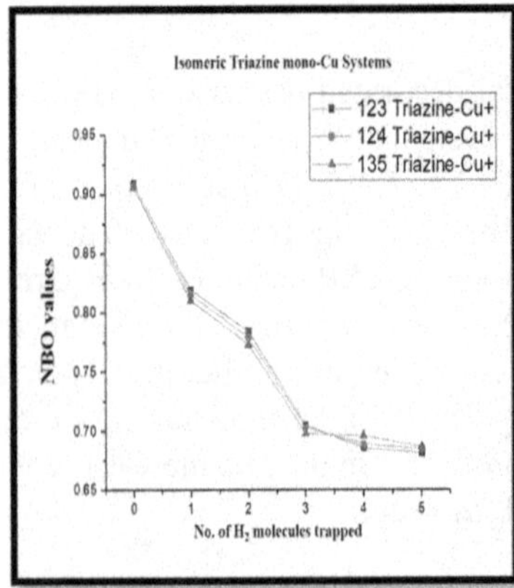

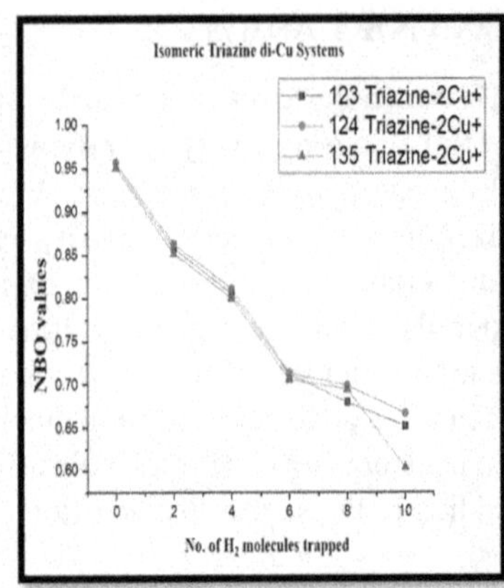

FIGURE 12.5 Variation of NBO charges in hosts and their sequential hydrogen adsorbed systems.

12.4 BONDING NATURE ANALYSIS

12.4.1 Electron Localization Function (ELF)

To know the binding nature between a heterocyclic atom (N) of the triazine ring to the Cu center and Cu center to H_2 molecules, a detailed investigation has been done by means of a shaded surface map for all the hydrogenated metal-doped systems. Figure 12.6 represents the hydrogenated mono- and di-metal-decorated isomeric triazine systems. It is found that the electron density accumulation occurs at a high electronegative N atom but not between N and the Cu center or between Cu and molecular H_2, indicating that the bonding between N and the Cu center, as well as between Cu and molecular H_2, is noncovalent in nature. The plot of the electron localization function (ELF) indicates the information about the localized electron. In this case, there are nearly no localized bonding electrons between N and the Cu center or between Cu and molecular H_2.

12.4.2 Noncovalent Interaction (NCI)

The noncovalent interaction (NCI) method, which is also known as the reduced density gradient (RDG) method, is very popular for analyzing weak bonding interactions among atoms, and it plays a crucial role for the identification of the different types of interaction. In the NCI plot, reduced density gradient (RDG) and sign $(\lambda_2)\rho$ are plotted to express the precise area of interaction. The sign $(\lambda_2)\rho$ is used to distinguish strong and weak noncovalent interactions. In the NCI plot, sign $(\lambda_2)\rho$ is placed on the *X*-axis, and RDGs are displayed in the *Y*-axis.

We have generated the NCI plot of hydrogenated and nonhydrogenated mono- and di-metal-decorated isomeric triazine systems. It is seen that the spikes appeared in the sign $(\lambda_2)\rho \approx 0$ zone after the adsorption of the H_2 molecule, whereas, in the absence of the H_2 molecule, no spikes appear in the sign $(\lambda_2)\rho \approx 0$ zone in the NCI plot (Figure 12.7.). Therefore, the presence of weak interaction supports the weak adsorption between metal and molecular H_2, which favors the quasi-sorption process.

12.4.3 Energy Decomposition Analysis (EDA)

Energy decomposition analysis (EDA) using Shubin Liu's energy decomposition approach has also been carried out using Multiwfn software. Here, total molecular energy is decomposed as $E_{Total} = E_{steric} + E_{electrostatic} + E_{quantum}$. Here, we break the molecule in two fragments—one is H_2-molecule

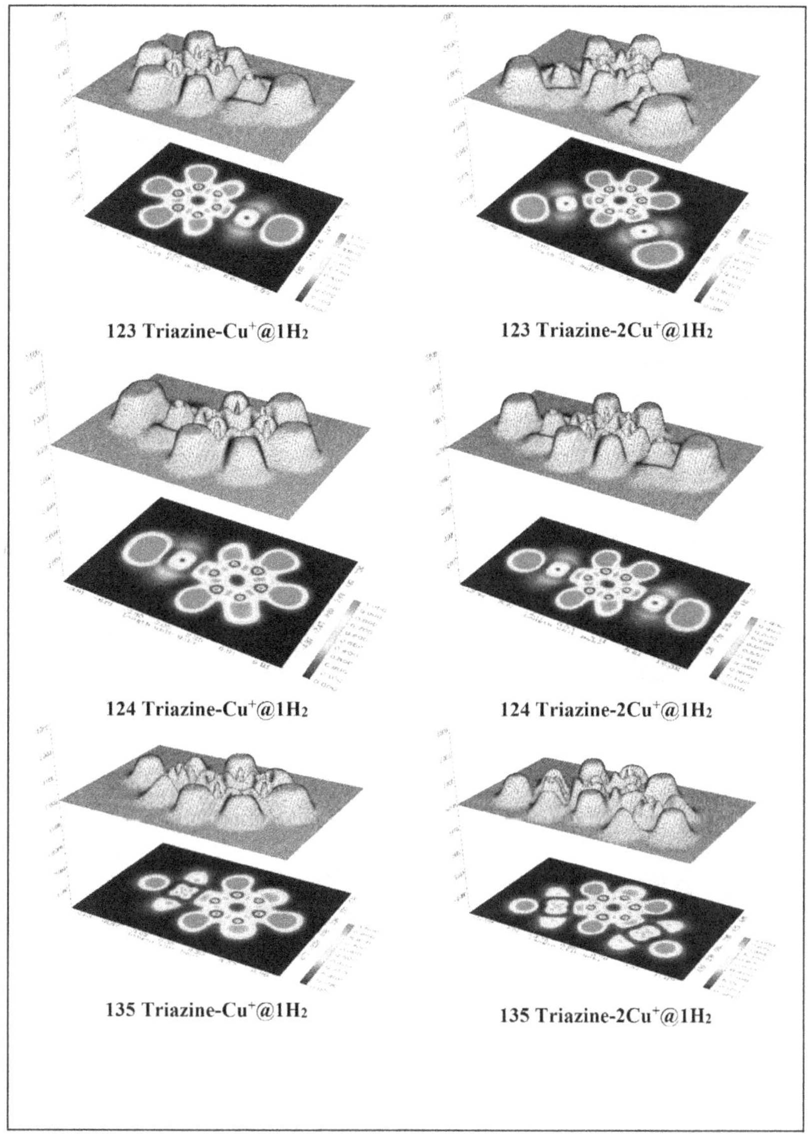

FIGURE 12.6 Shaded surface map with the projection of the electron density of the H_2-trapped mono- and di-Cu(I)-decorated isomeric triazine.

and another is metal-decorated isomeric triazine system. As can be seen from Table 12.4, each system has a negative E_{steric} value and a negative total energy value, which suggests that the model systems are stable.

12.4.4 Partial Density of State (PDOS) Analysis

To obtain the details analysis of the contribution pattern of hydrogen molecule, metal ion, and isomeric triazine toward framing the frontier molecular orbital (FMOs), we have performed the density of state (DOS) and partial density of state (PDOS) of all the studied systems, shown in the Figure 12.8.

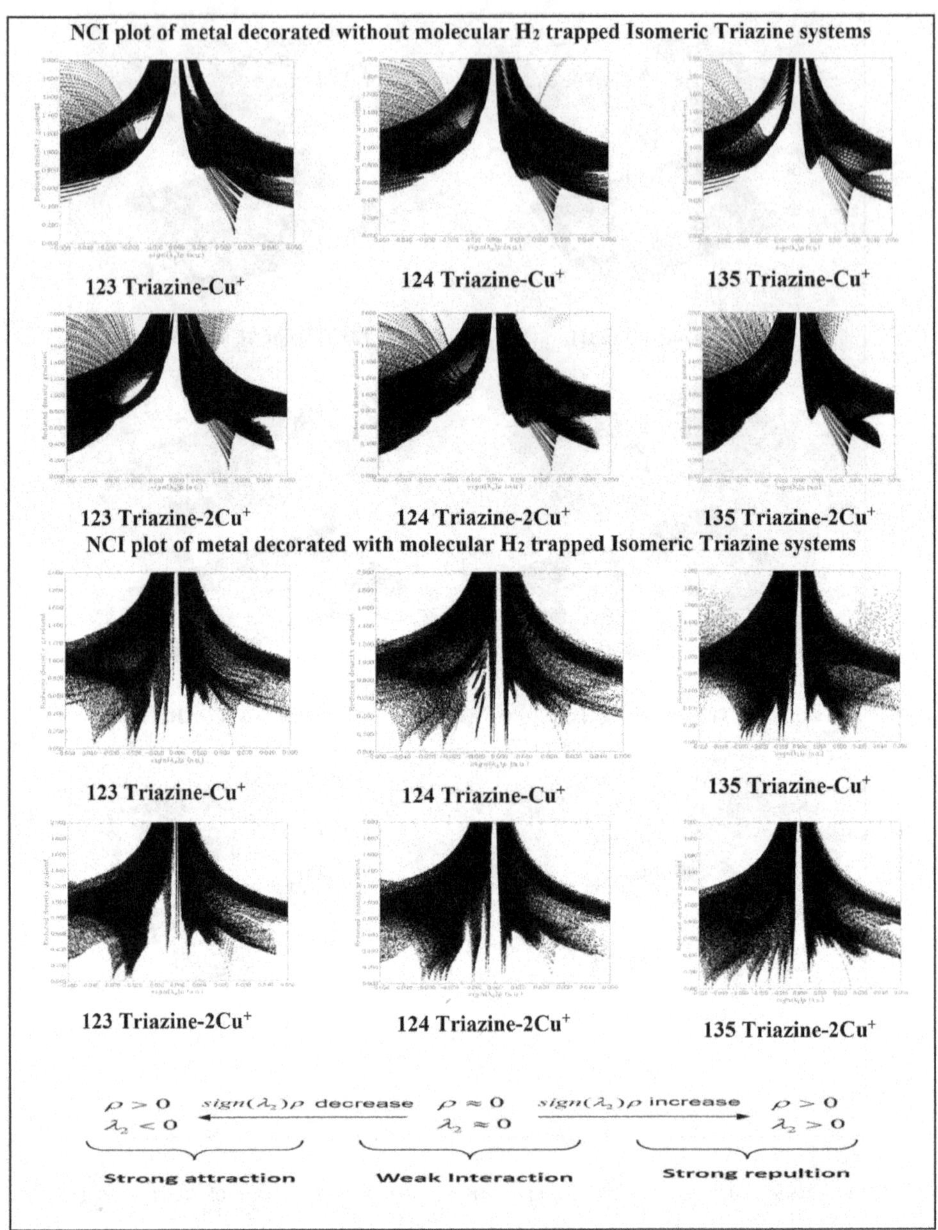

FIGURE 12.7 NCI plot of mono- and di-Cu(I)-decorated isomeric triazine with or without trapping of molecular H_2.

A study of frontier molecular orbitals (FMOs) of the studied complexes showed that the contribution of metal ion in di-metal-decorated systems toward HOMO is greater than that of mono-metal-decorated systems, except mono-metal-decorated 123-triazine. In the case of mono-metal-decorated systems, the contribution of isomeric triazine toward HOMO is greater than the di-metal-decorated systems except the 123-triazine system. In all cases, there is a significant contribution of H_2 molecules toward HOMO of all the complexes (except the 135-triazine system). At last, it is clear that, from Figure 12.8, the HOMO and LUMO of complexes are formed from the contribution of isomeric triazine, the metal ion, and molecular hydrogen.[69,70]

TABLE 12.4
EDA Results of Hydrogen-Trapped Cu(I)-Doped Isomeric Triazine Systems, Taking H_2 Molecule/s as One Fragment and Metal-Decorated Triazine Systems as Another Fragment

Mono-Metal-Decorated Systems				
Systems	ΔE_{steric}	$\Delta E_{electrostatic}$	$\Delta E_{quantum}$	ΔE_{total}
123 Triazine-Cu^+@$1H_2$	−15.37	8.27	6.33	−0.77
124 Triazine-Cu^+@$1H_2$	−15.93	8.24	6.91	−0.77
135 Triazine-Cu^+@$1H_2$	−15.43	8.74	5.90	−0.79

Di-Metal-Decorated Systems				
Systems	ΔE_{steric}	$\Delta E_{electrostatic}$	$\Delta E_{quantum}$	ΔE_{total}
123 Triazine-$2Cu^+$@$2H_2$	−31.77	18.34	11.73	−1.70
124 Triazine-$2Cu^+$@$2H_2$	−31.31	18.25	11.36	−1.70
135 Triazine-$2Cu^+$@$2H_2$	−31.94	18.68	11.53	−1.73

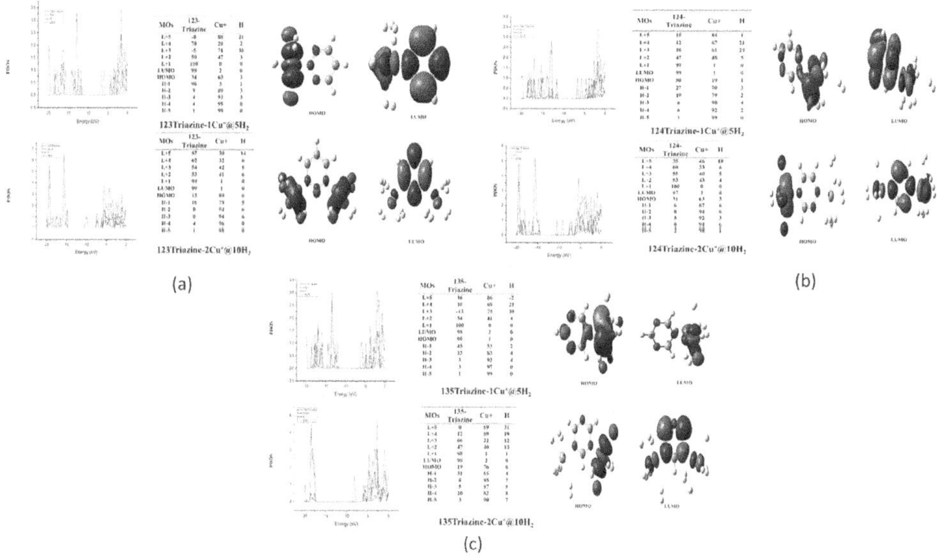

FIGURE 12.8 DOS and PDOS of H_2-trapped metal-decorated isomeric triazine (123(a), 124(b), and 135(c) with % of contribution of H, metal ion, and isomeric triazine toward FMOs.

12.5 EFFECT OF TEMPERATURE ON H_2 ADSORPTION

To investigate the spontaneity of the reversible H_2 adsorption and desorption process, Gibbs free energy change (ΔG) of mono and di-Cu(I)-decorated isomeric triazine has been investigated at different temperatures, which are shown in Figure 12.9. It is found that at 298 K, some systems show negative values of Gibbs free energy, and we want to know at which temperature all the systems show negative ΔG values. So we have decreased the temperature from 298 to 200 K. But, as we take different isomers of triazine, we want to investigate at which temperature for a particular isomer, negative ΔG appears. In Figure 12.9, it is observed that all the mono-Cu(I)- decorated isomer and di-Cu(I)-decorated 135-isomer show negative ΔG at very low temperatures, whereas the other two isomers of di-Cu(I)-decorated systems (123, 124) show negative ΔG at comparatively high temperatures.

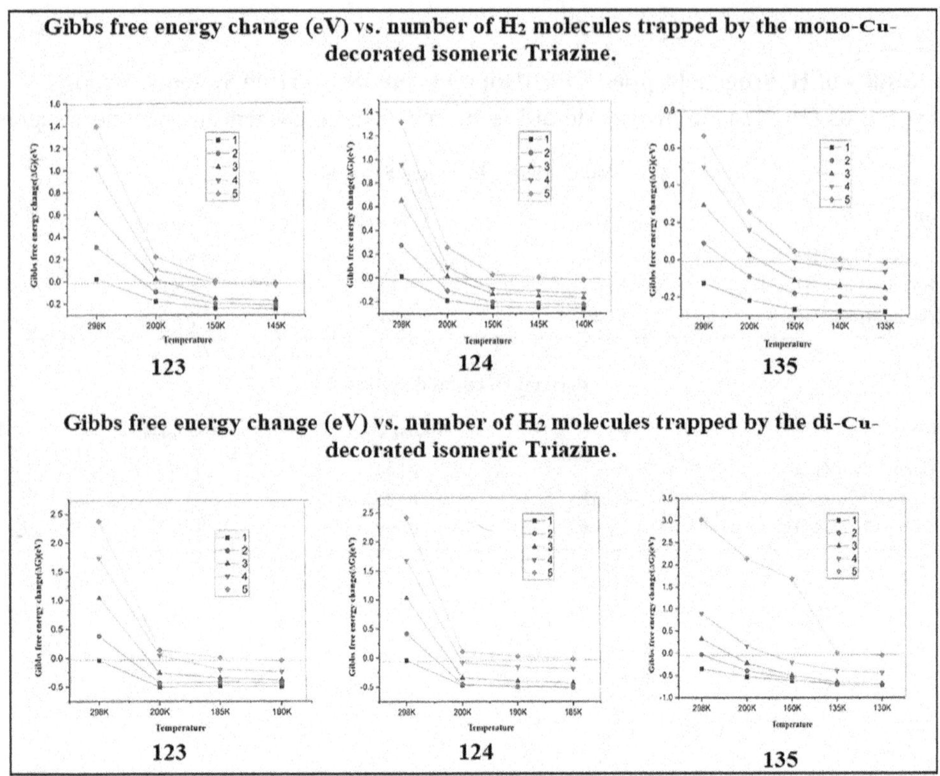

FIGURE 12.9 Gibbs free energy change (eV) vs. the number of H_2 molecules trapped with the mono- and di-Cu(I)-decorated isomeric triazine at different temperatures.

We calculated the gravimetric wt% of mono- and di-metal-decorated isomeric triazine systems by using equation (11.3). Then we found that the storage capacity of mono- and di-metal-decorated systems lies in the range of 6.5 and 8.8 wt% (gravimetric wt%), which satisfies the target set by U.S. Department of Energy (2025).[71]

12.6 CONCLUSION

In conclusion, we have reported our computational study on the H_2 storage capacity of mono- and di-metal-decorated isomeric triazine systems. For all the isomers of triazine, we put Cu(I) on the top and side of the different isomers of the triazine ring, but it is observed that, in all cases, after optimization, Cu(I) favors the side-on mode through the N center. We have introduced a maximum of two Cu(I) in each isomer of triazine. Geometrical parameters and CDFT descriptors confirmed the stability of the complexes. The average adsorption energy of adsorbed H_2 molecule over mono- and di-metal-decorated isomeric triazine has been observed to be in the range of 0.1–0.5 eV/H_2, confirming their quasi-binding interaction. This was further supported by the geometry analysis of the complexes. The ESP analysis confirmed the redistribution of electron density over the whole system. Also, NBO charge analysis supported the successive addition of the H_2 molecule on the Cu(I) center. Bonding nature analysis also has been performed by ELF and NCI analysis; both parameters support the weak interaction between Cu(I) to the ring and Cu(I) to the molecular H_2. Geometrical analysis showed that hydrogens are adsorbed in molecular form. Gibbs free energy change (ΔG) indicates that at a very low temperature, all the complexes can adsorb a maximum of five H_2 molecules (for mono-metal decoration) and ten H_2 molecules (for di-metal decoration).

The storage capacity of mono- and di-metal-decorated systems lies in the range of 6.5–8.8 wt% (gravimetric wt%), which satisfies the target set by the U.S. Department of Energy (2025). Hence our study concludes that the newly designed Cu(I)- decorated isomeric triazine can be considered a potential candidate for hydrogen storage systems. Our computational results might even inspire experimental research groups to develop hydrogen storage materials that can function under ambient thermodynamic conditions.

12.6.1 Acknowledgments

Gourisankar Roymahapatra sincerely thanks SERB for the financial support under the project TAR/2022/000162, dated January 27, 2023, and also thanks Haldia Institute of Technology for providing the research facilities.

REFERENCES

1. Shafiee S., Topal E. When will fossil fuel reserves be diminished? Energy Policy. 2009;37(1):181–189. https://doi.org/10.1016/j.enpol.2008.08.016
2. Johnsson F., Kjärstad J., Rootzén J. The threat to climate change mitigation posed by the abundance of fossil fuels. Clim Policy. 2019;19(2):258–274. https://doi.org/10.1080/14693062.2018.1483885
3. Singh S., Jain S., Venkateswaran P.S., Tiwari A.K., Nouni M.R., Pandey J.K., Goel S. Hydrogen: A sustainable fuel for future of the transport sector. Renew Sustain Energy Rev. 2015;51:623–633. https://doi.org/10.1016/j.rser.2015.06.040.
4. Schlapbach L., Züttel A. Hydrogen-storage materials for mobile applications. Materials for Sustainable Energy: A Collection of Peer-Reviewed Research and Review Articles from Nature Publishing Group. Nature. 2011;414:265–270. https://doi.org/10.1038/35104634.
5. Dresselhaus M., Crabtree G., Buchanan M. Basic research needs for the hydrogen economy. Technical report, Argonne National Laboratory, Basic Energy Sciences, US DOE, 2003. www.sc.doe.gov/bes/hydrogen.pdf.
6. U.S. Department of Energy's Energy Efficiency and Renewable Energy. 2010. https://www1.eere.energy.gov/hydrogen-andfuelcells/storage/current_technology.html.
7. Züttel A. Materials for hydrogen storage. Mater Today. 2003;6(9):24–33. https://doi.org/10.1016/S1369-7021(03)00922-2.
8. Jena P. Materials for hydrogen storage: Past, present, and future. J Phys Chem Lett. 2011;2(3):206–211. https://doi.org/10.1021/jz1015372.
9. Züttel A. Hydrogen storage methods. Naturwissenschaften. 2004;91(4):157–172. https://doi.org/10.1007/s00114-004-0516-x.
10. Eckert J. Hydrogen storage materials with binding intermediate between physisorption and chemisorption. Annual Progress Report, DOE Hydrogen Program, 2007.
11. Biniwale R.B., Rayalu S., Devotta S., Ichikawa M. Chemical hydrides: A solution to high-capacity hydrogen storage and supply. Int J Hydrog Energy. 2008;33:360. https://doi.org/10.1016/j.ijhydene.2007.07.028.
12. Zhu Q.L., Xu Q. Liquid organic and inorganic chemical hydrides for high-capacity hydrogen storage. Energy Environ Sci. 2015;8:478–512. https://doi.org/10.1039/C4EE03690E.
13. Li Y., Mi Y., Suk Chung J., Sung G.K. First-principles studies of K1-xMxMgH3 (M = Li, Na, Rb, or Cs) perovskite hydrides for hydrogen storage. International. J Hydrogen Energy. 2018;43:2232. https://doi.org/10.1016/j.ijhydene.2017.10.175.
14. Baysal M.B., Surucu G., Deligoz E., Ozısık H. The effect of hydrogen on the electronic, mechanical and phonon properties of LaMgNi4 and its hydrides for hydrogen storage applications. Int J Hydrog Energy. 2018;43:23397 https://doi.org/10.1016/j.ijhydene.2018.10.183.
15. Billur S., Farida L., Michael H. Metal hydride materials for solid hydrogen storage: A review. Int J Hydrog Energy 2007;32:1121. https://doi.org/10.1016/j.ijhydene.2006.11.022.
16. Orimo S.I., Nakamori Y., Eliseo J.R., Zuttel A., Jensen C.M. Complex hydrides for hydrogen storage. Chem Rev. 2007;107:4111. https://doi.org/10.1021/cr0501846.
17. Ley M.B., Jepsen L.H., Lee Y.S., Cho Y.W., von Colbe J.M.B., Jensen T.R. Complex hydrides for hydrogen storage – new perspectives. Mater Today. 2014;17:122. https://doi.org/10.1016/j.mattod.2014.02.013.
18. Yildirim T., Ciraci S. Titanium-decorated carbon nanotubes as a potential high-capacity hydrogen storage medium. Phys Rev Lett. 2005;94:175501. https://doi.org/10.1103/PhysRevLett.94.175501.

19. Dag S., Ozturk Y, Ciraci S., Yildirim T. Adsorption and dissociation of hydrogen molecules on bare and functionalized carbon nanotubes. Phys Rev B. 2005;72:155404. https://doi.org/10.1103/PhysRevB.72.155404.
20. Naghshineh N., Hashemianzadeh M. First-principles study of hydrogen storage on Si atoms decorated C60. Int J Hydrog Energy. 2009;34:2319. https://doi.org/10.1016/j.ijhydene.2008.12.064.
21. Kim D., Lee T.B., Choi S.B., Yoon J.H., Kim J., Choi S.H. A density functional theory study of a series of functionalized metal-organic frameworks. Chem Phys Lett. 2006;420:256. https://doi.org/10.1016/j.cplett.2005.12.083.
22. Rowsell J.L.C., Millward A.R., Park K.S., Yaghi O.M. Hydrogen sorption in functionalized metal–organic frameworks. J Am Chem Soc. 2004;126:5666. https://doi.org/10.1021/ja049408c.
23. Sun Q., Jena P., Wang Q., Marquez M. First-principles study of hydrogen storage on Li12C60. J Am Chem Soc. 2006;128:9741. https://doi.org/10.1021/ja058330c.
24. Veluswamy H.P., Kumar R., Linga P. Hydrogen storage in clathrate hydrates: Current state of the art and future directions. App Energy. 2014;122:112–132. https://doi.org/10.1016/j.apenergy.2014.01.063.
25. Davoodabadi A., Mahmoudi A., Ghasemi H. The potential of hydrogen hydrate as a future hydrogen storage medium. iScience. 2021;24:101907. https://doi.org/10.1016/j.isci.2020.101907.
26. Mondal S., Giri S., Chattaraj P.K. Possibility of having hf-doped hydrogen hydrates. J Phys Chem C. 2013;117(22):11625–11634. https://doi.org/110.1021/jp401342r.
27. Mondal S, Giri S, Chattaraj P.K. Methane hydrates and their HF doped analogues. Chem Phys Lett. 2013;578:110–114. https://doi.org/10.1016/j.cplett.2013.06.017.
28. Bandaru S., Chakraborty A., Giri S., Chattaraj P.K. Toward analyzing some neutral and cationic boron–lithium clusters (BxLiy x = 2–6; y = 1, 2) as effective hydrogen storage materials: A conceptual density functional study. Int J Quantum Chem. 2012;112:695–702. https://doi.org/10.1002/qua.23055.
29. Pan S., Giri S., Chattaraj P.K. A computational study on the hydrogen adsorption capacity of various lithium—doped boron hydrides. J Comput Chem. 2012;33(4):425–434. https://doi.org/10.1002/jcc.21985.
30. Yin Z., Linga P. Methane hydrates: A future clean energy resource. Chin J Chem Eng. 2019;27:2026–2036.
31. Lee S.Y., Holder G.D. Methane hydrates potential as a future energy source. Fuel Process Technol. 2001;71:181–186. https://doi.org/10.1016/S0378-3820(01)00145-X.
32. Targets for on-board hydrogen storage systems: Current R&D focus is on. Target. 2010. http://www1.eere.energy.gov/hydrogenandfuel.cells/pdfs/.freedomcar_targets_explanations.pdf. [Accessed 7 March 2012].
33. Dixit M., Maark T.A., Pal S. Ab initio and periodic DFT investigation of hydrogen storage on light metal-decorated MOF-5. Int J Hydrog Energy. 2011;36:10816. https://doi.org/10.1016/j.ijhydene.2011.05.165.
34. Deshmukh A., Chen Y.W., Kuo J.L. Tetrahedral silsesquioxane framework: A feasible candidate for hydrogen storage. J Phys Chem C. 2015;119:23820. https://doi.org/10.1021/acs.jpcc.5b06514.
35. Chattaraj S., Srinivasu K., Mondal S., Ghosh S.K. Hydrogen trapping ability of the pyridine–lithium+ (1:1) complex. J Phys Chem. 2015;119:3056. https://doi.org/10.1021/jp5129029.
36. Konda R., Deshmukh A., Titus E., Chaudhari A. Alkali, alkaline earth and transition metal doped B6H6 complexes for hydrogen storage. Int J Hydrog Energy. 2017;42:23723. https://doi.org/10.1016/j.ijhydene.2017.05.023.
37. Kumar A., Vyas N., Ojha A. Hydrogen storage in magnesium decorated boron clusters (Mg2Bn, n = 4–14): A density functional theory study. Int J Hydrog Energy 2020;45:12961. https://doi.org/10.1016/j.ijhydene.2020.03.018.
38. Dash M.K., Das S., Giri S., De G.C. Roymahapatra G. Comprehensive in silico study on lithiated Triazine isomers and its H_2 storage efficiency. J Indian Chem Soc. 2021;98:100134. https://doi.org/10.1016/j.jics.2021.1001340.
39. Dash M.K., Roymahapatra G., Chowdhury S.D., Chatterjee R., Maity S.P., Huang M., Bandyopadhyay M.N., Guo Z. Computational investigation on lithium fluoride for efficient hydrogen storage system. Eng Sci. 2022;18. https://doi.org/10.30919/es8d658.
40. Dash M.K., Sinha S., Das H.S., De G.C., Giri S., Roymahapatra G. H_2 storage capacity of Li-doped five member aromatic heterocyclic superalkali complexes; An in silico study. Sustain Energy Technol Assessments. 2022;52:102235. https://doi.org/10.1016/j.seta.2022.102235.
41. Parida R., Dash M.K., Roymahapatra G., Giri S. Aromatic N-heterocyclic superalkali M@C4H4N2 complexes (M=Li, Na, K); A promising potential hydrogen storage system. J Indian Chem Soc. 2021;98:100065. https://doi.org/10.1016/j.jics.2021.100065.
42. Dash M.K., Das S., Giri S. Comprehensive in silico study on lithiated triazine isomers and its H2 storage efficiency. J Indian Chem Soc. 2021;98(9):100134.

43. Roymahapatra G., Dash M.K., Sinha S., De G.C., Guo Z. Theoretical investigation of hydrogen adsorption efficiency of [Oxadiazole- xLi+] Complexes (x = 1, 2): In pursuit of green fuel storage. Eng Sci. 2022;19:114–124. https://doi.org/10.30919/es8d671.
44. Roymahapatra G., Dash M.K., Ghosh A., Bag A., Mishra S., Maity S. Hydrogen storage on lithium chloride/lithium bromide surface at cryogenic temperature. ES Energy Environ. 2022;18:90–100. https://doi.org/10.30919/esee8c756.
45. Sun Q, Wang Q, Jena P, Kawazoe Y. Clustering of Ti on a C60 surface and its effect on hydrogen storage. J Am Chem Soc. 2005;127:14582. https://doi.org/10.1021/ja0550125.
46. Parr R.G., Yang W. Density-Functional Theory of Atoms and Molecules. New York: Oxford University Press, 1989. https://doi.org/10.1007/978-94-009-9027-2_2.
47. Parr R.G. Absolute hardness: Companion parameter to absolute electronegativity. J Am Chem Soc. 1983;105:7512. https://doi.org/10.1021/ja00364a005.
48. Mardirossian N., Head-Gordon M. Thirty years of density functional theory in computational chemistry: An overview and extensive assessment of 200 density functionals. Mol Phys 2017;19:115. https://doi.org/10.1080/00268976.2017.1333644.
49. Makkar P., Ghosh N.N. A review on the use of DFT for the prediction of the properties of nanomaterials. RSC Adv. 2021;11:27897–27924. https://doi.org/10.1039/d1ra04876g.
50. Chakraborty D., Chattaraj P.K. Conceptual density functional theory based electronic structure principles. Chem Sci. 2021;12:6264–6279. https://doi.org/10.1039/d0sc07017c.
51. Geerlings P., Chamorro E., Chattaraj P.K., Proft F.D., Gázquez J.L., Liu S., Morell C., Toro-Labbé A., Vela A., Ayers P. Conceptual density functional theory: Status, prospects, issues. Theor Chem Acc. 2020;139:36. https://doi.org/10.1007/s00214-020-2546-7.
52. Sen K.D., Jorgensen C.K. Electronegativity; Structure and Bonding, 66. Berlin: Springer; 1987. https://doi.org/10.1002/bbpc.19890930424.
53. Parr R.G., Szentpaly L.V., Liu S. Electrophilicity index. J Am Chem Soc. 1999;121:1924. https://doi.org/10.1021/ja983494x.
54. Chattaraj P.K. Chemical Reactivity Theory: A Density Functional View. Boca Raton, FL: Taylor and Francis, CRC Press, 2009. https://doi.org/10.1201/9781420065442.
55. Stanger A. NICS – past and present. Eur J Chem. https://doi.org/10.1002/ejoc.201901829.
56. Chen Z., Wannere C.S., Corminboeuf C., Puchta R., Schleyer P.V.R. Nucleus-independent chemical shifts (NICS) as an aromaticity criterion. Chem Rev. 2005;105:3842. https://doi.org/10.1021/cr030088+.
57. Parida R., Rath M., Sinha S., Roymahapatra G., Giri S. Organic and inorganic benzene transform to superalkalis: An in silico study. J Indian Chem Soc. 2020;97(12a):2689–2697. http://indianchemicalsociety.com/portal/uploads/journal/A-Dec-16.pdf.
58. Sinha S., Das S., Roymahapatra G., Giri S. Stability, reactivity, and aromaticity of benzene, borazine, and hexasilabenzene. Int J Hit Transc Eccn. 2020;6(1a):7–13. www.researchgate.net/profile/GourisankarRoymahapatra/publication/351126056_Stability_Reactivity_and_Aromaticity_of_Benzene_Borazine_and_Hexasilabenzene/links/60e10897a6fdccb74503815c/Stability-Reactivity-and-Aromaticity-of-Benzene-Borazine-and-Hexasilabenzene.pdf
59. Parida R., Das S., Karas L.J., Wu J.-C., Roymahapatra G., Giri S. Superalkali ligands as a building block for aromatic trinuclear Cu(i)–NHC complexes. Inorg Chem Front 2019;6:3336–44. https://doi.org/10.1039/C9QI00873J
60. Lu T., Chen F.W. Multiwfn: A multifunctional wavefunction analyzer. J Comput Chem. 2012;33:580. https://doi.org/10.1002/jcc.22885
61. Ebrahimia A., Izadyar M., Khavanib M. Theoretical design of a new hydrogen storage based on the decorated phosphorene nanosheet by alkali metals. SSRN. 2022;30. http://dx.doi.org/10.2139/ssrn.4236115.
62. https://en.wikipedia.org/wiki/Non-covalent_interaction
63. Mondal S. Structure, bonding, and interaction with molecular hydrogen of the β-D-glucopyranose-silver+ (1:1) complex. J Phys Org Chem. 2022. https://doi.org/10.1002/poc.4448.
64. Hitler Louis H.O.E., Benjamin I., Gideon M., Unimuke T.O., Adalikwu S.A., Nwagu A.D., Adeyink A.S. Hydrogen storage capacity of C12X12 (X = N, P, and Si). Chem Phys Impact. 2022;5:100107. https://doi.org/10.1016/j.chphi.2022.100107.
65. Chen G., Chong W. The theoretical research of hydrogen storage capacities of Cu3Bx (X=1–4) compounds under ambient conditions. Int J Hydrog Energy. 2020;6:89. https://doi.org/10.1016/j.ijhydene.2020.06.089.

66. Jaiswal A., Chakraborty B., Sahu S. High capacity reversible hydrogen storage in Si substituted and Li decorated C20 fullerene: Acumen from density functional theory simulations. Int J Energy Res. 2022;1–17. https://doi.org/10.1002/er.8524.
67. Pearson R.G. The principle of maximum hardness. Acc Chem Res. 1993;26(5):250–255. https://doi.org/10.1021/ar00029a004.
68. Chamorro E., Chattaraj P.K., Fuentealba P. Variation of the electrophilicity index along the reaction path. Chem A Eur J. 2003;107(36):7068–7072. https://doi.org/10.1021/jp035435y.
69. Parida R., Reddy G.N., Ganguly A., Roymahapatra G., Chakraborty A., Giri S. On the making of aromatic organometallic superalkali complexes. Chem Commun. 2018. https://doi.org/10.1039/C8CC01170B.
70. Parida R., Rath M., Sinha S., Roymahapatra G., Giri S. Organic and inorganic benzene transform to superalkalis: An in silico study. J Indian Chem Soc. 2020;97(12a):2689–2697.
71. Sahoo R.K., Sahu S. Reversible hydrogen storage capacity of Li and Sc doped novel C_8N_8 cage: Insights from density functional theory. Int J Energy Res. 2022. https://doi.org/10.1002/er.8562.

13 Quantum Chemical Study on Pure and Silicon-Doped Activated Carbon Sheets

Ratnesh Kumar, Abhishek Kumar, Ambrish Kumar Srivastava, and Neeraj Misra

13.1 INTRODUCTION

Carbon-based materials come in a variety of forms, including diamond, graphene, carbon nanotubes, nanofibers, amorphous carbon, activated carbon, and others. Among these, activated carbon has a broader research scope due to its numerous applications in liquid and gas treatment,[1] metal selection,[2] gold purification,[3] medicine,[4] waste treatments,[5] electrodes, sensors,[6] and energy storage and conversion systems.[7–10] An activated carbon sheet is a porous material with a high surface area that makes it attractive for the adsorption of a wide range of substances. It has many potential applications such as in air and water purification, energy storage devices, and catalysis.[11] Activated carbon sheets have a wide range of applications due to their highly porous and adsorptive properties.[12] Some of the common applications include activated carbon sheets, which are commonly used for air and water purification. The porous structure of the sheets allows them to effectively adsorb and remove pollutants, odors, and contaminants from the air and water. They are used in air filters, water filters, respirators, and masks. It is also used in chemical filtration processes to remove impurities and contaminants from liquids and gases and is commonly used in the pharmaceutical, food and beverage, and chemical industries. Activated carbon sheets are used in environmental remediation to clean up contaminated soil and water. They can absorb and remove organic and inorganic pollutants, including heavy metals, pesticides, and volatile organic compounds (VOCs). They can be used as electrodes in supercapacitors and batteries such as fuel cells and electrolyzers due to their high surface area and electrical conductivity. They provide a large surface area for electrochemical reactions and can store energy efficiently.

Activated carbon sheets are used for adsorbing gases and vapors in various applications. They are used in gas masks and respirators to remove toxic gases and fumes. They are also used in industrial processes to capture and recover volatile organic compounds (VOCs) and other gases. These are also used in medical healthcare and industrial applications. They are used in wound dressings to absorb wound exudates and control odor. They are also used in filters for medical gases to remove impurities and are also utilized in various industrial processes, such as solvent recovery, gas purification, and decolonization of liquids.

In the 1950s, Rosalind Franklin produced the structure of carbon by the pyrolysis of organic materials and classified it in a graphitizing and nongraphitizing model.[13] Graphitization is observed with hexagonal rings of carbon; in nongraphitization, many hexagonal rings join together along with pentagonal and heptagonal rings to give activated carbon. The ratio of pentagons/heptagons to hexagons is approximately 1:50, and this structure is relatively hard but extremely stable as compared to other carbons.[14–17] Among all the carbon materials, activated carbon exhibits the largest specific surface area with increased high porosity to adsorbents having a low cost.[1, 9, 12, 18–20] It removes wastewater contaminants and pollutants from liquids and gases due to its a large amorphous structure for adsorbent, high surface activity, and chemical stability.[21, 22] Activated carbon is produced from organic precursors such as wood, rice husk, fruit shells, polymers, lignite, anthracite, brown

DOI: 10.1201/9781003441328-13

coal, and petrochemical products,[23, 24] by either physical activation or chemical activation.[25] The physical methods are typically thermal processes conducted at temperatures below 700 °C using oxidizing gases, the porous AC is produced when the oxidant converts carbon materials to form CO and CO_2.[26]

In chemical activation, initially, materials are activated by an activating agent such as hydrochloric acid,[27] sulfuric acid,[28] phosphoric acid,[29] potassium hydroxide,[30] sodium hydroxide,[31] and some others. It has been seen that rich carbon content, low inorganic content, and low-cost materials are used as raw materials for activation.[32, 33] To enlarge the diameter of pores, i.e., create new porosity, activation and carbonation are simultaneously performed during a chemical activation process.[34, 35] Chemical activation processes are preferred over physical activation due to its simplicity, lower temperature, and shorter activation time.[36, 37] An activated carbon sheet is a porous material with a high surface area that makes it attractive for the adsorption of a wide range of substances. It has many potential applications such as in air and water purification, energy storage devices, and catalysis. In a DFT study on activated carbon sheets, the electronic structure and properties of the material can be analyzed at the atomic and molecular levels. This can provide insights into the adsorption mechanism, surface chemistry, and reactivity of the material.

Activated carbon sheets possess some unique electronic properties, allowing for efficient energy transfer between the electrodes and the material and making it suitable for electrical applications.[6] Activated carbon sheets have a large surface area, which allows for increased charge transfer between the material and the surrounding environment.[8, 9] This property makes them ideal for use in supercapacitors, batteries, and other electronic devices. Some other properties include[38]:

- **High chemical stability:** Activated carbon sheets are highly chemically stable, which means they can withstand exposure to a variety of corrosive materials and environments without incurring damage to their electronic properties.
- **Low dielectric constant:** Activated carbon sheets have a low dielectric constant, which means they are not easily polarized by an electric field. This property makes them ideal for use in high-frequency electronic devices, such as antennas and capacitors.
- **Good thermal conductivity:** Activated carbon sheets have high thermal conductivity due to the presence of carbon, which allows for the efficient transfer of heat within the material. This property makes them useful for thermal management applications, such as heat sinks and insulation materials.

By using density functional theory (DFT), scientists can predict the adsorption capacity of activated carbon sheets for various substances, investigate the effect of different functional groups on their interactions with the material, and understand the impact of pore size and shape on their properties. In this chapter, we mainly focus on the DFT-based study of two kinds of structural analysis of activated carbon sheets by replacing carbon atoms and doping silicon atoms. Here, we have also studied the effect of the band gap due to the vacancies in the activated carbon sheet,[39] as well as the doping of silicon atoms in the activated carbon sheet and analysis of the conversion of semiconducting to conducting nature. Experimentally, these properties can be studied in the presence of omission in 2D activated carbon sheets.

13.2 COMPUTATIONAL METHODS

All calculations are performed by employing a computational program SIESTA (Spanish Initiative for Electronic Simulations with Thousands of Atoms),[40] which is based on the density functional theory.[41] A generalized gradient approximation (GGA) and norm-conserving pseudopotential are considered for atomic orbitals.[42, 43] Double zeta polarized basis sets (DZP) are taken for valence orbital $2p^2$ of carbon atoms.[44, 45] To run a program for such a system, there is a need to obtain the optimized values of cutoff, *k*-points, and the lattice constant of a system; therefore, flexible data format (FDF) input files are prepared for all systems for the optimization of these values. The final optimization is performed for all systems until the Hellmann–Feynman forces acting at all components of each atom are smaller than 0.001 eV/Å.

To proceed, an extensive simulation-based study of activated carbon (AC), 2D activated carbon sheets with a single pentagonal ring surrounded with hexagonal rings without and with single vacancy, double vacancy, and triple vacancy, is carried out with the help of density functional theory. The structural and electronic properties of these systems are analyzed and represented as AC, AC_1, AC_2, and AC_3, respectively. In the next section, we have modeled four structures as previously: activated carbon sheet (AC), $ACSi_1$ sheet with a single silicon atom, $ACSi_2$ sheet with two silicon atoms, and $ACSi_3$ sheet with three silicon atoms.[46]

13.3 RESULTS AND DISCUSSIONS

13.3.1 Geometrical Properties

The equilibrium configurations of the simulated structures of 2D-activated carbon sheets along with the vacancies are shown in Figure 13.1 Activated carbon (AC) is a sheet of 45 carbon atoms. Here, hexagonal rings (benzene) cover a pentavalent carbon ring at the center with little depth, so the pentavalent carbon atoms in this cluster are unsaturated in order to interact with the reactive portion. Vacancies are created in the pentagonal ring by removing carbon atoms (C1, C2, and C5) and with the increase in the number of vacancies the depth at the center increases gradually along with a variation in the width of a sheet. Si atoms are placed in the pentagonal ring by removing carbon atoms, and, with an increasing number of Si atoms, the depth at the center is gradually increased, and the total width of the sheet changes.

The variations in the bond lengths of C-C atoms at the defective region of different systems are 1.33–1.35 Å in AC, 1.37–1.40 Å in AC_1, 1.38–1.42 Å in AC_2, and 1.37–1.47 Å in AC_3. It is found that there is a larger variation in the bond lengths of AC_3 than of AC. Presumably, the availability of free space at the center increases, relative to the number of vacancies, and, as a result, the bond length between C-C atoms increases.

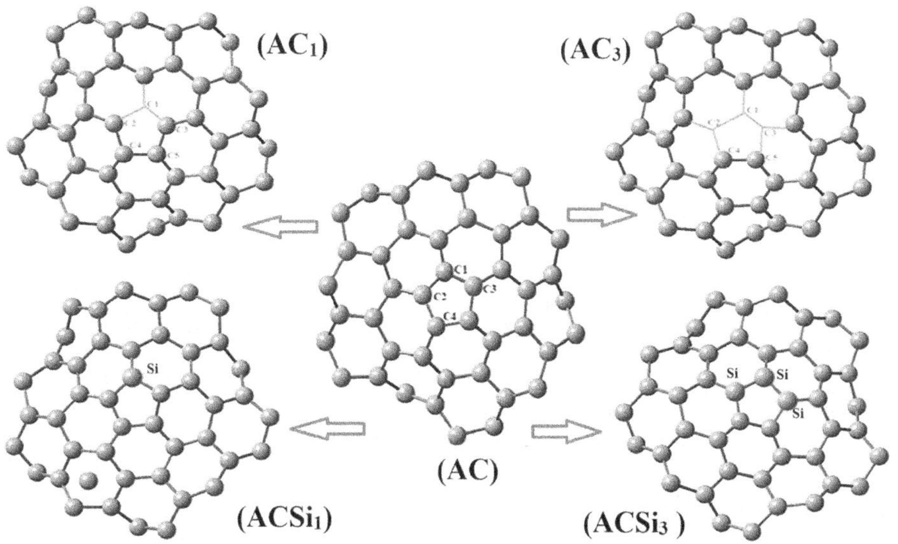

FIGURE 13.1 Equilibrium configurations of AC, AC1, AC3, ACSi1, and ACSi3.

Sources: From Ref. [39, 46].

TABLE 13.1
Bond Lengths, Fermi Energy, and Binding Energy of Systems

S.N.	System	Bond Length (Å)	Fermi Energy (eV)	Binding Energy (eV)
1	AC	1.33–1.35	−6.243	8.39
2	AC_1	1.37–1.40	−6.298	8.164
3	AC_2	1.38–1.42	−6.381	8.268
4	AC_3	1.37–1.47	−6.240	7.613
5	$ACSi_1$	1.69–2.98	−6.148	1.34
6	$ACSi_2$	1.78–3.54	−5.959	3.68
7	$ACSi_3$	1.86–3.63	−5.947	5.99

Sources: From Ref. [39, 46].

The bond length values, along with bond angles, are shown in Table 13.1. The observed bond angles at C2, C3, and C4 atoms in AC, AC_1, AC_2, and AC_3 structures are 109.0°, 122.0°, 113.0°, and 119.4°, respectively; i.e., there is an irregular variation. The bond length and bond angle of AC_3 are in good agreement with the experimental data.[47]

The binding energy (BE) is calculated using the following equation:

$$\textbf{BE}\left(\textbf{per atom}\right) = [NE(C) - E]/N$$

where $E(C)$ and E represent the total energy of the carbon atom and activated carbon system, respectively, and N is the number of carbon atoms in the respective system.

The Fermi energy is higher in AC_3 due to the less tightly held carbon atoms. Consequently, the average velocity of electrons in the AC_3 system is also higher with less binding energy, as given in Table 13.1. Binding energy is higher in AC and smaller in AC_3; i.e., a small amount of energy is required for the splitting of AC_3 than AC. In other words, higher energy is required in the formation of AC than AC_3. The calculated BE of AC_3 is found to be nearly equal to a previous study for a graphene sheet.[48] The vacancies can be created experimentally by providing the energy equal to the calculated binding energy, as shown in Table 13.1.[49]

In Si doping, the variation in bond lengths of C-C atoms and Si-C atoms increases, as shown in Figure 13.1 The bond length values of optimized structures are also mentioned in Table 13.1. The bond lengths of Si-Si atoms are 2.06 Å in the $ACSi_2$ sheet and 2.39Å, 2.40 Å, and 2.69 Å in the $ACSi_3$ sheet, respectively. It is observed that due to the placing of Si atoms in the activated carbon sheet, bond length as well as the depth at the center increases. There is also a variation in the total width and height of the system, as shown in Figure 13.1.

The energy of the highest occupied quantum state (Fermi energy) of the system is higher in $ACSi_3$; i.e., the average velocity of electrons is higher with a binding energy of 5.99 eV. The values of the binding energy of systems have also been presented in Table 13.1. The binding energy value is higher in activated carbon (AC) and smaller in $ACSi_1$; i.e., a small amount of energy is required to split $ACSi_1$ into its basic constituent carbon and silicon atoms. The binding energy of silicon-doped activated carbon sheets is calculated by the following equation:

$$\textbf{Binding Energy}\left(E_b\right) = E\left(AC\right) - E\left(AC\text{-}Si_n\right) - nE\left(Si\right)$$

where $E(AC)$, $E(AC\text{-}Si_n)$, and $E(Si)$ are the total equilibrium energies of an activated carbon sheet, pristine sheet doped with silicon atoms, and individual silicon atoms, respectively, and n represents the number of doped silicon atoms.

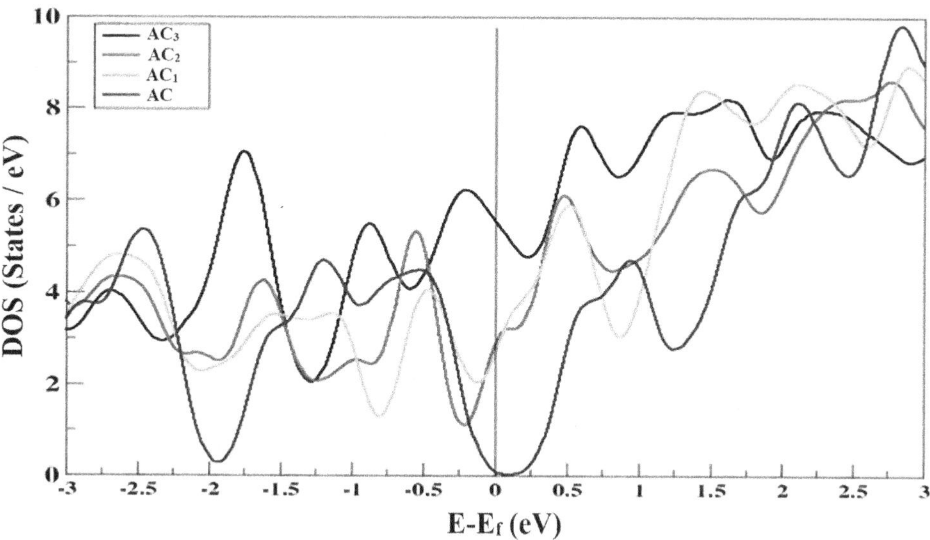

FIGURE 13.2 Density of states of AC, AC_1, AC_2, and AC_3.

Source: From Ref. [39].

13.3.2 Electronic Properties

The density of states exhibits the semiconducting nature of the AC system with a very small forbidden energy gap of 0.2 eV, which is equal to the experimentally obtained values,[50–52] and with one to three vacancies, the density of the state is shifted upward. This shifting increases with the increase in the number of vacancies in the activated carbon, as is evident in Figure 13.2. This shifting also justifies the outcome of the binding energy: with higher binding energy, the density of states lies at lower energy levels near the nucleus, and with the high number of vacancies, lower binding energy reflects the shifting of the density of states toward higher energy levels, i.e., far toward the atomic nucleus.

The COOP (crystal orbital overlap population) analyses also justify the outcome of the structural analysis, binding energy, and DOS. Higher peaks of antibonding states give lower binding energy, and the outermost valence electrons are weakly bound with the nuclei. They can be easily detached from the atom, and, as a result, the system has a higher concentration of electrons in the conduction band, thereby causing higher mobility and ultimately higher electron conductivity,[53] as observed in the AC_3 system rather than in $AC/AC_1/AC_2$.

The Mulliken population of defected carbon for the AC sheet is 0.461; AC_1 is 0.458; AC_2 is 0.445; and AC_3 is 0.377. The bond population is also reported here for an indication of the bond's nature as well as charge determination. The bond population is calculated by using Mullikan population analysis. The positive and negative values of the bond population show that atoms are bonded or antibonded, respectively. A larger positive value indicates that atoms are closely covalent, whereas a zero value means no interaction between the two atoms.[54] Relative values of the bond population are used for calculation because absolute population values give little physical meaning.

In Si doping, the result of the projected densities of states (PDOS) in Figure 13.3 shows high peaks of 2*p* orbital electrons of carbon atoms in the valence and conduction bands and 2*s* orbital electrons with high peaks around the Fermi level (0 eV) and in the conduction band. In Figure 13.3, the density of states is symmetrically found around the Fermi level. There is a high peak of 2*p* orbital electrons, then 2*s* electrons of carbon atoms, high peaks of 3*p* orbital electrons of the Si atom at the Fermi level, and the smaller peak of 3*s* orbital electrons. A similar kind of distribution of orbital electrons is found in Figure 13.3 with 2Si/3Si doped atoms in an activated carbon sheet. There is

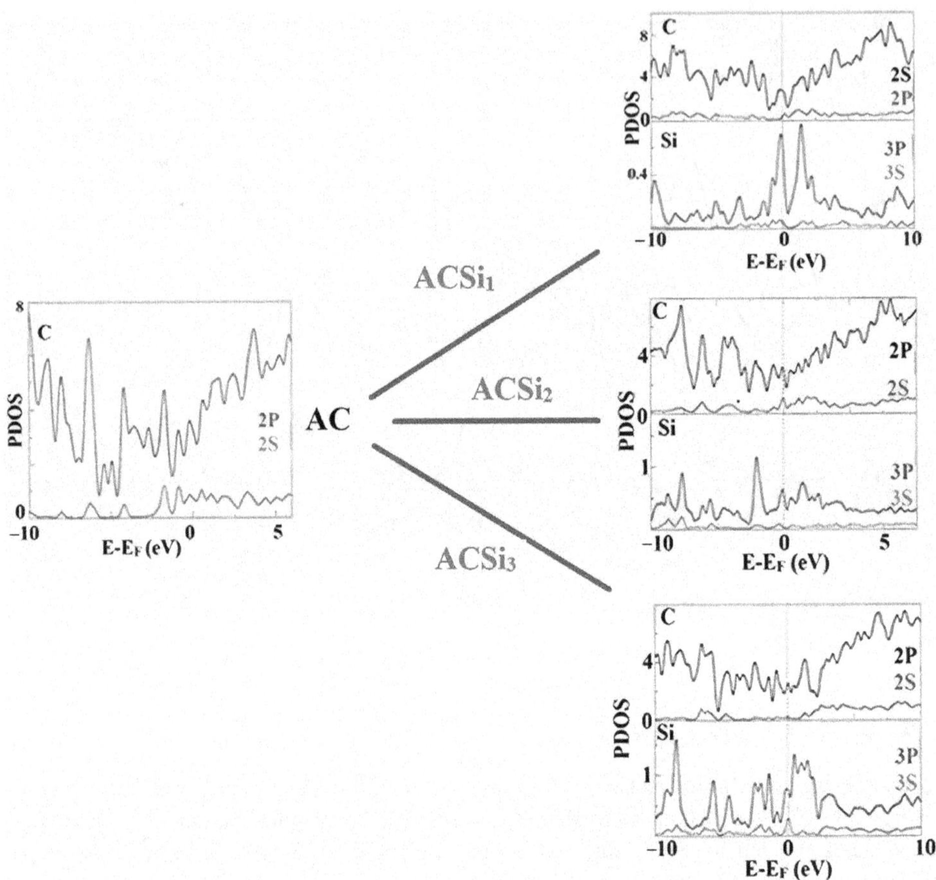

FIGURE 13.3 Projected density of states (PDOS) of AC, $ACSi_1$, $ACSi_2$, and $ACSi_3$.

Sources: From Ref. [39, 46].

s-p hybridization in all systems without and with Si atoms. The charge density distribution analysis also indicates the covalent bond between C-C atoms due to the overlapping of valence orbital wave functions.

The crystal orbital overlap population (COOP) analysis is shown in Figure 13.4. Here, the negative *Y*-axis shows antibonding and the positive *Y*-axis, bonding molecular orbital states. The existence of resultant antibonding states is found due to the overlapping of C2*s*-C2*s*, C2*s*-C2*p*, C2*p*-C2*p*, Si3*p*-C2*p*, Si3*p*-C2*s*, Si3*s*-C2*p*, Si3*s*-C2*s* atomic orbitals. High peaks of antibonding states are found in AC due to the overlapping of orbital wave functions of the C2*s*-C2*s* atomic orbital, as shown in Figure 13.4.

In $ACSi_2$, the orbital Si3*p*-C2*s* also indicates a high peak, but its value is less than the peak found in the AC system. On comparison with Si-doped systems, the resulting strength of antibonding states is greater than bonding orbitals states; hence it reduces the crystallinity of $ACSi_3$ and increases amorphicity. Since antibonding electrons are weakly bound to the molecules, they possibly can contribute a lot to the transport properties of materials (i.e., concentration of electrons, mobility, diffusion, and conduction of electrons) than the bonding electrons.[55]

The difference between the highest occupied molecular orbital (HOMO) and lowest unoccupied molecular orbital (LUMO) is known as the forbidden energy gap (E_g), and it plays a significant role in the conduction process.[56, 57] The charge transport properties of molecules depend on the

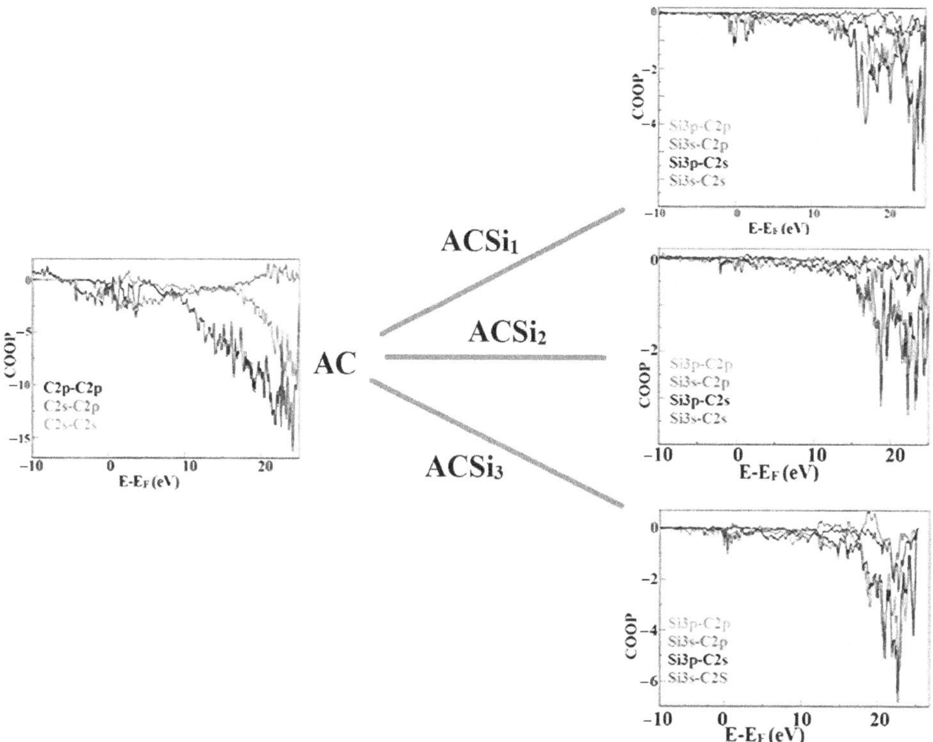

FIGURE 13.4 COOP analysis of AC, $ACSi_1$, $ACSi_2$, and $ACSi_3$.

Sources: From Ref. [39, 46].

forbidden energy gap; hence it is essential to examine the variations in the forbidden energy gap and molecular orbital energy level.[58, 59] The Fermi energy levels of all the systems are located at 0 eV along the *Y*-axis.

The energy band diagrams of AC and silicon-doped systems along the G-Y direction are shown in Figure 13.5. In Figure 13.5a, the lowest energy level of a conduction band and the highest energy level of a valence band appear between points G and R, which is observed as a 0.58 eV indirect energy gap for an activated carbon sheet. When a single carbon atom is replaced by doping a single silicon atom, it gets a small 0.29 eV indirect energy gap between the points near U and V.

Similarly, in $ACSi_2$, E_g = 0.19 eV is observed between points U and V by doping two silicon atoms. In $ACSi_3$, there is an overlapping of valence and conduction bands, which causes higher electrical conductivity than the AC sheet because there is an indirect zero band gap between points G and Y (Figure 13.5). Thus it is clear that when the number of doping silicon atoms increases in the AC sheet, it reduces the energy band gap by 0.58–0 eV, and electronic properties are enhanced. In the literature, too, it has been reported that activated carbon (AC) behaves as a semiconductor material, and the band gap energy (E_g) values reduce (<4 eV) in the presence of UV and gamma radiation.

13.4 CONCLUSIONS

The structural properties of the activated carbon sheet show that, by increasing the number of vacancies, the depth and interatomic distance at the center of activated carbon sheets gradually increase, and the total width changes. An activated carbon sheet with three vacancies has a lower binding energy and is without vacancy, and it has higher binding energy. AC shows a semiconducting nature

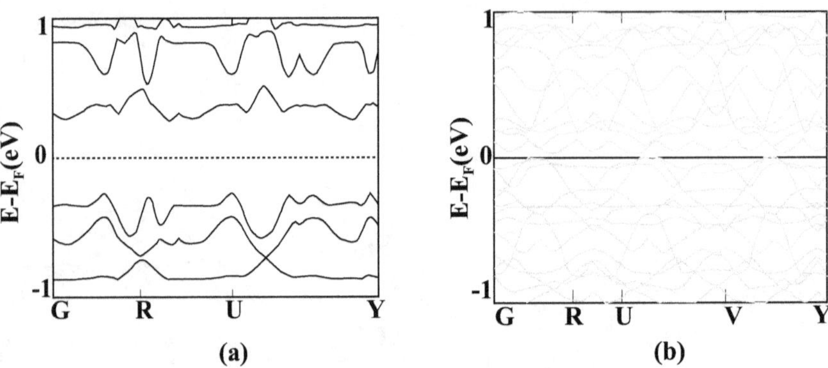

FIGURE 13.5 Energy band diagrams of (a) AC and (b) $ACSi_3$.

Sources: From Ref. [39, 46].

without vacancy; with vacancies, it exhibits a metallic nature. In all systems, we found resultant peaks of antibonding states having higher strength in the AC_3 system and smaller strength in AC, thereby exhibiting an amorphous structure. Consequently, it causes an increase in electron concentration, mobility, and finally electron conductivity with the gradual increase in the number of vacancies in the activated carbon.

The structural and electronic properties of activated carbon sheets with the subsequent doping of silicon atoms are also discussed in detail. The structural properties show that with the replacement of carbon atoms by Si atoms at the center of the activated carbon sheet, the depth at the center of the sheet gradually increases, and consequently the total width and bond lengths are also changed. AC and $ACSi_3$ have higher binding energies, but $ACSi_1$ and $ACSi_2$ systems have lower binding energies. The result of the projected density of states explains *s-p* hybridization in all the systems. The COOP analysis exhibits the strength of antibonding states in all systems. Hence, there is a probability of a little higher value of transport properties in $ACSi_3$ sheets than in $ACSi_1$ and $ACSi_2$sheets. The energy band diagrams reveal the metallic nature of $ACSi_3$ sheets and the semiconducting nature of $ACSi_1$ and $ACSi_2$ sheets with very small forbidden energy gaps.

The present study can possibly pave the way for experimentalists to work in the area of industrial pollution. In the future, activated carbon, due to its unique properties and versatile nature, is likely to find numerous applications in environmental remediation, water purification, air filtration, the food and beverage industry. It can remove harmful gases, volatile organic compounds, and unpleasant odors from indoor environments, making it valuable for residential, commercial, and industrial applications.

REFERENCES

[1]. L. Cermakova, I. Kopecka, M. Pivokonsky, L. Pivokonska, V. Janda, Removal of cyanobacterial amino acids in water treatment by activated carbon adsorption. Sep Purif Technol (2017) 173:330

[2]. H. Zhang, I.M. Ritchie, S.R. La Brooy, The adsorption of gold thiourea complex onto activated carbon. Hydrometallurgy (2004) 72:3

[3]. D. Fischer, D. Dennis, US Patent Issued (1976) 3:935

[4]. Y. Wang, W. Shi, S. Wang, M. Qian, J. Chen, R. Huang, Facile incorporation of dispersed fluorescent carbon nanodots into mesoporous silica nanosphere for pH-triggered drug delivery and imaging. Carbon (2016) 108:146

[5]. H. Yuan, Q. Jiao, S. Zhang, Y. Zhao, Q. Wu, H. Li, In situ chemical vapor deposition growth of carbon nanotubes on hollow CoFe2O4 as an efficient and low-cost counter electrode for dye-sensitized solar cells. J Power Sources (2016) 325:417

[6]. K. Qi, R. Selvaraj, T. Al. Fahdi, S. Al-Kindy, Y. Kim, G.C. Wang, M. Sillanpää, Enhanced photocatalytic activity of anatase- TiO2 nanoparticles by fullerene modification: A theoretical and experimental study. Appl Surf Sci (2016) 387:750
[7]. F. Ding, W. Xu, G.L. Graff, J. Zhang, M.L. Sushko, X. Chen, X. Liu, Dendrite-free lithium deposition via self-healing electrostatic shield mechanism. J Am Chem Soc (2013) 135:11
[8]. W.K. Chee, H.N. Lim, Z. Zainal, N.M. Huang, I. Harrison, Y. Andou, Flexible graphene-based supercapacitors: A review. J Phys Chem C (2016) 12:8
[9]. G. Sethia, A. Sayari, Activated carbon with optimum pore size distribution for hydrogen storage. Carbon (2016) 99:289
[10]. X. Yue, W. Sun, J. Zhang, F. Wang, Y. Yang, C. Lu, K. Sun, Macro-mesoporous hollow carbon spheres as anodes for lithium-ion batteries with high rate capability and excellent cycling performance. J Power Sources (2016) 331:10
[11]. S.Y. Kim, B.H. Kim, Silica decorated on porous activated carbon nanofiber composites for high-performance supercapacitors. J Power Sources (2016) 328:219
[12]. S. Gao, L. Zhu, Liu L, A. Gao, F. Liao, Shao M Improved energy storage performance based on gamma-ray irradiated activated carbon cloth. Electro Chim Acta (2016) 191:908
[13]. M.H. Ohenb, R.F. Rodriguez, Activated carbon. Elsevier, Oxford (2006)
[14]. S. Furmaniak, P. Kowalczyk, A.P. Terzyk, P.A. Gauden, P.J. Harris, Synergetic effect of carbon nanopore size and surface oxidation on carbon dioxide capture from its mixtures with methane. J Colloid Interface Sci (2013) 397:144
[15]. P.J.F. Harris, Z. Liu, K. Suenaga, Imaging the atomic structure of activated carbon. J Phys Condens Matter (2008) 20:362201–362211
[16]. P.J.F. Harris, A. Burian, S. Duber, High-resolution electron microscopy of microporous carbon. Philos Mag Lett (2000) 80:381
[17]. P.J.F. Harris, New perspectives on the structure of graphitic carbons. Crit Rev Solid State Mater Sci (2005) 30:235
[18]. X. Yue, W. Sun, J. Zhang, F. Wang, Y. Yang, C. Lu, Z. Wang, D. Rooney, K. Sun, Macro-mesoporous hollow carbon spheres as anodes for lithium-ion batteries with high rate capability and excellent cycling performance. J Power Sources. (2016) 331:10
[19]. S.Y. Kim, B.H. Kim, Silica decorated on porous activated carbon nanofiber composites for high-performance supercapacitors. J Power Sources. (2016) 328:219
[20]. C. Liu, B.B. Koyyalamudi, L. Li, S. Emani, C. Wang, L.L. Shaw, Improved capacitive energy storage via surface functionalization of activated carbon as cathodes for lithium-ion capacitors. Carbon (2016) 163:109
[21]. D. Mohan, K.P. Singh, V.K. Singh, Wastewater treatment using low cost activated carbons derived from agricultural byproducts. J Hazard Mater (2008) 152:1045
[22]. O.S. Amuda, A.A. Giwa, I.A. Bello, Removal of heavy metals from industrial wastewater using modified activated carbon coconut shell. Bio Chem Eng J (2007) 36:174
[23]. F. Derbyshire, M. Jagtoyen, M. Thwaites, Application of active carbons for gas separation and respiratory protection. In: Patrick JW (ed.), Porosity in carbons: Characterization and applications. Edward Arnold, London, (1995) 209
[24]. W. Tongpoothorn, M. Sriuttha, P. Homchan, S. Chanthai, C. Ruangviriyachai, Preparation of activated carbon derived from jatropha curcas fruit shell by simple thermo-chemical activation and characterization of their physicochemical properties. Chem Eng Res Des (2011) 89:3
[25]. A.W. Verla, M.J. Horsfall, E.N. Verla, A.I. Spiff, O.A. Ekpete, Preparation and characterization of activated carbon from fluted pumpkin (Telfairia occidentalis Hook. F) seed shell. Asian J Nat Appl Sci (2012) 39:1
[26]. M.L. Sekirifa, M. Hadj-Mahammed, S. Pallier, L. Baameur, D. Richard, A. H. Al-Dujaili,' Preparation and characterization of an activated carbon from a date stones variety by physical activation with carbon dioxide. J Anal Appl Pyrol (2013) 155:99
[27]. D. Savova, E. Apak, E. Ekinci, F. Yardim, N. Petrov, T. Budinova, V. Minkova, Biomass conversion to carbon adsorbents and gas. Biomass Bioenerg (2001) 21:2
[28]. K. Li, Y. Li, Z. Zheng, Kinetics and mechanism studies of p-nitroaniline adsorption on activated carbon fibers prepared from cotton stalk by NH4H2PO4 activation and subsequent gasification with steam. J Hazard Mater (2010) 178:1
[29]. F. Rozada, L. Calvo, A. Garcıa, J.M. Villacorta, M. Otero, Dye adsorption by sewage sludge-based activated carbons in batch and fixed-bed systems. Bioresour Technol (2003) 87:221

[30]. N. Petrov, T. Budinova, M. Razvigorova, J. Parra, P. Galiatsatou, Conversion of olive wastes to volatiles and carbon adsorbents. Biomass Bioenerg (2008) 32:12
[31]. T. Lee, C.H. Ooi, R. Othman, F.Y. Yeoh, Activated carbon fiber-the hybrid of carbon fiber and activated carbon. Rev Adv Mater Sci (2014) 36:2
[32]. W. Bae, J. Kim, J. Chung, Production of granular activated carbon from food-processing wastes (walnut shells and jujube seeds) and its adsorptive properties. J Air Waste Manag Assoc (2014) 64:8
[33]. O. Ioannidou, A. Zabaniotou, Agricultural residues as precursors for activated carbon production-a review. Renew Sustain Energy Rev (2007) 11:9
[34]. Y. Xu, D.D.L. Chung, Silane-treated carbon fiber for reinforcing cement. Carbon (2001) 39:13
[35]. M.Z. Alam, E.S. Ameem, S.A. Muyibi, N.A. Kabbashi, The factors affecting the performance of activated carbon prepared from oil palm empty fruit bunches for adsorption of phenol. Chem Eng J (2009) 155:1
[36]. M.J. Ahmed, S.K. Theydan, Physical and chemical characteristics of activated carbon prepared by pyrolysis of chemically treated date stones and its ability to adsorb organics. Powder Technol (2012) 229:237
[37]. P. Pragya, S. Sripal, K.Y. Mahesh, Preparation and study of properties of activated carbon produced from agricultural and industrial waste shells. Res J Chem Sci (2013)12:3
[38]. S. De, A.M. Balu, J.C. Van Der Waal, R. Luque, Biomass-derived porous carbon materials: Synthesis and catalytic applications. Chem Cat Chem (2015) 7(17):1608–1629.
[39]. R. Kumar, A. Kumar, B.K. Rao, A.K. Srivastava, N. Misra, Structural and electronic properties of 2D-activated carbon sheet. Carbon Letters (2021) 31:483–488.
[40]. E. Artacho, J.M. Cela, J.D. Gale, A. Garcia, J. Junquera, R.M. Martin, P. Ordejon, D.S. Portal, J.M. Soler, The SIESTA method; developments and applicability. J Phys Condens Matter (2011) 20(6):064208.
[41]. P. Hohenberg, W. Kohn, Phys Rev 136: B864. Kohn W, Sham LJ Phys Rev (1964) 140:A1133.
[42]. N. Troullier, J.L. Martins, Efficient pseudo-potentials for plane-wave calculations. Phys Rev B (1991) 43:3
[43]. R.Q. Zhang, Q.Z. Zhang, M.W. Zhao, A scheme for the economical use of numerical basis sets in calculations with SIESTA. Theor Chem Acc (2004) 112:3
[44]. J. Junquera, O. Paz, D. Sánchez-Porta, E. Artacho, Numerical atomic orbitals for linear-scaling calculations. Phys Rev B (2001) 64:23
[45]. J.M. Seminario, Modern density functional theory: A tool for chemistry. Elsevier, Amsterdam (1995) (ISBN: 9780080536705)
[46]. R. Kumar, A. Kumar, N. Misra, In-silico investigation of silicon-doped 2D-activated carbon sheet. Pramana (2022) 96(1):32.
[47]. Q. Hu, Q. Wu, G. Sun, X. Luo, Z. Liu, B. Xu, Y. Tian, First principle study of atomic oxygen adsorption on boron-substituted graphite. Surf Sci (2008) 602:1
[48]. R. Faccio, W.L. Fernández, H. Pardo, C. Goyenola, O.N. Ventura, A.W. Mombrú, Electronic and structural distortions in graphene induced by carbon vacancies and boron doping. J Phys Chem C (2010) 11:44
[49]. W. Tian, W. Li, W. Yu, X. Liu, A review on lattice defects in graphene: Types, generation, effects and regulation. Micromachines (2017) 8(5):163
[50]. W.M. Daud, M. Badri, H. Mansor, Possible conduction mechanisms in coconut-shell activated carbon. J Appl Phys (1990) 67(4):1915
[51]. L.J. Kennedy, J.J. Vijaya, G. Sekaran, Electrical conductivity study of porous carbon composite derived from rice husk. Mater Chem Phys (2005) 91(2–3):471
[52]. B.A. Barroso, F.M. Alexandre, G.C. Fernández, G.A. Macías, S.V. Gómez, Temperature dependence of the electrical conductivity of activated carbons prepared from vine shoots by physical and chemical activation methods. Microporous Mesoporous Mater (2015) 209:90
[53]. M. Kobayashi, T. Sakuma, H. Takahashi (eds), Physics of solid state ionics (1st ed.). Research Signpost, Thiruvananthapuram (2006)
[54]. T. Minami, A. Hayashi, M. Tatsumisago, Recent progress of glass and glass-ceramics as solid electrolytes for lithium secondary batteries. Solid State Ionics (2006) 177(26–32):2715–2720
[55]. K. Selvaraju, M. Jothi, P. Kumaradhas, A charge density analysis on quarter thiophene molecular nanowire under applied electric field: A theoretical study. J. Comput Theor Nanosci (2013) 10(2):357–367.
[56]. A. Shafiee, M.M. Salleh, M. Yahaya, Determination of HOMO and LUMO of [6, 6]-phenyl C61-butyric acid 3-ethylthiophene ester and poly (3-octyl-thiophene-2, 5-diyl) through voltametry characterization. Sains Malaysiana (2011) 40(2):173–176.

[57]. P.I. Djurovich, E.I. Mayo, S.R. Forest, M.E. Thompson, Measurement of the lowest unoccupied molecular orbital energies of molecular organic semiconductors. Organic Electronics (2009) 10(3):515–520.
[58]. T. Rangel, G.M. Rignanese, V. Olevano, Beilstein J. Nanotechnol (2015) 6:1247.
[59]. V. Balachandran, A. Lakshmi, A. Janaki, Ab initio, DFT, HOMO–LUMO and Natural Bond Orbital analyses of the electronic structure of 2-mercapto-1-methylimidazole. J Mol Struct (2011) 1006(1–3):395–401.

14 Quantum Computing in Materials

A Perspective

Ruby Srivastava and Sravani Joshi

14.1 INTRODUCTION

As quantum advantage has gained more attention using boson sampling, the potential applications of quantum computing in material design and drug development are now being explored. Unfortunately, it is difficult to find exact or nearly exact solutions for chemically relevant systems for more than ~30 electrons.[1] There are efficient and accurate algorithms that exist for larger systems, and they deliver exact quantum mechanical solutions in a cost-effective manner. However most of the computational methods can be easily implemented on classical computers. The simulation scale of quantum computers depends on the available quantum resources, and the critical portion of computations are left to quantum computers. Fast evolving quantum computing technology promises to completely shift the computational capabilities in materials and drug research by finding solutions for currently impossible calculations. Researchers are working in the field of quantum information science and technology (QIST) and exploiting quantum properties and behavior to establish new capabilities in communications and related fields. The potential of other leading technologies such as cloud computing, artificial intelligence (AI), and machine learning (ML) are used in quantum computing for significantly faster, more efficient, and remarkably less expensive computational calculations. A few studies showed that quantum algorithms provide exponential speedups over their classical counterparts.[2] As it is difficult to solve the Schrodinger equation exactly, modern quantum chemistry aimed at finding approximate solutions such as *ab initio* methods[3] (Hartree–Fock, Moller–Plesset, coupled cluster, Green's function, configuration interaction, etc.), semiempirical methods (extended Huckel, CNDO, INDO, AM1, PM3, etc.),[4] density functional methods (LDA, GGA, hybrid models, etc.),[5] density matrix methods,[6] algebraic methods (Lie groups, Lie algebras, etc.),[7] quantum Monte Carlo methods (variational, diffusion, Green's function forms, etc.),[8] and dimensional scaling methods.[9] These methods also faced challenges as they have high computational costs and are accurate only for larger systems. Due to the exponential speedup, quantum computers can accomplish simulation tasks within only a polynomial amount of time. Since quantum computers use quantum methods, the quantum mechanical simulation problems can run without memory loss with an exponentially faster rate. Various chemical systems can benefit from quantum computers with accurate quantum methods in material sciences and pharmaceutical sciences.[10,11] For larger disruptive industrial quantum simulations, advance quantum computers are needed. It has been reported that significant experimental developments in quantum computers has occurred in the past several years. Higher efficiencies have been observed for recent trapped ion quantum devices with an average of a two-qubit gate.[12] Quantum computers with 50–60 qubits have been used by Google and USTC to solve a classically unmanageable problem in superconducting technologies.[13,14]

The basic architecture of quantum computing is as follows. An electronic structure problem is mapped onto a quantum computer and compared with the scaling of quantum algorithms to find the ground state energy of an electronic Hamiltonian variational quantum eigensolver (VQE) and quantum phase estimation (QPE). Then the more favorable algorithm is used to discuss the aspects of error correction to run the algorithms with an estimated requirement of quantum resources. These

DOI: 10.1201/9781003441328-14

algorithms implement efficient unitary operators (Trotterization and qubitization) that are related to the Hamiltonian. Finally, the estimated results are compared, and the accurate and the efficient ones are applied for the material design or pharmaceutical system.

14.2 QUANTUM COMPUTATION

Quantum computational devices use the laws of quantum mechanics to perform calculations. The theory of quantum computing was first developed in the early 1980s by Paul Benioff, Richard Feynman, David Deutsch, and Peter Shor.[15] Quantum computing has the potential to perform polynomial calculations efficiently within the desired run time in dealing with problems of such a size that it cannot be performed efficiently on a digital computer. This initial work led in development of Shor's algorithm,[16] and then with the development of an algorithm in 1996 to search an unsorted database of size N in $O(\sqrt{N})$ time, compared to the $O(N)$ runtime required by classical algorithms.[17] High-level quantum mechanical computational models are used to accurately predict the reaction barriers and conformation stability. In principle, DFT is widely used in chemistry for predicting and interpreting complex system behaviors at an atomic scale with the exact ground state energy and electron density of a quantum system. The accuracy of the DFT approach depends on the exchange-correlation functions. As the exact forms of the functional are unknown, approximate exchange correlations functionals are used with low computational complexity as DFT is able to provide quantitative descriptions for very large-scale simulations of systems containing millions of atoms. One successful application of the divide and conquer (DC) approach was observed when the ground state potential energy surface of a ring containing 10 hydrogen atoms was estimated using the density matrix embedding theory (DMET) on a trapped-ion quantum computer in which a 20-qubit problem is decomposed into ten 2-qubit problems.[18]

A quantum computer consists of a register of quantum bits (qubits). These qubits can be in a state $|0\rangle$ or $|1\rangle$. Qubits can either be on or off or both (superposition). A quantum gate is any allowed operation on a system of qubits. The qubit is a two-state system whose states are labeled as $|0\rangle$ and $|1\rangle$. $|.\rangle$ is identified as the Dirac notation for the quantum state. Qubits can be in any superposition of the states $|0\rangle$ and $|1\rangle$:

$$|\Psi\rangle = \alpha\,|0\rangle \text{ and } \beta\,|1\rangle \quad \alpha,\beta \varepsilon C. \quad |\alpha|^2 + |\beta|^2 = 1. \tag{14.1}$$

where α, β are the complex coefficients known as the amplitudes of the respective states. They are related to the effect of physical measurement. The difference between classical bit and qubit is given in Table 14.1.

TABLE 14.1
Comparison between Classical Bit and Qubit

Classical Bit	Qubit
State 0 or 1	$|0\rangle$, $|1\rangle$, or superposition
Measurement does not change the state of the bit.	Measurement changes the system.
Deterministic result	Quantum state itself remains the same, and it is deterministic. Different results only occur with measurement when the quantum state collapses.
Can make a copy of bit (eavesdrop)	Cannot clone the qubit (security)
One number for a string bit	A qubit can represent a state vector with 2 degrees of freedom; thus, it can store one complex number

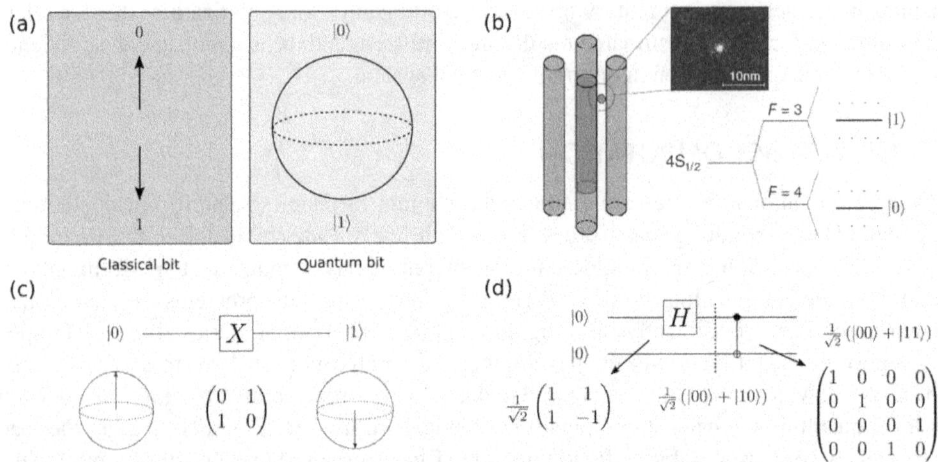

FIGURE 14.1 (a) Comparison between a classical bit and a qubit. (b) Scheme of an ion trap qubit approach used for experimental quantum computing. (c) Diagram of a quantum circuit implementing the X or quantum-NOT gate in matrix representation, and the change in the Bloch sphere. (d) Quantum circuit to generate a Bell state using the Hadamard (H) gate and the controlled-NOT gate. The dotted line in the middle represents the states of the circuit after applying the Hadamard gate.

Sources: (b) Adapted from *Phys Rev Lett.* 2014;113(22):220501; (d) adapted from *WIREs Comput Mol Sci.* 2020; e1481.

The comparison between the classical bit and the quantum bit is given in Figure 14.1, which shows that the classical bit can only take one of the two states, 0 or 1, while quantum bit can take any state of the form $|\Psi\rangle = \cos\frac{\theta}{2}|0\rangle + e^{-i\varphi}\sin\frac{\theta}{2}|1\rangle$. Single qubits are represented as Bloch sphere, where 1θ and 1φ are considered the azimuthal and polar angles in the sphere of unit radius.

Qubits and quantum gates are the basic components of any algorithm and used as the variables and functions of a programming language. The combination of great speed with probabilistic solutions to multiple calculations with higher accuracy at one time fits well in applications such as optimization, simulation of chemicals, and artificial intelligence (AI). Today's quantum computers known as noisy intermediate-scale quantum (NISQ) have limited computing power. They are able to assist in the first stage of drug discovery involving molecular simulation, wave function optimization, and ML. Even a small increase in simulation accuracy could result in more accurate predictions of potential drugs long before clinical trials, with reduced time and lowered cost. Computation using qubits rather than the classical bits are different in the presence of quantum entanglement, which occurs due to the correlation between multiple qubits. In this way, quantum computing strengthens quantum parallelism and quantum entanglement to offer an algorithmic speedup over the few classical computing evidences.[19–23] Quantum qubit provides an efficient representation of the many-body states of any interacting quantum system in quantum simulations of stationary/time-independent processes. Quantum qubits also provide real and imaginary time evolution[24, 25] in quantum dynamics either through Lie–Trotter–Suzuki expansion[26] or through variational frameworks.[27]

14.3 QUANTUM GATES AND QUANTUM-CIRCUIT-BASED PARADIGM

The quantum states of single and many qubits are transformed or manipulated with quantum gates. In the gate model of the quantum computing paradigm, transformations between states are achieved using unitary matrices that are represented as quantum gates. Since all quantum gates are unitary, the inverse of such gates necessarily exists, and hence transformations using quantum gates are always reversible. The irreversibility in the quantum-based circuit is incorporated by making projective measurements as it disturbs the state vector irreversibly, making it losing its present memory. The environmental interactions also induce irreversibility in the form of qubit decoherence.[28] A quantum circuit is a group of quantum gates that transform the initial state of a multiqubit system to the final desired state. The set of quantum gates represents a user-defined unitary transformation, which is followed by measurement either on a computational basis or on the basis of an operator that has the desired prepared statistical state.[28]

Various sources of error affect the quantum processor. For example, the coupling of a qubit to its environment can cause the system to collapse to one of its classical states. This process is known as decoherence. Small fluctuations transform the quantum gate to apply into a similar one, which produces a different output than expected, and this imperfect control mechanism always introduces an error. In previous studies, the least error was reported on the gates in a trapped ion processor for single-qubit gates and two-qubit gates.[29] Google reported an error rate of 0.1% for single-qubit gates and 0.3% for two-qubit gates in a superconducting processor.[30] So it can be easily anticipated that computation can become meaningless even with a small sequence of gates.

14.4 QUANTUM ALGORITHMS

In this section, we outline the variational quantum eigensolver (VQE) and quantum phase estimation (QPE) algorithms as quantum algorithms provide an exponential speedup in solving quantum chemistry problems.

14.4.1 Variational Quantum Eigensolver (VQE)

VQE is a hybrid algorithm that is used to explore the potential application of quantum computing. It uses both classical and quantum computational resources. A classical optimizer searches some set of quantum states with the smallest Hamiltonian expectation value. With variational principle, any such expectation value is necessarily greater than or equal to the ground state energy. The set of explored states is known as an ansatz. These states are prepared through some parametrized quantum circuit that is run to prepare a particular quantum state with some initialized parameters. Then the measurements of the state are made. The ansatz circuit must be applied many times to obtain the necessary information to estimate the expected Hamiltonian on the quantum state with some accuracy. This expected Hamiltonian is used to update the parameter values by the classical optimizer, and the expectation estimation process is repeated until some convergence criteria are met. The wave function ansatzes in the VQE are used to find an exact solution to the Schrödinger equation. The skeleton of VQE algorithms indicating which parts occur on the quantum computer and which are left to the classical ones are given in Figure 14.2.

An example is the unitary coupled cluster (UCC) with single and double excitations (UCCSD) wave function ansatzes. The depth of its circuit scales quadratically with the size of the system, though UCCSD ansatz is not able to describe a strong correlation effect such as the triple bond dissociation of molecular nitrogen. Studies have reflected the use of adaptive VQE algorithms to reduce circuit depth.[31–35] Another adaptive derivative assembled pseudo-Trotter (ADAPT) ansatz[31] also represents electronic wave functions with high accuracy. The electronic wave function of the qubit coupled cluster (QCC) is also represented in qubit, and QCC maintains a constant circuit depth for electronic structure simulations.[35]

14.4.2 Quantum Phase Estimation (QPE)

In 1995, Alexei Kitaev introduced quantum phase estimation (QPE) to study the Abelian stabilizer problem.[36] In 1998, Cleve et al. extended this QPE approach to estimate the phase of an arbitrary unitary operator.[37] The QPE method can be applied to find the eigenvalues of a chemical

Hamiltonian to a given accuracy with a runtime that scales polynomially to the system size. For this reason, it is said that quantum computers can perform accurate chemical calculations beyond the reach of classical devices. QPE is used in chemical and material simulations.[38] QPE algorithm has exponential acceleration while solving quantum chemistry problems. The probability of achieving a polynomial acceleration is also significantly present in quantum chemistry.[39] QPE algorithms are used to calculate the energies of chemical systems. It requires deeper circuits than VQE, but these calculations are not performed so frequently. QPE does not require an ansatz (see Figure 14.3). QPE

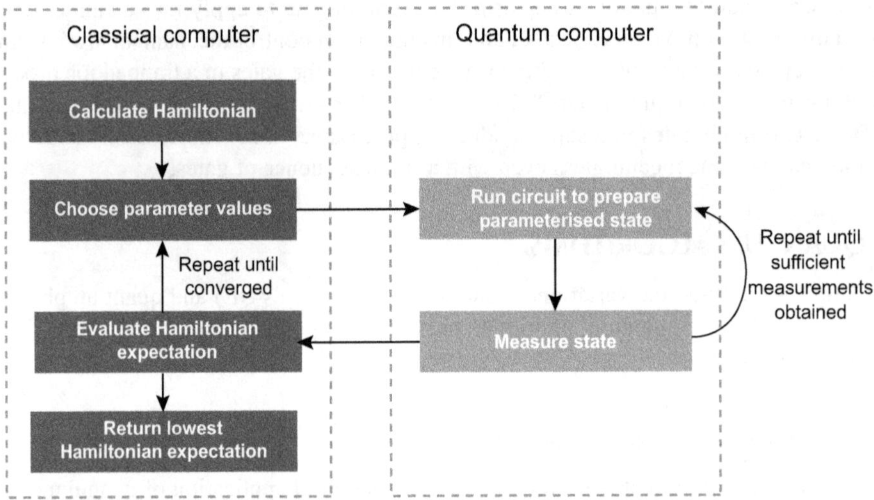

FIGURE 14.2 The schematic representation of VQE algorithms, indicating which parts occur on quantum and classical computers.

Source: Adapted from *J. Chem. Theory Comput.* 2022, 18, 12, 7001–7023.

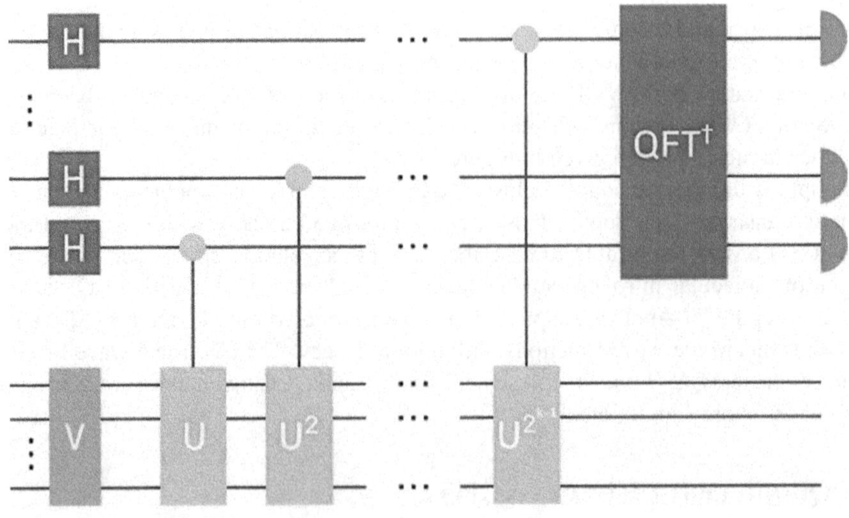

FIGURE 14.3 Outline of the circuit used to execute QPE.

Source: Adapted from *J. Chem. Theory Comput.* 2022, 18, 12, 7001–7023.

calculates the eigenvalues of the Hamiltonian directly with some precision. The quantum computers used to simulate large chemical systems will likely be fault tolerant. So these systems are able to run randomly deep algorithms without error. Quantum computers simulate a chemical system by mapping the behavior of its electrons to the behavior of its qubits and quantum gates.

Near-term quantum computers may be able to simulate small molecules that are beyond the reach of supercomputers. Hybrid quantum-classical computation is the universal approach for quantum computing, implying that it can carry out any task even if it is with remarkable overhead. In quantum computing, instead of achieving fault tolerance, it is better to reduce the errors with minimum effort in order to run bigger quantum calculations on larger circuits. Various approaches apply extra operations to discard computational runs with errors[40] or manipulate the error rate to deduce the correct result.[41,42] Quantum computing requires very large, fault-tolerant quantum computers to solve useful problems[43] in the coming decades. Quantum simulations of quantum chemistry problems involve tens or hundreds of orbitals. Fault-tolerant quantum computers scale up to hundreds of logical qubits with long-term quantum error correction. So it is better to use potential strategy that is customized for NISQ devices.[44–48]

14.5 MULTISCALE QUANTUM COMPUTING

Multiscale models are used to study the complex chemical reactions and biological processes. Classical computers are computationally not efficient in estimating total energies with chemical accuracy as the exact solutions to quantum mechanical problems are exponentially complicated. One of the popular methods of multiscale scheme to deal with complex systems at the electronic scale is to combine QM models with MM models. In this case the total energy of a QM/MM system can be written as:

$$E_{total} = E_{QM} + E_{MM} + E_{QM-MM} \tag{14.2}$$

with E_{QM-MM} being the interaction energy between the QM and MM regions. Even from this hybrid QM/MM method, it is impossible to carry out an exact screening due to numerical errors in material and drug design. Quantum computers provide the hope of coming up with a potential solution for accurate simulations of material properties or system behavior. These models include low-level computational models (EFF approaches and effective fragment potential [EFP] models) and high-level QM computational models (single-reference and multireference) approaches. These approaches have unreasonable computational complexity for accurate simulations of complex systems.

14.5.1 Divide and Conquer (DC) Approach

The two divide and conquer (DC) schemes used in quantum chemistry simulations are fragment-based approaches[49] and quantum embedding theories.[50] *Fragmentation strategy* fragments a complicated quantum chemistry problem into as many parts as possible according to the size and depth of the available quantum circuit. Fragment-based approaches divide a system according to chemical information, such as chemical functional groups and chemical bonds. Many other fragment-based approaches, such as the fragment molecular-orbital method,[51] molecular fragmentation with conjugated caps method,[52,53] and (generalized) many-body expansion,[54–56] are also used to study complex systems. The accuracy of these approaches depends on molecular fragmentation schemes and empirical corrections to high-order many-body interactions.[57] The size of subsystems can adapt to available quantum resources. These subsystems determine the number of qubits used for quantum computing. Then the massively parallel quantum simulations of subsystems can be performed on NISQ devices. Finally, by combining the properties of subsystems, the corresponding quantities of the system is attained. In this way, these quantum computational resources are utilized in practical applications. The depth of circuit in each subsystem simulation is remarkably reduced as compared to the simulation of the full system. Low-depth circuits are effectively combined with error reduction approaches to reduce the total error rate.

Quantum embedding theories have a flexible approach for partitioning large systems. The subsystem interactions with environment include bath orbitals in the calculations for the subsystem. However, applying analytical corrections to the density matrix embedding theory (DMET) is not an irrelevant task as it involves a self-consistent iteration method. DMET with VQE was implemented for a 10-hydrogen-atom ring on a trapped-ion quantum computer.[18] DMET-VQE method was applied for the numerical simulation of C18 with the cc-pVDZ basis (up to 144 qubits).[58] These studies showed the advantage of the DC[59] strategy in quantum computing. Still, whether to apply DC approaches within the VQE framework for quantum chemistry problems remains uncertain.

14.5.2 Correlation Energy Decomposition

The Hartree–Fock (HF) method is used to describe the coulomb and to exchange electron–electron interactions that are efficiently used in classical computing methods. The major task of quantum computing is to accurately recover the correlation energies (dynamic and static correlations). The correlation energy is performed according to the partitions of the orbital space, which include core, active, and virtual spaces. The static correlation is considered for the active space model. As quantum computers have computational complexity, which increases exponentially as a function of N, it is selected for finding a solution to this problem. Once the static correlation is acquired, dynamic correlation is obtained by classical *a posteriori* corrections.[60] This method is used in the wave function theory (WFT) and CASPT2 (CASSCF with second-order perturbation theory). Quantum computing also uses this method for improving the accuracy of calculations beyond the minimal basis set. A few other methods with corrections have been used in quantum computing[61] to solve the basis set convergence problem.[62,63] However, the requirements of quantum resources in material and (bio) molecular simulations are beyond the capacity of some quantum devices.

Quantum computers implement advanced algorithms for solving complex problems that are limited by available qubit counts, coherence time, and gate accuracy. Precise quantum simulations without error correction are feasible only for low-depth quantum algorithms. Recent advances in error reduction techniques require a dozen qubits, circuit sizes, and depths, which are still in the infant stages. Superconducting devices with 127 qubits are developed by IBM's Eagle processor.[64] These devices can run noisy intermediate-scale quantum algorithms (VQE)[65] with the largest experimental efforts by simulating the binding energy of up to 12-atom hydrogen chains.[66–69] Various experimental groups have implemented quantum error correction using multiple physical qubits to protect a smaller number of logical qubits from errors, thus increasing the efficiency of quantum computing results: for example, superconducting devices,[70] trapped ions,[71–74] and nuclear-spin qubits in diamond.[75]

14.6 CONCLUSION

These experimental results show a great deal of progress in the field of quantum computing. Quantum computers facilitate more effective algorithms with the help of operations that are not possible with classical computers. Quantum computers do not work faster than classical computers but operate in a different manner, so that they achieve unparallel speedups by avoiding irrelevant computation. The computation of full electronic wave function of an average drug molecule took longer than the age of the universe on any supercomputer with regular algorithms, but it took a time scale of days to solve this with the modest-sized quantum computers. Though many technical challenges of building quantum computers have emerged during the last decade, the developments of quantum computers have not stopped. The algorithms can be analyzed mathematically in the absence of hardware also. So further investigations need to be carried out in the future as quantum computers have immense potential in the applications of materials and drug development.

14.6.1 Acknowledgments

RS acknowledges the financial assistance from DST WOSA project (SR/WOS-A/CS-69/2018). RS is thankful to her mentor, Dr. Shrish Tiwari, Bioinformatics, CSIR-Centre for Cellular and Molecular Biology, and Dr. G. Narahari Sastry, Director, CSIR-NEIST for their technical support.

14.6.2 Conflict of Interest

The authors declare that they have no conflict of interest, financial and/or otherwise.

REFERENCES

1. Hermann, J.; Schätzle, Z.; Noé, F. Deep-neural-network solution of the electronic Schrödinger equation. Nat. Chem. **2020**, 12 (10), 891–897.
2. Kais, S. Introduction to Quantum Information and Computation for Chemistry. **2014**. www.chem.purdue.edu/kais/docs/QICC_Vol%20154_Chap1.pdf.
3. Jordan, S. The quantum Algorithm Zoo. **2012**. Available at http://math.nist.gov/quantum/zoo/.
4. Szabo, A.; Ostlund, N. S. Modern Quantum Chemistry—Introduction to Advanced Electronic Structure Theory, Dover Publications Inc., Mineola, NY, **1982**.
5. Hanson, D. M.; Harvey, E.; Sweeney, R.; Zielinksi, T. J. Quantum States of Atoms and Molecules. (Department of Education Open Textbook Pilot Project, the UC Davis Office of the Provost, the UC Davis Library, the California State University. **2022**.
6. Parr, R. G.; Yang, W. Density-Functional Theory of Atoms and Molecules, Oxford Science Publications, New York, **1989**.
7. Mazziotti, D. A. Reduced-density-matrix mechanics: With Application to many-electron atoms and molecules. Adv. Chem. Phys. **2007**, 134, 1.
8. Iachello, I.; Levine, R. D. Algebraic Theory of Molecules, Oxford University Press, New York, **1995**.
9. Nightingale, M. P.; Umrigar, C. J. Quantum Monte Carlo Methods in Physics and Chemistry, NATO Science Series(Vol. 525), Springer, **1998**.
10. What Happens When ‘If’ Turns to ‘When’ in Quantum Computing?. Boston Consulting Group (BCG), **2021**; www.bcg.com/publications/2021/building-quantum-advantage (accessed 2022–03–02)
11. Pharma’s digital Rx: Quantum computing in drug research and development. McKinsey, **2021**; www.mckinsey.com/industries/life-sciences/our-insights/pharmas-digital-rx-quantumcomputing-in-drug-research-and-development (accessed 2022–03–02).
12. Quantinuum Announces Quantum Volume 4096 Achievement. Quantinuum, **2022**; www.quantinuum.com/pressrelease/quantinuum-announces-quantum-volume-4096-achievement (accessed 2022–03–02).
13. Arute, F.; Arya, K.; Babbush, R.; Bacon, D.; Bardin, J. C.; Barends, R.; Biswas, R.; Boixo, S.; Brandao, F. G. S. L.; Buell, D. A.; Burkett, B.; Chen, Y.; Chen, Z.; Chiaro, B.; Collins, R.; Courtney, W.; Dunsworth, A.; Farhi, E.; Foxen, B.; Fowler, A.; Gidney, C.; Giustina, M.; Graff, R.; Guerin, K.; Habegger, S.; Harrigan, M. P.; Hartmann, M. J.; Ho, A.; Hoffmann, M.; Huang, T.; Humble, T. S.; Isakov, S. V.; Jeffrey, E.; Jiang, Z.; Kafri, D.; Kechedzhi, K.; Kelly, J.; Klimov, P. V.; Knysh, S.; Korotkov, A.; Kostritsa, F.; Landhuis, D.; Lindmark, M.; Lucero, E.; Lyakh, D.; Mandrá, S.; McClean, J. R.; McEwen, M.; Megrant, A.; Mi, X.; Michielsen, K.; Mohseni, M.; Mutus, J.; Naaman, O.; Neeley, M.; Neill, C.; Niu, M. Y.; Ostby, E.; Petukhov, A.; Platt, J. C.; Quintana, C.; Rieffel, E. G.; Roushan, P.; Rubin, N. C.; Sank, D.; Satzinger, K. J.; Smelyanskiy, V.; Sung, K. J.; Trevithick, M. D.; Vainsencher, A.; Villalonga, B.; White, T.; Yao, Z. J.; Yeh, P.; Zalcman, A.; Neven, H.; Martinis, J. M. Quantum supremacy using a programmable superconducting processor. Nature **2019**, 574, 505– 510.
14. Wu, Y.; Bao, W.-S.; Cao, S.; Chen, F.; Chen, M.-C.; Chen, X.; Chung, T.-H.; Deng, H.; Du, Y.; Fan, D.; Gong, M.; Guo, C.; Guo, C.; Guo, S.; Han, L.; Hong, L.; Huang, H.-L.; Huo, Y.-H.; Li, L.; Li, N.; Li, S.; Li, Y.; Liang, F.; Lin, C.; Lin, J.; Qian, H.; Qiao, D.; Rong, H.; Su, H.; Sun, L.; Wang, L.; Wang, S.; Wu, D.; Xu, Y.; Yan, K.; Yang, W.; Yang, Y.; Ye, Y.; Yin, J.; Ying, C.; Yu, J.; Zha, C.; Zhang, C.; Zhang, H.; Zhang, K.; Zhang, Y.; Zhao, H.; Zhao, Y.; Zhou, L.; Zhu, Q.; Lu, C.-Y.; Peng, C.-Z.; Zhu, X.; Pan, J.-W. Strong quantum computational advantage using a superconducting quantum processor. Phys. Rev. Lett. **2021**, 127, 180501.
15. Shor, P. W. The Early Days of Quantum Computation. arXiv Preprint (Quantum Physics), **2022**. arXiv:2208.09964v1. https://doi. org/10.48550/arXiv.2208.09964.

16. Shor, P. Algorithms for quantum computation: discrete logarithms and factoring. Proceedings 35th Annual Symposium on Foundations of Computer Science; IEEE, **1994**, 124–134. https://doi.org/10.1109/SFCS.1994.365700.
17. Grover, L. K. A Fast Quantum mechanical algorithm for database search. Proceedings of the Twenty-Eighth Annual ACM Symposium on Theory of Computing. New York, NY, USA, **1996**, 212–219.
18. Kawashima, Y.; Lloyd, E.; Coons, M. P.; Nam, Y.; Matsuura, S.; Garza, A. J.; Johri, S.; Huntington, L.; Senicourt, V.; Maksymov, A. O.; Nguyen, J. H. V.; Kim, J.; Alidoust, N.; Zaribayan, A.; Yamazaki, T. Optimizing electronic structure simulations on a trapped-ion quantum computer using problem decomposition. Commun. Phys. **2021**, 4, 245.
19. Bennett, C. H.; Brassard, G.; Crépeau, C.; Jozsa, R.; Peres, A.; Wootters, W. K. Teleporting an unknown quantum state via dual classical and Einstein-Podolsky-Rosen channels. Phys. Rev. Lett. **1993**, 70, 1895.
20. Harrow, A.; Hayden, P.; Leung, D. Superdense coding of quantum states. Phys. Rev. Lett. **2004**, 92, 187901.
21. Abrams, D. S.; Lloyd, S. Quantum algorithm providing exponential speed increase for finding eigenvalues and eigenvectors. Phys. Rev. Lett. **1999**, 83, 5162–5165.
22. Shor, P. W. Polynomial-time algorithms for prime factorization and discrete logarithms on a quantum computer. SIAM Rev. **1999**, 41, 303–332.
23. Grover, L. K. A fast quantum mechanical algorithm for database search. Proceedings of the Twenty-Eighth Annual ACM Symposium on Theory of Computing, **1996**, pp. 212–219.
24. Lin, S. H.; Dilip, R.; Green, A. G.; Smith, A.; Pollmann, F. Real- and imaginary-time evolution with compressed quantum circuits. PRX Quantum. **2021**, 2, 010342.
25. Yeter-Aydeniz, K.; Moschandreou, E.; Siopsis, G. Quantum imaginary-time evolution algorithm for quantum field theories with continuous variables. Phys. Rev. A. **2022**, 105, 012412.
26. Yi, C. Success of digital adiabatic simulation with large Trotter step. Phys. Rev. A, 2021, 104, 052603.
27. Endo, S.; Sun, J.; Li, Y.; Benjamin, S. C.; Yuan, X. Variational quantum simulation of general processes. Phys. Rev. Lett. **2020**, 125, 010501.
28. Nielsen, M. A.; Chuang, I. L. Quantum Computation and Quantum Information: 10th Anniversary Edition (10th ed.), Cambridge University Press, **2011**.
29. Harty, T.P.; Allcock, D.T.C.; Ballance, C.J.; Guidoni, L.; Janacek, H.A.; Linke, N.M.; Stacey, D.N.; Lucas, D.M. High-fidelity preparation, gates, memory, and readout of a trapped-ion quantum bit. Phys Rev Lett. **2014**, 113(22), 220501.
30. Arute, F.; Arya, K.; Babbush, R.; Bacon, D.; Bardin, J. C.; Barends, R.; Biswas, R.; Boixo, S.; Brandao, F. G. S. L.; Buell, D. A.; Burkett, B.; Chen, Y.; Chen, Z.; Chiaro, B.; Collins, R.; Courtney, W.; Dunsworth, A.; Farhi, E.; Foxen, B.; Fowler, A.;, Gidney, C.; Giustina, M.; Graff, R.; Guerin, K.; Martinis, J. M. Quantum supremacy using a programmable superconducting processor. Nature. **2019**, 574(7779), 505–510.
31. Grimsley, H. R.; Economou, S. E.; Barnes, E.; Mayhall, N. J. An adaptive variational algorithm for exact molecular simulations on a quantum computer. Nat. Commun. **2019**, 10, 3007.
32. Tang, H. L.; Shkolnikov, V.; Barron, G. S.; Grimsley, H. R.; Mayhall, N. J.; Barnes, E.; Economou, S. E. Qubit-ADAPT-VQE: An adaptive algorithm for constructing hardware-efficient ansätze on a quantum processor. PRX Quantum. **2021**, 2, 020310.
33. Yordanov, Y. S.; Armaos, V.; Barnes, C. H. W.; Arvidsson-Shukur, D. R. M. Qubit-excitation-based adaptive variational quantum eigensolver. Commun. Phys. **2021**, 4, 228.
34. Liu, J.; Li, Z.; Yang, J. Analytic evaluation of energy first derivatives for spin–orbit coupled-cluster singles and doubles augmented with noniterative triples method: General formulation and an implementation for first-order properties. J. Chem. Phys. **2021**, 154, 244112.
35. Fan, Y.; Liu, J.; Li, Z.; Yang, J. Equation-of-motion theory to calculate accurate band structures with a quantum computer. J. Phys. Chem. Lett. **2021**, 12, 8833–8840.
36. Kitaev, A. Y. Quantum measurements and the Abelian Stabilizer problem. arXiv Preprint (Quantum Physics), **1995**. arXiv:quant-ph/9511026. https://doi.org/10.48550/arXiv.quant-ph/9511026.
37. Cleve, R.; Ekert, A.; Macchiavello, C.; Mosca, M. Quantum algorithms revisited. Proc. R. Soc. London, A **1998**, 454, 339–354.
38. Reiher, M.; Wiebe, N.; Svore, K. M.; Wecker, D.; Troyer, M. Elucidating reaction mechanisms on quantum computers. Proc. Natl. Acad. Sci. U. S. A. **2017**, 114, 7555–7560.
39. Genin, S. N.; Ryabinkin, I. G.; Paisley, N. R.; Whelan, S. O.; Helander, M. G.; Hudson, Z. M. Estimating Phosphorescent Emission Energies in IrIII Complexes Using Large-Scale Quantum Computing Simulations. Angew. Chem., Int. Ed. **2022**, 134, e202116175.
40. McArdle, S, Yuan, X, Benjamin, S. Error-mitigated digital quantum simulation. Phys Rev Lett. **2019**, 122(18):180501.

41. Li, Y.; Benjamin, S. C. Efficient variational quantum simulator incorporating active error minimization. Phys Rev X. **2017**, 7(2), 021050.
42. Temme, K.; Bravyi, S.; Gambetta, J. M. Error mitigation for short-depth quantum circuits. Phys Rev Lett. **2017**;119(18):180509.
43. Preskill, J. Quantum computing in the NISQ era and beyond. Quantum. **2018**, 2:79.
44. Bauer, B.; Wecker, D.; Millis, A. J.; Hastings, M. B.; Troyer, M. Phys. Rev. X, **2016**, 6, 031045.
45. Kreula, J. M.; García-Alvarez, L.; Lamata, L.; Clark, S. R.; Solano, E.; Jaksch, D. EPJ Quantum Technology, **2016**, 3, 11.
46. Yamazaki, T.; Matsuura, S.; Narimani, A.; Saidmuradov, A.; Zaribafiyan, A. Towards the Practical Application of Near-Term Quantum Computers in Quantum Chemistry Simulations: A Problem Decomposition Approach, **2018**.
47. Bharti, K.; Cervera-Lierta, A.; Kyaw, T. H.; Haug, T.; Alperin-Lea, S.; Anand, A.; Degroote, M.; Heimonen, H.; Kottmann, J. S.; Menke, T.; Mok, W.-K.; Sim, S.; Kwek, L.-C.; Alán, A.-G., Noisy intermediate-scale quantum algorithms. Rev. Mod. Phys. **2022**, 94(1), 015004.
48. Li, W.; Huang, Z.; Cao, C.; Huang, Y.; Shuai, Z.; Sun, X.; Sun, J.; Yuan, X.; Lv, D. Toward practical quantum embedding simulation of realistic chemical systems on near-term quantum computers. Chem. Sci. **2022**, 13, 8953–8962.
49. Akimov, A. V.; Prezhdo, O. V. Large-scale computations in chemistry: A bird's eye view of a vibrant field. Chem. Rev. **2015**, 115, 5797–5890.
50. Sun, Q.; Chan, G. K. L. Quantum embedding theories. Acc. Chem. Res. **2016**, 49, 2705–2712.
51. Fedorov, D. G.; Nagata, T.; Kitaura, K. Exploring chemistry with the fragment molecular orbital method. Phys. Chem. Chem. Phys. **2012**, 14, 7562–7577.
52. Zhang, D. W.; Zhang, J. Z. H. Molecular fractionation with conjugate caps for full quantum mechanical calculation of protein–molecule interaction energy. J. Chem. Phys. **2003**, 119, 3599–3605.
53. Wang, X.; Liu, J.; Zhang, J. Z. H.; He, X. Electrostatically embedded generalized molecular fractionation with conjugate caps method for full quantum mechanical calculation of protein energy. J. Phys. Chem. A. **2013**, 117, 7149–7161.
54. Dahlke, E. E.; Truhlar, D. G. Electrostatically embedded many-body expansion for large systems, with applications to water clusters. J. Chem. Theory Comput. **2007**, 3, 46–53.
55. Richard, R. M.; Herbert, J. M. A generalized many-body expansion and a unified view of fragment-based methods in electronic structure theory. J. Chem. Phys. **2012**, 137, 064113.
56. Liu, J.; Herbert, J. M. Pair–pair approximation to the generalized many-body expansion: an alternative to the four-body expansion for ab initio prediction of protein energetics via molecular fragmentation. J. Chem. Theory Comput. **2016**, 12, 572–584.
57. Vorwerk, C.; Sheng, N.; Govoni, M.; Huang, B.; =Galli, G. Green's function formulation of quantum defect embedding theory. Nat. Comput. Sci. **2022**, 2, 424–432.
58. Li, W.; Huang, Z.; Cao, C.; Huang, Y.; Shuai, Z.; Sun, X.; Sun, J.; Yuan, X.; Lv, D. Toward practical quantum embedding simulation of realistic chemical systems on near-term quantum computers. Chem. Sci. **2022**, 13, 8953–8962.
59. Yang, W. Electron density as the basic variable: A divide-and-conquer approach to the ab initio computation of large molecules. J. Mol. Struct. **1992**, 255, 461–479.
60. Ryabinkin, I. G.; Izmaylov, A. F.; Genin, S. N. A posteriori corrections to the iterative qubit coupled cluster method to minimize the use of quantum resources in large-scale calculations. Quantum Science and Technology. **2021**, 6, 024012.
61. Bauman, N. P.; Bylaska, E. J.; Krishnamoorthy, S.; Low, G. H.; Wiebe, N.; Granade, C. E.; Roetteler, M.; Troyer, M.; Kowalski, K. Downfolding of many-body Hamiltonians using active-space models: Extension of the sub-system embedding sub-algebras approach to unitary coupled cluster formalisms. J. Chem. Phys. **2019**, 151, 014107.
62. Kutzelnigg, W.; Klopper, W. Wave functions with terms linear in the interelectronic coordinates to take care of the correlation cusp. I. General theory. J. Chem. Phys. **1991**, 94, 1985–2001.
63. Kumar, A.; Asthana, A.; Masteran, C.; Valeev, E. F.; Zhang, Y.; Cincio, L.; Tretiak, S.; Dub, P. A. Quantum simulation of molecular electronic states with a transcorrelated Hamiltonian: Higher accuracy with fewer qubits. J. Chem. Theory Comput. **2022**, 18, 5312–5324.
64. Collins, H.; Easterly, K. IBM Unveils Breakthrough 127-Qubit Quantum Processor. IBM, **2021**; https://newsroom.ibm.com/2021-11-16-IBM-Unveils-Breakthrough-127-Qubit-Quantum-Processor [accessed 2022–03–02].
65. Peruzzo, A.; McClean, J.; Shadbolt, P.; Yung, M.-H.; Zhou, X.- Q.; Love, P. J.; Aspuru-Guzik, A.; O'Brien, J. L. A variational eigenvalue solver on a quantum processor. Nat. Commun. **2014**, 5, 4213.

66. Hempel, C.; Maier, C.; Romero, J.; McClean, J.; Monz, T.; Shen, H.; Jurcevic, P.; Lanyon, B. P.; Love, P.; Babbush, R.; Aspuru-Guzik, A.; Blatt, R.; Roos, C. F. Quantum chemistry calculations on a trapped-ion quantum simulator. Phys. Rev. X. **2018**, 8, 031022.
67. Chen, M.-C.; Gong, M.; Xu, X.-S.; Yuan, X.; Wang, J.-W.; Wang, C.; Ying, C.; Lin, J.; Xu, Y.; Wu, Y.; Wang, S.; Deng, H.; Liang, F.; Peng, C.-Z.; Benjamin, S. C.; Zhu, X.; Lu, C.-Y.; Pan, J.-W. Demonstration of adiabatic variational quantum computing with a superconducting quantum coprocessor. Phys. Rev. Lett. **2020**, 125, 180501.
68. Arute, F.; Aryae, K.; Babbush, R.; Bacon, D.; Bardin, J. C.; Barends, R.; Boixo, S.; Broughton, M.; et al. Google A. I. Quantum and collaborators; Hartree-Fock on a superconducting qubit quantum computer. Science. **2020**, 369, 1084–1089.
69. Roffe, J. Quantum error correction: An introductory guide. Contemp. Phys. **2019**, 60, 226–245.
70. Chen, Z.; Satzinger, K. J.; Atalaya, J.; Korotkov, A. N.; Dunsworth, A.; Sank, D.; Quintana, C.; McEwen, M.; Barends, R.; Klimov, P. V.; et al. Exponential suppression of bit or phase errors with cyclic error correction. Nature. **2021**, 595, 383–387.
71. Nguyen, N. H.; Li, M.; Green, A. M.; Huerta Alderete, C.; Zhu, Y.; Zhu, D.; Brown, K. R.; Linke, N. M. Demonstration of Shor Encoding on a Trapped-Ion Quantum Computer. Phys. Rev. Appl. **2021**, 16, 024057.
72. Egan, L.; Debroy, D. M.; Noel, C.; Risinger, A.; Zhu, D.; Biswas, D.; Newman, M.; Li, M.; Brown, K. R.; Cetina, M.; Monroe, C. Fault-Tolerant Operation of a Quantum Error-Correction Code. arXiv Preprint (Quantum Physics), **2020**. arXiv:2009.11482. https://doi.org/10.48550/arXiv.2009.11482.
73. Pino, J. M.; Dreiling, J. M.; Figgatt, C.; Gaebler, J. P.; Moses, S. A.; Allman, M. S.; Baldwin, C. H.; Foss-Feig, M.; Hayes, D.; Mayer, K.; Ryan-Anderson, C.; Neyenhuis, B. Demonstration of the trappedion quantum CCD computer architecture. Nature. **2021**, 592, 209–213.
74. Postler, L.; Heußen, S.; Pogorelov, I.; Rispler, M.; Feldker, T.; Meth, M.; Marciniak, C. D.; Stricker, R.; Ringbauer, M.; Blatt, R.; Schindler, P.; Müller, M.; Monz, T. Demonstration of fault-tolerant universal quantum gate operations. Nature. **2022**, 605, 675–680.
75. Abobeih, M. H.; Wang, Y.; Randall, J.; Loenen, S. J. H.; Bradley, C. E.; Markham, M.; Twitchen, D. J.; Terhal, B. M.; Taminiau, T. H. Fault-tolerant operation of a logical qubit in a diamond quantum processor. arXiv Preprint (Quantum Physics), **2021**. arXiv:2108.01646. https://doi.org/10.48550/arXiv.2108.01646.

Index

Note: Page numbers in *italics* indicate figures; page numbers in **bold** indicate tables.

For Product Safety Concerns and Information please contact our EU
representative GPSR@taylorandfrancis.com
Taylor & Francis Verlag GmbH, Kaufingerstraße 24, 80331 München, Germany

www.ingramcontent.com/pod-product-compliance
Lightning Source LLC
LaVergne TN
LVHW081316110826
845149LV00006B/1516

* 9 7 8 1 0 3 2 5 7 8 5 8 3 *